THEORETICAL AERODYNAMICS

BY

L. M. MILNE-THOMSON, C.B.E.

EMERITUS PROFESSOR OF APPLIED MATHEMATICS, UNIVERSITY OF ARIZONA
EMERITUS PROFESSOR OF MATHEMATICS IN THE ROYAL NAVAL COLLEGE
PROFESSOR OF APPLIED MATHEMATICS IN BROWN UNIVERSITY
PROFESSOR IN THE MATHEMATICS RESEARCH CENTER AT THE UNIVERSITY OF WISCONSIN
VISITING PROFESSOR AT THE UNIVERSITIES OF ROME, QUEENSLAND, CALGARY, OTAGO

FOURTH EDITION

Revised and enlarged

DOVER PUBLICATIONS, INC.
NEW YORK

The time will come when thou shalt lift thine eyes
To watch a long drawn battle in the skies
While aged peasants too amazed for words
Stare at the flying fleets of wondrous birds

THOMAS GRAY
Luna Habitabilis, 1737

Published in Canada by General Publishing Company, Ltd., 30 Lesmill Road, Don Mills, Toronto, Ontario.
Published in the United Kingdom by Constable and Company, Ltd., 10 Orange Street, London WC 2.

This Dover edition, first published in 1973, is an unabridged and unaltered republication of the fourth edition (1966) of the work originally published by Macmillan and Company Limited in 1958.

International Standard Book Number: 0-486-61980-X
Library of Congress Catalog Card Number: 73-85109

Manufactured in the United States of America
Dover Publications, Inc.
180 Varick Street
New York, N. Y. 10014

PREFACE TO THE FIRST EDITION

THE airflow round an aircraft is a phenomenon of high complexity. To study it, in the present state of our knowledge, demands simplifying assumptions. These must be largely based on experimental observation of what actually happens ; that is one aspect of the practical side of aerodynamics. To make mathematical deductions and predictions belong to the theoretical side and it is the theoretical* side with which this book is concerned.

The aim is therefore to lay bare the assumptions, to bring them to explicit statement so that the reader may be consciously aware of what is assumed, and then to examine what can be deduced from the assumptions as a first approximation.

The treatment is based on my lectures to junior members of the Royal Corps of Naval Constructors at the Royal Naval College during the past ten years.

The mathematical equipment of the reader is presumed not to extend beyond the elements of the differential and integral calculus. What further is needed is mostly developed in the course of the exposition, which is thus reasonably self-contained. It is therefore hoped that the book will provide a solid introduction to the theory which is the indispensable basis of practical applications.

Since the use of vectors, or in two-dimensions the complex variable, introduces such notable simplifications of physical outlook and mathematical technique, I have had no hesitation in using vector methods. On the other hand the subject has been presented in such fashion that the reader who prefers cartesian notations should encounter little difficulty in adapting the vector arguments to a cartesian presentation. Chapter XXI on vectors has been added for the benefit of those with little or no previous acquaintance with vector methods. This chapter may be read first, or just before Chapter IX, or merely used as a compendium for reference.

Apart from Chapters I and II which are of a preliminary general character, and Chapter XXI on vectors, the work falls into four fairly well-defined parts. Chapters III to VIII contain the theory of two-dimensional, Chapters IX to XIV that of three-dimensional aerofoils, including propellers and wind tunnel corrections. Chapters XV, XVI, XVII deal with the effect of the compressibility of air in subsonic and supersonic flow. Chapters XVIII to XX are concerned with the aircraft as a whole.

The chapters are divided into sections numbered in the decimal notation. The equations are numbered in each section independently. Thus 7·14(3) refers to equation (3) in section 7·14 which, as the integer before the decimal point indicates, occurs in Chapter VII. Backward and forward references are

---

* To "the uninstructed and popular world" *practical* and *theoretical* are antonyms; a palpably false proposition.

freely used to aid the reader in following an argument or in comparing similar situations. Each diagram, of which there are 260, bears the number of the section to which it belongs and may therefore be traced without delay or exasperation.

About 300 exercises have been provided, collected into sets of Examples at the ends of the chapters. Some of these are very easy, others quite difficult, and many supplement the text. The majority were composed specially for this book.

References to literature are given where they appear to me to be appropriate or useful, but no attempt has been made to give systematic citations. I have made absolutely no endeavour to settle or assign priority of discovery. The coupling of a particular name with a theorem or method simply indicates an association in my own mind. The proper historical setting I must leave to those who have the time and the taste for such research.

This book was intended to appear long since, but other preoccupations during the war years prevented that. The delay has, however, allowed the presentation of matter which has appeared in the interim, and has also given me the fortunate opportunity of having the whole book read in manuscript by my colleague Mr B. M. Brown, to whom I owe a great debt for his criticisms and improvements. Another friend, Mr A. C. Stevenson, has likewise rendered invaluable service by his diligent help in proof reading and by important suggestions. To both these friends I wish to express my lively gratitude and appreciation. I also take this opportunity of expressing my thanks to the officials of the Glasgow University Press for the care and attention which they have given to the typography, and for maintaining a standard of excellence which could scarcely have been surpassed in the pre-war years.

L. M. MILNE-THOMSON

Royal Naval College
Greenwich
May 1947

PREFACE TO THE FOURTH EDITION

THE gratifying reception accorded to this work has encouraged me to strive for improvements. Opportunity has been taken to make several corrections, to revise certain passages and to carry out extensive rearrangements. Additional matter has been introduced particularly in connection with supersonic flow. Moreover a new circle theorem, here called the " second circle theorem ", to deal with flow of constant vorticity is given for the first time.

L. M. M.-T.

Mathematics Department
The University of Arizona
Tucson, Arizona
April 1966

CONTENTS

CHAPTER I

PRELIMINARY NOTIONS

CHAPTER II

BERNOULLI'S THEOREM

CONTENTS

CHAPTER III

TWO-DIMENSIONAL MOTION

CHAPTER IV

RECTILINEAR VORTICES

CHAPTER V

THE CIRCULAR CYLINDER AS AN AEROFOIL

CHAPTER VI

JOUKOWSKI'S TRANSFORMATION

CHAPTER VII

THEORY OF TWO-DIMENSIONAL AEROFOILS

CHAPTER VIII

THIN AEROFOILS

CHAPTER XII

LIFTING SURFACE THEORY

CHAPTER XIII

PROPELLERS

CHAPTER XIV

WIND TUNNEL CORRECTIONS

CHAPTER XV

SUBSONIC FLOW

CHAPTER XVI

SUPERSONIC FLOW

CHAPTER XVII

SUPERSONIC SWEPTBACK AND DELTA WINGS

CHAPTER XVIII

SIMPLE FLIGHT PROBLEMS

CHAPTER XIX

MOMENTS

CHAPTER XX

STABILITY

CHAPTER XXI

VECTORS

SYMBOLS

THE following list is intended to show the most frequent conventional meaning with which certain symbols are used in this book.

The list is not exhaustive nor does it preclude some symbols being used in other senses (which are always defined).

The numbers in brackets indicate the section where the meaning is first used or explained.

A   aerodynamic force (1·01)

A aspect ratio (1·13)

a, a' slope of the (C_L, α) graph for finite aspect-ratio (11·24) ; interference factors (13·4)

a_0 slope of the (C_L, α) graph for two-dimensional motion (7·13)

b span (1·1)

c chord (1·11, 1·12) ; speed of sound (1·5)

C cross-sectional area of a wind tunnel (14·4)

C_L lift coefficient, C_D drag coefficient, etc. (1·73)

D drag (1·02)

g, g acceleration due to gravity and its magnitude (20·01, 2·5)

i square root of -1 (3·4)

i, j, k   unit vectors along the x-, y-, z-axes (11·1)

J rate of advance coefficient (13·42)

K circulation (5·5)

L lift (1·02)

l typical length (1·71) : l^2 in Joukowski transformation (6·1)

M pitching moment (1·73) ; Mach number (1·71)

n   unit normal vector (9·31)

p pressure (1·4) ; aerodynamic pressure (2·13) ; angular velocity of rolling (19·2)

Q engine torque (13·1)

q, q air velocity and speed (1·21); angular velocity of pitching (19·0)

R Reynolds' number (1·71)

r   position vector (1·21)

r angular velocity of yawing (19·2)

S plan area of wings (1·13)

T absolute temperature (2·5) ; propeller thrust (13·1)

t time

u, v, w	components of air speed (3·11)
$\mathbf{V}, V$	velocity of aircraft, and aircraft speed (1·01, 1·71)
W	weight of aircraft (1·01)
w	complex potential (3·7); velocity of downwash (11·21); wing; loading (18·31)
X, Y, Z	components of aerodynamic force (5·4)
z	$x + iy$, complex variable (3·4)
α	incidence, angle of attack (1·15)
β	absolute incidence (7·13); angle of side-slip (18·33); $\sqrt{(1 - M_0{}^2)}$ (15·4)
γ	angle between Axes I and II (7·14); gliding angle (1·02) dihedral angle (19·6); ratio of specific heats (15·01)
Γ	circulation (11·2)
δ	aerofoil characteristic (11·53)
ϵ	angle of downwash (11·24); complex variable (17·12)
ϵ_I	wind tunnel interference angle (14·4)
ζ	vorticity vector (9·3)
ζ	$\xi + i\eta$ in conformal mapping (3·6)
η	efficiency of a propeller (15·1)
κ	strength of a vortex (4·11); propeller characteristic (13·42)
μ	Mach angle (16·1); relative aircraft density (20·2)
ν	kinematic viscosity (1·6)
π	ratio of length of circumference of a circle to its diameter
ϖ	impulsive pressure (3·31)
Π	air pressure at infinity (5·32)
ρ	air density (1·3)
Σ	summation (1·71)
τ	aerofoil characteristic (11·51); propeller characteristic (13·42); unit of time (20·21)
ϕ	velocity potential (3·31)
ψ	stream function (3·1)
Ω	angular velocity of aircraft (20·01)
Ω	angular speed of propeller (13·1); force potential (2·11)
$\boldsymbol{\omega}$	surface vorticity (9·6); angular velocity (20·1)
ω	magnitude of vorticity (3·21); angular speed
∇	$\mathbf{i}\,\dfrac{\partial}{\partial x} + \mathbf{j}\,\dfrac{\partial}{\partial y} + \mathbf{k}\,\dfrac{\partial}{\partial z}$ (9·1)

REFERENCES TO LITERATURE

The following abbreviations are used :

Proc. Camb. Phil. Soc.	*Proceeding of the Cambridge Philosophical Society*
Proc. Roy. Soc. (A)	*Proceedings of the Royal Society (Series A)*
Proc. Roy. Soc. Edin. (A)	*Proceedings of the Royal Society of Edinburgh (Series A)*
R. and M.	*Reports and Memoranda of the Aeronautical Research Committee*
N.A.C.A.	*Reports of the National Advisory Committee for Aeronautics* (U.S.A.)
Z.a.M.M.	*Zeitschrift für angewandte Mathematik und Mechanik*

Milne-Thomson *Theoretical Hydrodynamics.* References are to the 4th edition.

THE GREEK ALPHABET

alpha	α	A		nu	ν	N
beta	β	B		xi	ξ	Ξ
gamma	γ	Γ		omicron	o	O
delta	δ	Δ		pi	π $\tilde{\omega}$	Π
epsilon	ϵ	E		rho	ρ	P
zeta	ζ	Z		sigma	σ	Σ
eta	η	H		tau	τ	T
theta	θ	Θ		upsilon	υ	Υ
iota	ι	I		phi	ϕ	Φ
kappa	κ	K		chi	χ	X
lambda	λ	Λ		psi	ψ	Ψ
mu	μ	M		omega	ω	Ω

CHAPTER I

PRELIMINARY NOTIONS

1·0. The science of aerodynamics is concerned with the motion of air and of bodies moving through air. In particular it is concerned with the motion of aircraft.

1·01. Aerodynamic force. Air is a fluid, and in accordance with the hydrostatical theorem known as the principle of Archimedes an aircraft will be buoyed up by a force equal to the weight of air displaced by the aircraft. Thus

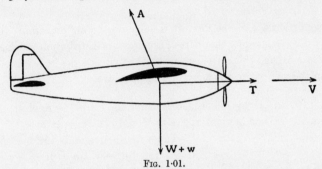

Fig. 1·01.

if we denote by the vector * **W** the weight of the aircraft and by the vector **w** the force of buoyancy, the total force due to gravity and buoyancy is **W** + **w**. We observe that the weight **W** is a force of magnitude W whose direction is vertically downwards, while the force of buoyancy **w** is a force of magnitude w whose direction is vertically upwards, so that the magnitude of the force **W** + **w** is $W - w$. This force **W** + **w** will act whether the aircraft is at rest or in motion.

To fix our ideas let us suppose that the aircraft is moving with constant velocity **V** in a horizontal direction through air which is otherwise at rest, that is to say any motion of the air is due solely to the motion of the aircraft. Let this motion be maintained by a tractive force **T** exerted by the propeller.†

Newton's first law of motion asserts that the resultant force on the aircraft must be zero, for the motion is unaccelerated. It follows that there must be an additional force **A**, say, such that the vector sum

$$\mathbf{T} + (\mathbf{W} + \mathbf{w}) + \mathbf{A} = 0.$$

---

* We shall denote vectors by letters in clarendon (heavy) type while their magnitudes will be denoted by the corresponding italic letter. See 21·0.

† We typify the propulsive system by the term propeller. The actual mechanism may be other than an airscrew, for example, jet propulsion.

This force **A** is called the *aerodynamic force* exerted on the aircraft. One of the major problems of aerodynamics is the investigation of this force.

The above definition of aerodynamic force is based on a state of motion in which the aircraft advances with constant velocity **V** in air otherwise at rest. If we imagine superposed on the whole system, aircraft and air, a uniform velocity − **V**, the aircraft may be considered at rest with the air streaming past it, the air velocity at points distant from the aircraft being − **V**.

It is important to observe that the aerodynamic force is theoretically the same in both cases, and therefore we may adopt whichever point of view may be the more convenient in any particular case. The use of wind tunnels to measure forces on aircraft is based on this principle.* We shall always refer to the direction of **V** as the *direction of motion*, and to the direction of − **V** as the *direction of the air stream*, or *relative wind*.

The whole surface of the aircraft is subjected to air pressure. The aerodynamic force is caused by that part of the pressure distribution which is due to motion, see 2·13, and this statement will serve to define the force in the general case when the velocity of the aircraft is not uniform.

1·02. Lift and drag. The aerodynamic force can be resolved into two component forces, one at right angles to **V** and one opposite to **V**.

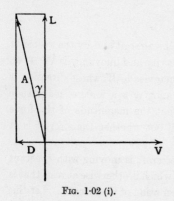

FIG. 1·02 (i).

These forces **L** and **D** are called respectively the *lift* and *drag*. If γ is the angle between **L** and **A**, we have

$$L = A \cos \gamma, \quad D = A \sin \gamma, \quad \tan \gamma = \frac{D}{L}.$$

The angle γ is called the *gliding angle* (see 18·3). Inasmuch as drag is an undesirable (but unavoidable) feature, for it entails expenditure of energy to maintain flight, the gliding angle should be as small as possible. An aircraft with a small gliding angle is said to be *streamlined*, and proper streamlining is another problem of aerodynamics.

In order to avoid any false impression that lift and drag are related to vertical and horizontal directions we give two formal definitions.

Def. Lift *is the component of aerodynamic force perpendicular to the direction of motion.*

---

* Corrections have to be made to measured values, see, for example, Ch. XIV. It may also be observed that natural air is but little turbulent whereas artificially generated wind streams are usually markedly turbulent. Thus the above mentioned theoretical equivalence of the two states has practical limitations.

Def.　Drag *is the component of aerodynamic force opposite to the direction of motion.*

This point is illustrated in fig. 1·02 (ii) which shows the various directions of the lift as an aircraft " loops the loop ".　In particular it can be seen that the lift can be directed horizontally and even vertically downwards.

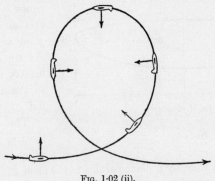

Fig. 1·02 (ii).

1·1. Monoplane aircraft.　Fig. 1·1 (*a*) shows in diagram form the essential features of a monoplane aircraft.　The propellers are omitted.　The

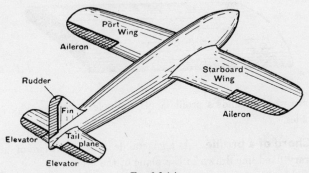

Fig. 1·1 (*a*).

main lifting system consists of two *wings* which together constitute the *aerofoil*. The tail-plane also exerts lift.　According to the design the aerofoil may, or may not, be interrupted by the fuselage.　The designer will subsequently allow for the effect of the fuselage as a disturbance or perturbation of the properties of the aerofoil.　For the present purpose we shall ignore the fuselage, and treat the aerofoil as one continuous surface.

The control surfaces consist of the elevators, ailerons, and rudder.

When the ailerons and rudder are in their *neutral* positions the aircraft has a median *plane of symmetry* which divides the whole machine into two parts each of which is the optical image of the other in this plane considered as a mirror.

The wings are then the portions of the aerofoil on either side of the plane of symmetry.

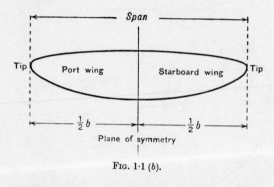

Fig. 1·1 (b).

The *wing tips* consist of those points of the wings which are at the greatest distance from the plane of symmetry. Thus the tip can be a point, a line, or an area according to the design of the aerofoil.

The distance between the tips is called the *span* which will be denoted by b.

The section of a wing by a plane parallel to the plane of symmetry is called a *profile*. The shape and general orientation of a profile will usually depend on its distance from the plane of symmetry. In the case of a cylindrical wing fig. 1·1 (c), the profiles are the same at every distance.

Fig. 1·1 (c).

The curve which bounds a profile is of " tadpole " outline. Fig. 7·0 shows several profiles.

1·11. Chord of a profile. As a general definition the *chord* of any profile is an arbitrary fixed line drawn in the plane of the profile.

The chord has direction, position, and length. The main requisite is that in each case the chord should be precisely defined, since the chord enters into the constants which describe the aerodynamic properties of the profile.

The official definition is the line which joins the centres of the circles of curvature of minimum radius at the nose and tail.

Fig. 1·11 (a).

Another definition is the longest line which can be drawn to join two points of the profile.

A third definition which is sometimes convenient is the projection of the profile on the double tangent to its lower surface (i.e. the tangent which touches the profile at two distinct points).

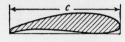

FIG. 1·11 (b).

This definition fails if there is no such double tangent.

1·12. Chord of an aerofoil. In the case of a cylindrical aerofoil, fig. 1·1 (c), the *chord of the aerofoil* is taken to be the chord of the profile in which the plane of symmetry cuts the aerofoil.

In all other cases we define the chord of the aerofoil as a mean or average chord located in the plane of symmetry.

Take any convenient (see 11·1) axes of reference x, y, z, the origin being situated in the plane of symmetry and the y-axis being perpendicular to that plane. Consider a profile whose distance * from the plane of symmetry is $|y|$. Let c be the length of the chord of this profile, θ the inclination of the chord to the xy plane, and (x, y, z) the coordinates of the quarter point of the chord, that is to say the point of the chord at distance $\frac{1}{4}c$ from the leading edge of the chord (see 7·31). Since the profile is completely defined when y is given, all these quantities are functions of y. The chord of the aerofoil is defined by averaging across the span. Thus if c_m is the length of the mean chord, $(x_m, 0, z_m)$ its quarter point, and θ_m its inclination, we take

$$c_m = \frac{1}{b}\int_{-b/2}^{b/2} c\,dy, \quad \theta_m = \frac{1}{b}\int_{-b/2}^{b/2} \theta\,dy, \quad x_m = \frac{1}{b}\int_{-b/2}^{b/2} x\,dy, \quad z_m = \frac{1}{b}\int_{-b/2}^{b/2} z\,dy.$$

These mean values completely define the chord of the aerofoil in length, direction, and position.

1·13. Aspect ratio. Consider a cylindrical aerofoil, fig. 1·1 (c). Let us imagine this to be projected on to the plane which contains the chords of all the sections. (This plane is perpendicular to the plane of symmetry and contains the chord of the aerofoil.) The projection in this case is a rectangle of area S, say, which is called the *plan area* of the aerofoil.

The plan area is quite distinct from the total surface area of the aerofoil. The simplest cylindrical aerofoil would be a rectangular plate and the plan area would then be half the total area.

The *aspect ratio* of the cylindrical aerofoil is then defined by

$$A = \frac{b}{c} = \frac{b^2}{S}.$$

---

* $|y|$ means the numerical or absolute value of y. Thus, for example, $|3| = 3$, $|-4| = 4$.

In the case of an aerofoil which is not cylindrical the plan area is defined to be the area of the projection on the plane through the chord of the aerofoil (mean chord) perpendicular to the plane of symmetry, and the aspect ratio is then defined to be

$$A = \frac{b^2}{S}.$$

A representative value of the aspect ratio is 6.

1·14. Camber.

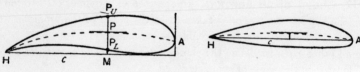

FIG. 1·14.

Consider a profile and its chord. Let y_U, y_L be the ordinates of points on the upper and lower parts of the profile respectively for the same value of x and let c be the chord, taken as x-axis. We then define

$$upper\ camber = (y_U)_{max} : c,$$
$$lower\ camber = (y_L)_{max} : c,$$

where the notation refers to that ordinate which is numerically greatest. Camber has sign, positive or negative, according to the sign of $(y_U)_{max}$ or $(y_L)_{max}$. We observe that the abscissae of the points which correspond to $(y_U)_{max}$ and $(y_L)_{max}$ may be different.

We also define the *camber line* of the profile as the locus of the point $(x, \frac{1}{2}(y_U + y_L))$. In the case of a symmetrical profile $y_U + y_L = 0$ so that the camber line is straight and coincides with the chord. Denoting the numerically greatest ordinate of the camber line by $(y)_{max}$, we define

$$mean\ camber = (y)_{max} : c.$$

Observe that mean camber is not, in general, the same as the mean of upper and lower camber ; also that the mean camber of a symmetrical profile is zero. The word camber, without qualification, usually refers to mean camber.

The *thickness ratio* is the ratio of the maximum thickness (measured perpendicularly to the chord) to the chord.

1·15. Incidence.
When an aircraft advances in the plane * of symmetry, the angle between the direction of motion and the direction of the chord of a profile is called the *geometrical incidence* of the profile and will be denoted by the

---

* This is to be understood to mean that the direction of motion of the aircraft is parallel to the plane of symmetry.

letter α. An alternative term is *angle of attack* (fig. 1·15). For the aeroplane as a whole the geometrical incidence will be defined as the angle between the direction of motion and the chord of the aerofoil. When the chords of the

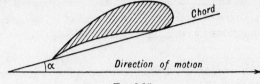

Chord

α Direction of motion

FIG. 1·15.

various profiles of an aerofoil are parallel the incidence is the same at each section. When the chords are not parallel the incidence varies from section to section and the wing has *twist*.

The value of the geometrical incidence would be altered if a different line were chosen as chord.

1·2. Fluids.

All materials * exhibit *deformation* under the action of forces ; *elasticity* when a given force produces a definite deformation, which vanishes if the force is removed ; *plasticity* if the removal of the forces leaves permanent deformation ; *flow* if the deformation continually increases without limit under the action of forces, *however small*.

A *fluid* is material which flows.

Actual fluids fall into two categories, namely gases and liquids.

A *gas* (such as atmospheric air) will ultimately fill any closed space to which it has access and is therefore classified as a (highly) *compressible fluid*.

A liquid at constant temperature and pressure has a definite volume and when placed in an open vessel will take under the action of gravity the form of the lower part of the vessel and will be bounded above by a horizontal free surface. All known liquids are to some slight extent compressible. For most purposes it is, however, sufficient to regard liquids as *incompressible fluids*.

It may be observed that for speeds which are sufficiently small fractions of the speed of sound, the effect of compressibility on atmospheric air can be neglected, and in many experiments which are carried out in wind tunnels the air is treated as incompressible.†

Actual liquids (and gases) in common with solids exhibit *viscosity* arising from internal friction in the substance. Our definition of a fluid distinguishes a viscous fluid, such as treacle or pitch, from a plastic solid, such as putty or clay,

---

* In this summary description the materials are supposed to exhibit a macroscopic continuity, and the forces are not great enough to cause rupture. Thus a heap of sand is excluded, but the individual grains are not.

† In this sense we may use the convenient term " incompressible air ".

since the former cannot permanently resist any shearing stress, however small, whilst in the case of the latter, stresses of a definite magnitude are required to produce deformation. Pitch is an example of a very viscous liquid, water is an example of a liquid which is but slightly viscous. For the present, in order to render the subject amenable to exact mathematical treatment, we shall follow the course adopted in other branches of mechanics and make simplifying assumptions by defining an ideal substance known as an *inviscid* or *ideal* fluid.

Definition. An inviscid fluid is a continuous fluid substance which can exert no shearing stress, *however small*.

The continuity is postulated in order to evade the difficulties inherent in the conception of a fluid as consisting of a granular structure of discrete molecules. The inability to exert any shearing stress, however small, will be shown later to imply that the pressure at any point is the same for all directions at that point.

Moreover, the absence of tangential stress between the fluid on the two sides of any small surface imagined as drawn in the fluid implies the entire absence of internal friction, so that no energy can be dissipated from this cause. A further implication is that, when a solid moves through the fluid or the fluid flows past a solid, the solid surface can exert no tangential action on the fluid, so that the fluid flows freely past the boundary and no energy can be dissipated there by friction. In this respect the ideal fluid departs widely from the actual fluid which, as experimental evidence tends to show, adheres to the surface of solid bodies immersed in it.

1·21. Velocity.

Since our fluid is continuous, we can define a *fluid particle* as consisting of the fluid contained within an infinitesimal volume, that is to say, a volume whose size may be considered so small that for the particular purpose in hand its linear dimensions are negligible. We can then treat a fluid particle as a geometrical point for the particular purpose of discussing its velocity and acceleration.

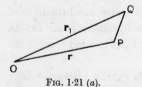

FIG. 1·21 (a).

If we consider, fig. 1·21 (a), the particle which at time t is at the point P, defined by the vector *

$$\mathbf{r} = \overrightarrow{OP},$$

at time t_1 this particle will have moved to the point Q, defined by the vector

$$\mathbf{r}_1 = \overrightarrow{OQ}.$$

---

* The subject of vectors and the notations here used is explained in Chapter XXI.

The velocity of the particle at P is then defined by the vector *

$$\mathbf{q} = \lim_{t_1 \to t} \frac{\mathbf{r}_1 - \mathbf{r}}{t_1 - t} = \frac{d\mathbf{r}}{dt}.$$

Thus the velocity $\mathbf{q}$ is a function of $\mathbf{r}$ and t, say

$$\mathbf{q} = f(\mathbf{r}, t).$$

If the form of the function f is known, we know the motion of the fluid. At each point we can draw a short line to represent the vector $\mathbf{q}$, fig. 1·21 (b).

To obtain a physical conception of the velocity field defined by the vector $\mathbf{q}$, let us imagine the fluid to be filled with a large (but not infinitely large) number of luminous points moving with the fluid.

A photograph of the fluid taken with a short time exposure would reveal the tracks of the luminous points

FIG. 1·21 (b).

as short lines, each proportional to the distance moved by the point in the given time of the exposure and therefore proportional to its velocity. This is in fact the principle of one method of obtaining pictorial records of the motion of an actual fluid. In an actual fluid the photograph may reveal a certain regularity of the velocity field in which the short tracks appear to form parts of a regular system of curves. The motion is then described as *streamline* motion. On the other hand, the tracks may be wildly irregular, crossing and recrossing, and the motion is then described as *turbulent*. The motions of our ideal inviscid fluid will always be supposed to be of the former character. An exact mathematical treatment of turbulent motion has not yet been achieved.

1·22. Streamlines and paths of the particles. A line drawn in the fluid so that its tangent at each point is in the direction of the fluid velocity at that point is called a *streamline*.

When the fluid velocity at a given point depends not only on the position of the point but also on the time, the streamlines will alter from instant to instant. Thus photographs taken at different instants will reveal a different system of streamlines. The aggregate of all the streamlines at a given instant constitutes the *flow pattern* at that instant.

When the velocity at each point is independent of the time, the flow pattern will be the same at each instant and the motion is described as *steady*. In this connection it is useful to describe the type of motion which is *relatively steady*. Such a motion arises when the motion can be made steady by super-

---

* The symbol $\lim_{t_1 \to t}$ is to be read as " the limit when t_1 tends to the value t ". This is the usual method of defining differential coefficients, whose existence we shall infer on physical grounds. The symbol $\to$ alone is read " tends to ".

posing on the whole system a constant velocity. Thus when an aircraft flies on a straight course with constant speed in air otherwise undisturbed, to an observer in the aircraft the flow pattern which accompanies him appears to be steady and could in fact be made so by superposing the reversed velocity of the aircraft on the whole system consisting of the aircraft and air.

If we fix our attention on a particular particle of the fluid, the curve which this particle describes during its motion is called a *path line*. The direction of motion of the particle must necessarily be tangential to the path line, so that the path line touches the streamline which passes through the instantaneous position of the particle as it describes its path.

Thus the streamlines show how each particle is moving at a given instant The path lines show how a given particle is moving at each instant.

When the motion is steady, the path lines coincide with the streamlines.

1·23. Stream tubes and filaments.
If we draw the streamline through each point of a closed curve we obtain a *stream tube*.

A *stream filament* is a stream tube whose cross-section is a curve of infinitesimal dimensions.

When the motion is dependent on the time, the configuration of the stream tubes and filaments changes from instant to instant, but the most interesting applications of these concepts arise in the case of the steady motion of incompressible air, which we shall now discuss.

In the steady motion, a stream tube behaves like an actual tube through which the air is flowing, for there can be no flow into the tube across the walls since the flow is, by definition, always tangential to the walls. Moreover, these walls are fixed in space since the motion is steady, and therefore the motion of the air within the walls would be unaltered if we replaced the walls by a rigid substance.

Consider a stream filament of air in steady motion. We can suppose the cross-sectional area of the filament so small that the velocity is the same at each point of this area, which can be taken perpendicular to the direction of the velocity.

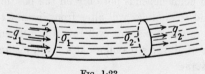

FIG. 1·23.

Now let q_1, q_2 be the speeds of the flow at places where the cross-sectional areas are σ_1 and σ_2. Since the air is incompressible, in a given time the same volume must flow out at one end as flows in at the other. Thus

$$q_1 \sigma_1 = q_2 \sigma_2.$$

This is the simplest case of the equation of conservation of mass, or the *equation of continuity*, which asserts in the general case that the rate of generation of mass within a given volume must be balanced by an equal net outflow of mass from the volume. The above result can be expressed in the following theorem.

The product of the speed and cross-sectional area is constant along a stream filament of incompressible air in steady motion.

It follows from this that a stream filament is narrowest at places where the speed is greatest and is widest at places where the speed is least.

A further important consequence is that a stream filament cannot terminate at a point within the fluid unless the velocity becomes infinite at that point. Leaving this case out of consideration, it follows that in general stream filaments are either closed or terminate at the boundary. The same is of course true of streamlines, for the cross-section of the filament may be considered as small as we please.

1·3. Density. If M is the mass of the air within a closed volume V, we can write

(1) $$M = V\rho_1,$$

and ρ_1 is then the average density of the air within the volume at that instant. In a hypothetical medium continuously distributed we can define the density ρ as the limit of ρ_1 when $V \to 0$.

1·4. Pressure. Consider a small plane of infinitesimal area $d\sigma$, whose centroid is P, drawn in the fluid, and draw the normal PN on one side of the area which we shall call the positive side. The other side will be called the negative side.

We shall make the hypothesis that the mutual action of the fluid particles on the two sides of the plane can, at a given instant, be represented by two equal but opposite forces of magnitude $p\,d\sigma$ applied at P, each force being a push not a pull, that is to say, the fluid on the positive side pushes the fluid on the negative side with a force of magnitude $p\,d\sigma$.

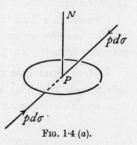

Fig. 1·4 (a).

Experiment shows that in a fluid at rest these forces act along the normal. In a real fluid in motion these forces make an angle ϵ with the normal (analogous to the angle of friction). When the viscosity is small, as in the case of air and water, ϵ is very small. In an inviscid fluid which can exert no tangential stress $\epsilon = 0$, and in this case p is called the *pressure* at the point P.

In the above discussion there is nothing to show that the pressure p is independent of the orientation of the element $d\sigma$ used in defining p. That this independence does in fact exist is proved in the following theorem.

Theorem. The pressure at a point in an inviscid fluid is independent of direction.

Proof. Let P, Q be two neighbouring points, and consider a cylinder of fluid, whose generators are parallel to PQ, bounded by a cross-section $d\sigma_1$ and

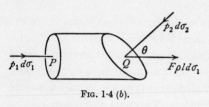

an oblique section $d\sigma_2$, the centroids of these sections being P and Q respectively. Let the pressures at P and Q, defined by the sections $d\sigma_1$ and $d\sigma_2$, be p_1 and p_2, and let the normal at Q make an angle θ with PQ. The volume of fluid within the cylinder is $l\,d\sigma_1$,

Fig. 1·4 (b).

where l is infinitesimal. Let F be the component in the direction of PQ of the external force per unit mass of fluid, and let f be the acceleration of the cylinder in the direction of PQ. Then if ρ is the density, the second law of motion gives

$$p_1\,d\sigma_1 - p_2\,d\sigma_2\cos\theta + F\,\rho l\,d\sigma_1 = f\,\rho l\,d\sigma_1.$$

Now, $d\sigma_2\cos\theta = d\sigma_1$. Therefore dividing by $d\sigma_1$,

$$p_1 - p_2 = l\rho(f - F).$$

If we let Q approach P, l will tend to zero and therefore $p_1 - p_2$ tends to zero. Thus when Q coincides with P we get $p_1 = p_2$. Since the direction of the normal to the section at Q is quite arbitrary, we conclude that the pressure at P is the same for all orientations of the defining element of area. Q.E.D.

Pressure is a *scalar* quantity, i.e. independent of direction. The dimensions of pressure (see 1·7) in terms of measure ratios M, L, T of mass, length and time are indicated by $ML^{-1}T^{-2}$.

The *thrust* on an area $d\sigma$ due to pressure p is a force, of magnitude $p\,d\sigma$, that is, a vector quantity whose complete specification requires direction as well as magnitude.

Pressure in a fluid in motion is a function of the position of the point at which it is measured and of the time. When the motion is steady the pressure may vary from point to point, but at a given point it is independent of the time.

1·41. Thrust due to pressure. Consider a cylinder of ideal fluid of infinitesimal length ds and of cross-sectional area σ, the dimensions of the cross-section being small compared with ds.

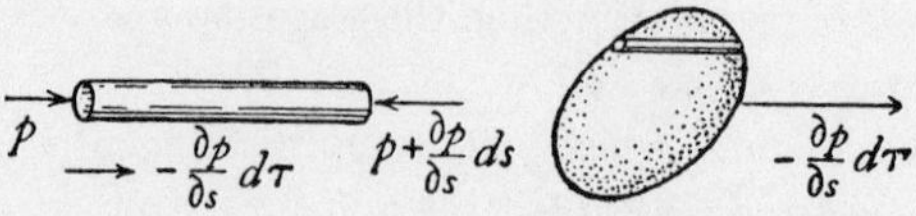

$$P \xrightarrow{\quad} \quad -\frac{\partial p}{\partial s}d\tau \qquad p+\frac{\partial p}{\partial s}ds \qquad -\frac{\partial p}{\partial s}d\tau$$

Fig. 1·41.

The pressure thrusts on the ends are $p\sigma$ and $(p + (\partial p/\partial s)\,ds)\sigma$ in opposite senses and the pressure thrusts on the curved surface of the cylinder form a system of forces in equilibrium. Thus the net pressure thrust on the cylinder is in the direction of its length and is of magnitude

$$(1) \qquad\qquad -\frac{\partial p}{\partial s}\, d\tau,$$

where $d\tau = \sigma\, ds$ is the volume of the contained fluid.

This result can be applied to a volume $d\tau$ of any shape, provided its dimensions are infinitesimal, the thrust in the direction of a line element ds being given by (1). To see this, observe that such a volume can be divided into slender cylinders, of the type just considered, whose generators are all parallel to the direction of ds.

In vector notation the thrust is $-\,(\nabla p)\, d\tau$, see 21·3.

1·5. The speed of sound.

We shall suppose that sound is propagated in air by small to-and-fro motions of the air whereby the disturbance passes rapidly from place to place without causing a transference of the air itself. This view is supported by " dust tube " experiments in which fine particles suspended in air show no appreciable motion when sound passes through the air in which they float.

The basic assumptions are as follows :

(i) The variations of the pressure, density and velocity caused by the passage of sound are infinitesimal quantities of the first order, that is to say their squares and products may be neglected.

(ii) The pressure is a function of the density alone.

With regard to (ii), experiment shows that the adiabatic law (15·01)

$$(1) \qquad\qquad p = \kappa\rho^{\gamma} = f(\rho)$$

is best suited to give agreement between theory and observation. Here γ is the ratio of the specific heats at constant volume and constant pressure, and has for air the value 1·405 approximately.

To fix our ideas, consider a long horizontal tube of small cross-section A containing air at rest and let us study the portion of air, density ρ_0, pressure p_0, between the sections at distance x and $x + dx$ from a fixed point in tube. So long as the air is undisturbed p_0 and ρ_0 are independent of x.

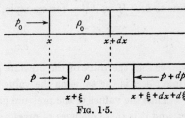

Fig. 1·5.

When the sound is travelling through the air the bounding sections will become displaced, say to $x + \xi$ and $x + \xi + dx + d\xi$ at time t, and the pressure on these sections will be perturbed, say to p and $p + dp$, while the density will change from ρ_0 to ρ. Observe that x is not a function of t but merely identifies the section which we are considering. On the other hand, ξ measures the displacement of that section at time t and therefore ξ depends both on t and x.

By the second law of motion

$$- A\, dp = \rho_0\, A\, dx\, \frac{\partial^2 \xi}{\partial t^2},$$

for $\partial^2 \xi / \partial t^2$ is the acceleration of all the air particles at $x + \xi$ and by (i) the accelerations of the remaining particles of air here considered will differ but infinitesimally from this. Thus

(2) $$- \frac{\partial p}{\partial x} = \rho_0\, \frac{\partial^2 \xi}{\partial t^2}.$$

Again, the mass of air concerned is the same in both cases so that

$$\rho (dx + d\xi)\, A = \rho_0\, dx\, A$$

and therefore $\rho(1 + \partial \xi / \partial x) = \rho_0$ so that (see 1·9 (1))

(3) $$\rho = \rho_0 \left(1 - \frac{\partial \xi}{\partial x} \right),$$

for ρ and ρ_0 differ infinitesimally and therefore $(\partial \xi / \partial x)^2$ is negligible. We proceed to eliminate ρ and ξ from the three equations (1), (2), (3). From (1) and (3)

$$\frac{\partial p}{\partial x} = f'(\rho)\, \frac{\partial \rho}{\partial x} = - \rho_0 f'(\rho)\, \frac{\partial^2 \xi}{\partial x^2} = - \rho_0 f'(\rho_0)\, \frac{\partial^2 \xi}{\partial x^2}$$

to our order of approximation (see 1·9). Combining this with (2) we get

(4) $$\frac{\partial^2}{\partial t^2} \left(\frac{\partial p}{\partial x} \right) = c_0^2\, \frac{\partial^2}{\partial x^2} \left(\frac{\partial p}{\partial x} \right), \text{ where}$$

(5) $$c_0^2 = f'(\rho_0) = \left(\frac{dp}{d\rho} \right)_0 = \frac{\gamma p_0}{\rho_0}.$$

It is readily verified that (4) is satisfied if we equate $\partial p / \partial x$ to any arbitrary (differentiable) function of $x - c_0 t$ or $x + c_0 t$, or the sum of any two such functions. If we take, for example,

(6) $$\frac{\partial p}{\partial x} = F(x - c_0 t),$$

we see that the value of the pressure gradient $\partial p / \partial x$ at the section $x + c_0 \tau$ at the time $t + \tau$ is the same as the value of $\partial p / \partial x$ at the section x at time t.

Thus the values of the pressure gradient move along the tube with the speed c_0, and it is these changes of pressure gradient which the ear detects in audible sound. Thus we may identify the value of c_0 given by (5) as the speed of sound in air whose unperturbed pressure and density are p_0, ρ_0.

In standard air at sea-level (2·5) we have approximately

$$c_0 = 1120 \text{ ft./sec.} = 764 \text{ mi./hr.}$$

In an ideal incompressible fluid any change of pressure is propagated instantaneously or, as we may say, with "infinite speed". Thus the greater the speed of sound in a given fluid the more we should expect it to exhibit the properties characteristic of incompressibility. Therefore for motions of air in which the maximum speed involved is a sufficiently small fraction of the speed of sound we should expect to be able, as a first approximation, to treat the air as incompressible.

1·6. Maxwell's definition of viscosity.

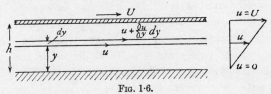

FIG. 1·6.

A horizontal plate moves forwards with velocity U over fluid which is in contact with a fixed horizontal plane. The fluid in contact with the plate is at rest relatively to the plate and moves with it, while the fluid in contact with the plane is at rest. Thus the fluid is urged forwards, i.e. in the direction of U, from above and retarded from below. If we consider the fluid between two planes at heights y and $y + dy$ and denote by F the tractive force per unit area of the surface of the plane at height y, the assumption is made that

$$F = \mu \frac{\partial u}{\partial y},$$

where u is the velocity of the fluid in the plane at height y and μ is a constant. The corresponding tractive force per unit area on the fluid in the plane at height $y + dy$ will then be

$$F + dF = \mu \left(\frac{\partial u}{\partial y} + \frac{\partial^2 u}{\partial y^2} dy \right).$$

When a steady state has been reached we must have $dF = 0$ otherwise the fluid would be accelerated. Therefore $\partial^2 u/\partial y^2 = 0$ and so $u = Uy/h$ where h is the height of the plate, since $u = 0$ when $y = 0$ and $u = U$ when $y = h$; so that $F = \mu U/h$, which is constant. Thus if $U = 1$ and $h = 1$, we get $F = \mu$, and we have Maxwell's definition of the coefficient of viscosity μ.

The coefficient of viscosity is the tangential force per unit area on either of two parallel plates at unit distance apart, one fixed and the other moving with unit velocity.

The dimensions (see 1·7) of μ are those of

$$\frac{\text{force per unit area}}{\text{velocity per unit length}} = \frac{ML}{T^2} \times \frac{1}{L^2} \div \left(\frac{L}{T} \times \frac{1}{L} \right) = \frac{M}{LT}.$$

In practice the kinematic coefficient of viscosity

$$\nu = \frac{\mu}{\rho}$$

is more useful. The dimensions of ν are therefore $ML^{-1}T^{-1} \times L^3M^{-1} = L^2T^{-1}$. Thus in the c.g.s. system ν is measured in cm.^{2}/sec. ; in the British system in ft.^{2}/sec.

For air at 15° C., $\nu = 1·59 \times 10^{-4}$ ft.^{2}/sec.

1·7. Physical dimensions. Physics deals with the measurable properties of physical quantities, certain of which, as for example, length, mass, time and temperature, are regarded as fundamental, since they are independent of one another, and others, such as velocity, acceleration, force, thermal conductivity, pressure, energy are regarded as derived quantities, since they are defined ultimately in terms of the fundamental quantities. Mathematical physics deals with the representation of the measures of these quantities by numbers and deductions therefrom. These measures are all of the nature of ratios of comparison of a measurable magnitude with a standard one of like kind, arbitrarily chosen as the unit, so that the number representing the measure depends on the choice of unit.

Consider a *dynamical system*, i.e. one in which the derived quantities depend only on length, mass and time, and change the fundamental units from, say, foot, pound, second, to mile, ton, hour. Let l_1, m_1, t_1 and l_2, m_2, t_2 be the measures of the same length, mass and time respectively in the two sets of units. Then we have

(1) $$l_1 = \frac{l_1}{l_2} \times l_2 = Ll_2, \quad m_1 = Mm_2, \quad t_1 = Tt_2,$$

where L, M, T are numbers independent of the particular length, mass or time measured, but depending only on the choice of the two sets of units. Thus in this case, we have $L = 5280, M = 2240, T = 3600$. These numbers L, M, T we call the respective *measure-ratios* of length, mass, time for the two sets of units, in the sense that measures of these quantities in the second set are converted into the corresponding measures in the first set by multiplication by L, M, T.

The measure-ratios V, A, F of the derived quantities, velocity v, acceleration a, and force f, are then readily obtained from the definitions of these quantities as

$$V = L/T, \quad A = V/T, \quad F = MA,$$

so that ultimately the measure ratio of a force is given by $F = ML/T^2$. And in general if n_1, n_2 are the measures of the same physical quantity n in the two sets of units, we arrive at the measure-ratio

(2) $$\frac{n_1}{n_2} = N = L^x M^y T^z,$$

and we express this conventionally by the statement that the quantity is of *dimensions* $L^x M^y T^z$ (or is of dimensions x in length, y in mass, and z in time). If $x = y = z = 0$, then $n_1 = n_2$, so the quantity in question is independent of any units which may be chosen, as for example, the quantity defined as the ratio of the mass of the pilot to the mass of the aircraft. In such a case we say the quantity is dimensionless and is represented by *a pure number*, meaning that it does not change with units.

Now consider a definitive relation

(3) $a = bc$

between the measures a, b, c of physical quantities in a dynamical system, i.e. a relation which is to hold whatever the sets of units employed, and which is not merely an accidental relation * between numbers arising from measurement in one particular set of units. Suppose the dimensions of a, b, c are respectively (p, q, r), (s, t, u), and (x, y, z), so that

(4) $a_1 = a_2 L^p M^q T^r,\quad b_1 = b_2 L^s M^t T^u,\quad c_1 = c_2 L^x M^y T^z.$

Then (3) would become $a_1 = b_1 c_1$, and (4) would then give by substitution

$$a_2 L^p M^q T^r = b_2 L^s M^t T^u c_2 L^x M^y T^z.$$

Now $a_2 = b_2 c_2$, since the form of (3) is independent of units, and therefore

$$L^p M^q T^r = L^{s+x} M^{t+y} T^{u+z}, \quad \text{or} \quad p = s + x, \quad q = t + y, \quad r = u + z.$$

In other words, each fundamental measure-ratio must occur with the same index on each side of (3), i.e. each side of (3) must be of the same physical dimensions.

In systems involving temperature as well as length, mass, and time as fundamental quantities (*thermodynamical systems*) a measure-ratio (say D) of temperature must be introduced (cf. 2·5).

1·71. Aerodynamic force; dimensional theory.

In 1·01 the aerodynamic force was defined as the force on an aircraft caused by that part of the pressure distribution which is due to motion. Thus gravity does not enter into the specification of this force. Restricting our consideration to steady motion without rotation the aerodynamic force on an aerofoil or on a complete aircraft may be expected to depend on the following quantities whose physical dimensions are given :

Quantity	Symbol	Dimensions
Typical length	l	L
Forward speed	V	L/T
Air density	ρ	M/L^3
Velocity of sound	c	L/T
Kinematic viscosity	ν	L^2/T

---

* For an example of this see the period of the phugoid oscillation, 18·51.

Let us denote by A the magnitude of the aerodynamic force. The dimensions of A are then those of force, namely ML/T^2.

To begin with, let us suppose the air incompressible and inviscid. Then $c = \infty$ and $\nu = 0$, and A does not depend on either of these quantities, and we should be led to assume that A, depending only on l, V, ρ would be given by a formula such as

(1) $$A = \tfrac{1}{2}k\rho^p V^q l^r,$$

where $\tfrac{1}{2}k$ is a dimensionless number, or by a sum of terms like that on the right. Since each side of (1) is of the same dimensions, we must have

$$\frac{ML}{T^2} \equiv \left(\frac{M}{L^3}\right)^p \left(\frac{L}{T}\right)^q L^r \equiv \frac{M^p L^{q+r-3p}}{T^q}.$$

Thus $p = 1$, $q = 2$, $q + r - 3p = 1$, and therefore (1) becomes

(2) $$A = \tfrac{1}{2}k\rho V^2 l^2.$$

It now appears that (1) only requires a single term.

If we wish to take account of compressibility and viscosity, c and ν should also appear and (1) will be replaced by *

(3) $$A = \Sigma \tfrac{1}{2}k_{st}\rho^p V^q l^r c^s \nu^t,$$

where $\tfrac{1}{2}k_{st}$ is a dimensionless number, and each term must have the dimensions of a force. Therefore

$$\frac{ML}{T^2} \equiv \left(\frac{M}{L^3}\right)^p \left(\frac{L}{T}\right)^q L^r \left(\frac{L}{T}\right)^s \left(\frac{L^2}{T}\right)^t.$$

Equating the indices of M, L, T on the two sides we get

$$p = 1, \quad q + r + s + 2t - 3p = 1, \quad q + s + t = 2,$$

whence $\quad q = 2 - s - t, \quad r = 2 - t, \quad$ and therefore

$$\rho^p V^q l^r c^s \nu^t = \rho V^{2-s-t} l^{2-t} c^s \nu^t = \rho V^2 l^2 \left(\frac{V}{c}\right)^{-s} \left(\frac{Vl}{\nu}\right)^{-t}.$$

The dimensionless number $M = V/c$ is called the *Mach number*,† or the *Rayleigh number*. It arises from taking account of compressibility. For an incompressible fluid $M = 0$.

The dimensionless number $R = Vl/\nu$ is called the *Reynolds' number*. For an inviscid fluid $R = \infty$, and for air, since ν is small (see 1·6), R is large unless Vl is also small.

Thus (3) becomes

$$A = \tfrac{1}{2}\rho V^2 l^2 \Sigma k_{st} M^{-s} R^{-t} = \tfrac{1}{2}\rho V^2 S f(M, R),$$

---

* Σ denotes the sum of all allowable terms such as the specimen which follows it.

† There should be no occasion to confuse the Mach number M with the measure-ratio M.

where l^2 has been replaced by the plan area S, a proportional number of the same dimensions, and $f(M, R)$ is a function, whose form is not determined by the present method, with values which are independent of physical units.

The dimensionless number

$$C_A = \frac{A}{\frac{1}{2}\rho V^2 S} = f(M, R)$$

is called the (dimensionless) coefficient of the aerodynamic force A. The effect of compressibility can usually be neglected if $M < \frac{1}{2}$ (see 2·32), so that in this case $C_A = F(R)$ a function of the Reynolds' number only.

1·72. Similar systems; scale effect. This last result gives rise to some remarks concerning the inferences to be drawn as to the behaviour of the full-scale machine from experiments made on a geometrically similar model. If the model tests give an aerodynamic coefficient C_{Am} for a test conducted at a Reynolds' number R_m, the *scale effect* on the coefficient is given by $C_A : C_{Am} = F(R) : F(R_m)$.

The model tests will give the aerodynamic coefficient $C_A = C_{Am}$ directly if $R = R_m$.

Since $R = Vl/\nu = Vl\rho/\mu$ and since μ and ρ are the same in the machine and its model, while V and l are both greater for the machine than for the model, the model experiments are necessarily conducted for a smaller Reynolds' number than that for which the machine will be used. Thus the above correction will have to be made in calculating the actual C_A from the values obtained by model experiments. There is, however, a way out, namely by using a compressed air wind tunnel which has the effect of increasing ρ and therefore the Reynolds' number.

In conclusion it should be noted that in giving the Reynolds' number a statement should always be made as to what particular length is taken for the typical length l.

1·73. Coefficients. We have just seen how to define a dimensionles coefficient C_A of aerodynamic force. In exactly the same way we define the coefficient of any component. Thus if L and D are the lift and drag components we have *lift and drag coefficients*

$$C_L = \frac{L}{\frac{1}{2}\rho V^2 S}, \qquad C_D = \frac{D}{\frac{1}{2}\rho V^2 S}.$$

Fig. 1·73 shows typical curves of C_L and C_D plotted against geometrical incidence α. We shall consider the properties of such curves later.

In addition to force coefficients we have moment coefficients. The moment of the aerodynamic force about an axis perpendicular to the plane of symmetry,

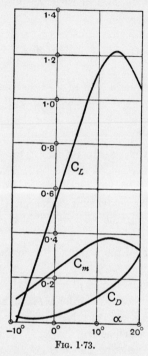

FIG. 1·73.

called the pitching moment, will depend on the particular axis chosen. Denoting this moment about the chosen axis by M, we define the *coefficient of pitching moment* by

$$C_m = \frac{M}{\frac{1}{2}\rho V^2 Sc}.$$

A typical graph is shown in fig. 1·73.

It may be noted that if we choose $\frac{1}{2}\rho V^2 S$ as the *unit of pressure* and the chord c as the *unit of length* the above aerodynamic forces reduce to the lift C_L, the drag C_D and the moment C_m.

1·8. The boundary layer.

Consider a flat plate of length l at rest over which a stream of fluid passes with general velocity V.

Consider a point P of the plate and a normal PN erected there. Let us draw vectors at points of this normal to represent the fluid velocity parallel to the plate. At P this velocity is zero (see 1·6). If the Reynolds' number Vl/ν is large, say of the order 10^5, it is found that the velocity rapidly attains the value V as we recede from P. If we denote by δ the height at which the velocity attains the value $99V/100$, say, the maximum value of δ/l is about 0·02. Fig. 1·8 (i) shows the state of affairs, the vertical scale being *greatly exaggerated*. If $PQ = \delta$ the locus of the points Q is a surface which passes through the leading

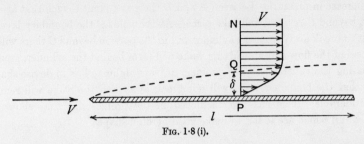

FIG. 1·8 (i).

edge of the plate, and the fluid between this surface and the plate constitutes the *boundary layer*. Inside the boundary layer the effect of viscosity is important. Outside the boundary layer the effect is negligible. The greater the Reynolds' number the thinner becomes the boundary layer and we have practically the case of an inviscid fluid flowing past the plate. There is, however, this difference. However small the viscosity the plate is subjected to a tangential traction or drag urging it in the direction of V. This force is known as the *friction drag*, and this force can never be entirely eliminated. On the other hand the fluid outside the boundary layer behaves like an inviscid fluid.

Now consider the steady flow past a circular cylinder at rest.

If V is the speed of the stream in an inviscid fluid, it can be proved that the speed at the top A is $2V$ and gradually falls off to zero at B. There is a

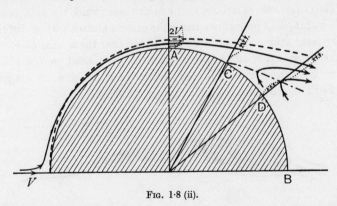

FIG. 1·8 (ii).

corresponding increase of pressure as we pass round from A to B. Now, if the fluid is viscous, the fluid in contact with the cylinder is at rest and the fluid in the immediate neighbourhood of the cylinder has a very small speed and cannot force its way round to B against the increasing pressure. Thus, if we perform the same construction of vectors on the normals as before, fig. 1·8 (ii),* the diagram at A resembles that at P in fig. 1·8 (i), but just above a point such as D the component of velocity parallel to the tangent is actually reversed owing to the increase in pressure.   Between A and D there is a point C such that at any point beyond C reversal will take place.   At the point C the boundary layer is about to break away from the cylinder and in the portion beyond C there will be reversal of the flow and an eddying wake will form behind the cylinder, greatly increasing the resistance.   The problem of " streamlining " is to devise shapes such that the boundary layer will not break away and the wake will remain inconsiderable.   This has been achieved in the profiles like that shown in fig. 1·8 (iii) which are found to make good aerofoil shapes.

Fig. 1·8 (iii).

For those there is a narrow ribbon wake but, to a first approximation, the problem of flow past such a shape is that of inviscid fluid flowing smoothly past the body.

The above considerations give rise to some general observations :

(1) It is found that to delay the breaking away of the boundary layer the region where the fluid is moving against increasing pressure should curve as gradually as possible, i.e. should have a large radius of curvature.

(2) The importance of smoothing the surface over which the fluid passes lies in the fact that small projections above the general surface may so disturb the boundary layer as to cause a breaking away too near the leading edge.   Apart from other reasons it is easy to see that a rivet whose head projects above the boundary layer may entirely alter the character of the flow, an exaggerated picture being shown in fig. 1·8 (iv).

Fig. 1·8 (iv).

(3) Good streamline shapes should be such that the breaking away point is as near as possible to the trailing edge.

---

* The firm lines are streamlines ; – – – – is the boundary layers frontier; between · — · — · — · and the cylinder back flow is taking place.   The diagram is purely schematic.

1·9. Approximations. In the applications of mathematics to physical measurements it is frequently desirable and often necessary to make approximations in which certain numbers are neglected on the grounds that their inclusion would not affect significantly the accuracy of the calculation. The standard of approximation which is adopted in any given case is a matter of convention. Thus, if an error of ten per cent. is regarded as permissible, the approximate formula $(1 + x)^2 = 1 + 2x$ might be used to yield $(1·4)^2 = 1·8$, the percentage error in defect of the true value $1·96$ being less than 10 per cent.

Without entering into individual numerical cases, the *linear approximation* will be defined as that approximation in which the squares (and higher powers) and products of the numbers which are to be regarded as small (in comparison with those retained) are neglected. On this understanding one of the commonest approximations is the binomial formula

$$(1) \qquad (1 + x)^n = 1 + nx.$$

Similarly, when x and y are small we can write

$$(2) \quad (1 + x)(1 + y) = 1 + x + y, \quad \frac{1 + x}{1 + y} = (1 + x)(1 + y)^{-1}$$
$$= (1 + x)(1 - y) = 1 + x - y.$$

Taylor's theorem (or, when $a = 0$, Maclaurin's theorem) in the form *

$$f(a + x) = f(a) + xf'(a) + \tfrac{1}{2}x^2 f''(a) + \cdots$$

yields the approximation

$$(3) \qquad f(a + x) = f(a) + xf'(a)$$

when x is regarded as small.

An application of this was made in 1·5, where in effect,

$$f'(\rho) = f'[\rho_0 + (\rho - \rho_0)] = f'(\rho_0) + (\rho - \rho_0)f''(\rho_0) + \cdots,$$

and therefore if $(\rho - \rho_0)$ and $\partial^2 \xi / \partial x^2$ are small, we have approximately

$$f'(\rho) \frac{\partial^2 \xi}{\partial x^2} = f'(\rho_0) \frac{\partial^2 \xi}{\partial x^2}.$$

It may happen that $f'(a) = 0$, and then Taylor's theorem yields the approximation

$$f(a + x) = f(a) + \tfrac{1}{2}x^2 f''(a).$$

This is not a linear approximation in the sense defined above (the linear approximation is $f(a)$), but it is the *first approximation* in the sense that it is the first approximation which differs from a simple constant. The idea of a

---

* $f'(a)$ is always to be obtained by first differentiating $f(x)$ with respect to x, and then putting $x = a$.

first approximation is thus more general than the idea of a linear approximation.

To illustrate this latter point, the expansion

$$\tan\left(\tfrac{1}{2}\pi - x\right) = \frac{1}{x} - \frac{x}{3} - \cdots$$

shows that when x is small, the first approximation to $\tan\left(\tfrac{1}{2}\pi - x\right)$ is $1/x$, a number which increases as x decreases. Taking $\tfrac{1}{2}\pi = 1\cdot5708$ we get the approximation $\tan 1\cdot4 = 5\cdot851$, which exceeds the correct value $5\cdot798$ by less than 1 per cent.

A few examples which will be useful later are given :

(i) Expand $2\lambda e^{\lambda}/(e^{2\lambda} - 1)$ as far as the term in λ^2. Since

$$e^{\lambda} = 1 + \lambda + \tfrac{1}{2}\lambda^2 + \tfrac{1}{6}\lambda^3 + \cdots ,$$
$$e^{2\lambda} = 1 + 2\lambda + 2\lambda^2 + \tfrac{4}{3}\lambda^3 + \cdots ,$$

the given expression is

$$\frac{1 + \lambda + \tfrac{1}{2}\lambda^2 + \cdots}{1 + \lambda + \tfrac{2}{3}\lambda^2 + \cdots} = 1 - \tfrac{1}{6}\lambda^2 + \cdots ,$$

by long division of the numerator by the denominator.

(ii) If $\operatorname{cn} x = 1 - \tfrac{1}{2}x^2 + \dfrac{1 + 4m}{24}\,x^4 - \cdots$, find the first approximation, when x is small, to

$$1 - \tfrac{1}{2}x\left[\sqrt{\frac{1 + \operatorname{cn} x}{1 - \operatorname{cn} x}} - \sqrt{\frac{1 - \operatorname{cn} x}{1 + \operatorname{cn} x}}\right].$$

Here $1 - \operatorname{cn} x = \tfrac{1}{2}x^2\left(1 - \dfrac{1 + 4m}{12}\,x^2\right)$, $1 + \operatorname{cn} x = 2\left(1 - \tfrac{1}{4}x^2\right)$.

Using (2) we get

$$\frac{1 - \operatorname{cn} x}{1 + \operatorname{cn} x} = \tfrac{1}{4}x^2\left(1 + \frac{1 - 2m}{6}\,x^2\right).$$

Therefore from (1)

$$\sqrt{\frac{1 - \operatorname{cn} x}{1 + \operatorname{cn} x}} = \tfrac{1}{2}x\left(1 + \frac{1 - 2m}{12}\,x^2\right), \qquad \sqrt{\frac{1 + \operatorname{cn} x}{1 - \operatorname{cn} x}} = \frac{2}{x}\left(1 - \frac{1 - 2m}{12}\,x^2\right),$$

and therefore the given expression is equal to

$$1 - 1 + \frac{1 - 2m}{12}\,x^2 + \tfrac{1}{4}x^2 = \frac{2 - m}{6}\,x^2.$$

Observe that here the term in x^2 must be retained throughout.

(iii) If $a = x - \tfrac{1}{6}x^3$, find a second approximation to x when a is small.

Clearly a first approximation is $x = a$. We therefore put $x = a + \epsilon$, which gives

$$a = a + \epsilon - \tfrac{1}{6}(a^3 + 3a^2\epsilon + 3a\epsilon^2 + \epsilon^3).$$

Now ϵ is necessarily small since a is itself small. Neglect all powers of ϵ except the first. We then get $\epsilon(1 - \frac{1}{2}a^2) = \frac{1}{6}a^3$, so that from (1),

$$\epsilon = \frac{1}{6}a^3(1 + \frac{1}{2}a^2) = \frac{1}{6}a^3,$$

retaining only the most important term. Thus to a second approximation

$$x = a + \frac{1}{6}a^3.$$

This process could be continued as often as required.

Indeterminate ratios of the form $0/0$ may be evaluated by l'Hospital's theorem, namely, if $f(a) = 0$ and $\phi(a) = 0$, then

$$\lim_{x \to a} \frac{f(x)}{\phi(x)} = \frac{f'(a)}{\phi'(a)}.$$

Proof. If h is small, we have, approximately, from (3),

$$f(a + h) = f(a) + hf'(a) = hf'(a), \quad \phi(a + h) = \phi(a) + h\phi'(a) = h\phi'(a),$$
$$\lim_{x \to a} \frac{f(x)}{\phi(x)} = \lim_{h \to 0} \frac{f(a + h)}{\phi(a + h)} = \frac{f'(a)}{\phi'(a)}.$$

If $f'(a) = 0$ and $\phi'(a) = 0$, the theorem can be applied again.

Now consider $f(a + x, b + y)$ when x, y are small. The extended form of Taylor's theorem gives the linear approximation

$$f(a + x, \ b + y) = f(a, b) + x\left[\frac{\partial f(a + x, \ b + y)}{\partial x}\right]_0 + y\left[\frac{\partial f(a + x, \ b + y)}{\partial y}\right]_0,$$

where suffix 0 denotes that, after the differentiation, we put $x = y = 0$. Denoting these coefficients by A and B, we get

(4) $$f(a + x, \ b + y) = f(a, \ b) + Ax + By.$$

This result embodies the very important principle of the *superposition by addition of small changes*. If y were zero, we should have

$$f(a + x, \ b) = f(a, \ b) + Ax,$$

so that Ax is the increment in $f(a, b)$ when b remains fixed and a is increased to $a + x$. Similarly By is the increment in $f(a, b)$ when a remains fixed and b is increased to $b + y$. When both a and b vary slightly we add the increments, each of which can be obtained independently of the other. The principle applies whatever the number of variables.

Thus, for example, if an aircraft is given a small rotation and a small translation, the change of position in space of its centre of gravity can be got by considering the changes due to rotation without translation, and translation without rotation, and then adding the results.

EXAMPLES I

1. An aircraft has a mass of 6 tons and its solid structure displaces 250 ft.^{3} Calculate in lb. wt. the force of buoyancy, and the vertical component of the aerodynamic force when the aircraft is moving horizontally at constant speed. Take the density of air to be 0·002378 slug/ft.^{3} See 2·5.

2. An aircraft is travelling horizontally at constant speed of 300 mi./hr. If the drag is 70 lb. wt., calculate the least horse-power which the engines must exert to overcome this.

3. The following table gives corresponding values of lift and drag on a certain aircraft ; calculate the gliding angles and exhibit your results as a graph.

L	1820	3540	5290	7020	8760	10,480
D	120	165	243	361	506	666

Draw also a graph of the efficiency L/D.

4. An aircraft of mass 3000 lb. is "looping the loop". At the top of the loop the speed is 90 mi./hr. and the radius of curvature of the path is 500 ft. Calculate the lift at the top of the loop.

5. A flat aerofoil is in the shape of the quadrilateral $ABCD$ where A, B, C, D are successive vertices of a regular hexagon. If the span is 30 ft., calculate the length of the mean chord, the plan area, and the aspect ratio.

6. The aerofoil of Ex. 5 has the port and starboard wings rotated upwards through 5° about the line of symmetry so that the aerofoil is no longer plane. Find the mean chord and calculate the percentage change in the aspect ratio.

7. A symmetrical profile is formed by drawing direct common tangents to two circles of radii 1 and 20, the nose and tail thus consisting of circular arcs. Find the length of the chord c, taken as the longest line in the profile, in terms of d the distance between the centres of the circles, and draw a graph to show the relation between c and d. Find the upper camber and the thickness ratio in terms of c.

8. A profile is in the form of a segment of a circle of radius a cut off by a chord of length c. Calculate the mean camber, and obtain an approximation if c^4/a^4 is negligible.

9. A profile is in the form of the segment of the parabola $x^2 = l(h - y)$ cut off by a chord of length c parallel to the x-axis. Show that the mean camber of the profile is $c/8l$, and express this in terms of the angle which the tangent to the profile at the leading edge of the chord makes with the chord.

10. Plot the profile of the aerofoil Clarke YH from the following values of (x, y) where x, y are percentages of the chord length, and x is measured from the leading edge of the chord ; y_U, y_L refer to the upper and lower surfaces.

x	0	1·25	2·5	5	7·5	10	15	20	30
y_U	3·50	5·45	6·50	7·90	8·85	9·60	10·68	11·36	11·70
y_L	3·50	1·93	1·47	0·93	0·63	0·42	0·15	0·03	0

x	40	50	60	70	80	90	95	100
y_U	11·40	10·51	9·15	7·42	5·62	3·84	2·93	2·05
y_L	0	0	0	0·06	0·38	1·02	1·40	1·85

Draw the camber line, estimate the upper, lower and mean cambers, and the thickness ratio.

11. Incompressible air flows steadily through a conical tube whose cross-sectional diameter decreases linearly in the direction of flow from $2R$ to R. If V is the mean speed where the diameter is $2R$, and v the mean speed where the diameter is $2r$, draw a graph to show v/V as a function of R/r at different points of the tube.

12. Show that atmospheric pressure at sea-level is about one ton weight per square foot. See 2·5.

13. Assuming the speed of sound in air at sea-level to be

$$331 \cdot 1 + 0 \cdot 6T$$

metres per second when the temperature is $T°$ C., draw a graph to give the speed in ft./sec. when the temperature is $T°$ F., in the range $-40°$ F. to $20°$ F.

14. The following table gives the kinematic viscosity of air at various Centigrade temperatures :

$T°$ C.	0	20	40	60	80
$1000\,\nu$ ft.²/sec.	0·142	0·161	0·181	0·202	0·225

Exhibit these results graphically and estimate ν for $T = 56°$, $30°$, $15°$.

Assuming that ν is increased by 4% when the pressure is increased to 25 atmospheres, draw a corresponding graph of $(\nu,\,T)$ for the range of temperature $0°$ to $80°$.

15. Assuming that a mass m falling vertically under gravity experiences a drag kmv^2 when its speed is v, prove that when the mass has fallen a vertical distance y from rest

$$v\frac{dv}{dy} + kv^2 = g.$$

Prove, or verify, that this problem is solved by

$$v^2 = \frac{g}{k}(1 - e^{-2ky})$$

and that the terminal speed of the body is $\sqrt{(g/k)}$. (The terminal speed is the lowest speed which cannot be exceeded. At this speed the drag is equal to the weight.)

16. Assuming the resistance to vertical descent of a passenger-carrying parachute of area S to be $Sv^2/54$ when the speed is v, find an expression for the terminal speed in terms of the total mass M of parachute and passenger.

If $S = 600$ ft.², the parachute weighs 30 lb., and the passenger 10 stone, calculate the terminal speed in ft./sec. and mi./hr.

17. An aircraft is flying at 200 mi./hr., in air at temperature $15°$ C. Taking the wing span 30 ft. as the typical length, calculate the Reynolds' number R and the Mach number M.

18. A sphere of radius a moves with speed V in incompressible air (i.e. in air whose compressibility is neglected.) Prove that the resistance is of the form

$$\tfrac{1}{2}\pi\rho a^2 V^2 f\left(\frac{\nu}{Va}\right),$$

where ν is the kinematic viscosity.

19. Show that two spheres will have the same drag if their speeds are inversely proportional to their radii.

20. If the aircraft rolls with angular velocity Ω, show that the coefficient of aerodynamic force is a dimensionless function of the type

$$f(R,\,M,\,l\Omega/V),$$

where V is the forward speed, l the typical length and R and M are the Reynolds' and Mach numbers.

21. The following table gives corresponding values of incidence, lift coefficient, and drag coefficient for the aerofoil profile Clark YH, aspect ratio 6, Reynolds' number $6\cdot83 \times 10^6$:

α°	$-2\cdot9$	$-1\cdot7$	$+0\cdot6$	$2\cdot8$	$5\cdot1$	$7\cdot4$	$9\cdot6$	$11\cdot8$	$14\cdot0$
C_L	$-0\cdot011$	$+\cdot076$	$\cdot250$	$\cdot420$	$\cdot590$	$\cdot760$	$\cdot924$	$1\cdot084$	$1\cdot224$
C_D	$0\cdot009$	$\cdot009$	$\cdot012$	$\cdot018$	$\cdot027$	$\cdot041$	$\cdot058$	$\cdot081$	$\cdot103$

α°	$16\cdot2$	$17\cdot3$	$18\cdot4$	$19\cdot3$	$20\cdot3$	$22\cdot3$	$25\cdot3$	$28\cdot4$
C_L	$1\cdot366$	$1\cdot426$	$1\cdot474$	$1\cdot304$	$1\cdot252$	$1\cdot102$	$0\cdot912$	$\cdot854$
C_D	$\cdot126$	$\cdot138$	$\cdot151$	$\cdot196$	$\cdot220$	$\cdot277$	$\cdot330$	$\cdot396$

Draw the (C_L, α), (C_D, α) graphs and estimate the maximum value of C_L and the incidence at which this occurs.

22. Referring to fig. 1·8 (i), show that the velocity u given by

$$u = V \sin \frac{\pi y}{2h},$$

where y is the distance from the plate, is a possible distribution in a boundary layer of thickness h.

Taking
$$h^2 = \frac{2\pi^2}{4 - \pi} \frac{vx}{V},$$

where x is the distance from the leading edge, sketch the outline of the boundary layer as x increases from 0 to l.

CHAPTER II

BERNOULLI'S THEOREM

2·0. In this chapter we consider Bernoulli's theorem * and some of its consequences, and, in particular, derive the justification for ignoring the compressibility of air in a first approximation to flight phenomena at sufficiently low speeds.

2·1. Bernoulli's theorem

In its most general form the theorem is as follows.

In the steady motion of an inviscid fluid the quantity

$$\frac{p}{\rho} + K$$

is constant along a streamline, where p is the pressure, ρ is the density and K is the energy per unit mass of the fluid.

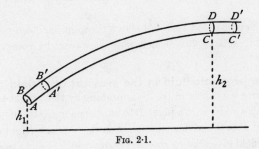

FIG. 2·1.

Proof. Consider the fluid body bounded by the cross-sections AB and CD of a stream filament.

We shall denote by suffixes 1 and 2 the values of quantities at AB and CD respectively. Thus $p_1, q_1, \rho_1, \sigma_1, K_1$ will denote the pressure, fluid speed, density, cross-sectional area and energy per unit mass at AB.

After a short time δt the above fluid body will have moved and will now occupy the portion of the filament bounded by the cross-sections $A'B'$ and $C'D'$ where

---

* Discovered by Daniel Bernoulli, 1700–1783.

(1) $$AA' = q_1 \,\delta t, \quad CC' = q_2 \,\delta t.$$

Since the motion is steady, the mass m of fluid between AB and $A'B'$ will be the same as that between CD and $C'D'$ so that

(2) $$m = \sigma_1 q_1 \,\delta t \,\rho_1 = \sigma_2 q_2 \,\delta t \,\rho_2$$

Let H denote the total energy of the portion of the fluid between $A'B'$ and CD. Then the increase of energy of the fluid body in time δt is

(3) $$(mK_2 + H) - (mK_1 + H) = m(K_2 - K_1)$$

This increase of energy is due to the work done by the pressure thrusts at AB and CD, namely

(4) $$p_1 \sigma_1 q_1 \,\delta t - p_2 \sigma_2 q_2 \,\delta t = \frac{mp_1}{\rho_1} - \frac{mp_2}{\rho_2}$$

Equating (3) and (4) we find that

$$\frac{p_1}{\rho_1} + K_1 = \frac{p_2}{\rho_2} + K_2$$

which shows that

$$\frac{p}{\rho} + K$$

has the same value at any two points of a streamline and is therefore constant along it.

Q.E.D.

It should be emphasised that the above theorem has been proved only for *steady motion* of *inviscid fluid* which may, however, be compressible or incompressible.

2·11. Incompressible fluid in the gravitational field.

The gravitational field of force is a conservative field, meaning by this that the work done by gravity in taking a body from a point P to another point Q is independent of the path taken from P to Q and depends solely on the vertical height of Q above P. A conservative field gives rise to *potential energy* which is measured by the work done in taking the body from one standard position to any other position. In the case of the gravitational field the potential energy per unit mass is gh, where h is the height above a fixed horizontal datum plane. Thus in fig. 2·1 the potential energy per unit mass is gh_1 at A and gh_2 at C.

In addition to potential energy per unit mass the fluid has kinetic energy $\tfrac{1}{2}q^2$ per unit mass. Also ρ is constant so that there is no energy due to compressibility, therefore in 2·1 we have

$$K = \tfrac{1}{2}q^2 + gh,$$

and Bernoulli's theorem is

$$\frac{p}{\rho} + \tfrac{1}{2}q^2 + gh = \text{constant along a streamline.}$$

2·12. The constant in Bernoulli's theorem.

If we fix our attention on a particular streamline, 1, Bernoulli's theorem states that

$$\frac{p}{\rho} + \tfrac{1}{2}q^2 + gh = C_1,$$

where C_1 is constant for that streamline. If we take a second streamline, 2, we get

$$\frac{p}{\rho} + \tfrac{1}{2}q^2 + gh = C_2,$$

where C_2 is constant along the second streamline. We have not proved (and in the general case it is false) that $C_1 = C_2$. When, however, the motion is irrotational, a term which will be explained later (3·3), it is true that the constant is the same for all streamlines, so that

$$\frac{p}{\rho} + \tfrac{1}{2}q^2 + gh = C,$$

where C has the same value at each point of the fluid. It will also be shown later (3·31) that this case arises whenever an inviscid fluid is set in motion by ordinary mechanical means, such as by moving the boundaries suddenly or slowly, by opening an aperture in a closed vessel, or by moving a body through the fluid.

2·13. Aerodynamic pressure.

In the steady motion of an incompressible fluid Bernoulli's theorem enables us to elucidate the nature of pressure still further. In a fluid at rest there exists at each point a hydrostatic pressure p_H and the principle of Archimedes states that a body immersed in the fluid is buoyed up by a force equal to the weight of the fluid which it displaces. The particles of the fluid are themselves subject to this principle and are therefore in equilibrium under the hydrostatic pressure p_H and the force of gravity. It follows at once that $p_H/\rho + gh$ is constant throughout the fluid. If we write

$$p = p_H + p_D,$$

Bernoulli's theorem gives

$$\frac{p_D}{\rho} + \tfrac{1}{2}q^2 + \frac{p_H}{\rho} + gh = C,$$

and therefore

(1) $$\frac{p_D}{\rho} + \tfrac{1}{2}q^2 = C',$$

where $C' = C - (p_H/\rho + gh)$ is a new constant.

Note that (1) is the form which Bernoulli's theorem would assume if the force of gravity were non-existent.

The quantity p_D may be called the *aerodynamic pressure*, or the pressure due to motion. This pressure p_D measures the force with which two air particles are pressed together (for both are subject to the same force of buoyancy). It will be seen that the knowledge of the aerodynamic pressure will enable us to calculate the *total* effect of the air pressure on a body, for we have merely to work out the effect due to p_D and then add the effect due to p_H, which is known from the principles of hydrostatics. This is a very important result, for it enables us to neglect the external force of gravity in investigating many problems, due allowance being made for this force afterwards.

It is often felt that aerodynamic problems in which external forces are neglected or ignored are of an artificial and unpractical nature. This is by no means the case. The omission of external forces is merely a device for avoiding unnecessary complications in our analysis.

It should therefore be borne in mind that when we neglect external forces we in effect calculate the aerodynamic pressure.

We also see from (1) that the aerodynamic pressure is greatest where the speed is least, and also that the greatest aerodynamic pressure occurs at points of zero velocity i.e. stagnation points.

2·2. The Pitot tube. Fig. 2·2 (*a*) shows a tube $ABCD$ open at A, where it is drawn to a fine point, and closed, at D containing mercury in the **U**-shaped part.

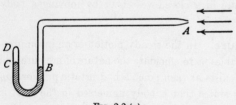

FIG. 2·2 (*a*).

If this apparatus is placed with the open end upstream in steadily flowing air, the axis of the horizontal part in the figure will form part of the streamline which impinges at A. Hence if p_1 is the pressure just inside the tube at A, and p is the pressure ahead of A, we shall have, by Bernoulli's theorem,

$$\frac{p_1}{\rho} = \frac{p}{\rho} + \tfrac{1}{2}q^2,$$

since the air inside the tube is at rest. The pressure p_1 is measured by the difference in levels of the mercury at B and C, assuming a vacuum in the part CD. This is the simplest form of Pitot tube for determining the quantity $p + \tfrac{1}{2}\rho q^2$.

In applications it is often required to measure the speed q. In order to do this we must have a means of measuring p.

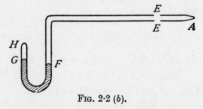

FIG. 2·2 (*b*).

This measurement can be made by means of the apparatus shown in fig. 2·2 (*b*), which differs from the former only in having the end A closed and holes

in the walls of the tube at E slightly downstream of A. The streamlines now follow the walls of the tube from A, and the air within the tube being at rest and the pressure being necessarily continuous, the pressure just outside the tube at E is equal to the pressure just in-side the tube at E, and this is measured by the difference in the levels of the mercury at G and F. In practice it is usual

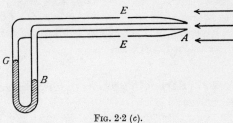

FIG. 2·2 (c).

to combine both tubes into a single apparatus as shown in fig. 2·2 (c).

In this apparatus the difference in levels of the mercury at B and G measures $p_1 - p = \frac{1}{2}\rho q^2$.

The above description merely illustrates the principle of speed measurements with the Pitot tube. The actual apparatus has to be very carefully designed, to interfere as little as possible with the fluid motion. With proper design and precautions in use, the Pitot tube can give measurements within one per cent. of the correct values in an actual fluid, such as air or water.

2·3. The work done by air in expanding.

Let S and S' be the surfaces of a unit mass of air before and after a small expansion.

Let the normal displacement of the element dS of the surface S be dn.

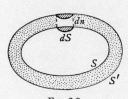

FIG. 2·3.

Suppose the pressure of the air to be p. Then the work done by the air is

$$p \, \Sigma \, dS \, . \, dn = p \times \text{increase in volume} = p \, dv,$$

where v is the volume within S. But since the mass is unity, $v\rho = 1$.

Hence the work done by the air $= pd\left(\dfrac{1}{\rho}\right)$,

and if the expansion is from density ρ to density ρ_0,

$$\text{the work done} = \int_{\rho}^{\rho_0} pd\left(\frac{1}{\rho}\right).$$

We suppose that the pressure is a function of the density only.[*]

We shall call *intrinsic energy* per unit mass the work which a unit mass of the air could do as it expands under the assumed relation between p and ρ from its actual state to some standard state in which the pressure and density are p_0 and ρ_0. Calling E the intrinsic energy per unit mass, we get

$$E = \int_{\rho}^{\rho_0} pd\left(\frac{1}{\rho}\right) = \frac{p_0}{\rho_0} - \frac{p}{\rho} - \int_{p}^{p_0} \frac{dp}{\rho}$$

---

[*] When the pressure is a function of the density the flow is called *barotropic*.

on integrating by parts. Thus

$$(1) \qquad E = \frac{p_0}{\rho_0} - \frac{p}{\rho} + \int_{p_0}^{p} \frac{dp}{\rho}.$$

Note that intrinsic energy is a form of potential energy analogous to that of a stretched elastic string.

2·31. Bernoulli's theorem for compressible flow.

Here the energy per unit mass is

$$K = \tfrac{1}{2}q^2 + gh + E$$

where E is the intrinsic energy given by 2·3 (1). Substituting in 2·1 we find that

$$\frac{p}{\rho} + \tfrac{1}{2}q^2 + gh + \frac{p_0}{\rho_0} - \frac{p}{\rho} + \int_{p_0}^{p} \frac{dp}{\rho}$$

is constant along a streamline and therefore that

$$\int_{p_0}^{p} \frac{dp}{\rho} + \tfrac{1}{2}q^2 + gh = \text{constant along a streamline.}$$

If we neglect gravity, that is if we deal with aerodynamic pressure only (2·13), we have

$$\int_{p_0}^{p} \frac{dp}{\rho} + \tfrac{1}{2}q^2 = \text{constant along a streamline.}$$

The differential form of this, which is useful in considering compressible flow about an aerofoil is

$$(1) \qquad dp = -\rho q \, dq$$

2·32. Application of Bernoulli's theorem to adiabatic expansion.

When air expands adiabatically (that is to say without gain or loss of heat, 15·01), the pressure and the density are connected by the relation

$$(1) \qquad p = \kappa \rho^{\gamma},$$

where κ and γ are constants. For dry air, $\gamma = 1·405$. Therefore

$$\int_{p_0}^{p} \frac{dp}{\rho} = \kappa \gamma \int_{\rho_0}^{\rho} \rho^{\gamma-2} \, d\rho = \frac{\kappa \gamma}{\gamma - 1} [\rho^{\gamma-1} - \rho_0{}^{\gamma-1}] = \frac{\gamma}{\gamma - 1} \left(\frac{p}{\rho} - \frac{p_0}{\rho_0} \right).$$

Since p_0/ρ_0 refers to a standard state, this is constant, and therefore Bernoulli's theorem gives

$$\frac{\gamma}{\gamma - 1} \frac{p}{\rho} + \tfrac{1}{2}q^2 + gh = C.$$

If we take p_0 to be the pressure when the velocity is zero * and neglect the effect of gravity, we obtain

$$\frac{\gamma}{\gamma - 1}\frac{p}{\rho} + \tfrac{1}{2}q^2 = \frac{\gamma}{\gamma - 1}\frac{p_0}{\rho_0},$$

so that

(2)
$$q^2 = \frac{2\gamma}{\gamma - 1}\frac{p_0}{\rho_0}\left(1 - \frac{p\,\rho_0}{p_0\,\rho}\right).$$

Now
$$\frac{p\,\rho_0}{p_0\,\rho} = \frac{\rho^{\gamma-1}}{\rho_0^{\gamma-1}} = \left(\frac{p}{p_0}\right)^{\frac{\gamma-1}{\gamma}} \quad \text{from (1).}$$

Also, from the theory of sound waves, it is known (1·5) that the speed of sound c_0 when the pressure is p_0 is given by

$$c_0^2 = \frac{\gamma\,p_0}{\rho_0}.$$

Therefore we obtain from (2)

$$\left(\frac{p}{p_0}\right)^{\frac{\gamma-1}{\gamma}} = 1 - \frac{\gamma - 1}{2}\left(\frac{q}{c_0}\right)^2,$$

and therefore

$$\frac{p}{p_0} = \left[1 - \frac{\gamma - 1}{2}\left(\frac{q}{c_0}\right)^2\right]^{\frac{\gamma}{\gamma-1}} = 1 - \frac{\gamma}{2}\left(\frac{q}{c_0}\right)^2 + \frac{\gamma}{8}\left(\frac{q}{c_0}\right)^4 + \dots$$

$$= 1 - \frac{1}{2}\frac{\rho_0\,q^2}{p_0} + \frac{\gamma}{8}\left(\frac{q}{c_0}\right)^4 + \dots.$$

The ratio of the third term to the second in this expansion is $q^2/4c_0^2$, so that even when the speed q is equal to half the speed of sound this ratio is $1/16$. Thus it appears that we may, to a good approximation, neglect the third term, unless q is a considerable fraction of c_0.

Bernoulli's theorem for air will then take the form

$$\frac{p}{\rho_0} + \tfrac{1}{2}q^2 = \frac{p_0}{\rho_0},$$

which means that the air may be treated as incompressible within a very considerable range of speeds. In particular, for speeds of 300 miles per hour, the error in speed measurements made by the use of the Pitot tube (see 2·2) will only be about 2 per cent.

---

* It is not asserted that zero velocity is attained.   The pressure p_0 is nevertheless uniquely defined by the equation which follows.

2·4. The Venturi tube. The principle of the Venturi tube is illustrated in fig. 2·4. The apparatus is used for measuring the flow in a pipe and

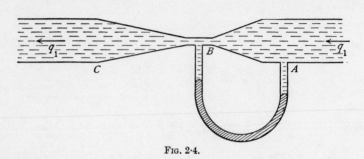

FIG. 2·4.

consists essentially of a conical contraction in the pipe from the full bore at A
to a constriction at B, and a gradual widening of the pipe to full bore again
at C. To preserve the streamline flow, the opening from B to C has to be very
gradual. A **U**-tube manometer containing mercury joins openings at A and
B, and the difference in level of the mercury measures the difference in pressures
at A and B. Let p_1, q_1, p_2, q_2 be the pressures and speeds at A and B respectively. Then

$$\frac{p_1}{\rho} + \tfrac{1}{2} q_1^2 = \frac{p_2}{\rho} + \tfrac{1}{2} q_2^2,$$

by Bernoulli's theorem.

Let S_1, S_2 be the areas of the cross-sections at A and B.

Then
$$q_1 S_1 = q_2 S_2,$$

since the same volume of fluid crosses each section in a given time. Therefore

$$q_1 = \sqrt{\frac{2(p_1 - p_2)}{\rho \left(\dfrac{S_1^2}{S_2^2} - 1\right)}},$$

$p_1 - p_2$ is given by observation and the value of q_1 follows.

If h is the difference in level of the mercury in the two limbs of the manometer and σ is the density of mercury, the formula becomes

$$q_1 = \sqrt{\frac{2 g h \sigma}{\rho \left(\dfrac{S_1^2}{S_2^2} - 1\right)}} = K\sqrt{h},$$

K being a constant for the apparatus.

2·41. Flow of air measured by the Venturi tube. Assuming adiabatic changes in the air from the entrance to the throat, we obtain from Bernoulli's theorem and the equation of continuity

$$\frac{\gamma}{\gamma - 1}\frac{p_1}{\rho_1} + \tfrac{1}{2}q_1{}^2 = \frac{\gamma}{\gamma - 1}\frac{p_2}{\rho_2} + \tfrac{1}{2}q_2{}^2,$$

$$\rho_1\, q_1\, S_1 = \rho_2\, q_2\, S_2,$$

whence we easily obtain

$$q_1{}^2 = \frac{\dfrac{2\gamma}{\gamma - 1}\left(\dfrac{p_1}{\rho_1} - \dfrac{p_2}{\rho_2}\right)}{\left(\dfrac{\rho_1}{\rho_2}\right)^2 \dfrac{S_1{}^2}{S_2{}^2} - 1}.$$

Now, $\dfrac{p_1}{p_2} = \left(\dfrac{\rho_1}{\rho_2}\right)^{\gamma}$, and therefore

$$q_1{}^2 = \frac{\dfrac{2\gamma}{\gamma - 1}\dfrac{p_1}{\rho_1}\left[1 - \left(\dfrac{p_2}{p_1}\right)^{\frac{\gamma - 1}{\gamma}}\right]}{\left(\dfrac{p_1}{p_2}\right)^{\frac{2}{\gamma}}\left(\dfrac{S_1}{S_2}\right)^2 - 1}.$$

To use this formula we must know p_1, p_2 and ρ_1. The instrument must therefore be modified so that A and B in fig. 2·4 are connected to separate manometers, thereby obtaining measures of the actual pressures p_1, p_2 and not their difference, as in the case of incompressibility. For speeds not comparable with the speed of sound, the ordinary formula and method may be used (see 2·32).

2·5. Standard atmosphere. Since the density, pressure and temperature of air depend on many circumstances of date, position, humidity and so on, it is usual in aerodynamics to postulate certain arbitrary standard values for these fundamental quantities, afterwards making any necessary corrections for local conditions.

The combined gas laws of Boyle and Charles may be expressed in the form

$$(1) \qquad\qquad p = R\,\rho\,T,$$

where T is the absolute temperature and R is a constant,* whose physical dimensions are $L^2 T^{-2} D^{-1}$, where D is the measure-ratio of temperature. The absolute zero of temperature is taken to be $-273°$ C. If we assume that, at normal temperature $15°$ C., and normal atmospheric pressure 760 mm. of mercury, or 2116 lb.wt./ft.², the density of air is 0·07651 lb./ft.³, then $T = 273° + 15° = 288°$ C., and $R = 3090$ ft.²/(sec.² degree C.).

---

* There should be no occasion to confuse this with Reynolds' number.

If, however, we adopt the practical or British engineering units in which the fundamental quantities are length, time and *force*, with the foot, second and lb. wt. respectively as their units, then *mass* is a quantity derived from Newton's second law of motion, force = mass × acceleration, whose unit is termed the *slug*. With this new set of three fundamental quantities, of measure-ratios L, T, F, the dimensions of mass are $FL^{-1}T^2$. Hence, since 1 lb. wt. = g poundals, we find, on changing from practical units to the ft.-lb.-sec. system by the method of dimensions of 1·7, that 1 slug = g lb.

If we take $g = 32·174$, the density of air becomes 0·002378 slug/ft.^{3} at normal temperature and pressure.

Consider a small cylinder of air whose axis, of length dh, is vertical and whose cross-sectional area is σ. The weight of the contained air is, in equilibrium, just balanced by the difference in pressure thrusts on the ends so that $\sigma\, dp + \sigma\, dh\, g\rho = 0$. Thus

$$(2) \qquad \frac{dp}{dh} = -g\rho.$$

The *international standard atmosphere* is defined by the assumption that the temperature is a particular linear function of the height h above sea-level up to 36,093 ft.* and is thereafter constant. Using the centigrade absolute scale

$$(3) \qquad T = 288 - 0·001981h, \quad h < 36{,}093 \text{ ft.}$$
$$(4) \qquad T = 273 - 56·5, \qquad\quad h > 36{,}093 \text{ ft.}$$

The part of the atmosphere below 36,093 ft. is called the *troposphere*, the part above that level is the *stratosphere*. The height which divides the two parts is arbitrarily laid down as a reasonable average representation of the conditions.

Considering the troposphere and eliminating ρ and T between (1), (2), and (3), we get

$$\frac{1}{p}\frac{dp}{dh} = -\frac{g}{R(288 - 0·001981h)}$$

which yields on integration

$$(5) \qquad \frac{p}{p_0} = (1 - 0·00000688h)^{5·256},$$

where p_0 is the pressure at sea-level.

If ρ_0 and T_0 are the density and absolute temperature at sea-level, (1) gives

$$\frac{p}{p_0} = \frac{\rho}{\rho_0} \times \frac{T}{T_0},$$

and therefore

$$(6) \qquad \frac{\rho}{\rho_0} = (1 - 0·00000688h)^{4·256}.$$

Corresponding relations for the stratosphere are obtained from (1), (2) and (4).

---

* The odd 93 ft. have no real significance. The numbers here given are simply adjusted to be consistent.

Fig. 2·5 gives a diagrammatic representation of aerial pressures and heights attained up to 1943.   Since then many successive height records have been

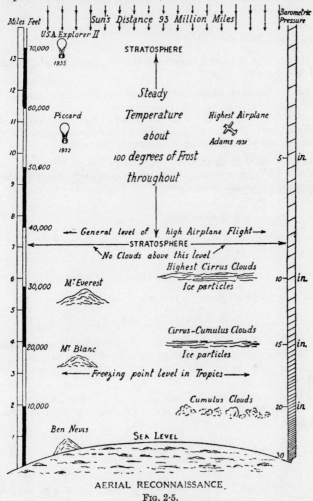

AERIAL RECONNAISSANCE.
FIG. 2·5.

established, among which are the ascent of Malcolm Ross in a balloon to 113,500 ft. in 1961, and the reaching of a height of 351,000 ft. in 1963, by Joseph Walker in an X-15 rocket plane.   Heights of satellites are excluded.

EXAMPLES II

1. Petrol is led steadily through a pipe line which passes over a hill of height h into the valley below, the speed at the crest being v.   Show that by properly adjusting the ratio of the cross-sections of the pipe at the crest and in the valley the pressure may be equalised at these two places.

2. Incompressible air of density ρ flows through a Venturi tube. The points where the cross-sections are σ_1, σ_2 are connected by a differential pressure gauge which indicates the pressure $p_1 - p_2$. Prove that the mass of air flowing through the tube per unit time is

$$\sigma_1 \, \sigma_2 \sqrt{\frac{2\rho\,(p_1 - p_2)}{\sigma_1^{\,2} - \sigma_2^{\,2}}}.$$

3. If air pressure is halved under adiabatic expansion, find the ratio of the initial and final densities.

4. If $pv^{1\cdot4} = $ constant, show that the work done by air expanding from the state (p_1, v_1) to the state (p_2, v_2) is $2\cdot5\,(p_1v_1 - p_2v_2)$.

An air compressor takes 4 ft.^{3} of air per stroke at the pressure of standard air at sea-level and compresses it to 4 times that pressure according to the above law. Find the least horse-power required at 60 strokes per minute.

5. One cubic metre of air at $20°$ C. and pressure 950 gm. wt./cm.^{2} is compressed to a pressure of 9,500 gm. wt./cm.^{2} in such a way that $pv^{1\cdot2} = $ constant throughout the compression. Find the work done. Take the density of air at $20°$ C. to be $0\cdot0013$ gm./cm.^{3}

6. With notation of 2·32 show that

$$\frac{\rho}{\rho_0} = 1 - \tfrac{1}{2} M_0^{\,2} + \frac{2 - \gamma}{8} M_0^{\,4} - \dots\,,$$

where M_0 is the Mach number q/c_0.

If $M_0^{\,4}$ is neglected, exhibit graphically the relation between ρ/ρ_0 and M_0.

7. Establish the formula for the pressure along a streamline

$$\frac{p}{p_0} = 1 - \tfrac{1}{2}\gamma\, M_0^{\,2} + \tfrac{1}{8}\gamma\, M_0^{\,4},$$

where $M_0^{\,6}$ and higher powers are neglected, M_0 being the Mach number q/c_0.

An air-speed indicator is graduated on the basis of the formula $p_0 - p = \tfrac{1}{2}\rho_0\, q^2$. Draw a graph to show the percentage error in the indicated air-speed for values of M_0 up to $0\cdot6$.

8. If air flows out adiabatically, from a large closed vessel in which the pressure is n times the atmospheric pressure p, through a thin pipe, show that the speed V of efflux is given by

$$V^2 = \frac{2\gamma\, p}{(\gamma - 1)\rho}\, [n^{1-1/\gamma} - 1],$$

ρ being the density of the atmosphere.

9. Air flows along a tube of small variable cross-section σ at the point whose distance in arc from a fixed cross-section is s. Use the equation of continuity to prove that

$$\frac{d}{ds} \log \rho + \frac{d}{ds} \log q = -\frac{d}{ds} \log \sigma.$$

10. Air flows along a tube of small variable cross-section σ at the point whose distance in arc from a fixed cross-section is s. Prove that, if the expansion is adiabatic,

$$\frac{d}{ds} \log \sigma + \left(1 - \frac{q^2}{c^2}\right)\frac{d}{ds} \log q = 0,$$

where c is the speed of sound at the point considered.

11. If q, σ, ρ, p are corresponding values of speed, cross-section, density and pressure at any point of a stream filament, prove that

$$\text{(i)} \quad q + \frac{c^2}{\rho} \frac{d\rho}{dq} = 0,$$

$$\text{(ii)} \quad \frac{d\sigma}{dq} = -\frac{\sigma}{q}\left(1 - \frac{q^2}{c^2}\right),$$

where $c^2 = \gamma p/\rho$ gives the local speed of sound.

12. In the preceding example show that the cross-sectional area and the speed increase together if the speed exceeds the local speed of sound. Prove also that the cross-sectional area of the filament has a minimum value.

13. If c_m is the speed of sound at the minimum cross-section in Ex. 12, prove that there is an upper limit to the value of q given by

$$q_{max} = c_m \times \sqrt{\frac{\gamma + 1}{\gamma - 1}} = 2{\cdot}45\, c_m.$$

14. Show that the density of standard air at sea-level is $0{\cdot}07651$ lb./ft.^{3}

15. Draw graphs to show how relative pressure p/p_0, and relative density ρ/ρ_0 are related to height in the standard atmosphere.

16. Express the (temperature, height) relation for standard air in the Fahrenheit scale of temperature.

Find the temperature, density and pressure of standard air at 10,000 ft.

17. Calculate the pressure and density of standard air at 50,000 ft.

18. Compare the work done by 3 ft.^{3} of air at a pressure of 120 lb. wt./in^2. expanding to 10 ft.^{3} in the two cases where the air obeys (i) Boyle's law, (ii) the adiabatic law.

19. If dry atmosphere air ($\gamma = 1{\cdot}4$) at 14° C. is suddenly compressed adiabatically to one-tenth of its original volume, find its final temperature and pressure, taking the barometric height to be 76 cm. of mercury.

CHAPTER III

TWO-DIMENSIONAL MOTION

3·0. Motion in two dimensions. Motion of a fluid is said to be two-dimensional when the velocity at every point is parallel to a fixed plane and is the same at every point of any given normal to that plane.

We shall in particular consider the two-dimensional motion of air regarded as an ideal inviscid fluid whose compressibility is neglected ; incompressible air.

It is often useful in order to form a vivid mental picture of the phenomenon to suppose the fluid to be confined between two planes parallel to the plane of the motion and at unit distance apart, the fluid being supposed to glide freely over those planes without encountering any resistance of a frictional nature.

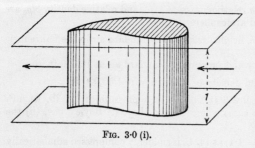

This idea corresponds with the case of a two-dimensional wind tunnel. Here the aerofoil on which experiments are to be performed is cylindrical and stretches from one wall to the opposite, fig. 3·0 (i).

FIG. 3·0 (i).

To complete the picture we choose as representative plane of the motion the parallel plane midway between these hypothetical fixed planes.

We shall call this plane the x, y plane and the section of the aerofoil by this plane will be a profile, fig. 3·0 (ii). We can then use the language of plane geometry and speak of the flow past the curve. In using this terminology the state of affairs conventionally depicted in fig. 3·0 (ii)

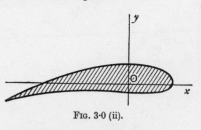

FIG. 3·0 (ii).

should always be mentally referred to the state depicted in fig. 3·0 (i).

3·1. Stream function. In the two-dimensional motion of air regarded as an incompressible fluid, let A be a fixed point in the plane of the motion, and ABP, ACP two curves also in the plane joining A to an arbitrary point P. We suppose that no air is created or destroyed within the region R bounded by these curves. Then the condition of continuity may be expressed in the following form.

The rate at which air flows into the region R from right to left across the curve ABP is equal to the rate at which it flows out from right to left across the curve ACP. We shall use the convenient term *flux* to denote rate of flow.

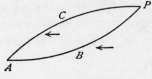

FIG. 3·1 (i).

The term from right to left is relative to an observer who proceeds along the curve from the fixed point A in the direction in which the arc s of the curve measured from A is increasing.

Thus the flux from right to left across ACP is equal to the flux from right to left across any curve joining A to P.

Once the *base point* A has been fixed this flux therefore depends solely on the position of P, and the time t. If we denote the flux by ψ, it is a function of the position of P and the time. In cartesian coordinates, for example,

$$\psi = \psi(x, y, t).$$

The function ψ is called the *stream function*.

The existence of this function is merely a consequence of the assertion of the continuity of *incompressible* air.

Now take two points P_1, P_2, and let ψ_1, ψ_2 be the corresponding values of the stream function.

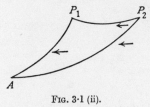

FIG. 3·1 (ii).

Then, from the same principle, the flux across AP_2 is equal to the flux across AP_1 plus that across $P_1 P_2$. Hence the flux across $P_1 P_2$ from right to left $= \psi_2 - \psi_1$.

It follows from this that if we take a different base point, A' say, the stream function merely changes by the flux from right to left across $A'A$.

Moreover, if P_1 and P_2 are points of the same streamline, the flux from right to left across P_1P_2 is equal to the flux from right to left across the part of the streamline between P_1 and P_2. Thus $\psi_1 - \psi_2 = 0$. Therefore

the stream function is constant along a streamline.

The equations of the streamlines are therefore obtained from $\psi = c$, by giving arbitrary values to the constant c.

When the motion is steady, the streamline pattern is fixed. When the motion is not steady, the pattern changes from instant to instant.

The dimensions of the stream function are represented by L^2T^{-1}. See 1·7.

3·11. Velocity derived from the stream function.

Let $P_1 P_2 = \delta s$ be an infinitesimal arc of a curve, so short that it may be considered as straight.

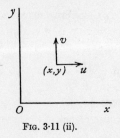

The fluid velocity across this arc can be resolved into components along and perpendicular to δs. The component along δs contributes nothing to the rate of flow across. The component at right angles to δs

$$= \text{flux across divided by } \delta s$$
$$= (\psi_2 - \psi_1)/\delta s,$$

FIG. 3·11 (i).

where ψ_1, ψ_2 are the values of the stream functions at P_1, P_2. Thus the velocity from right to left across δs becomes in the limit $\partial \psi / \partial s$.

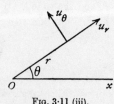

FIG. 3·11 (ii).   FIG. 3·11 (iii).

In cartesian coordinates, by considering infinitesimal increments δx, δy, the components u, v of velocity parallel to the axes are given by

$$u = -\frac{\partial \psi}{\partial y}, \quad v = \frac{\partial \psi}{\partial x}.$$

In polar coordinates, we get similarly

$$u_r = -\frac{\partial \psi}{r \, \partial \theta}, \quad u_\theta = \frac{\partial \psi}{\partial r},$$

for the radial and transverse components, fig. 3·11 (iii), since δs will be δr and $r \, \delta \theta$ for the radial and transverse increments respectively.

3·12. Rankine's theorem.

If the stream function ψ can be expressed as the sum of two stream functions in the form $\psi = \psi_1 + \psi_2$, the streamlines can be drawn when the streamlines $\psi_1 = \text{constant}$, $\psi_2 = \text{constant}$ are known.

Taking a small constant ω, we draw the streamlines $\psi_1 = \omega, 2\omega, 3\omega \ldots$, $\psi_2 = \omega, 2\omega, 3\omega, \ldots$, and so obtain a network of streamlines as shown in fig. 3·12.

At the points marked 3, $\psi = 3\omega$, at the points marked 4, $\psi = 4\omega$, and so on. If we join the points with the same numeral we obtain lines along which $\psi = \text{constant}$, the dotted lines in the figure.

The meshes of the network can be made as small as we please by taking ω small enough, and the meshes can be regarded as parallelograms (of different sizes). The streamlines are then obtained by drawing the diagonals of the meshes. The streamlines which pass through the corners of a mesh are approximately parallel in the neighbourhood of the mesh.

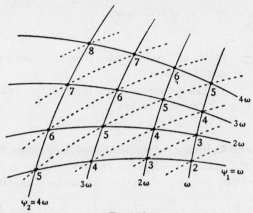

FIG. 3·12.

3·13. The stream function of a uniform wind.
Suppose every air particle to move with the constant speed U parallel to the x-axis.

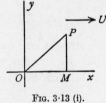

FIG. 3·13 (i).

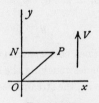

FIG. 3·13 (ii).

If P is the point (x, y), the flux from right to left across OP is the same as the flux from right to left across PM, where PM is perpendicular to Ox. Thus the flux is $- Uy$, and therefore

$$\psi = - Uy$$

is the stream function for this motion. In polar coordinates,

$$\psi = - Ur \sin \theta.$$

Similarly, for a uniform wind in the direction Oy of speed V we get

$$\psi = Vx = Vr \cos \theta.$$

If we superpose the two winds, we get a wind of speed $\sqrt{U^2 + V^2}$ inclined to the x-axis at the angle $\alpha = \tan^{-1} V/U$, and for this wind

$$\psi = - Uy + Vx.$$

Writing $U = Q\cos\alpha$, $V = Q\sin\alpha$, we obtain the stream function for a uniform wind Q, making an angle α with the x-axis, namely

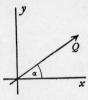

$$\psi = Q(x\sin\alpha - y\cos\alpha),$$

or, in polar coordinates,

$$\psi = -Qr\sin(\theta - \alpha),$$

and in all these cases the streamlines are straight lines, as is indeed obvious.

FIG. 3·13 (iii).

The streamline which passes through the origin corresponds to $\psi = 0$ and is therefore the line

$$\theta = \alpha.$$

3·14. Circular Cylinder. Consider the following stream function, which gives the flow of a uniform wind past a circular cylinder, as can readily be verified, for $\psi = 0$ on the circle $r = a$, and the motion is irrotational (3·311)

$$\psi = V\left(r\sin\theta - \frac{a^2}{r}\sin\theta\right) = Vy\left(1 - \frac{a^2}{r^2}\right) = \psi_1 + \psi_2,$$

where
$$\psi_1 = Vy, \quad \psi_2 = -\frac{a^2\,Vy}{x^2 + y^2}.$$

Putting $\psi_1 = m\,Va$, $\psi_2 = -n\,Va$, we get

$$y = ma, \quad x^2 + \left(y - \frac{a}{2n}\right)^2 = \frac{a^2}{4n^2},$$

so that the lines corresponding to ψ_1 and ψ_2 are straight lines parallel to the x-axis and circles touching the x-axis at the origin. By giving m, n the values $0·1, 0·2, 0·3, \ldots$, the streamlines can be readily plotted with the aid of Rankine's theorem.

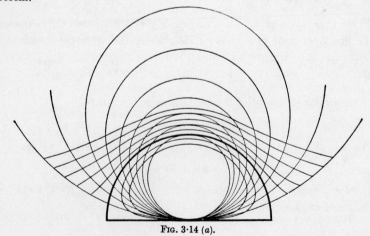

FIG. 3·14 (a).

The streamlines are symmetrical with respect to the y-axis, for changing the sign of x does not alter the equation.

The streamlines above the x-axis are the reflection in that axis of the streamlines below it, as is obvious from symmetry. If the velocity V is reversed the streamline pattern is unaltered.

Writing $\psi = kVa$, the equation of the streamlines is

$$ka = y\left(1 - \frac{a^2}{r^2}\right),$$

so that when $r \to \infty$, $y \to ka$ and therefore $y = ka$ is the asymptote of the streamline. Also if $k > 0$, then $y > ka$ and therefore the streamline approaches its asymptote from above.

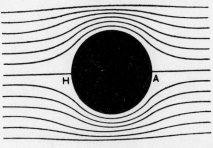

Fig. 3·14 (b).

3·15. The dividing streamline. In the flow past any cylinder the contour of the cylinder must itself form part of a streamline. Since the stream function for the circular cylinder (3·14) is

$$\psi = Uy\left(1 - \frac{a^2}{r^2}\right),$$

and since on the cylinder we have $r = a$, it follows that the contour is part of the streamline $\psi = 0$. The complete streamline $\psi = 0$ consists therefore of the circle $r = a$ and that part of $y = 0$ which is external to the cylinder, see fig. 3·14 (b).

Thus the stream advances towards the cylinder along the x-axis until the point A is reached, then divides and proceeds in opposite senses round the cylinder, joins up again at H and moves off along the x-axis. This streamline which divides on the contour is called the *dividing streamline*. The dividing line is important, for a knowledge of its position at once enables us to draw the general form of the flow pattern by successive lines at first nearly coincident with it, and then becoming less and less influenced by its shape. A study of fig. 3·14 (b) will make this clear.

3·2. Circulation. Consider a closed curve C imagined to lie entirely in the fluid. Note that such a closed curve, or *circuit*, is a purely geometrical concept; it is not a boundary interfering with the flow.

Let A be a fixed point on the curve and let P be any other point of C. The position of P is determined if we know the length s of the arc AP and the sense, say counterclockwise, in which this arc is to be measured. The fluid velocity **q** at P will make an angle α, say, with the tangent at P drawn in the sense in which s increases. Then

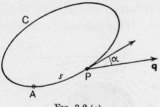

$$q_s = q \cos \alpha$$

will be the component along the tangent of the velocity **q**.

Def. The circulation *in the closed circuit C is the line integral of the tangential component of the velocity taken round the circuit in the sense in which the arc s increases.*

Fig. 3·2 (a).

Thus we can write

(1) $$\operatorname{circ} C = \int_{(C)} q \cos \alpha \, ds = \int_{(C)} q_s \, ds = \int_{(C)} \mathbf{q} \, d\mathbf{r},$$

the last being the vector form of the statement (see 21·12).

If the curve lies in the x, y plane the tangential component of the velocity is

$$u \frac{dx}{ds} + v \frac{dy}{ds},$$

Fig. 3·2 (b).

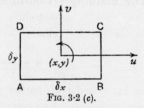

Fig. 3·2 (c).

and in this case $$\operatorname{circ} C = \int_{(C)} (u \, dx + v \, dy).$$

If in particular we take for C an infinitesimal rectangle $ABCD$ whose centre is the point (x, y) and whose sides AB, BC are parallel to the axes and of lengths δx, δy, respectively,

$$\operatorname{circ} ABCD = u_{AB} . AB + v_{BC} . BC - u_{CD} . CD - v_{DA} . AD,$$

where u_{AB} means the average value of u on AB, with similar meanings for v_{BC}, etc. Thus

$$\operatorname{circ} ABCD = \delta x \, \delta y \left(\frac{v_{BC} - v_{DA}}{\delta x} - \frac{u_{CD} - u_{AB}}{\delta y} \right).$$

Now to the first order, by Taylor's theorem for two variables,*

$$v_{BC} = v + \tfrac{1}{2} \delta x \frac{\partial v}{\partial x}, \quad v_{DA} = v - \tfrac{1}{2} \delta x \frac{\partial v}{\partial x}.$$

---

* At any point $(x + \tfrac{1}{2}\delta x,\ y + \eta)$ of BC the v-component is

$$v + \tfrac{1}{2}\delta x \frac{\partial v}{\partial x} + \eta \frac{\partial v}{\partial y}$$

and the average value of the last term is zero as η goes from $-\tfrac{1}{2}\delta y$ to $\tfrac{1}{2}\delta y$.

Thus we can write

(2) $\qquad \operatorname{circ} ABCD = \delta x\, \delta y \left(\dfrac{\partial v}{\partial x} - \dfrac{\partial u}{\partial y} \right) = ABCD \left(\dfrac{\partial v}{\partial x} - \dfrac{\partial u}{\partial y} \right).$

Thus the circulation in an infinitesimal rectangle will be proportional to its area.

Now any simple plane circuit C can be divided by lines parallel to the axes into infinitesimal rectangles, and the circulation round C is clearly the sum of the circulations in the infinitesimal meshes of the net so obtained, for the contributions to

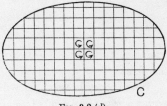

Fig. 3·2 (d).

the circulations of any boundary common to two meshes cancel. Thus

(3)

$$\operatorname{circ} C = \iint \left(\frac{\partial v}{\partial x} - \frac{\partial u}{\partial y} \right) dx\, dy \quad \text{or} \quad \int_{(C)} (u\, dx + v\, dy) = \iint \left(\frac{\partial v}{\partial x} - \frac{\partial u}{\partial y} \right) dx\, dy,$$

the surface integral being over the whole surface contained by C.

It is proper to remark that the above proof assumes that the fluid is distributed over the whole of the interior of C.

If, for example, the circuit C were to embrace a cylinder or aerofoil C', the above demonstration would now prove only that

Fig. 3·2 (e).

(4) $\qquad \operatorname{circ} C - \operatorname{circ} C' = \iint \left(\dfrac{\partial v}{\partial x} - \dfrac{\partial u}{\partial y} \right) dx\, dy,$

the surface integral being taken over the area between C and C'.

We also observe that if P and Q are any two continuous and differentiable functions of x, y defined over the whole area inside C a similar argument shows that

(5) $\qquad \displaystyle\int_{(C)} (P\, dx + Q\, dy) = \iint \left(\dfrac{\partial Q}{\partial x} - \dfrac{\partial P}{\partial y} \right) dx\, dy,$

and on writing $-Q$ for P and P for Q that

(6) $\qquad \displaystyle\int_{(C)} (P\, dy - Q\, dx) = \iint \left(\dfrac{\partial P}{\partial x} + \dfrac{\partial Q}{\partial y} \right) dx\, dy.$

These results constitute the two-dimensional form of Stokes's theorem (21·7).

3·21. Vorticity. We have seen in 3·2 that the circulation in an infinitesimal plane circuit is proportional to the area of the circuit.

Def. In two-dimensional motion the vector at a point P which is perpendicular to the plane of the motion and whose magnitude is equal to the limit of the ratio of the *circulation in an infinitesimal circuit embracing P to the area of the circuit is called the vorticity at P.*

Thus if we denote the magnitude of the vorticity by ω, we see from 3·2 (2) that

$$\omega = \frac{\partial v}{\partial x} - \frac{\partial u}{\partial y},$$

while the vorticity is the vector $\omega\mathbf{k}$ where $\mathbf{k}$ is a unit vector perpendicular to the plane.

In the general case of three-dimensional motion the vorticity (see 9·3) is the vector whose components (ξ, η, ζ) parallel to the x, y, z-axes are given by

$$\xi = \frac{\partial w}{\partial y} - \frac{\partial v}{\partial z}, \quad \eta = \frac{\partial u}{\partial z} - \frac{\partial w}{\partial x}, \quad \zeta = \frac{\partial v}{\partial x} - \frac{\partial u}{\partial y}.$$

In the two-dimensional case $w = 0$ and u, v are by definition independent of z. Thus $\xi = 0$, $\eta = 0$, $\zeta = \omega$.

3·22. Motion of a fluid element.   Consider a circular drop of fluid of infinitesimal radius, whose centre is at the point (x, y) and has velocity (u, v).

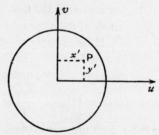

FIG. 3·22 (a).

Let (u', v') be the velocity of any point P of the drop whose coordinates are $(x + x', y + y')$. Since x', y' are infinitesimal, Taylor's theorem gives

$$u' = u + x'\frac{\partial u}{\partial x} + y'\frac{\partial u}{\partial y}, \quad v' = v + x'\frac{\partial v}{\partial x} + y'\frac{\partial v}{\partial y},$$

Let $\qquad \chi = -\tfrac{1}{2}\frac{\partial u}{\partial x}x'^2 - \tfrac{1}{2}\left(\frac{\partial u}{\partial y} + \frac{\partial v}{\partial x}\right)x'y' - \tfrac{1}{2}\frac{\partial v}{\partial y}y'^2.$

The vorticity at the centre of the drop is of magnitude

$$\omega = \frac{\partial v}{\partial x} - \frac{\partial u}{\partial y}.$$

Thus, as can be easily verified,

$$u' = u - \frac{\partial \chi}{\partial x'} - \tfrac{1}{2}y'\omega, \quad v' = v - \frac{\partial \chi}{\partial y'} + \tfrac{1}{2}x'\omega.$$

This shows that the drop moves like a rigid body with velocity of translation of its centre (u, v), and angular speed of rotation $\frac{1}{2}\omega$ about an axis perpendicular to its plane, and that on this motion is superposed a velocity of deformation in which the point (x', y') moves relatively to the centre with a velocity whose components are $-\partial\chi/\partial x'$, $-\partial\chi/\partial y'$.

The foregoing result is known as *Helmholtz's first theorem*. It is equally true for the three-dimensional motion of a spherical drop and the analysis is substantially the same.

It is precisely this velocity of deformation which is characteristic of the " fluidity " of the medium. If the drop were suddenly frozen solid without change of angular momentum, it would begin to rotate with angular velocity $\frac{1}{2}\omega$.

It may also be observed that if the vorticity were zero, the drop, so frozen, would move with a velocity of translation only.

In an ideal fluid the pressure thrust on the boundary of a circular drop is normal to the boundary and therefore passes through the centre. Thus pressure can neither increase nor decrease the angular momentum of the drop.

Def. Motion in which the vorticity is different from zero is called rotational,

In an ideal fluid rotational motion persists, for its rotational character cannot be altered by pressure, and therefore fluid which is once moving rotationally will continue so to move and must have been so moving in all its past history.

As a simple example of rotational motion in an actual fluid consider the motion described in 1·6. Here

$$\omega = -\frac{\partial u}{\partial y} = -U/h.$$

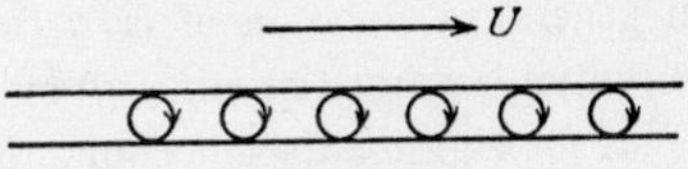

FIG. 3·22 (b).

In this motion there is constant vorticity throughout, the sense of rotation being clockwise in fig. 1·6.

Fig. 3·22 (b) indicates the vorticity between two adjacent planes parallel to the direction of U.

3·3. Irrotational motion.
Motion in which the vorticity is zero is said to be *irrotational*.

In an inviscid fluid irrotational motion is *permanent* in the sense that fluid which at any instant has no vorticity can never acquire vorticity, nor can it have lost vorticity at a previous time.

Since fluid at rest has no vorticity it follows that inviscid fluid set into motion from rest will move irrotationally. The same is true of the *initial motion* of fluids of small viscosity such as air and water. Thus, for example, in the case of streaming past a circular cylinder, photographs show that the initial stages of the motion conform to the flow pattern of an inviscid fluid. The reason for this is accounted for by the boundary layer theory (1·8). The only fluid appreciably affected by viscosity is that which has passed near the boundary of the cylinder, and when the coefficient of viscosity is small a correspondingly large volume of fluid must have passed the boundary before a visible effect can be built up.

3·31. Velocity potential.
Referring to fig. 3·1 (i), let us suppose that ABP, ACP are two curves joining A to P, *these curves and the region between them* being in fluid in which the vorticity is everywhere zero, i.e.

$$\frac{\partial v}{\partial x} - \frac{\partial u}{\partial y} = 0.$$

From 3·2 (3) it follows that circ $ABPCA = 0$ and therefore

$$\int_{(ABP)} q_s \, ds = \int_{(ACP)} q_s \, ds.$$

Thus the line integral from A to P is independent of the particular path taken from A to P, so that if A is a fixed point, the value of the integral depends on P alone. If we write

$$\phi = \phi_P = - \int_{(AP)} q_s \, ds,$$

the function ϕ is called the *velocity potential*. If we wish to emphasise the particular point P at which the velocity potential is to be evaluated we use the notation ϕ_P instead of ϕ. Like the stream function the velocity potential is, in general, a function of (x, y, t). When the motion is steady the velocity potential is a function of (x, y) only.

If P_1, P_2 are adjacent points (fig. 3·11 (i)), we have

$$\phi_{P_2} - \phi_{P_1} = - \int_{(P_1 P_2)} q_s \, ds = - P_1 P_2 \, q_s \text{ approximately.}$$

Thus the velocity component in the direction $P_1 P_2$ is

$$q_s = - \lim_{P_2 \to P_1} \frac{\phi_{P_2} - \phi_{P_1}}{P_2 P_1} = - \frac{\partial \phi}{\partial s}.$$

The negative sign used in the above definition of ϕ has no essential significance. Its conventional adoption here means that the fluid moves in the direction in which the velocity potential decreases, and agrees with the convention adopted for the derivation of other quantities from potential functions in mathematical physics.

Referring to figs. 3·11 (ii), (iii), we have for cartesian coordinates

(1) $$u = - \frac{\partial \phi}{\partial x}, \quad v = - \frac{\partial \phi}{\partial y},$$

and for polar coordinates

(2)
$$u_r = -\frac{\partial \phi}{\partial r}, \quad u_\theta = -\frac{\partial \phi}{r\,\partial \theta}.$$

It is important to observe that, provided the motion is irrotational, a velocity potential exists whether the fluid is compressible or incompressible.

On the other hand the stream function, as defined in 3·1, exists only when compressibility is negligible, irrespectively of whether the motion is rotational or irrotational (see 15·44 (1)).

In the case of the irrotational motion of an incompressible fluid both ϕ and ψ exist, and if we equate the values of the velocity components (u, v) derived from them, we obtain the relations

(3)
$$\frac{\partial \phi}{\partial x} = \frac{\partial \psi}{\partial y}, \quad \frac{\partial \phi}{\partial y} = -\frac{\partial \psi}{\partial x}.$$

The velocity potential, when one-valued, has an interesting physical interpretation. Suppose the existing irrotational motion of an incompressible fluid to be generated instantaneously from rest by the application of impulsive pressure ϖ. Then the dynamical law of impulse applied to a small volume $d\tau$ gives

$$-\frac{\partial \varpi}{\partial s}\,d\tau = \rho\,d\tau\,q_s,$$

for the left side gives the resultant impulsive pressure thrust (see 1·41) in the direction of ds and the right side gives the momentum generated. Therefore·

$$q_s = -\frac{\partial}{\partial s}\left(\frac{\varpi}{\rho}\right),$$

so that the velocity potential is ϖ/ρ. Thus $\rho\phi$ is the impulsive pressure required to generate the motion instantaneously from rest.

Conversely, a motion generated from rest by impulsive pressure only is necessarily irrotational, the velocity potential being ϖ/ρ. Irrotationality must characterise any motion started from rest, as for example when an aircraft starts in still air. The argument is true even for a viscous fluid as regards the *initial* motion, but vortex sheets (see 10·2) may form even in inviscid air due to the bringing together of layers of air which were previously separated, and which are moving with different velocities. The presence of even slight viscosity may cause these sheets to roll up and form concentrated vortices (see 10·4).

3·311. Laplace's equation.　In the case of irrotational motion it follows from 3·31 (3) that

$$\frac{\partial^2 \phi}{\partial x^2} = \frac{\partial^2 \psi}{\partial x\,\partial y} = -\frac{\partial^2 \phi}{\partial y^2},$$

and therefore that
$$\frac{\partial^2 \phi}{\partial x^2} + \frac{\partial^2 \phi}{\partial y^2} = 0,$$

which is known as Laplace's equation * (in its two-dimensional form). We prove similarly that ψ also satisfies Laplace's equation.

3·32. Cyclic motion. The existence of a velocity potential is not incompatible with the coexistence of circulation. Referring to fig. 3·2 (e), if the air between the circuit C and the aerofoil C' is devoid of vorticity, there is a velocity potential ϕ, and

$$\operatorname{circ} C = \int_{(C)} - \frac{\partial \phi}{\partial s}\, ds = [-\phi]_C,$$

where the notation means the change in $-\phi$, i.e. the decrease in ϕ, when we go once round the circuit C. On the other hand, from 3·2 (4), we see that

$$\operatorname{circ} C = \operatorname{circ} C',$$

so that ϕ decreases by the same amount when we go round any circuit embracing the aerofoil once. In this case ϕ is a many-valued function.

For example, suppose that

$$\phi = - \kappa \tan^{-1} \frac{y}{x} = - \kappa \theta.$$

Then $$u = \frac{-\kappa y}{x^2 + y^2}, \quad v = \frac{\kappa x}{x^2 + y^2} \quad \text{and} \quad \frac{\partial v}{\partial x} - \frac{\partial u}{\partial y} = 0,$$

but if we go round a circuit C which embraces the origin once, θ increases by 2π and ϕ decreases by $2\pi\kappa$, so that there is a circulation of this amount in the circuit.

3·4. Complex numbers. Let x, y be *real numbers* positive or negative. Let i be a symbol which obeys the ordinary laws of algebra, and in addition satisfies the relation

(1) $$i^2 = -1.$$

The combination

(2) $$z = x + iy$$

is then called a *complex number*.

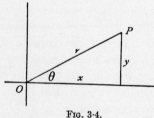

FIG. 3·4.

Such a complex number can be represented by the point P whose cartesian coordinates are (x, y).

The resulting picture in which the number is so represented is called the *Argand diagram*. With this representation we may talk of "the point z", meaning thereby the point (x, y) or P.

---

* The three-dimensional form is $\nabla^2 \phi = 0$, or in cartesian coordinates,

$$\frac{\partial^2 \phi}{\partial x^2} + \frac{\partial^2 \phi}{\partial y^2} + \frac{\partial^2 \phi}{\partial z^2} = 0 \quad (19·5).$$

The numbers x and y in (2) are called respectively the real and imaginary parts of the complex number z ;

$$x = \text{Real part of } z, \quad y = \text{Imaginary part of } z.$$

When $y = 0$ the number z is said to be purely real, $z = x$. When $x = 0$ the number z is said to be purely imaginary, $z = iy$.

Two complex numbers which differ only in the sign of i are said to be *conjugate*.

We turn a number into its conjugate by writing a bar above it. Thus z, $\bar{z}$ are conjugate,

(3) $z = x + iy, \quad \bar{z} = x - iy.$

Since $z + \bar{z} = 2x$, and $z - \bar{z} = 2iy$ we have two seemingly trivial but in fact important theorems.

Theorem A. The imaginary part of the sum of two conjugate complex numbers is zero.

Theorem B. The real part of the difference of two conjugate complex numbers is zero.

Moreover equations (3) may be regarded as equations of transformation from two real variables x, y to two complex variables z, $\bar{z}$.

Since the point P can also be described by polar coordinates (r, θ), in which r is necessarily positive, we have, using Euler's theorem * $\cos\theta + i\sin\theta = e^{i\theta}$,

(4) $z = x + iy = r\cos\theta + ir\sin\theta = r(\cos\theta + i\sin\theta) = re^{i\theta}$,

$z^n = r^n(\cos n\theta + i\sin n\theta) = r^n e^{ni\theta}.$

Note also that $2\cos\theta = e^{i\theta} + e^{-i\theta}, \quad 2i\sin\theta = e^{i\theta} - e^{-i\theta}.$

When polar coordinates are used the positive number r is called the *modulus* of z, written

$$r = \text{mod } z = |z| = \sqrt{(x^2 + y^2)} = \sqrt{(z\,\bar{z})}.$$

Thus the product of two conjugate complex numbers is the square of the modulus of either.

3·41. The Argument.   The angle θ is called the *argument* of z written

$$\theta = \arg z.$$

Clearly all complex numbers whose moduli are the same and whose arguments differ by an integral multiple of 2π are represented by the same point P in the Argand diagram. We call the principal value of $\arg z$ that angle θ which lies between $-\pi$ and $+\pi$. Denoting by $P[\arg z]$ the principal value of $\arg z$, the precise definition is

$$- \pi < P[\arg z] \leqslant \pi.$$

The principal value of the argument of a positive real number is zero, and of a negative real number is π.

---

* Milne-Thomson, *Theoretical Hydrodynamics*, 5·13. See also Ex. III, 7.

Referring to fig. 3·41, the curve C encircles the origin, the curve C_1 does not. If θ is the initial value of arg z and if z is represented by the point P, it is clear

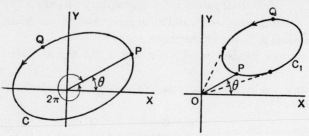

Fig. 3·41.

that when a point Q originally coinciding with P is moved round C in the counterclockwise sense, the corresponding value of its argument increases, and when we finally return to P after one excursion round C, we now have arg $z = \theta + 2\pi$. On the other hand, if we go round C_1, the argument of Q decreases at first (for the position shown in fig. 3·41) until OQ becomes a tangent to C_1, then increases until OQ again becomes a tangent and finally decreases to its original value. Thus we conclude that if arg z has a given value at one point of a curve such as C_1 which does not encircle the origin, the value of arg z is *one-valued* at every point inside and on C_1, provided we assume that arg z varies continuously with z.

Now consider

$$(1) \qquad \log z = \log (re^{i\theta}) = \log r + i\theta.$$

If we take z once round C, θ increases by 2π and therefore $\log z$ increases by $2\pi i$. Thus $\log z$ is a many-valued function if z moves inside or upon a curve which encircles the origin. On the other hand, $\log z$ can be regarded as a one-valued function if z is restricted to the interior of a curve such as C_1 which does not encircle the origin.

As to the multiplication of complex numbers, if $z_1 = r_1 e^{i\theta_1}$, $z_2 = r_2 e^{i\theta_2}$, then $z_1 z_2 = r_1 r_2 e^{i(\theta_1 + \theta_2)}$.

Thus the modulus of the product is the product of the moduli, while the argument of the product is the sum of the arguments, in symbols

$$(2) \qquad |z_1 z_2| = |z_1| \, |z_2| \quad \text{and} \quad \arg(z_1 z_2) = \arg z_1 + \arg z_2.$$

In applying this last result it is well to bear in mind that each of the arguments may be many-valued and therefore the right-hand member is only one of the possible values of arg $(z_1 z_2)$.

Clearly with similar limitations

$$(3) \qquad \arg\left(\frac{z_1}{z_2}\right) = \arg z_1 - \arg z_2.$$

3·42. Differentiation. Let n be a positive integer. Then we define, as in the case of real variables,

$$\frac{d}{dz} z^n = \lim_{z_1 \to z} \frac{z_1^n - z^n}{z_1 - z} = \lim_{z_1 \to z} (z_1^{n-1} + z_1^{n-2} z + \ldots + z^{n-1}) \times \frac{z_1 - z}{z_1 - z} = nz^{n-1}.$$

Clearly if $f(z)$ is a polynomial, say

$$f(z) = a_0 + a_1 z + a_2 z^2 + \ldots + a_n z^n,$$

then $\qquad f'(z) = \dfrac{df(z)}{dz} = a_1 + 2a_2 z + 3a_3 z^2 + \ldots + n\, a_n z^{n-1}.$

More generally we shall assume, for a rigorous proof would lead too far afield, that if $f(z)$ can be represented by an infinite *power series,*

$$f(z) = a_0 + a_1 z + a_2 z^2 + \ldots + a_n z^n + \ldots$$

convergent in a certain region, then

$$f'(z) = a_1 + 2a_2 z + \ldots + n\, a_n z^{n-1} + \ldots,$$

where the new power series is convergent in the *same* region.

3·43. Holomorphic functions. Consider a simple closed curve C and a function $f(z)$. The function $f(z)$ is said to be *holomorphic when z is within C* if

(i) to every value of z within C there corresponds one and only one value of $f(z)$ and that value is finite (i.e. its modulus is not infinite) ;

(ii) for each value of z within C the function has a one-valued finite derivative (differential coefficient), defined by

$$f'(z) = \lim_{z_1 \to z} \frac{f(z_1) - f(z)}{z_1 - z},$$

where $z_1 \to z$ by any path all of whose points lie within C.

A function is said to be holomorphic inside and upon C, if it is holomorphic inside a larger curve C' to which every point of C is interior.

Examples. The functions z^n (n a positive integer), e^z, $\sin z$, $\cos z$, $\sinh z$, $\cosh z$ are holomorphic in any finite region.

The function z^{-n} (n a positive integer) is holomorphic in every region which does not include the origin.

The function $\log z$ is holomorphic in any finite region which does not enclose the origin, provided that the determination (see 3·41) of $\log z$ is prescribed at one point of the region.

3·44. Conjugate functions. The real and imaginary parts of a holomorphic function $f(z)$ are functions of x and y which are called *conjugate functions.* Thus we can write

$$(1) \qquad f(z) = f(x + iy) = \phi(x, y) + i\psi(x, y) = \phi + i\psi.$$

Hence
$$\frac{\partial \phi}{\partial x} + i\frac{\partial \psi}{\partial x} = \frac{\partial f(z)}{\partial x} = f'(z)\frac{\partial z}{\partial x} = f'(z),$$

$$\frac{\partial \phi}{\partial y} + i\frac{\partial \psi}{\partial y} = \frac{\partial f(z)}{\partial y} = f'(z)\frac{\partial z}{\partial y} = if'(z).$$

These yield
$$i\left(\frac{\partial \phi}{\partial x} + i\frac{\partial \psi}{\partial x}\right) = \frac{\partial \phi}{\partial y} + i\frac{\partial \psi}{\partial y} \quad \text{or}$$

(2)
$$\frac{\partial \phi}{\partial x} = \frac{\partial \psi}{\partial y}, \quad \frac{\partial \phi}{\partial y} = -\frac{\partial \psi}{\partial x}.$$

These two relations embody the characteristic property of conjugate functions and are known as the Cauchy-Riemann equations. They express the geometrical fact that if the two families of curves

$$\phi(x, y) = \text{constant}, \quad \psi(x, y) = \text{constant}$$

are drawn, then they intersect everywhere at right angles.

To prove this observe that the gradients dy/dx of the two families of curves are given by, respectively,

$$\frac{\partial \phi}{\partial x} + \frac{\partial \phi}{\partial y}\frac{dy}{dx} = 0, \quad \frac{\partial \psi}{\partial x} + \frac{\partial \psi}{\partial y}\frac{dy}{dx} = 0,$$

and the Cauchy-Riemann equations express that the product of these two gradients is -1, i.e. the tangents are at right angles.

Another proof is as follows. From (1)

$$f'(z)\, dz = d\phi + i\, d\psi.$$

Therefore
$$(\arg dz)_{\phi=\text{constant}} = \tfrac{1}{2}\pi + (\arg dz)_{\psi=\text{constant}},$$

so that the elements of arc of the curves $\phi = \text{constant}$, $\psi = \text{constant}$, are at right angles.

We state here, without proof, that if equations (2) are satisfied, and if all the partial derivatives are continuous, then $f(z)$ is a holomorphic function of z. See also Ex. III, 32.

3·45. The function $\bar{f}(z)$. Given a holomorphic function $f(z)$ we can form the conjugate complex $\bar{f}(\bar{z})$ by replacing i by $-i$ wherever it occurs. Thus if

(1)
$$f(z) = (2 + 3i)z + e^{-4iz}$$

we shall have

(2)
$$\bar{f}(\bar{z}) = (2 - 3i)\bar{z} + e^{-4i\bar{z}}$$

Def. The function $\bar{f}(z)$ is formed from the function $f(z)$ by first forming $\bar{f}(\bar{z})$ and then in $\bar{f}(\bar{z})$ writing z instead of $\bar{z}$.

Thus in (1) above, we get from (2) $\bar{f}(z) = (2 - 3i)z + e^{-4iz}$

Again if $f(z) = i\kappa \log(3 - h)$ where h is a complex constant and κ is real the steps are

$$\bar{f}(\bar{z}) = - i\kappa \log (\bar{z} - \bar{h}), \quad \bar{f}(z) = - i\kappa \log (z - \bar{h})$$

We shall have frequent occasion to use this process and after a little practice it will be found quite simple to go straight to the final result.

The function $f(\bar{z})$ which is distinct from both $f(z)$ and $\bar{f}(\bar{z})$ is formed by writing $\bar{z}$ for z in $f(z)$. Thus from (1)

$$f(\bar{z}) = (2 + 3i)\bar{z} + e^{4i\bar{z}}$$

Thus starting with $f(z)$

(i) to form $f(\bar{z})$ we change i into $-i$ in z only.

(ii) to form $\bar{f}(\bar{z})$ we change the signs of all the i's.

(iii) to form $\bar{f}(z)$ we change the sign of all the i's except those in z

3·47. The coordinates z **and** $\bar{z}$. Since $z = x + iy$, $\bar{z} = x - iy$, we have $x = \frac{1}{2}(z + \bar{z})$ and $y = \frac{1}{2}i(z - \bar{z})$ so that any function of x, y can be expressed as a function of z, $\bar{z}$ and vice-versa. Also

$$\frac{\partial z}{\partial x} = \frac{\partial \bar{z}}{\partial x} = 1, \quad \frac{\partial z}{\partial y} = i, \quad \frac{\partial \bar{z}}{\partial y} = - i$$

Let us regard the stream function ψ as expressed as a function of x, y and again as a function of z, $\bar{z}$. Then from 3·11

$$u = - \frac{\partial \psi}{\partial y} = - \frac{\partial \psi}{\partial z}\frac{\partial z}{\partial y} - \frac{\partial \psi}{\partial \bar{z}}\frac{\partial \bar{z}}{\partial y} = - i\frac{\partial \psi}{\partial z} + i\frac{\partial \psi}{\partial \bar{z}}$$

$$v = \frac{\partial \psi}{\partial x} = \frac{\partial \psi}{\partial z}\frac{\partial z}{\partial x} + \frac{\partial \psi}{\partial \bar{z}}\frac{\partial \bar{z}}{\partial x} = \frac{\partial \psi}{\partial z} + \frac{\partial \psi}{\partial \bar{z}}$$

Combining these we get

(1) $$u - iv = - 2i\frac{\partial \psi}{\partial z}, \quad u + iv = 2i\frac{\partial \psi}{\partial \bar{z}}$$

Since $$u - iv = - \frac{\partial \psi}{\partial y} - i\frac{\partial \psi}{\partial x} = - i\left(\frac{\partial \psi}{\partial x} - i\frac{\partial \psi}{\partial y}\right)$$

we have, on comparison of the two expressions for $u - iv$ the equivalence of operators

(2) $$2\frac{\partial}{\partial z} = \frac{\partial}{\partial x} - i\frac{\partial}{\partial y}, \quad 2\frac{\partial}{\partial \bar{z}} = \frac{\partial}{\partial x} + i\frac{\partial}{\partial y}$$

the second result following from the expression of $u + iv$, or more simply as the conjugate complex of the first.

From 3·21 we have for the vorticity

$$\omega = \frac{\partial v}{\partial x} - \frac{\partial u}{\partial y} = \frac{\partial^2 \psi}{\partial x^2} + \frac{\partial^2 \psi}{\partial y^2} = \left(\frac{\partial}{\partial x} - i\frac{\partial}{\partial y}\right)\left(\frac{\partial}{\partial x} + i\frac{\partial}{\partial y}\right)\psi$$

(3) Thus, using (2), $$\omega = \nabla^2 \psi = \frac{4 \, \partial^2 \psi}{\partial z \, \partial \bar{z}}$$

3·5. Cauchy's integral theorem.

Let C be a simple closed contour such that the function $f(z)$ is holomorphic at every point of C and in the interior of C.* Then

$$\int_{(C)} f(z)\, dz = 0.$$

This is Cauchy's integral theorem.

Proof. Let $f(z) = \phi + i\psi$.

Then

$$\int_{(C)} f(z)\, dz = \int_{(C)} (\phi + i\psi)(dx + i\, dy) = \int_{(C)} (\phi\, dx - \psi\, dy) + i \int_{(C)} (\psi\, dx + \phi\, dy)$$

$$(1) \qquad = -\int_{(S)} \left(\frac{\partial \psi}{\partial x} + \frac{\partial \phi}{\partial y} \right) dS + i \int_{(S)} \left(\frac{\partial \phi}{\partial x} - \frac{\partial \psi}{\partial y} \right) dS,$$

using Stokes's theorem (3·2, (5), (6)).

From the Cauchy-Riemann equations (3·44),

$$\frac{\partial \psi}{\partial x} + \frac{\partial \phi}{\partial y} = 0, \quad \frac{\partial \phi}{\partial x} - \frac{\partial \psi}{\partial y} = 0.$$

Therefore
$$\int_{(C)} f(z)\, dz = 0. \qquad\qquad \text{Q.E.D.}$$

The proof here given is of course based on the assumption pointed out in 3·44 that sufficient conditions of holomorphy are satisfied. A complete proof would be long and difficult and the conditions here assumed are satisfied in the applications.

3·51. Singularities.

A point at which a function ceases to be holomorphic is called a *singular point*, or *singularity* of the function.

Thus the function $f(z) = (z - a)^{-1}$ is holomorphic in any region from which the point $z = a$ is excluded (e.g. by drawing a small circle round it). At $z = a$ the function ceases to be finite and therefore does not satisfy the first part of the definition of holomorphy.

More generally, if near the point $z = a$ the function can be expanded in positive and negative powers of $z - a$, say

$$f(z) = \ldots + A_2(z - a)^2 + A_1(z - a) + A_0 + \frac{B_1}{z - a} + \frac{B_2}{(z - a)^2} + \ldots,$$

the point $z = a$ is a singular point.

If only a finite number of terms contain negative powers of $z - a$, the point $z = a$ is called a *pole*; a *simple pole* if B_1 alone is different from zero.

Again, consider the function $f(z) = \log z$. This function ceases to be holomorphic at $z = 0$. We have seen in 3·41 that $\log z$ is many-valued. If we

---

*This means that C and its interior lie wholly within a larger contour inside which the function is holomorphic.

choose one particular determination, say that which reduces to zero when $z = 1$, and allow z to describe a closed curve which does not encircle the point $z = 0$, $\log z$ will return to its starting value and will be holomorphic inside the curve.

3·52. Residues. We have seen that a function, which in the neighbourhood of $z = a$ has an expansion which contains negative powers of $z - a$, is singular at $z = a$.

In this case the coefficient of $(z - a)^{-1}$ is called the *residue* of the function at $z = a$.

Let us consider

$$\int (z - a)^n \, dz$$

taken round a circle of radius R whose centre is at the point $z = a$. On the circumference of this circle $z - a = Re^{i\theta}$, and therefore

$$\int (z - a)^n \, dz = \int_0^{2\pi} R^{n+1} e^{(n+1)i\theta} i \, d\theta$$

$$= \frac{R^{n+1}}{(n + 1)} \left[e^{(n+1)i\theta} \right]_0^{2\pi} = 0, \text{ if } n \neq -1.$$

If, however, $n = -1$, we get

$$\int \frac{dz}{z - a} = \int_0^{2\pi} i \, d\theta = 2\pi i.$$

Now, suppose that $f(z)$ has an expansion in the neighbourhood of $z = a$ of the form

$$\ldots + A_2(z - a)^2 + A_1(z - a) + A_0 + \frac{B_1}{(z-a)} + \frac{B_2}{(z - a)^2} + \ldots .$$

If we integrate round a small circle surrounding $z = a$, we get

$$\int f(z) \, dz = 2\pi i \, B_1,$$

for all the integrals vanish except that of $B_1(z - a)^{-1}$.

Thus we see the importance of the residues, for they form the only contributions to the integral of a function which is holomorphic at all points except singularities of the kind described above.

3·53. Cauchy's residue theorem. Let C be a closed contour inside and upon which the function $f(z)$ is holomorphic, except at a finite number of singular points within C at which the residues are $a_1, a_2, \ldots, a_n$. Then

$$\int_{(C)} f(z) \, dz = 2\pi i (a_1 + a_2 + \ldots + a_n).$$

Proof. Suppose there are three singularities. Surround them by small circles, C_1, C_2, C_3. Then, from 3·5,

$$\int_{(C)} f(z)\,dz = \int_{(C_1)} f(z)\,dz + \int_{(C_2)} f(z)\,dz + \int_{(C_3)} f(z)\,dz$$
$$= 2\pi i a_1 + 2\pi i a_2 + 2\pi i a_3,$$

from 3·52. This proves the theorem in the case of three singularities. The proof for any finite number is the same.   Q.E.D.

3·6. Conformal mapping.

Let us take two complex variables

$$\zeta = \xi + i\eta, \quad z = x + iy$$

and the corresponding Argand diagrams which we shall call the ζ-plane and the z-plane respectively. Let $f(\zeta)$ be a holomorphic function of ζ in the region R

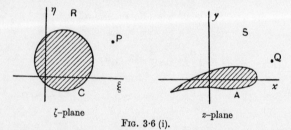

ζ-plane z-plane

FIG. 3·6 (i).

exterior to a simple closed curve C, that is to say, exterior to the region shaded in fig. 3·6 (i). We can then establish a correspondence between points of this region and points of the z-plane by means of the relation

$$(1) \qquad\qquad z = f(\zeta).$$

Since $f(\zeta)$ is holomorphic, it is one-valued and therefore to each point ζ there corresponds a unique point z, and as ζ describes the curve C, z will describe a corresponding curve A in the z-plane which is the *map* in the z-plane, of the curve C, given by the *mapping function* $f(\zeta)$.

We shall assume that the mapping function is so chosen that A is a simple closed curve and that the points of the region R of the ζ-plane exterior to C map into points of the region S of the z-plane exterior to A.

Moreover, we assume that the mapping is biuniform, that is to say, that there is a one-to-one correspondence between the points exterior to C and the points exterior to A. These conditions will be satisfied in all the applications which we shall make.*

Now referring to fig. 3·6 (ii), let $d\sigma$, ds be infinitesimal arcs of any two corresponding curves in the two planes and put

$$d\zeta = d\sigma\, e^{i\omega}, \quad dz = ds\, e^{i\theta}.$$

---

* For further details and proofs the reader is referred to Milne-Thomson, *Theoretical Hydrodynamics.* Chapter V.

From (1) we have

$$dz = d\zeta \cdot f'(\zeta)$$

and so if $f'(\zeta) = me^{i\alpha}$, we have

$$ds\, e^{i\theta} = m\, d\sigma\, e^{i(\omega+\alpha)},$$

and therefore

(2) $$ds = m\, d\sigma, \quad \theta = \omega + \alpha.$$

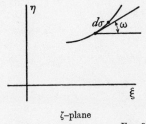 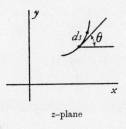

ζ–plane z–plane

FIG. 3·6 (ii).

These relations show (a) that the angle at ζ between any two curves passing through ζ is equal to the angle at z between their maps, and (b) that the element of arc of the map in the z-plane is m times the element of arc of the curve in the ζ-plane. It follows that any small region round ζ maps into a geometrically similar small region round z, the *linear scale*, or *magnification*, of the mapping being $m = |f'(\zeta)|$, (and therefore being in general variable from one part of the plane to another). For this reason the mapping given by (1) is said to be *conformal*.

Observe that the conformal character of the mapping breaks down at any point for which $f'(\zeta)$ is zero or infinite.

3·7. Complex potential. Let ϕ, ψ be the velocity potential and stream function of the irrotational two-dimensional motion of air regarded as incompressible. Then equating the velocity components, we get

(1) $$\frac{\partial \phi}{\partial x} = \frac{\partial \psi}{\partial y}, \quad \frac{\partial \phi}{\partial y} = -\frac{\partial \psi}{\partial x}.$$

We define the *complex potential* of the motion by the relation

$$w(z) = w = \phi + i\psi.$$

We see from 3·44 that, on account of (1), w is a holomorphic function of the complex variable $z = x + iy$ in any region where ϕ and ψ are one-valued.

Conversely, if we assume for w any holomorphic function of z, the corresponding real and imaginary parts give the velocity potential and stream function of a possible two-dimensional irrotational motion, for they satisfy (1) and Laplace's equation (see 3·311).

Thus, for example, $w = z^2$ gives $\phi = x^2 - y^2$, $\psi = 2xy$.

Since iw is likewise a function of z, it follows that $-\psi$ and ϕ are the velocity potential and stream function of another motion in which the streamlines and lines of equal velocity potential are interchanged.

It will be found that the mathematical analysis is very considerably simplified by working with the complex potential instead of ϕ and ψ separately. The simplification is of the same nature as that attained by using one vector equation instead of three cartesian equations. In two dimensions we work with one equation in z instead of two in x and y.

The dimensions of the complex potential are those of a velocity multiplied by a length, i.e. $L^2 T^{-1}$.

We give a few simple illustrations.

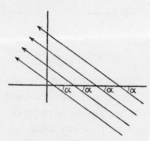

FIG. 3·7 (i).

(i) $w = Ve^{i\alpha} z.$

Here $\psi = V(y \cos \alpha + x \sin \alpha)$ and the streamlines are therefore straight. We have a uniform wind V at incidence α to the x-axis.

(ii) $w = -m \log z; \quad \phi = -m \log r, \quad \psi = -m\theta.$

The streamlines are straight and radiate from the origin. This motion is due to a *simple source*, or point of outward radial flow, at the origin. Such a motion would result were a circular cylinder to expand uniformly in still air. Similarly, $w = m \log z$ is the complex potential for a *sink* or point of inward radial flow.

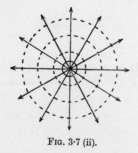

FIG. 3·7 (ii).

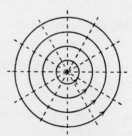

FIG. 3·7 (iii).

(iii) $w = i\kappa \log z; \quad \phi = -\kappa\theta, \quad \psi = \kappa \log r.$

The streamlines are concentric circles with centre at the origin. This motion is due to a rectilinear *vortex* at the origin (see 4·1).

(iv) $w = \dfrac{\mu}{z}$; $\psi = -\dfrac{\mu y}{x^2 + y^2}$.

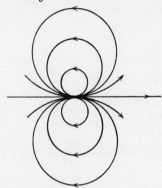

Fig. 3·7 (iv).

The streamlines are circles touching the x-axis. The motion is due to a *doublet* at the origin, directed along the x-axis. Such a doublet arises as a limiting case from the juxtaposition of a source and an equal sink.

3·71. The complex velocity.

From the complex potential $w = \phi + i\psi$ we get

$$\frac{\partial \phi}{\partial x} + i\,\frac{\partial \psi}{\partial x} = \frac{dw}{dz}\,\frac{dz}{dx} = \frac{dw}{dz},$$

and since

$$u = -\,\partial\phi/\partial x, \quad v = \partial\psi/\partial x,$$

(1)
$$u - iv = -\frac{dw}{dz}.$$

We shall call $u - iv$ the *complex velocity* and note that the complex velocity is obtained directly from the complex potential as shown in (1). Graphically, the vector representing the complex velocity is the reflection in the line, through the point considered, parallel to the x-axis, of the vector representing the actual velocity $u + iv$.

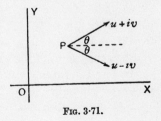

Fig. 3·71.

The relation is shown in fig. 3·71. It is very important to note that $-dw/dz$ gives $u - iv$ and not $u + iv$. Thus, for example, if $w = iz^2$, we have $\quad u - iv = -2iz, \quad u + iv = 2i\bar{z},$ either of these leading to $\quad u = 2y, \quad v = 2x.$

At a *stagnation point* the velocity is zero. Thus $u = 0$ and $v = 0$. Therefore the stagnation points are given by $dw/dz = 0$. Thus, for example, if

(2)
$$w = Vz + \frac{Va^2}{z},$$

the stagnation points are given by $1 - a^2/z^2 = 0$, i.e. $z = a$ and $z = -a$
Now (2) is the complex potential of a circular cylinder in a uniform wind (see
5·2 (i) and fig. 3·14 (b)). The stagnation points are therefore where the
wind impinges directly on the cylinder and the diametrically opposite point.

To calculate the air speed q at any point we have

$$(3) \qquad\qquad q = \sqrt{(u^2 + v^2)} = \left| \frac{dw}{dz} \right|.$$

An alternative method is as follows :

$$(4) \quad q^2 = u^2 + v^2 = (u - iv)(u + iv) = \frac{dw(z)}{dz} \times \frac{d\bar{w}(\bar{z})}{d\bar{z}} = \frac{dw}{dz} \times \frac{d\bar{w}}{d\bar{z}}.$$

As an example, taking (2) above,

$$q^2 = \left(V - \frac{Va^2}{z^2} \right) \left(V - \frac{Va^2}{\bar{z}^2} \right) = V^2 \left(1 - \frac{a^2 e^{-2i\theta}}{r^2} \right) \left(1 - \frac{a^2 e^{2i\theta}}{r^2} \right)$$

$$= V^2 \left(1 - \frac{2a^2 \cos 2\theta}{r^2} + \frac{a^4}{r^4} \right).$$

3·8. Application of conformal mapping.

Consider a mapping of the
ζ-plane on the z-plane by

$$(1) \qquad\qquad z = f(\zeta),$$

such that the region R exterior to C in the ζ-plane maps into the region S
exterior to A in the z-plane, see fig. 3·6 (i). Then the contour C maps into the
contour A.

Let a fluid motion in the region R of the ζ-plane be given by the complex
potential

$$(2) \qquad\qquad w(\zeta) = w = \phi + i\psi.$$

Then at corresponding points ζ and z given by (1), w and therefore ϕ and ψ
take the same values.

Now C is a boundary, and so a streamline, and therefore $\psi = k$, a constant,
at all points of C. Since A corresponds point by point with C, $\psi = k$ at all
points of A. Therefore A is a streamline of the motion given by (2) and (1)
together in the z-plane.

The actual form of the complex potential in terms of z would be got by
eliminating ζ between (1) and (2), but it is often preferable to look on ζ as a
parameter and forgo the elimination.

Thus to find the velocity at Q in the z-plane corresponding with P in the
ζ-plane, we have

$$\frac{dw}{dz} = \frac{dw}{d\zeta} \times \frac{d\zeta}{dz},$$

and therefore

$$u_Q - iv_Q = (u_P - iv_P)/f'(\zeta).$$

EXAMPLES III

1. Draw the streamlines of the motion given by $\psi = xy$ for $x \geqslant 0$, $y \geqslant 0$.

2. Determine the condition that the velocity components

$$u = ax + by, \quad v = cx + dy$$

may satisfy the equation of continuity, and show that the magnitude of the vorticity is $c - b$.

3. Considering a circular drop of infinitesimal radius in two-dimensional motion, prove that if the drop were suddenly frozen solid without change of angular momentum, it would begin to rotate with angular velocity equal to half the vorticity before solidification.

4. Prove that $u = 2cxy$, $v = c(a^2 + x^2 - y^2)$ are the velocity components of a possible fluid motion. Determine the stream function and sketch the streamlines. Prove that the motion is irrotational and find the velocity potential.

5. Represent on an Argand diagram the number $3 + 4i$ and its square roots.

6. If $\phi + i\psi = f(z)$ and $f(z)$ is real when $y = a$, show that $\psi = 0$ when $y = a$.

7. If we define $e^{i\theta}$ by putting $x = i\theta$ in the exponential series

$$e^x = 1 + x + \frac{x^2}{2!} + \frac{x^3}{3!} + \ldots,$$

prove that $de^{i\theta}/d\theta = ie^{i\theta}$.

Hence show that the differential equation $du/d\theta = iu$ is satisfied by

$$u_1 = e^{i\theta}, \quad u_2 = \cos\theta + i\sin\theta.$$

From the fact that $u_1 = u_2 = 1$ when $\theta = 0$, deduce Euler's theorem

$$\cos\theta + i\sin\theta = e^{i\theta}.$$

8. Obtain the expansions

$$\cos\theta = 1 - \frac{\theta^2}{2!} + \frac{\theta^4}{4!} - \frac{\theta^6}{6!} + \ldots, \quad \sin\theta = \theta - \frac{\theta^3}{3!} + \frac{\theta^5}{5!} - \frac{\theta^7}{7!} + \ldots.$$

9. Prove the following results :

(i) $e^{i\pi/2} = i$,   (ii) $e^{i\pi} = -1$,   (iii) $e^{2i\pi} = 1$,

and hence show that, if n is an integer

$$e^{ni\pi/2} = (i)^n$$

can take only four different values i, $-i$, -1, 1.

10. Prove that $\cos 2\theta + i\sin 2\theta = (\cos\theta + i\sin\theta)^2$, and hence express $\cos 2\theta$ and $\sin 2\theta$ in terms of $\cos\theta$, $\sin\theta$.

11. Prove that

(i) $\cosh\theta = \cos i\theta$, $i\sinh\theta = \sin i\theta$.   (ii) $\cosh i\theta = \cos\theta$, $\sinh i\theta = i\sin\theta$.

Hence expand $\cosh\theta$, and $\sinh\theta$, in ascending powers of θ.

12. If $z = x + iy$, $z' = x' + iy'$, prove that $z = z'e^{i\alpha}$ turns the axes of reference (x, y) through the angle α, and that $z' = ze^{-i\alpha}$ turns the axes (x', y') through the angle α in the opposite sense. Hence express the formulae for rotation of axes (x, y) in terms of x', y' and (x', y') in terms of x, y.

13. Separate the real and imaginary parts of z^3, $\dfrac{1}{z}$, $\cos z$, $\sin z$, $\cosh z$, $\sinh z$.

14. Prove that the equation of any circle which passes through the points z_1, z_2 can be written in the form

$$(z - z_1)(\bar{z} - \bar{z}_2) + \lambda(\bar{z} - \bar{z}_1)(z - z_2) = 0.$$

Deduce the equation of the circle which circumscribes the triangle whose vertices are the points z_1, z_2, z_3.

15. If a is real and $(z - a)/(z + a)$ is purely imaginary, prove that z describes a circle on the line joining $(-a, 0)$ to $(a, 0)$ as diameter.

16. If $z = c \cosh \zeta$, where $\zeta = \xi + i\eta$, show that the curve for which ξ has a constant value ξ_0 is an ellipse whose semi-axes are $c \cosh \xi_0$, $c \sinh \xi_0$, that η is the eccentric angle and that all such ellipses are confocal.

Find ξ_0 if the semi-axes of the ellipse are a, b.

17. If $z = c \sin \zeta$, where $\zeta = \xi + i\eta$, show that the curves for which η has the constant value η_0 is an ellipse whose semi-axes are $c \cosh \eta_0$, $c \sinh \eta_0$, and that all such ellipses are confocal. Interpret the meaning of ξ.

Find η_0 if the semi-axes of the ellipse are a, b.

18. If $x + iy = c \cos(\phi + i\psi)$, show that

$$\frac{x^2}{c^2 \cosh^2 \psi} + \frac{y^2}{c^2 \sinh^2 \psi} = 1.$$

Hence prove that the streamlines are confocal ellipses.

Prove that the circulation round any one of these ellipses is 2π.

19. If $\phi + i\psi = a/z$, where a is real, show that the curves $\phi = $ constant, $\psi = $ constant, are circles and verify that they intersect orthogonally.

20. If $\phi + i\psi = \log z$, prove that ϕ has a constant value on concentric circles, and that ψ has a constant value on radial lines.

21. Form the functions $\overline{f}(z)$ and $\overline{f}(\bar{z})$ when $f(z)$ is any one of the functions

$$Vze^{i\alpha}, \quad i\kappa \log z, \quad -m \log z, \quad i\kappa \log \frac{z - a}{z + a}, \quad e^{(m+in)z},$$

and in each case verify that $f(z) + \overline{f}(\bar{z})$ is real and $f(z) - \overline{f}(\bar{z})$ is purely imaginary.

22. Find the poles and the corresponding residues of the functions

$$\frac{z}{z + 1}, \quad \frac{z + 2}{z^2 - 1}, \quad \frac{z + z^2}{z^2 + 1}, \quad \frac{4z^2}{z^4 - 1}.$$

Calculate the integral of each of these functions round the circle $z\bar{z} = 2$.

23. By writing $\zeta = (a+b)\frac{1}{2}e^{i\theta}$, or otherwise, prove that the transformation

$$z = \zeta + \frac{a^2 - b^2}{4\zeta}$$

maps the region exterior to the circle $|\zeta| = \frac{1}{2}(a + b)$ in the ζ-plane on the region exterior to the ellipse

$$\frac{x^2}{a^2} + \frac{y^2}{b^2} = 1$$

in the z-plane.

24. Show that $z = \zeta^k$, $k > 1$, maps the infinite sector between $\theta = 0$ and $\theta = \pi/k$ in the ζ-plane, on the half-plane $y \geqslant 0$ in the z-plane.

25. Prove that $z = e^{\pi\zeta/a}$ maps the infinite strip $0 \leqslant \eta \leqslant a$ in the ζ-plane on the upper half ($y \geqslant 0$) of the z-plane.

26. Prove that the transformation $z = i\, e^{\pi \zeta/a}$ maps the infinite strip $-\tfrac{1}{2}a \leqslant \eta \leqslant \tfrac{1}{2}a$ in the ζ-plane on the upper half $(y \geqslant 0)$ of the z-plane.

27. Prove that $z = \cosh(\pi\zeta/a)$ maps the semi-infinite strip $0 \leqslant \eta \leqslant a,\ \xi \geqslant 0$ in the ζ-plane on the upper half $(y \geqslant 0)$ of the z-plane.

28. Taking the conformal transformations

$$\text{(i)}\ z = \zeta + a, \quad \text{(ii)}\ z = \zeta e^{i\alpha}, \quad \text{(iii)}\ z = b\zeta, \quad \text{(iv)}\ z = 1/\zeta,$$

where α is real and a, b, may be complex, prove that the first gives a translation, the second a rotation, the third a rotation and a magnification, the fourth an inversion followed by a reflection.

Prove that $z = (\alpha\zeta + \beta)/(\gamma\zeta + \delta)$ may be compounded of a succession of the above transformations and hence gives a mapping in which circles and straight lines map into circles or straight lines. Note that $\alpha,\ \beta,\ \gamma,\ \delta$ may be complex.

29. If the circle $|\zeta| \leqslant r$ is mapped on the region B of the z-plane by the relation $z = \zeta + a_2\zeta^2 + a_3\zeta^3 + \dots$, prove that the area of B is

$$\pi\{r^2 + 2\,|a_2|^2\, r^4 + 3\,|a_3|^2\, r^6 + \dots\},$$

and is therefore greater than the area of the given circle.

30. Prove that the complex potential

$$w = m \log z - m \log\left(z - \frac{a^2}{f}\right) - m \log(z - f)$$

represents the motion due to sources of strength m at $(f, 0)$, $(a^2/f, 0)$ and a sink of strength m at $(0, 0)$.

Prove that $\psi = $ constant on the circle $|z| = a$, and hence that this circle is a streamline and could therefore be replaced by a rigid boundary.

31. If $z = \tan w$, prove that

$$\coth 2\psi = \frac{x^2 + y^2 + 1}{2y}, \quad \cot 2\phi = \frac{1 - x^2 - y^2}{2x},$$

and hence draw the streamlines.

Discuss the possibility of stagnation points, and calculate the air speed at any point.

32. A function of the real variables x, y is transformed by 3·4 (3) into a function $F(z, \bar{z})$ of the complex variables, $z, \bar{z}$. Prove that

$$\frac{\partial F}{\partial \bar{z}} = \tfrac{1}{2}\left(\frac{\partial F}{\partial x} + i\,\frac{\partial F}{\partial y}\right).$$

If $F(z, \bar{z}) = \phi(x, y) + i\psi(x, y)$ and if $\phi,\ \psi$ satisfy the Cauchy-Riemann equations, prove that $\partial F/\partial \bar{z} = 0$, i.e. that the Cauchy-Riemann equations ensure that F shall be a function of z only.

33. Show that $w = Uz^2$ gives irrotational flow in a corner formed by the first quadrant in the xy-plane. Prove that the origin is a stagnation point. Find the streamlines and show the direction of motion of the fluid particles on these lines.

Show that a fluid drop, bounded by lines parallel to the axes of reference, remains rectangular throughout its motion.

CHAPTER IV

RECTILINEAR VORTICES

4·0. Two-dimensional vortices. In this chapter some aspects of two-dimensional vortex motion will be considered. The vorticity vector is by definition (3·21) perpendicular to the plane of the motion, so that the vortex lines (see 9·31) are straight and parallel. All vortex tubes (9·31) are therefore cylinders whose generators are perpendicular to the plane of the motion. Such vortices are known as *rectilinear vortices*. As usual we shall consider the fluid to be confined between parallel planes at unit distance apart and parallel to the plane of the motion, which is half-way between them, and we shall use the language of plane geometry.

4·1. Circular vortex. Let there be a single cylindrical vortex tube, whose cross-section is a circle of radius a, surrounded by unbounded fluid.

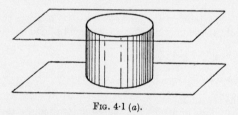

FIG. 4·1 (*a*).

The section of the vortex by the plane of the motion is a circle and the arrangement may therefore be referred to as a *circular vortex*.

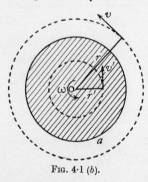

FIG. 4·1 (*b*).

We shall suppose that the vorticity over the area of this circle has the constant value ω. Outside the circle the vorticity is zero. Draw circles, concentric with the circle which bounds the vortex, of radii r' and r, where $r' < a < r$. Let v' and v be the speeds of fluid motion on the circles of radii r' and r respectively. It is clear from the symmetry that the speed at every point of the circle radius r' is the same, and that the velocity is tangential to this circle, for a radial component would entail a net flux across the circle and its centre O would then be a source or a sink. Similarly the velocity at any point of the circle of radius r is tangential to that circle.

Apply Stokes's circulation theorem (3·2) to these circles. Then

$$\int v' \, ds = \omega \, \pi r'^2, \quad r' < a \; ; \quad \int v \, ds = \omega \, \pi a^2, \quad r > a.$$

Since v' and v are constants on their respective circles we get

$$2\pi r' \, v' = \omega \, \pi r'^2, \quad 2\pi r \, v = \omega \, \pi a^2.$$

Thus $v' = \frac{1}{2}\omega r'$, $r' < a$; $v = \frac{1}{2}\omega a^2/r$, $r > a$. When $r' = r = a$ we have $v' = v = \frac{1}{2}a\omega$ so that the velocity is continuous as we pass through the circle.

From this it appears that the existence of a vortex such as we have described implies the co-existence of a certain distribution or *field of velocity*. This velocity field which co-exists with the vortex is known as the *induced velocity field* and the velocity at any point of it is called the *induced velocity*.

It is customary to refer to the velocity at a point of the field as the *velocity induced by the vortex*, but this must be understood merely as a convenient abbreviation of the fuller statement that were the vortex alone in the otherwise undisturbed field the velocity at the point would have the value in question. In this sense, when several vortices are present the field of each will contribute its proper amount to the velocity at a point.

Returning to the circular vortex the induced velocity at the extremity of any radius vector r joining the centre of the vortex to a point of the fluid external to the vortex is of magnitude inversely proportional to r and is perpendicular to r. Thus the induced velocity tends to zero at great distances.

As to the fluid within the vortex, its velocity is of magnitude proportional to r and therefore the fluid composing the vortex moves like a rigid body rotating about the centre O with angular velocity $\frac{1}{2}\omega$. The velocity at the centre is zero. This important fact may be stated in the following way.

A circular vortex induces no velocity at its centre. This is to be understood to mean that the centre of a circular vortex alone in the otherwise undisturbed fluid will not tend to move.

Still considering the fluid within the vortex, the velocities at the extremities of oppositely directed radii are of the same magnitude but of opposite sense so that the mean velocity of the fluid within the vortex is zero. Thus, if a circular vortex of small radius be " placed " in a field of flow at a point where the velocity is **u** the mean velocity at its centre will still be **u** and the fluid composing the vortex will move with velocity **u** ; it will " swim with the stream " carrying its vorticity with it.

The circular vortex is illustrated in nature on the grand scale by the tropical cyclone (hurricane, typhoon) which attains a diameter * of from 100 to 500 miles, and travels at a speed seldom exceeding 15 miles per hour. Within

---

* D. Brunt, *Weather Study*, London (1942).

the area the wind can reach hurricane force, while there is a central region, "the eye of the storm", of diameter 10 to 20 miles where conditions may be comparatively calm.

4·11. Velocity distribution.   If we introduce the *strength* κ of the circular vortex, defined by

$$2\pi\,\kappa = \text{circulation} = \pi a^2 \omega,$$

we have $\kappa = \frac{1}{2}a^2\omega$ and therefore

$$v' = \kappa\,\frac{r'}{a^2}, \quad r' < a\,; \quad v = \frac{\kappa}{r}, \quad r > a.$$

Since the velocities at all points of a diameter are perpendicular to that diameter the extremities of the velocity vectors at the different points of the diameter will lie on a curve which gives the velocity distribution as we go along the diameter from $-\infty$ to $+\infty$.

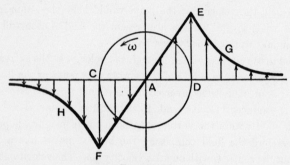

Fig. 4·11 (*a*).

This is shown in fig. 4·11 (*a*) where A is the centre of the vortex and CD is the diameter. For points between C and D the graph is a straight line EAF, for points on CD produced the graph is part of a rectangular hyperbola whose asymptotes are the diameter CD and the perpendicular diameter through A. The ordinates DE, CF each represent the velocity κ/a. Thus, if we keep κ constant and reduce the radius a of the vortex, DE will increase and DA will decrease. Therefore in the limit when $a \to 0$ the graph consists of the rectangular hyperbola and the asymptote perpendicular to CD.

Let us now introduce a second circular vortex also of radius a but with opposite vorticity $-\omega$ having its centre B on DC produced. If BA is sufficiently large compared with a, we can suppose, to a first approximation, that the vortices do not interfere, that is to say, that they remain circular and that their velocity fields may be compounded by the ordinary law of vector addition. The effect on the distribution graph of A will be to reduce all the velocities at points near A on the diameter CD by approximately $V = \kappa/BA$, by greater amounts on DC produced and by less amounts on CD produced.

The general shape of the distribution graph for the pair of vortices will now be that shown in fig. 4·11 (b).

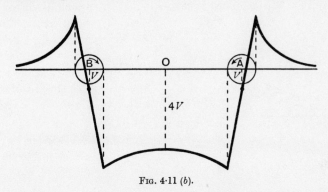

FIG. 4·11 (b).

It will be observed that the centre of each vortex is now in the field of velocity induced by the other and will therefore move with velocity V perpendicularly to AB. Thus the vortices are no longer at rest, but move with equal uniform velocity, remaining at constant distance apart. This is an application of the theorem that a vortex induces no velocity in itself.

The above diagram has its application to the study of the induced velocity due to the wake of a monoplane aerofoil at a distance behind the trailing edge (see 11·7).

4·12. Size of a circular vortex. It can be shown* that the pressure in the field of a circular vortex is least at the centre of the vortex, and that its value there is $\Pi - \kappa^2\rho/a^2$, where Π is the pressure at infinity. It follows that, if the pressure is to be nowhere negative, $a^2 \geqslant \kappa^2\rho/\Pi$, and the radius of the vortex cannot be less than the amount given by this relation, when κ and Π are assigned. In the following sections we shall be concerned with the case $a \to 0$, but the resulting point vortex must be regarded as a (very convenient) abstraction. We can of course make a as small as we please by making κ small enough, or Π big enough, but we shall still have a circular vortex and the induced velocity will be everywhere finite. The apparently infinite velocities which occur subsequently are therefore to be ascribed to the over-simplification of taking $a = 0$. If this is borne in mind, no difficulty need be felt at their occurrence.

A similar lower limit exists for the size of a point source (3·7) in two-dimensional motion, and is given by the same relation if κ is the strength of the source.

---

* Milne-Thomson, *Theoretical Hydrodynamics*, 13·11.

4·2. Point rectilinear vortex.

We have seen that at any point outside a circular vortex, at distance r from the centre, the velocity is κ/r at right angles to r where κ is the strength of the vortex.

If we let the radius a of the vortex tend to zero, κ remaining constant, the circle shrinks to a point and we have thus in the plane a point rectilinear vortex,

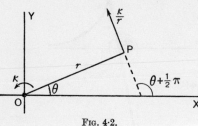

Fig. 4·2.

or simply a point vortex, of strength κ. The cylinder of fig. 4·1 (a) which represents the vortex tube shrinks to a straight line and the vortex is now a single rectilinear vortex represented by a point in the plane of the motion.

If we take the origin at the vortex, the velocity at the point $P(r, \theta)$ is represented by the complex number

$$\frac{\kappa}{r} e^{i(\theta + \frac{1}{2}\pi)} = \frac{i\kappa}{re^{-i\theta}},$$

and therefore the complex velocity (3·71) is

$$-\frac{dw}{dz} = u - iv = \frac{-i\kappa}{re^{i\theta}} = \frac{-i\kappa}{z}.$$

Hence, ignoring an added constant, which is irrelevant, the complex potential is given by

$$w = i\kappa \log z.$$

Observe that the motion is irrotational except at O where the vortex is situated and so a complex potential exists, with a logarithmic singularity at the vortex.

If the vortex were at the point z_0 instead of at the origin, we should have

$$w = i\kappa \log(z - z_0).$$

Observe also that the velocity derived from the complex potential is the velocity induced by the vortex.

4·21. Vortex pair.

A pair of point vortices of strengths κ and $-\kappa$ form a *vortex pair*.

If the vortices of the pair are at A and B respectively, each induces a velocity κ/AB in the other, perpendicular to AB and in the same sense. Thus the vortex pair moves perpendicularly to AB, remaining at the constant distance AB apart. The fluid

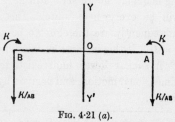

Fig. 4·21 (a).

velocity at O, the mid-point of AB, is $2\kappa/AB + 2\kappa/AB = 4\kappa/AB$, which is four times the velocity of either vortex (cf. fig. 4·11 (b)).

Taking O as origin and the x-axis along OA, if $AB = 2a$, we have the complex potential, at the instant when the vortices are on the x-axis,

(1) $$w = i\kappa \log (z - a) - i\kappa \log (z + a).$$

Thus $$u - iv = - i\kappa \left(\frac{1}{z - a} - \frac{1}{z + a} \right).$$

To find the velocity distribution along the x-axis we put $y = 0$, so that

$$u - iv = - \frac{2ai\kappa}{x^2 - a^2}.$$

Thus $u = 0$, $v = 2a\kappa/(x^2 - a^2)$. The graph of v against x is shown in fig. 4·21 (b).

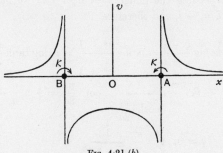

FIG. 4·21 (b).

The curve has the equation $v(x^2 - a^2) = 2a\kappa$, so that the asymptotes are $v = 0$, $x = a$, $x = -a$. This curve is the limit of that depicted in fig. 4·11 (b) with which it should be compared. The explanation of the occurrence of velocities which are apparently infinite is given in 4·12. The straight parts of fig. 4·11 (b) go over into the asymptotes $x = \pm a$ and this explains why we cannot read on fig. 4·21 (b) the velocity of A, although this is still one-quarter of the velocity at O.

4·22. Image of a vortex in a plane. Referring to fig. 4·21 (a), it is clear from the symmetry that there is no flux across YY' the perpendicular bisector of AB. Thus YY' is a streamline and could therefore be replaced by a rigid boundary, and the motion due to a vortex at A in the presence of this boundary is the same as the motion would be if the boundary were removed and an equal vortex of opposite rotation were introduced at B. The vortex at B is called the *image* of the actual vortex at A with respect to the plane boundary and the complex potential is still given by 4·21 (1).

4·3. Vortex between parallel planes.

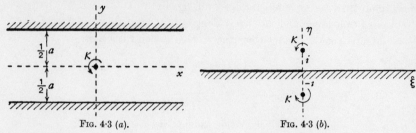

FIG. 4·3 (a).   FIG. 4·3 (b).

Let the vortex, strength κ, be midway between the planes $y = \pm a/2$ and at the origin.

The transformation $\zeta = ie^{\pi z/a}$, as is easily verified, maps the strip between the planes on the upper half of the ζ-plane (the heavy and thin lines in figs. (a), (b) indicate which parts of the boundaries correspond), and $z = 0$ corresponds with $\zeta = i$. Using the image system of 4·22 we have vortices κ at $\zeta = i$ and $-\kappa$ at $\zeta = -i$, and therefore

$$w = i\kappa \log \frac{\zeta - i}{\zeta + i} = i\kappa \log \frac{e^{\frac{\pi z}{a}} - 1}{e^{\frac{\pi z}{a}} + 1} = i\kappa \log \tanh \frac{\pi z}{2a},$$

$$u - iv = -i\kappa \cdot \frac{\pi}{2a} \cdot \frac{1}{\cosh^2 \frac{\pi z}{2a}} \times \coth \frac{\pi z}{2a} = -i\kappa \frac{\pi}{a} \frac{1}{\sinh \frac{\pi z}{a}}.$$

Thus when $\qquad y = 0, \quad v = \frac{\kappa \pi}{a} \frac{1}{\sinh \frac{\pi x}{a}}, \quad u = 0,$

and the velocity at points of the x-axis is given by this formula.

If the walls were absent we should have, on the x-axis, $v_0 = \kappa / x$.

Thus $\qquad\qquad \dfrac{v}{v_0} = \dfrac{\pi x}{a} \times \dfrac{1}{\sinh \dfrac{\pi x}{a}} < 1.$

Therefore the walls reduce the velocity v at points on the x-axis.

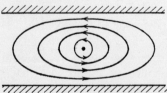

FIG. 4·3 (c).

Thus, for example, if $x = a$,

$$\frac{v}{v_0} = \frac{3 \cdot 14}{11 \cdot 53} = \frac{1}{4} \text{ roughly.}$$

The streamlines are somewhat as shown.

These are got from circular streamlines by observing that the walls increase u when $x = 0$, and decrease v when $y = 0$, so that the streamlines crowd and spread as shown.

4·4. Force on a vortex.

A rectilinear vortex may be regarded as the limit of a circular vortex which rotates about its centre as if rigid.

Let us suppose a small circular vortex inserted in a steady field of flow, so that its centre is at the point whose velocity is (u_0, v_0) before the vortex is

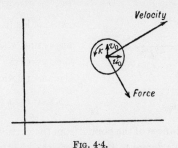

FIG. 4·4.

inserted. The vortex would then swim with the fluid (4·1), and begin to move with the velocity (u_0, v_0), so that the motion would no longer be steady. Let us imagine the vortex to be held fixed by the application of a suitable force (in the form of a pressure distribution). This force would be equal but opposite to that exerted by the fluid.

Since the motion is now steady, the force exerted by the fluid is the Kutta-Joukowski lift, which is investigated in 5·5. This is independent of the size and shape of the vortex and is given by

$$X + iY = -2\pi\kappa\rho i(u_0 + iv_0).$$

This force, being independent of size, is also the force exerted by the fluid on a point vortex.

The direction of the force (see fig. 4·4) is obtained by rotating the velocity vector through a right angle in the sense opposite to that of the circulation (vorticity) (see 5·5).

4·5. Mutual action of two vortices.

Consider two vortices of strengths κ and κ' at the points $(0, 0)$ and $(0, h)$, respectively.

Here for the vortex κ' we have

$$u_0 = -\kappa/h, \quad v_0 = 0$$

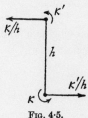

and therefore $X = 0$, $Y = 2\pi\kappa\kappa'\rho/h$, so that the two vortices repel one another if κ and κ' have the same sign, and attract if the signs are opposite. This result has its application to the action between the wings of a biplane.

FIG. 4·5.

4·6. Energy due to a pair of vortices. We consider two circular vortices each of radius a so small compared with the distance $2b$ between their centres that their circular form is preserved.

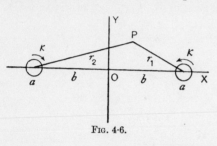

If κ is the strength of each vortex we can write

$$w = i\kappa \log (z - b) - i\kappa \log (z + b),$$

neglecting the interaction between the vortices as in 4·11, and the stream function is therefore

$$\psi = \kappa \log \frac{r_1}{r_2},$$

FIG. 4·6.

where r_1, r_2 are the distances of the point z from the vortices, as shown in fig. 4·6.

Then for the region *external* to the vortices the kinetic energy of the fluid is

$$T_e = \tfrac{1}{2}\rho \iint (u^2 + v^2)\, dx\, dy.$$

Now

$$u^2 + v^2 = - u \frac{\partial \psi}{\partial y} + v \frac{\partial \psi}{\partial x} = \frac{\partial (v\psi)}{\partial x} - \frac{\partial (u\psi)}{\partial y},$$

since in this region $\partial v/\partial x - \partial u/\partial y = 0$.

Hence by the two-dimensional form of Stokes's theorem (3·2) we have

$$T_e = \tfrac{1}{2}\rho \times 2 \int_{(C)} - (u\psi\, dx + v\psi\, dy)$$

where the integral is taken positively round C, the circumference of the vortex at $z = b$. The negative sign is accounted for by the fact that C is an internal boundary, and the factor 2 because each vortex must contribute the same amount to the energy.

Now $u\, dx + v\, dy = q_s\, ds$ where q_s is the tangential speed, so that

$$\int_{(C)} q_s\, ds = 2\pi\kappa, \quad \text{the circulation.}$$

Also on C, $r_1 = a$, and $r_2 = 2b$ nearly, so that we may write

$$T_e = - \rho \times 2\pi\kappa \times \kappa \log \frac{a}{2b} = 2\pi\rho\kappa^2 \log \frac{2b}{a}.$$

The fluid internal to C is rotating (4·11) with angular velocity κ/a^2 and moving as a whole with velocity $\kappa/2b$ induced by the other vortex. Thus its kinetic energy is

$$T_i = \pi a^2 \rho \left(\frac{1}{2} \frac{\kappa^2}{4b^2} + \frac{1}{2} \cdot \frac{a^2}{2} \frac{\kappa^2}{a^4} \right) = \tfrac{1}{4}\pi\rho\kappa^2,$$

neglecting a^2/b^2. Thus the total energy is, very nearly,

$$T = T_e + 2T_i = 2\pi\rho\kappa^2 \left(\tfrac{1}{4} + \log \frac{2b}{a} \right).$$

4·7. Continuous line of vortices. Let there be a straight line AH stretching from $(-\frac{1}{2}c, 0)$ to $(\frac{1}{2}c, 0)$ of continuous vortices, that is to say,

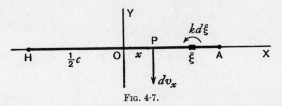

FIG. 4·7.

let the element $d\xi$ of the line at the point $(\xi, 0)$ behave like a point rectilinear vortex of strength $k\,d\xi$, where k may be constant or a function of ξ. This element taken by itself will induce at the point $P(x, 0)$ a velocity dv_x, in the negative sense of the y-axis, given by

$$dv_x = \frac{k\,d\xi}{\xi - x},$$

and therefore the whole line of vortices will induce at P the velocity

$$(1) \qquad v_x = \int_{-c/2}^{c/2} \frac{k\,d\xi}{\xi - x}.$$

Observe that in (1) ξ is *variable*, x is *fixed*. When $\xi = x$ the integrand is infinite. On the other hand, using the principle that a vortex induces no velocity at its own centre, we note that the point x must be omitted from the range of variation of ξ. To do this we define the " improper " integral (1) by its *principal value*, namely,

$$(2) \qquad v_x = \lim_{\epsilon \to 0} \left\{ \int_{-c/2}^{x-\epsilon} \frac{k\,d\xi}{\xi - x} + \int_{x+\epsilon}^{c/2} \frac{k\,d\xi}{\xi - x} \right\},$$

for in this way the point $(x, 0)$ is always the centre of the omitted portion between $x - \epsilon$ and $x + \epsilon$.

For subsequent use in the theory of aerofoils the type of integral (2) in which we shall be interested is that for which $\xi = -\frac{1}{2}c\cos\phi$ and $k = k_n \sin n\phi$ where k_n is independent of ϕ, and it is this problem that we shall investigate.

If, for convenience, we put $x = -\frac{1}{2}c\cos\theta$, where θ, like x, is fixed, we get from (1)

$$v_x = k_n \int_0^{\pi} \frac{\sin n\phi \sin\phi\,d\phi}{\cos\theta - \cos\phi} = \tfrac{1}{2}k_n \int_0^{\pi} \frac{[\cos(n-1)\phi - \cos(n+1)\phi]d\phi}{\cos\theta - \cos\phi}.$$

Our problem therefore reduces to calculating an improper integral of the type

$$I_n = \int_0^{\pi} \frac{\cos n\phi\,d\phi}{\cos\phi - \cos\theta}.$$

It will be proved in the next section that $I_n = \pi \sin n\theta / \sin\theta$. It therefore follows that

$$v_x = \tfrac{1}{2}k_n[I_{n+1} - I_{n-1}] = \tfrac{1}{2}\pi k_n \frac{\sin(n+1)\theta - \sin(n-1)\theta}{\sin\theta} = \pi k_n \cos n\theta.$$

4·71. Evaluation of the definite integral.

(1) $$I_n = \int_0^\pi \frac{\cos n\phi}{\cos \phi - \cos \theta}\, d\phi, \quad n \text{ an integer.}$$

As explained in 4·7 we define I_n by its *principal value*

$$I_n = \lim_{\epsilon \to 0}\left\{ \int_0^{\theta - \epsilon} \frac{\cos n\phi}{\cos \phi - \cos \theta}\, d\phi + \int_{\theta + \epsilon}^\pi \frac{\cos n\phi}{\cos \phi - \cos \theta}\, d\phi \right\}.$$

This is physically tantamount (see 4·7) to omitting the vorticity between $\theta - \epsilon$ and $\theta + \epsilon$ and then taking the limit $\epsilon \to 0$ so that θ remains the centre of the omitted portion.

If $n = 0$ we have, as may readily be verified, by differentiation,

$$\int_0^{\theta - \epsilon} \frac{d\phi}{\cos \phi - \cos \theta} = \left[\frac{1}{\sin \theta} \log \frac{\sin \frac{1}{2}(\theta + \phi)}{\sin \frac{1}{2}(\theta - \phi)} \right]_0^{\theta - \epsilon},$$

$$\int_{\theta + \epsilon}^\pi \frac{d\phi}{\cos \phi - \cos \theta} = \left[\frac{1}{\sin \theta} \log \frac{\sin \frac{1}{2}(\phi + \theta)}{\sin \frac{1}{2}(\phi - \theta)} \right]_{\theta + \epsilon}^\pi.$$

Hence $$I_0 = \lim_{\epsilon \to 0} \frac{1}{\sin \theta} \log \frac{\sin (\theta - \frac{1}{2}\epsilon)}{\sin (\theta + \frac{1}{2}\epsilon)} = 0.$$

It follows that we may write

(2) $$I_n = I_n - I_0 \cos n\theta = \int_0^\pi \frac{\cos n\phi - \cos n\theta}{\cos \phi - \cos \theta}\, d\phi.$$

In particular, putting $n = 1$, we have $I_1 = \pi$.

Now $\cos (n + 1)\phi - \cos (n + 1)\theta + \cos (n - 1)\phi - \cos (n - 1)\theta$

$$= 2 \cos n\phi (\cos \phi - \cos \theta) + 2 \cos \theta (\cos n\phi - \cos n\theta)$$

and $$\int_0^\pi \cos n\phi\, d\phi = 0. \quad \text{Therefore, from (2),}$$

$$I_{n+1} + I_{n-1} = 2 \cos \theta\, I_n.$$

To solve this difference equation * put $I_n = x^n$, which gives

$$x^2 - 2x \cos \theta + 1 = 0,$$

so that $$x = e^{i\theta} \quad \text{or} \quad e^{-i\theta}.$$

Thus $$I_n = A \sin n\theta + B \cos n\theta.$$

Since $I_0 = 0$ we have $B = 0$, and since $I_1 = \pi$ we have $A = \pi \operatorname{cosec} \theta$ and therefore

(3) $$I_n = \pi \frac{\sin n\theta}{\sin \theta},$$

which is valid for all positive integral values of n including zero.

---

* Milne-Thomson, *The Calculus of Finite Differences* (1965), 13·0.

EXAMPLES IV

1. If p' is the pressure at radius r' within a cylindrical vortex, show that

$$\frac{1}{\rho}\frac{dp'}{dr'} = \frac{\omega^2\,r'}{4} = \frac{\kappa^2\,r'}{a^4},$$

and deduce that

$$p' = \frac{\kappa^2\,r'^2\,\rho}{2a^4} + p_0,$$

where p_0 is the pressure at the centre of the vortex.

2. If p is the pressure at a point external to a cylindrical vortex, prove that

$$\frac{p}{\rho} + \frac{\kappa^2}{2r^2} = \frac{\Pi}{\rho},$$

where Π is the pressure at infinity.

3. Draw a graph to show the pressure distribution within and outside a cylindrical vortex.

4. Show that the pressure due to a cylindrical vortex is least at the centre and has the value there $\Pi(1 - k)$, where

$$k = \frac{\kappa^2\,\rho}{a^2\,\Pi}.$$

Hence show that if $k > 1$, the vortex has a concentric cylindrical hollow space. Deduce that, if $k = 2$, there is a completely hollow cylindrical space around which there is cyclic irrotational motion.

5. Show that the stream function for a cylindrical vortex is $\psi = -\frac{1}{4}\omega(a^2 - r^2)$, or $\psi = \frac{1}{2}\omega a^2 \log \dfrac{r}{a}$, according as $r < a$ or $r > a$.

6. Draw the pressure distribution for a rectilinear vortex filament in unbounded incompressible air.

7. If the vortices of a vortex pair are situated at A and B, show that the stream function at P is

$$\kappa \log \frac{PA}{PB}.$$

Deduce that the streamlines are coaxal circles.

8. For a vortex between parallel planes (4·3) show that when $x = 0$

$$u = -\frac{\kappa\pi}{a}\operatorname{cosec}\frac{\pi y}{a}.$$

Explain the significance of the negative sign.

9. Vortex filaments of strengths κ_1, κ_2 are placed at A_1, A_2. Show that the centroid of masses κ_1, κ_2 remains at rest if the masses are imagined to move with the vortices. Prove also that each vortex describes a circle.

10. Draw the streamlines for the combination of source and vortex the complex potential of which is

$$w = (-m + i\kappa)\log z.$$

11. Show that in steady two-dimensional motion the vorticity remains constant along a streamline.

12. A two-dimensional vortex filament of strength κ is near the corner of a large rectangular tank and is parallel to the edge of the corner. Show that the filament will trace out in the plane the curve $r \sin 2\theta = $ constant, and that the motion will be regulated by the equation

$$r \frac{d\theta}{dt} = \frac{-\kappa}{2r}.$$

13. Three parallel rectilinear vortices of the same strength κ and in the same sense meet any plane perpendicular to them in an equilateral triangle of side a. Show that all three vortices move round the same cylinder with uniform speed in time $2\pi a^2/(3\kappa)$.

14. The two-dimensional motion of incompressible air is such that the vorticity ω is uniform. Show that the stream function ψ is given by

$$\psi = \tfrac{1}{4}\omega(x^2 + y^2) + f(x + iy) + \bar{f}(x - iy),$$

where f is an arbitrary function.

15. Prove that a source of strength m is acted upon by the fluid with a force $X + iY = -2\pi\rho m(u_0 + iv_0)$.

16. Prove that two sources attract one another, but that a source and a sink repel one another.

17. Prove that a source and a vortex exert on one another a force perpendicular to the line joining them.

CHAPTER V

THE CIRCULAR CYLINDER AS AN AEROFOIL

5·0. In this chapter we shall consider the properties of two-dimensional airflow past a circular cylinder, the air being treated as inviscid and incompressible. The considerations adduced in 1·8 show clearly that a circular cylinder is an unsuitable shape for an aerofoil but, as we shall see later, it is easy to transform a circle into an aerofoil profile by a conformal mapping, and from the flow past the circle we can then deduce the corresponding flow past the aerofoil. For this reason a careful consideration of the circular profile is a useful preliminary.

5·1. The points z and a^2/z.

Let C be the circle $|z| = a$ in the Argand diagram of the z-plane.

Let P be the point $z = re^{i\theta}$. Then if Q is the point a^2/z, we have

$$\frac{a^2}{z} = \frac{a^2}{re^{i\theta}} = \frac{a^2}{r} e^{-i\theta}.$$

If we mark, on OP between O and P, the point S such that $OS = OQ = a^2/r$ we see that

$$(1) \qquad OS \cdot OP = a^2,$$

so that S and P are *inverse points* with respect to the circle C, and the point Q

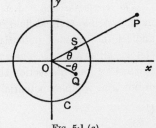

Fig. 5·1 (a).

is the optical reflection of S in the x-axis regarded as a mirror. It is clear from (1) that if P is *outside* the circle ($OP>a$), then S and therefore Q is *inside* the circle.

If, however, P is *on* the circumference of the circle, S coincides with P, and Q then also lies on the circumference : so that if z is on the circumference,

$$(2) \qquad \bar{z} = \frac{a^2}{z}.$$

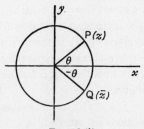

Fig. 5·1 (b).

Let $f(z)$ be a function of z which is holomorphic in the whole plane except at certain isolated singular points all of which are at a distance greater than a from the origin. We can then form the associated function $\bar{f}(z)$ described in 3·45. Consider the pair of functions

$$f(z), \quad \bar{f}\left(\frac{a^2}{z}\right),$$

the second being obtained from $\bar{f}(z)$ by writing a^2/z instead of z.

If the point z is outside the circle C, the point a^2/z is inside C. It follows that all the singular points of $\bar{f}(a^2/z)$ are inside C, since by hypothesis all those of $f(z)$ are outside C.

Also if z is on C, so is $\bar{z} = a^2/z$. Therefore the function

$$f(z) + \bar{f}\left(\frac{a^2}{z}\right)$$

is such that all its singularities outside the circle C are the same as those of $f(z)$ while if z is on the circle the value of the function becomes

$$f(z) + \bar{f}(\bar{z}),$$

which being the sum of two conjugate complex numbers has its imaginary part equal to zero, from Theorem A, 3·4.

5·2. The circle theorem. We now prove a general theorem * which will be of great use subsequently.

The circle theorem. Let there be irrotational two-dimensional flow of incompressible inviscid fluid in the z-plane. Let there be no rigid boundaries, and let the complex potential of the flow be $f(z)$, where the singularities of $f(z)$ are all at a distance greater than a from the origin. If a circular cylinder typified by its cross-section the circle C, $|z| = a$, be introduced into the field of flow, the complex potential becomes

$$(1) \qquad\qquad w = f(z) + \bar{f}\left(\frac{a^2}{z}\right).$$

Proof. Since there are no rigid boundaries, the flow given by $f(z)$ is determinate at every point of the z-plane, except perhaps at the singularities of $f(z)$ which arise from the vortices, sources, doublets, streams, etc., to which the flow is due.

After the cylinder is inserted, C must become a streamline $\psi = $ constant, and without loss of generality we may assume that it is the streamline $\psi = 0$.

We have seen in 5·1 that the singularities of w, given by (1), in the region external to C are the same as those of $f(z)$, and therefore no new singularities are introduced in this region. In particular, since by hypothesis $f(z)$ has no singularity at $z = 0$, $\bar{f}(a^2/z)$ has no singularity at infinity.

Now w is purely real on the cylinder, for there $w = f(z) + \bar{f}(\bar{z})$, and therefore $\psi = 0$ on C. Thus C is a streamline and all the conditions are satisfied. Q.E.D.

---

* Milne-Thomson, *Proc. Camb. Phil. Soc.*, 36 (1940).

Applications.

(i) A uniform wind in the negative direction of the x-axis. Here $f(z) = Vz$. When the cylinder is inserted $w = Vz + Va^2/z$ (cf. fig. 3·14 (b)).

(ii) A uniform wind at incidence α to the x-axis. From 3·7 (i) $f(z) = Ve^{i\alpha}z$ and therefore $w = Ve^{i\alpha}z + Va^2e^{-i\alpha}/z$.

(iii) The cylinder is inserted in the same uniform wind as in (ii), but the centre of the cylinder is at the point $z = z_0$. Then

$$w = Ve^{i\alpha}(z - z_0) + \frac{a^2\,Ve^{-i\alpha}}{z - z_0} = Ve^{i\alpha}z + \frac{a^2\,Ve^{-i\alpha}}{z - z_0} + \text{constant},$$

and the constant may be omitted for it contributes nothing to the velocity $- dw/dz$.

(iv) A vortex of strength κ at the point z_0. The centre of the cylinder is at the origin. Here $f(z) = i\kappa \log(z - z_0)$, and therefore, if $|z_0| > a$,

$$w = i\kappa \log(z - z_0) - i\kappa \log(a^2/z - \bar{z}_0).$$

Thus $\quad w = i\kappa \log z + i\kappa \log(z - z_0) - i\kappa \log\left(z - \frac{a^2}{\bar{z}_0}\right) + \text{constant}.$

The point $a^2/\bar{z}_0$ is the inverse of z_0 with respect to the circle.

(v) A vortex pair κ at z_0, $-\kappa$ at z_0', both outside the cylinder whose centre is the origin. Using (iv)

$$w = i\kappa \log(z - z_0) - i\kappa \log(z - z_0') - i\kappa \log\left(z - \frac{a^2}{\bar{z}_0}\right) + i\kappa \log\left(z - \frac{a^2}{\bar{z}_0'}\right).$$

(vi) A vortex pair *inside* the cylinder. The solution is the same as (v), for clearly if z_0, z_0' are inside the cylinder, $a^2/\bar{z}_0$, $a^2/\bar{z}_0'$ are outside it, and the cylinder is a streamline.

5·3. Circulation about a circular cylinder.

The complex potential of a vortex of strength κ at the origin is

(1) $w = i\kappa \log z = i\kappa \log r - \kappa\theta.$

Thus $\quad \phi = -\kappa\theta, \quad \psi = \kappa \log r.$

It follows that the streamlines $\psi = $ constant are the circles $r = $ constant. Any one of these circles, say $r = a$, can be taken as a rigid boundary. Now take a circuit which embraces the circle once. If we go once round this circuit, θ increases by 2π and therefore ϕ decreases by $2\pi\kappa$ which means (see 3·32) that there is a circulation $2\pi\kappa$ in any circuit which embraces

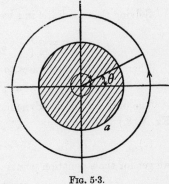

Fig. 5·3.

the circle once. Thus (1) is the complex potential of an irrotational circulatory motion round the cylinder of radius a.

5·31. Circular cylinder in a wind with circulation. Consider the uniform wind whose complex potential is Vz. If we insert the circular cylinder, typified by the circle $|z| = a$, the complex potential becomes (see 5·2)

$$Vz + V\frac{a^2}{z}.$$

To include circulation $2\pi\kappa$ we add the term $i\kappa \log z$, which gives finally

$$(1) \qquad\qquad w = Vz + V\frac{a^2}{z} + i\kappa \log z.$$

It is easy to verify that the circle is still a streamline, for putting $z = ae^{i\theta}$ we get

$$\psi = \kappa \log a,$$

which is constant.

To find the stagnation points we equate dw/dz to zero, which gives

$$V - \frac{Va^2}{z^2} + \frac{i\kappa}{z} = 0.$$

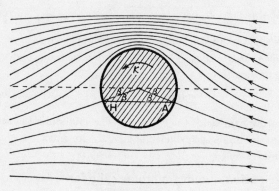

Fig. 5·31 (a).

Solving this quadratic in z we get

$$z = a\left\{- i\frac{\kappa}{2aV} \pm \sqrt{\left(1 - \frac{\kappa^2}{4a^2V^2}\right)}\right\}.$$

There are thus three cases to distinguish,

$$\kappa < 2aV, \quad \kappa = 2aV, \quad \kappa > 2aV.$$

The only case of aerodynamic interest is the first. (For the other two see Ex. V, 8.) Putting

$$(2) \qquad\qquad \frac{\kappa}{2aV} = \sin \beta,$$

we get for the stagnation points

$$z = a(\pm \cos \beta - i \sin \beta) = ae^{-i\beta}, \quad -ae^{i\beta}.$$

Thus the stagnation points lie on the cylinder and on a line parallel to the undisturbed wind stream (in this case parallel to the real axis).

Fig. 5·31 (a) shows the stagnation points * A and H, the interpretation of the angle β and the disposition of the streamlines. The general effect of the circulation is to increase the speed of the air at points of the major arc AH and to decrease the speed at points of the minor arc AH. Thus, by Bernoulli's theorem, the pressure above is diminished and the pressure below is increased so that there will be an upward force on the cylinder perpendicular to the wind, in other words a lift.

From (1) we get the stream function

(3)
$$\psi = Vy\left(1 - \frac{a^2}{r^2}\right) + \kappa \log r,$$

which is unaltered when $-x$ is written for x, and therefore the streamlines are symmetrical about the y-axis so that there will be no resultant force in the direction of the wind.

It should be observed that (2) gives the circulation which makes a given point on the cylinder a stagnation point. It also appears from (2) that when there is no circulation the stagnation points are at the ends of a diameter and (3) shows that in this case the streamlines are also symmetrical, fig. 3·14 (b), about the x-axis so that the lift vanishes.

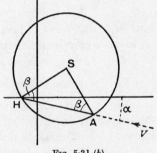

FIG. 5·31 (b).

If we take the centre of the cylinder at the point $z = s$ and the wind at incidence α, the complex potential (1) is replaced by (see 5·2 (iii)),

(4)
$$w = Vze^{i\alpha} + \frac{Va^2 e^{-i\alpha}}{z - s} + 2ai\,V \sin \beta \log (z - s).$$

In fig. 5·31 (b) the rear stagnation point H is shown on the real axis. This is the usual disposition but it does not affect the form of (4). The flow pattern is of course independent of any choice of axes.

5·311. Given stagnation point.

To find the circulation which will make a given point z_0 of the circular cylinder a stagnation point.

We use the circle theorem. Let $f(z)$ be the complex potential before the cylinder $|z| = a$ is inserted. After the cylinder is inserted with circulation $2\pi\kappa$ we have

$$w = f(z) + \bar{f}\left(\frac{a^2}{z}\right) + i\kappa \log z,$$

---

* Mnemonic ; A for anterior, H for hindmost stagnation point.

and, since z_0 is to be a stagnation point, $dw/dz = 0$ when $z = z_0$. Therefore

$$f'(z_0) - \frac{a^2}{z_0^2} \bar{f}' \left(\frac{a^2}{z_0} \right) + \frac{i\kappa}{z_0} = 0.$$

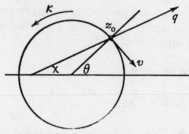

Fig. 5·311.

Now, since z_0 is on the cylinder $a^2/z_0 = \bar{z}_0$. Therefore

$$\kappa = iz_0 f'(z_0) - i\bar{z}_0 \bar{f}'(\bar{z}_0) = \text{real part of} - 2iz_0(u_0 - iv_0),$$

where $u_0 - iv_0 = -f'(z_0)$ is the complex velocity at z_0 in the undisturbed flow, i.e. *when the cylinder and circulation are absent.*

Let $z_0 = ae^{i\theta}$, $u_0 + iv_0 = qe^{i\chi}$. Then

$$\kappa = \text{real part of} - 2iaq \, e^{i(\theta-\chi)} = 2aq \sin(\theta - \chi) = 2av,$$

where $v = q \sin(\theta - \chi)$ is the component at z_0 of the *undisturbed velocity* along the tangent to the circle. Observe that κ and v are always in *opposite senses*. Equation 5·31 (2) could be deduced from this theorem. See also 8·7.

5·32. The pressure on the cylinder.

At points on the cylinder (5·31 (1)),

$$\frac{dw}{dz} = V(1 - e^{-2i\theta}) + \frac{i\kappa}{a} e^{-i\theta} = e^{-i\theta} \left[2i \, V \sin \theta + \frac{i\kappa}{a} \right].$$

Hence

$$q^2 = \left(2V \sin \theta + \frac{\kappa}{a} \right)^2 = 4V^2 (\sin \theta + \sin \beta)^2,$$

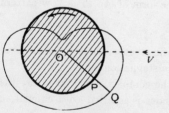

Fig. 5·32.

and therefore if Π is the pressure at a great distance from the cylinder Bernoulli's theorem gives

$$(1) \quad \frac{p}{\rho} + 2V^2(\sin \theta + \sin \beta)^2 = \frac{\Pi}{\rho} + \tfrac{1}{2}V^2,$$

so that

$$(2) \quad \frac{p}{\tfrac{1}{2}\rho V^2} = \frac{\Pi}{\tfrac{1}{2}\rho V^2} + 1 - 4(\sin \theta + \sin \beta)^2.$$

In fig. 5·32 we have a polar diagram in which the radius vector OQ represents the pressure at the point P of the cylinder.

To draw such a diagram we must know the values of $\Pi/\frac{1}{2}\rho V^2$ and $\sin \beta$; OQ is then drawn to represent $p/\frac{1}{2}\rho V^2$.

That the cylinder experiences a lift but no drag is evident from the diagram.

5·33. Force on the cylinder.

If X and Y denote the resultant thrusts in the positive directions of the x- and y-axes due to air pressure we have

$$X = - \int_0^{2\pi} p \cos \theta \, a \, d\theta, \quad Y = - \int_0^{2\pi} p \sin \theta \, a \, d\theta.$$

Now from 5·32 we see that

$$p = p_1 - 4\rho V^2 \sin \beta \sin \theta,$$

where

$$p_1 = \Pi + \tfrac{1}{2}\rho V^2 - 2\rho V^2 \sin^2 \theta - 2\rho V^2 \sin^2 \beta,$$

and it is clear that the resultant pressure thrusts due to p_1 vanish, for p_1 is unaltered when θ is replaced by $\theta + \pi$, so that the thrusts due to p_1 at diametrically opposite points cancel. Therefore

$$X = 4\rho a V^2 \sin \beta \int_0^{2\pi} \sin \theta \cos \theta \, d\theta, \quad Y = 4\rho a V^2 \sin \beta \int_0^{2\pi} \sin^2 \theta \, d\theta,$$

whence

$$X = 0, \quad Y = 4\pi\rho a V^2 \sin \beta = 2\pi\kappa\rho V.$$

Thus the aerodynamic force on the cylinder is a lift equal to

Circulation $\times$ air density $\times$ wind speed.

The foregoing calculation is simple owing to the form of the contour of the cylinder. In the general case of an aerofoil of other than circular section the direct integration of the components of pressure thrust can become exceedingly complicated. There is, however, a simple method which can be used in all such cases which will form the subject of the following section.

5·4. The theorem of Blasius.

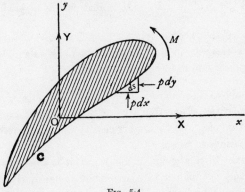

Fig. 5·4.

Consider a cylinder of any cross-section placed in steady irrotational air flow. If the origin O is taken as base point the aerodynamic force (per unit length of cylinder) may be reduced to a force (X, Y) and a pitching moment M about O. If w is the complex potential, the theorem of Blasius is as follows :

(1) $X - iY = \frac{1}{2} i\rho \int \left(\frac{dw}{dz}\right)^2 dz,$ $M = $ real part of $- \frac{1}{2}\rho \int z \left(\frac{dw}{dz}\right)^2 dz,$

where the integrals are taken round the contour C of the profile or round any contour reconcilable * with C without passing over a singularity of the integrands.

Proof. The pressure thrusts on an element ds of the contour give rise at O to the force system (dX, dY) and moment dM where

$$dX = - p\, dy, \quad dY = p\, dx, \quad dM = p(x\, dx + y\, dy).$$

Now the pressure equation gives $p = p_0 - \frac{1}{2}\rho q^2$ where p_0 is a constant pressure and the resultant effect of such a constant pressure is zero. Thus we can ignore p_0 and so write

$$d(X - iY) = \frac{1}{2}\rho q^2 (dy + i\, dx), \quad dM = - \frac{1}{2}\rho q^2 (x\, dx + y\, dy).$$

(2)   Now $dy + i\, dx = i\, d\bar{z},$ $x\, dx + y\, dy = $ real part of $z\, d\bar{z},$

and $q^2 = \frac{dw}{dz} \cdot \frac{d\overline{w}}{d\bar{z}}.$

Thus $d(X - iY) = \frac{1}{2}\rho i \dfrac{dw}{dz} d\overline{w},$ $dM = $ real part of $- \frac{1}{2} \rho z \dfrac{dw}{dz} d\overline{w}.$

Now on the cylinder the stream function ψ is constant and therefore $d\psi = 0$. Therefore

$$d\overline{w} = d(\phi - i\psi) = d(\phi + i\psi) = dw = \frac{dw}{dz} dz,$$

and so

$$d(X - iY) = \frac{1}{2}\rho i \left(\frac{dw}{dz}\right)^2 dz, \quad dM = \text{real part of } - \frac{1}{2}\rho z \left(\frac{dw}{dz}\right)^2 dz.$$

Integrating round C the theorem follows when C is the contour of integration, and by Cauchy's theorem (3·5) the contour can be enlarged or contracted provided no singularity of an integrand is crossed.

Q.E.D.

Notes on the above theorem :

(i) The singularities in question arise at those points where dw/dz becomes infinite and nowhere else. At such a singularity there is therefore a source or vortex or combinations of these.

(ii) Singularities of w are not in question; it is the behaviour of dw/dz which matters, and although w will have singularities at the same points as dw/dz it is the form of the latter which determines the aerodynamic force.

---

* This means that if we regard C as elastic, we can stretch it or contract it provided that no singularity is crossed in the process.

(iii) It is frequently advantageous to take as contour of integration a circle of large radius enclosing the origin. This can always be done when there are no sources or vortices in the fluid.

(iv) The calculation of the aerodynamic force is reduced to the calculation of residues (3·52).

5·41. Theorem of Blasius in terms of the stream function.

The theorem of Blasius stated in 5·4 applies only to irrotational motion. We now obtain a form in terms of the stream function. Since in two-dimensions a stream function always exists the new form will apply also to rotational motion.

We use the figure and notations of 5·4. From 5·4 (1)

$$d(X - iY) = \tfrac{1}{2}\rho q^2 i\, d\bar{z} \quad \text{and} \quad q^2 = u^2 + v^2 = (u - iv)(u + iv)$$

Therefore from 3·47 (1)

$$d(X - iY) = 2i\rho \frac{\partial \psi}{\partial z} \frac{\partial \psi}{\partial \bar{z}}\, d\bar{z}$$

Now ψ is constant on the boundary. Therefore

$$0 = d\psi = \frac{\partial \psi}{\partial z}\, dz + \frac{\partial \psi}{\partial \bar{z}}\, d\bar{z} \text{ so that}$$

$$d(X - iY) = -2i\rho \left(\frac{\partial \psi}{\partial z}\right)^2 dz$$

and therefore

(1)
$$X - iY = -2i\rho \int \left(\frac{\partial \psi}{\partial z}\right)^2 dz,$$

where the integral is taken round the boundary or any reconcilable contour.

Similarly we prove that for the moment M about the origin

(2)
$$M = \text{real part of } 2\rho \int z \left(\frac{\partial \psi}{\partial z}\right)^2 dz.$$

Observe also that on the boundary $\bar{z}$ is a function of z and so the above integrals may be evaluated by the residue theorem.

5·5. The theorem of Kutta and Joukowski.

An aerofoil at rest in a uniform wind of speed V, with circulation K round the aerofoil, undergoes a lift $K\rho V$ perpendicular to the wind. The direction of the lift vector is got by rotating the wind velocity vector through a right angle in the sense opposite to that of the circulation.

Proof. Since there is a uniform wind, the velocity at a great distance from the aerofoil must tend simply to the wind velocity, and therefore if $|z|$ is sufficiently large, we may write

(1)
$$-\frac{dw}{dz} = -Ve^{i\alpha} + \frac{A}{z} + \frac{B}{z^2} + \cdots,$$

where α is the incidence. Thus

$$w = Ve^{i\alpha} z - A \log z + \frac{B}{z} + \dots,$$

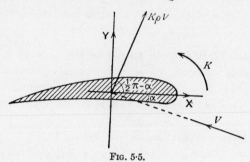

FIG. 5·5.

and since there is circulation K, we must have

(2) $$-A = \frac{iK}{2\pi},$$

for $\log z$ increases by $2\pi i$ when we go once round the aerofoil in the positive sense. From (1) and (2) we get

(3) $$\left(\frac{dw}{dz}\right)^2 = V^2 e^{2i\alpha} + \frac{iK\,Ve^{i\alpha}}{\pi z} - \frac{K^2 + 8\pi^2\,BVe^{i\alpha}}{4\pi^2\,z^2} - \dots.$$

If we now integrate round a circle whose radius is sufficiently large for the expansion (3) to be valid, the theorem of Blasius gives (see 3·52),

$$X - iY = \tfrac{1}{2}i\rho \times 2\pi i\left(\frac{iK\,Ve^{i\alpha}}{\pi}\right) = -iK\rho\,Ve^{i\alpha},$$

so that, changing the sign of i,

(4) $$X + iY = iK\rho\,Ve^{-i\alpha} = K\rho\,Ve^{i(\frac{1}{2}\pi-\alpha)}.$$

Comparison with fig. 5·5 shows that this force has all the properties stated in the enunciation.

Q.E.D.

Notes. (i) The theorem was discovered independently by Kutta (1902), and by Joukowski (1906).

(ii) *The lift is independent of the form of the profile.*

(iii) Observe that in applying the rule for the direction of the lift, the velocity vector must be drawn *from* the origin in the direction of the velocity.

(iv) If the aerofoil is regarded as moving in air otherwise at rest, the lift is got by rotating the velocity vector *of the aerofoil* through a right angle in the *same sense* as the circulation.

(v) The theorem of Blasius applied to (3) gives the moment about the origin

(5) $$M = \text{real part of } 2\pi i\rho BVe^{i\alpha}.$$

5·7. The second circle theorem.   The circle theorem of 5·2 applies to irrotational motion. The theorem about to be enunciated applies to motion in which the vorticity is constant.

The second circle theorem.   Let there be two-dimensional flow with constant vorticity ω in the z-plane, given by the stream function.

(1) $$\psi_0(z, \bar{z}) = F(z) + \bar{F}(\bar{z}) + \tfrac{1}{4}\omega z\bar{z}$$

Let there be no rigid boundaries and let all the singularities of $F(z)$ be a distance greater than a from the origin. If a circular cylinder typified by its cross-section of circumference C, $|z| = a$, be introduced into the field of flow, the stream function of the perturbed flow becomes

(2) $$\psi(z, \bar{z}) = F(z) - F\left(\frac{a^2}{\bar{z}}\right) + \bar{F}(\bar{z}) - \bar{F}\left(\frac{a^2}{z}\right) + \tfrac{1}{4}\omega z\bar{z}$$

Proof.   Since on C, $z\bar{z}=a^2$ the stream function $\psi(z, \bar{z})=\tfrac{1}{4}\omega a^2$ on C which is constant so that C is a streamline for the motion given by (2).

Since all the singularities of $F(z)$, and therefore also of $\bar{F}(\bar{z})$ are outside the circumference C, all the singularities of $F\left(\frac{a^2}{\bar{z}}\right)$ and of $\bar{F}\left(\frac{a^2}{z}\right)$ are inside C so that no new singularities are introduced at infinity and the motion given by (2) at infinity is the same as that given by (1).

The vorticity of the flow given by (2) is, 3·47 (2),

$$4\frac{\partial^2 \psi}{\partial z\, \partial \bar{z}} = \omega$$

Thus (2) satisfies all the conditions and is therefore the stream function of the perturbed motion.

Corollary.   If in (1) we replace $\tfrac{1}{4}\omega z\bar{z}$ by $\tfrac{1}{4}\omega z\bar{z}+\tfrac{1}{2}\kappa \log (z\bar{z})$ we get for the perturbed flow

$$\psi(z, \bar{z}) = F(z) - F\left(\frac{a^2}{\bar{z}}\right) + \bar{F}(\bar{z}) - \bar{F}\left(\frac{a^2}{z}\right) + \tfrac{1}{4}\omega z\bar{z} + \tfrac{1}{2}\kappa \log (z\bar{z})$$

This allows for circulation $2\pi\kappa$ about C.

Observe that save for an added constant (2) is the unique solution, for (2) solves the Dirichlet problem for the function $\psi(z, \bar{z}) - \tfrac{1}{4}\omega z\bar{z}$.

5·72. Uniform shear flow.   Let the x-axis be horizontal, say on ground level, and the y-axis vertically upwards. The velocity distribution

(1) $$u = -\omega y, \quad v = 0, \quad \omega \text{ constant}$$

is one in which the speed is proportional to the distance from the ground and decreases to zero as the ground is approached.

This type of velocity distribution is frequently exhibited by natural wind and is known as *uniform shear flow*.

More precisely we have the following definition.

Def. Two-dimensional flow in which the velocity at a point is parallel to a fixed line and proportional to the distance of the point from the line is called *uniform shear flow*.

The stream function for the flow (1) is

$$\tfrac{1}{2}\omega y^2 = -\tfrac{1}{8}\omega(z - \bar{z})^2.$$

Hence by a simple rotation of the axes through the angle β we find that if the velocity of the shear flow is parallel to the line $y \cos \beta - x \sin \beta = 0$, the stream function is

$$(2) \qquad \psi_0 = -\tfrac{1}{8}\omega(ze^{-i\beta} - \bar{z}e^{i\beta})^2.$$

It follows from 3·47 (3) that the vorticity is

$$(3) \qquad \frac{4\,\partial^2\psi_0}{\partial z\,\partial \bar{z}} = \omega$$

so that the vorticity in uniform shear flow is constant.

The stream function ψ_0 of (2) can also be written

$$(4) \qquad \psi_0 = -\tfrac{1}{8}\omega z^2 e^{-2i\beta} - \tfrac{1}{8}\omega\bar{z}^2 e^{2i\beta} + \tfrac{1}{4}\omega z\bar{z}$$

Comparing this with 5·7 (1), for the second circle theorem we see that

$$(5) \qquad F(z) = -\tfrac{1}{8}\omega z^2 e^{-2i\beta}$$

If in addition to shear flow we wish to have circulation also we write ψ_1 instead of ψ_0 in (4) where

$$(6) \qquad \psi_1 = -\tfrac{1}{8}\omega z^2 e^{-2i\beta} - \tfrac{1}{8}\omega\bar{z}^2 e^{2i\beta} + \tfrac{1}{4}\omega z\bar{z} + \tfrac{1}{2}\kappa \log (z\bar{z}).$$

5·74. Circular cylinder in uniform shear flow.

We consider the flow consisting of a uniform stream V at incidence α, uniform shear flow parallel to $y \cos \beta - x \sin \beta = 0$, circulation $2\pi\kappa$. The stream function for the undisturbed flow is

$$(1) \quad \psi_0 = -\tfrac{1}{2}iVze^{i\alpha} + \tfrac{1}{2}iV\bar{z}e^{-i\alpha} - \tfrac{1}{8}\omega z^2 e^{-2i\beta} - \tfrac{1}{8}\omega\bar{z}^2 e^{2i\beta} + \tfrac{1}{4}\omega z\bar{z} + \tfrac{1}{2}\kappa \log z\bar{z}$$

To this we apply the second circle theorem, 5·7, with

$$(2) \qquad F(z) = -\tfrac{1}{2}iVze^{i\alpha} - \tfrac{1}{8}\omega z^2 e^{-2i\beta}$$

so that the stream function for the perturbed motion is

$$(3) \quad \psi = -\tfrac{1}{2}iVze^{i\alpha} - \tfrac{1}{8}\omega z^2 e^{-2i\beta} - \tfrac{1}{2}iV\frac{a^2}{z}e^{-i\alpha} + \tfrac{1}{8}\omega\frac{a^4}{z^2}e^{2i\beta} + \tfrac{1}{2}iV\bar{z}e^{-i\alpha}$$

$$- \tfrac{1}{8}\omega\bar{z}^2 e^{2i\beta} + \tfrac{1}{2}iV\frac{a^2}{\bar{z}}e^{i\alpha} + \tfrac{1}{8}\omega\frac{a^2}{\bar{z}^2}e^{-2i\beta} + \tfrac{1}{4}\omega z\bar{z} + \tfrac{1}{2}\kappa \log (z\bar{z})$$

We note that this fulfills all the conditions and reduces to (1) as $z \to \infty$.

To find the force on the cylinder we use the theorem of Blasius in the form given in 5·41. Thus finding $\partial\psi/\partial z$ from (3) and then putting $\bar{z} = a^2/z$ we get, on C,

$$\frac{\partial\psi}{\partial z} = -\tfrac{1}{4}\omega e^{-2i\beta}z - \tfrac{1}{2}iVe^{i\alpha} + \frac{a^2\omega + 2\kappa}{4z} + \tfrac{1}{2}i\frac{Va^2e^{-i\alpha}}{z^2} - \tfrac{1}{4}\frac{a^4\omega e^{2i\beta}}{z^3}$$

Therefore from 5·41 (1), using the residue theorem,

(4) $X - iY = -\pi\rho Vi\{2\kappa e^{i\alpha} + a^2\omega(e^{\,i\alpha-2i\beta} + e^{i\alpha})\}$

This gives the Kutta-Joukowski lift when $\omega = 0$ but we note now that even when $\kappa = 0$ there is a force on the cylinder.

If the x-axis is parallel to the ground and α and β are both zero the force given by (4) is the lift

$$Y = 2\pi\rho V(\kappa + a^2\omega)$$

Thus the shear flow increases or decreases the lift $2\pi\rho V\kappa$ according as ω is positive or negative. Indeed if $\omega = -\kappa/a^2$, the lift vanishes thus revealing a possible danger in landing in a certain type of shear flow.

EXAMPLES V

1. Show that the image (4·22) of a vortex of strength κ at a point A outside a circular cylinder is an equal vortex at the centre of the cylinder and an equal vortex of the opposite rotation at B the inverse of A with respect to the cylinder.

Verify that with this system of vortices the boundary of the cylinder is in fact a streamline.

2. A vortex of strength κ is placed at the point $(f, 0)$ outside a circular cylinder, centre $(0, 0)$, of radius a. By calculating the forces (4·4) exerted on the image system prove that the cylinder is acted upon by a force of magnitude,

$$\frac{2\pi\kappa^2\rho a^2}{f(f^2 - a^2)}.$$

In what direction is the cylinder urged by this force?

3. Find the image system of a vortex pair inside a circular cylinder.

4. A vortex of strength κ is placed at the point A $(f, 0)$ inside a circular cylinder, centre the origin, of radius a. Prove that a vortex of strength $-\kappa$ placed at the inverse point $B(a^2/f, 0)$ will make the circle a streamline.

Prove that A begins to move with speed $\kappa f/(a^2 - f^2)$ perpendicularly to AB, and hence show that A will move round a circle concentric with the cylinder.

5. A rectilinear vortex of strength κ is situated outside a fixed solid circular cylinder of radius a. The vortex is parallel to and at distance f from the axis of the cylinder and there is no circulation in any circuit which does not enclose the vortex. Show that the vortex moves about the axis of the cylinder with constant angular velocity equal to $\kappa a^2/f^2(f^2 - a^2)$.

Find the velocity of the air at a point on the cylinder such that the axial plane through the point makes an angle θ with the axial plane through the vortex, and proceed to show how the resultant thrust on the cylinder may be calculated.

6. A vortex of strength κ is at a fixed distance R from the centre of a circle of radius a, round which there is a circulation of strength κ'. Prove that the force on the circle is

$$\frac{2\pi\rho}{R}\left\{\frac{\kappa^2 a^2}{R^2 - a^2} - \kappa\kappa'\right\}.$$

7. A column of air whose outer boundary is an infinitely long circular cylinder of radius b, is in cyclic irrotational motion and is under the action of a uniform pressure P over the external surface. Prove that there must be a concentric cylindrical hollow whose radius a is determined by the equation $2\pi a^2 b^2 P = M\kappa^2$, where κ is the strength of the circulation, and M is the mass of air per unit length of the column.

8. In the case of a circular cylinder in a wind stream with circulation (5·31) discuss the nature of the stagnation points (i) when $\kappa = 2aV$, (ii) $\kappa > 2aV$, showing that in the latter case part of the air circulates round the cylinder without ever joining the main stream.

9. Use Rankine's theorem to plot the streamlines given by 5·31 (3) in the three cases $\kappa <, =, > 2aV$.

10. The circle $|z| = a$ is placed in a wind V in the negative sense of the x-axis. Find the circulation $2\pi\kappa$ which will make the point $ae^{i\theta}$ a stagnation point, and draw a graph to show κ as a function of θ as θ varies from $\pi/2$ to $3\pi/2$.

11. Prove, with the usual notation, that the pressure on the boundary of the cylinder is everywhere positive if the speed of the oncoming air is less than the value given by

$$V^2 = \frac{2\Pi}{\rho}\frac{1}{(1 + 2\sin\beta)(3 + 2\sin\beta)}.$$

Plot the critical speed (cavitation speed) given by the above formula in the range $0 \leqslant \beta \leqslant \pi/2$.

Show that the critical speed is greatest when there is no circulation.

12. The circle $(x - a)^2 + y^2 = a^2$ is placed in an oncoming wind of velocity V and there is circulation $2\pi\kappa$. Find the complex potential and use the theorem of Blasius to show that the moment about the origin is $2\pi\kappa\rho aV$.

13. A source of strength m is placed at the point $(f, 0)$ outside the circle $|z| = a$. Prove that the complex potential is

$$w = -m\log(z - f) - m\log\left(z - \frac{a^2}{f}\right) + m\log z.$$

Use the theorem of Blasius to prove that there is no moment about the centre of the circle, and that the circle is urged towards the source by a force

$$\frac{2\pi\rho m^2 a^2}{f(f^2 - a^2)}.$$

Find the corresponding result when the source is replaced by a vortex.

14. Apply the circle theorem to show that the complex potential when the cylinder $|z| = a$ is in the presence of the doublet of 3·7 (iv) at $(f, 0)$ is

$$\frac{\mu}{z - f} - \frac{\mu z}{fz - a^2}.$$

Prove that the force on the cylinder is

$$\frac{4\pi\rho\,\mu^2 a^2 f}{(f^2 - a^2)^3}$$

and that there is no moment.

CHAPTER VI

JOUKOWSKI'S TRANSFORMATION

6·1. Joukowski's transformation. The simplest form of the transformation is

(1) $$z = \zeta + \frac{l^2}{\zeta}$$

where l is a real constant. By means of it we can map any selected region of the ζ-plane on the z-plane. In aerodynamic practice the region mapped is generally that exterior to a circle in the ζ-plane.

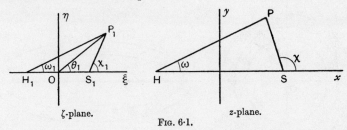

ζ-plane.   z-plane.

Fig. 6·1.

Let us denote corresponding points in the ζ- and z-planes by the same letter with, and without, suffix 1. In particular the points S_1, H_1 given by $\zeta = l$, and $\zeta = -l$, will map into the points S, H, given by $z = 2l$, $z = -2l$. These points play an important part in the geometry of the mapping.

From (1) we get at once by subtracting $2l$

(2) $$z - 2l = \frac{(\zeta - l)^2}{\zeta},$$

whence $\arg(z - 2l) = 2\arg(\zeta - l) - \arg\zeta$, $|z - 2l| = \dfrac{|\zeta - l|^2}{|\zeta|}$.

which in the notation of fig. 6·1 means that

(3) $$\chi = 2\chi_1 - \theta_1, \quad SP = S_1P_1^2/OP_1.$$

Thus, if χ_1 and θ_1 increase from 0 to 2π, so does χ.

Similarly, by adding $2l$ to each side of (1) we get

(4) $$z + 2l = \frac{(\zeta + l)^2}{\zeta}, \quad \omega = 2\omega_1 - \theta_1, \quad HP = H_1P_1^2/OP_1.$$

From (3) and (4) we see that

(5) $$\angle SPH = \chi - \omega = 2(\chi_1 - \omega_1) = 2\angle S_1P_1H_1;$$

(6) $$SP + HP = \frac{S_1P_1^2 + H_1P_1^2}{OP_1} = \frac{2OS_1^2 + 2OP_1^2}{OP_1}$$

by the theorem of Apollonius, since OP_1 is a median of the triangle $S_1P_1H_1$.

We also notice that for large values of $|\zeta|$ we have $z = \zeta$ nearly, so that the distant parts of the planes are undistorted by the mapping. This property is important, for it implies that a uniform wind in one plane will appear as the same uniform wind in the other.

The scale of the mapping is given by

$$\left|\frac{dz}{d\zeta}\right| = \left|1 - \frac{l^2}{\zeta^2}\right|,$$

which vanishes when $\zeta = l$ or $-l$. Thus the points S_1, H_1 are points where the mapping ceases to be conformal so that we must not map any region to which these points are interior, though they may appear on the boundary.

6·11. Circles with centre at the origin.

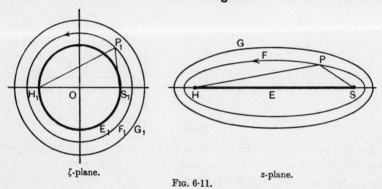

ζ-plane.   z-plane.

Fig. 6·11.

Let us apply Joukowski's transformation to circles whose centre is at the origin in the ζ-plane. We shall consider only circles to which the points S_1, H_1 are not external (see 6·1).

If P_1 is on one of the circles, say F_1, we have from 6·1 (6)

$$SP + HP = \text{constant},$$

since OP_1 is constant. Thus P describes an ellipse F whose foci are S and H. Similarly if P_1 describes a larger circle such as G_1, P will describe a larger ellipse G, which shows that points exterior to F_1 map into points exterior to F.

Thus the Joukowski transformation (1) maps circles in the ζ-plane whose centre is the origin into confocal ellipses in the z-plane.

As a particular case, the circle E_1 on S_1H_1 as diameter maps into the straight line SH. This is readily seen from 6·1 (5), for in this case $\angle S_1P_1H_1 = \pi/2$ and therefore $\angle SPH = \pi$, so that P moves on the line SH.

It now appears that the transformation will map the region external to any one of the circles on the region external to the corresponding ellipse.

In particular the region external to the circle E_1 which passes through S_1 and H_1 maps into the region external to the line SH. In the language of our subject the circular cylinder typified by the circle E_1 is transformed into the rectangular aerofoil, of chord $4l$, typified by the line SH.

6·2. Joukowski fins, rudders, and struts.

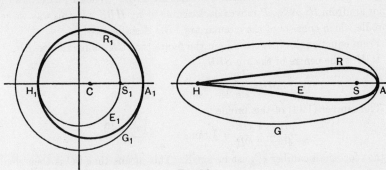

FIG. 6·2.

Instead of mapping a circle whose centre is the origin, let us take a circle R_1 whose centre C is on the real axis and which passes through H_1, and which encloses S_1. Then the circles E_1, G_1 with centres at the origin can be drawn to touch R_1 at H_1 and at A_1, the second point at which R_1 meets the real axis. Clearly the map of R_1 must lie between the maps of E_1 and G_1 and the map of R_1 will resemble the map of E_1 in the neighbourhood of H and the map of G_1 in the neighbourhood of A. Thus the circle R_1 maps into a symmetrical profile with a blunt nose at A and a cusp at H, the trailing edge. Such shapes are suited to form the profiles of fins, rudders and struts where symmetry is desirable.

6·3. Circular arc profiles.

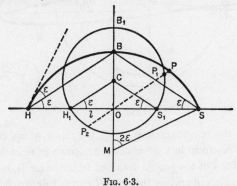

FIG. 6·3.

Let us transform a circle of radius a whose centre C is on the imaginary axis and which passes through S_1 and H_1.

Let us introduce the useful practice of marking the map on the *same* Argand diagram as the circle, so that the figure shows points of both the z- and ζ-planes.

If $\angle CH_1S_1 = \epsilon$, we have, from 6·1 (5), $\angle SPH = 2\angle S_1P_1H_1 = \pi - 2\epsilon$ since $\angle S_1CH_1$ at the centre is equal to $2\angle S_1P_1H_1$ at the circumference.

Thus when P_1 describes the major arc $S_1B_1H_1$ in fig. 6·3 the $\angle SPH$ remains constant and therefore P describes a circular arc SBH. When P_1 describes the minor arc from H_1 to S_1, P moves back again* along HBS. In this way we get a profile which consists of the circular arc SBH described twice.

From equal angles it is clear from the figure that CH_1 is parallel to BH. If M is the centre of the arc SBH,

$$SM = \frac{2l}{\sin 2\epsilon} = \frac{a}{\sin \epsilon} = \frac{a^2}{\sqrt{(a^2 - l^2)}}.$$

The camber (1·14) of this profile is

$$\frac{OB}{4l} = \frac{1}{2}\frac{OB}{OH} = \tfrac{1}{2}\tan \epsilon = \frac{1}{2}\frac{\sqrt{(a^2 - l^2)}}{l},$$

so that for small camber ϵ must be small. This means that OC is then small compared with l, in other words that the centre C of the circle to be transformed is near to O.

6·4. The general Joukowski profile.
This is obtained by transforming a circle of radius a which passes through the point H_1 but whose centre C is not on either the real or the imaginary axis.

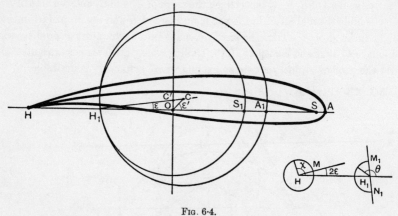

Fig. 6·4.

If CH_1 meets the imaginary axis at C', the circle whose centre is C' and radius $C'H_1$ will transform in a circular arc as shown in fig. 6·4. This circular arc forms the *skeleton* of the profile obtained from the circle, centre C. If C is near to C' this profile will enclose the arc SH and not depart far from it at any point ; the profile will be thin. The greater the distance CC' the thicker will be the profile. The actual construction of such a profile can easily be carried out

---

* For if P_2O meets the circle again at P_2, $OP_2 . OP_1 = l^2$ and therefore the map of P_2 is the reflexion in OB of P and thus lies on the arc HBS.

by the general method to be described in 6·51. Fig. 6·51 (c) shows the details of such a construction.

To examine the nature of the profile at H, observe that the transformation can be written

$$(1) \qquad \frac{z + 2l}{z - 2l} = \frac{(\zeta + l)^2}{(\zeta - l)^2}.$$

Near H_1 and H we can write

$$\zeta + l = re^{i\theta}, \quad z + 2l = Re^{i\chi},$$

where r and R are infinitesimal and therefore (1) gives

$$- Re^{i\chi} = r^2 e^{2i\theta}/l$$

approximately, so that taking arguments

$$(2) \qquad \chi + \pi = 2\theta.$$

If we draw a semicircle, centre H_1 radius r, on the tangent to the circle at H_1 and outside the circle, we can go round H_1 on this semicircle from M_1 to N_1 in fig. 6·4. In this passage θ increases by π and therefore χ increases by 2π. Thus there is a cusp at H, the two branches touching the same tangent. Also at M_1, $\theta = \frac{1}{2}\pi + \epsilon$ and therefore from (2) $\chi = 2\epsilon$, which is the inclination of the tangent at the cusp to $H_1 S_1$.

The existence of the cusp could also be explained on the basis of the reasoning of 6·2.

6·5. Geometrical construction.

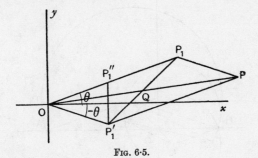

FIG. 6·5.

The Joukowski transformation

$$(1) \qquad z = \zeta + \frac{l^2}{\zeta}$$

can be replaced by the successive transformations

$$\zeta_1 = \frac{l^2}{\zeta}, \quad z = \zeta + \zeta_1.$$

Let us mark all the complex numbers ζ, ζ_1, z on the same Argand diagram, fig. 6·5. Let P_1 be the point ζ, P_1' the point ζ_1, and P the point z.

If $$\zeta = re^{i\theta}, \quad \text{then} \quad \zeta_1 = \frac{l^2}{r} e^{-i\theta}.$$

Draw $P_1'P_1''$ perpendicular to the real axis to meet OP_1 in P_1''. Then

$$OP_1'' = OP_1' = l^2/r, \quad OP_1 \,.\, OP_1'' = l^2.$$

Thus P_1, P_1'' are inverse points with respect to O, and to obtain P_1' we first find the inverse point P_1'' and then reflect OP_1'' in the real axis. The point P is then obtained by completing the parallelogram $OP_1'PP_1$. We also observe that if the diagonals of the parallelogram meet at Q then $OQ = \frac{1}{2}OP$ so that the locus of Q is similar and similarly situated to the locus of P but on half the scale.

In the majority of applications the point P_1 will be made to describe a circle. The point P_1'' will then describe the inverse of a circle which will be shown (6·51) to be a circle also and P_1' will describe the circle got by reflecting the locus of P_1'' in the real axis.

From (1) we have for the *scale of the mapping* (see 3·6)

$$(2) \qquad m = \left| \frac{dz}{d\zeta} \right| = \left| 1 - \frac{l^2}{\zeta^2} \right| = \left| \zeta - \frac{l^2}{\zeta} \right| \div |\zeta| = \frac{P_1 P_1'}{OP_1}.$$

This means that all lengths in a small region R round P will be m times the corresponding lengths in the region R_1 round P_1 of which R is the map.

6·51. Mapping a circle.

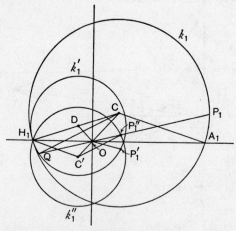

Fig. 6·51 (a).

Consider the circle k_1, centre C, radius a, which meets the real axis in H_1 and A_1 as shown in fig. 6·51 (a), and take $l = OH_1$.

Let P_1'' be the inverse of P_1 on the circle so that $OP_1'' \cdot OP_1 = OH_1^2 = l^2$. Let P_1O meet the circle k_1 again in Q and draw $P_1''C'$ parallel to CQ to meet CO produced at C'. By the rectangle property of the circle we have $OP_1 \cdot OQ = OH_1 \cdot OA_1$. Therefore, by division, $OP_1''/OQ = OH_1/OA_1$, a constant.

Since the triangles $OP_1''C'$, OQC are similar,

$$\frac{OC'}{OC} = \frac{C'P_1''}{CQ} = \frac{OP_1''}{OQ} = \frac{OH_1}{OA_1}.$$

Since $OC : OC'$ is constant, C' is a fixed point.

Since $C'P_1'' : CQ$ is constant, $C'P_1''$ is of constant length. Therefore P_1'' describes a circle k_1'' whose centre is C'. The point H_1 is its own inverse since $OH_1 \cdot OH_1 = l^2$ and therefore the locus of P_1'' passes through H_1.

Since $OC'/OC = OH_1/OA_1$, the triangles $OC'H_1$. OCA_1 are similar and similarly situated. Therefore CA_1 is parallel to $C'H_1$ and

$$\angle CH_1O = \angle CA_1H_1 = \angle C'H_1O.$$

Therefore the circle k_1', which is the reflection of the locus of P_1'' in the real axis, will have its centre D on CH_1 and will pass through H_1. It follows that the

circles k_1, k_1' touch at the point H_1. This circle k_1' is the locus of P_1' which is the reflection of P_1''. Since OD is the reflection of OC', it follows that OD and OC are equally inclined to the real axis.

The complete construction is shown in fig. 6·51 (b). Starting with the circle k_1 which passes through H_1, we find the point D on CH_1 such that OC, OD are equally inclined to the imaginary axis. The circle k_1', centre D, is then drawn. To map the point P_1 of k_1 we join OP_1 and find the point P_1' on k_1' such that OP_1,

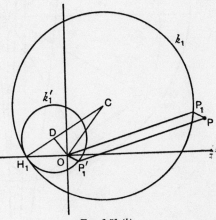

Fig. 6·51 (b).

OP_1' are equally inclined to the real axis. Completing the parallelogram $OP_1'PP_1$, the point P is the map of P_1.

Fig. 6·51 (c) shows the details of such a construction, the profile being sketched through the points obtained by drawing radii vectores at 30° intervals; corresponding points on the circle and aerofoil bear the same numerals.

The dotted circular arc shows the skeleton (6·4), got by transforming the

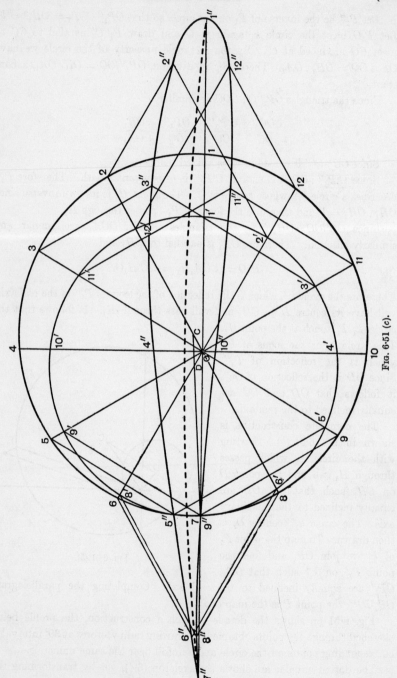

Fig. 6·51 (c).

circle through 7 whose centre is the point where CD cuts the y-axis, which can often conveniently replace the camber line in theoretical work.

Various simple link mechanisms have been devised for automatically describing the profiles arising from a given circle.

6·6. Reversal. To find the point P_1 of the ζ-plane which corresponds with a given point P of the z-plane, we consider the associated mapping

$$(1) \qquad Z = \tfrac{1}{2}z = \tfrac{1}{2}\left(\zeta + \frac{l^2}{\zeta}\right).$$

Referring to fig. 6·5, Z is the mid-point Q of P_1P_1' in fig. 6·6.

Let P_1O meet the circle k through S_1, P_1, H_1 at R. Then P_1' is the image of R in the imaginary axis, for $OP_1 \cdot OP_1' = l^2 = OH_1 \cdot OS_1 = OP_1 \cdot OR.$

Again, turning to fig. 6·3, when P_1, i.e. ζ, is on the arc $S_1B_1H_1$, the point P, i.e. z, is on the arc SBH and therefore Q, i.e. Z, is on the arc of the circle through S_1, C, H_1, for $OC = \tfrac{1}{2}OB$ and $Z = \tfrac{1}{2}z$.

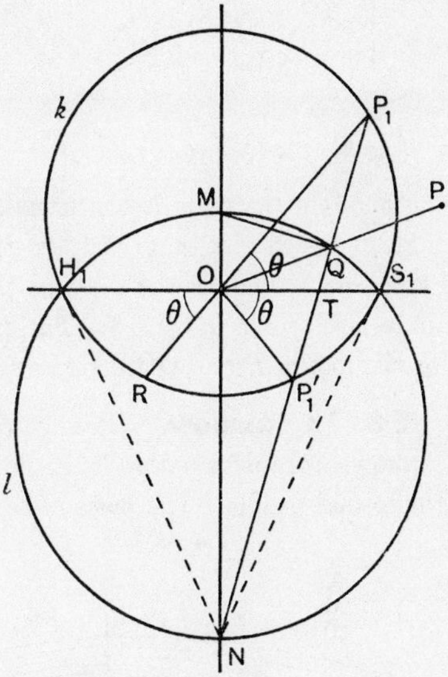

Therefore in fig. 6·6, as P_1 describes the upper arc of the circle k, the point Q will describe the upper arc of the circle l through S_1, M, H_1, where M is the centre of k.

Now MQ is the perpendicular bisector of P_1P_1', so that if P_1P_1' meets the imaginary axis in N, the $\angle MQN$ is a right angle. Therefore the point N is

FIG. 6·6.

also on l at the opposite end of the diameter through M.

From these facts we derive the following construction to find P_1 when P is given.

Find Q the mid-point of OP and draw the circle S_1QH_1. This determines M and N, M being on the same side of S_1H_1 as Q. The circle k, centre M, radius MS_1 can now be drawn. The line NQ then meets k in P_1 and P_1'. From the construction it appears that of the two points P_1, P_1' only one lies on the same side of S_1H_1 as P, and the choice of that one determines a bi-uniform (3·6) mapping.

In mapping the region outside any circle which encloses S_1 and passes through or encloses H_1, there is no ambiguity, for one of the points P_1, P_1' is always inside the circle and does not fall in the region to be mapped.

6·7. Construction of tangents.

To draw the tangent to a given profile we proceed as follows, still using the associated mapping 6·6 (1).

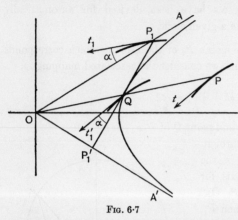

FIG. 6·7

Let the point P_1 move along a radius OA. Then from 6·5 we see that P_1' moves along a radius OA', the reflection of OA in the real axis. Also

$$OP_1 . OP_1' = l^2$$

and therefore Q, the mid-point of $P_1 P_1'$, describes a hyperbola* whose asymptotes are OA, OA' and $P_1 P_1'$ touches this hyperbola at Q. Let t_1 be the tangent at P_1 to the curve which maps into the profile described by P.

Since the mapping is conformal, the angle between t_1 and OA is equal to the angle between the tangent t_1' to the locus of Q and the tangent at Q to the hyperbola (which is the map in the Z-plane of the radius OA). The equal angles are shown in fig. 6·7. The tangent t at P to the profile is parallel to t_1', for the loci of P and Q are similar and similarly situated.

6·8. The airflow.

Having considered in some detail the geometry of the Joukowski transformation let us now examine its application to airflow round a Joukowski profile. The basis of the method is given in 3·8.

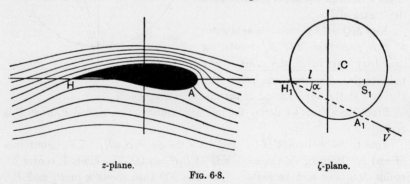

z-plane.   ζ-plane.

FIG. 6·8.

Let the profile of fig. 6·8 be obtained by transforming the circle $|\zeta - s| = a$ by the Joukowski transformation

$$(1) \qquad\qquad z = \zeta + \frac{l^2}{\zeta}.$$

---

* The triangle OP_1P_1' is of constant area, and therefore P_1P_1' touches a hyperbola whose asymptotes are OA, OA'. The point of contact of the tangent is the mid-point of P_1P_1', i.e. Q.

If there is a wind V at incidence α to the real axis in the ζ-plane, from 5·31 (4), the complex potential is

(2) $$w = V\zeta e^{i\alpha} + \frac{Va^2 e^{-i\alpha}}{\zeta - s} + 2aiV \sin \beta \log (\zeta - s),$$

and, from 3·8, (1) and (2) together define the complex potential in the z-plane, the plane of the profile. Observe that all quantities are given except the angle β. From (2)

(3) $$\frac{dw}{d\zeta} = Ve^{i\alpha} + \frac{2aiV \sin \beta}{\zeta - s} - \frac{Va^2 e^{-i\alpha}}{(\zeta - s)^2}$$
$$= V \left(e^{i\alpha} + \frac{ae^{i\beta}}{\zeta - s} \right) \left(1 - \frac{ae^{-i(\alpha+\beta)}}{\zeta - s} \right).$$

To find the velocity in the z-plane we have

(4) $$\frac{dw}{dz} = \frac{dw}{d\zeta} \times \frac{d\zeta}{dz} = \frac{dw}{d\zeta} \cdot \frac{\zeta^2}{\zeta^2 - l^2}.$$

Since $\zeta^2 - l^2$ vanishes at $\zeta = -l$ (and also at $\zeta = l$ which is inside the circle and is therefore not mapped), we see that the velocity at the cusp H will, in general, be infinite. If, however, we arrange that H_1 ($\zeta = -l$) is a stagnation point on the circle, dw/dz at H will assume the indeterminate form 0/0 and we shall prove that this yields a finite velocity at H.

To make H_1 a stagnation point, we see from fig. 5·31 (b) that $\angle CH_1S_1 = \beta - \alpha$ and therefore

(5) $$-l = s - ae^{i(\beta-\alpha)} \quad \text{or} \quad ae^{i\beta} = e^{i\alpha}(l + s).$$

This equation determines β and then

$$e^{i\alpha} + \frac{ae^{i\beta}}{\zeta - s} = e^{i\alpha} \left(1 + \frac{l + s}{\zeta - s} \right) = e^{i\alpha} \frac{\zeta + l}{\zeta - s},$$

so that (3) and (4) give, after an easy reduction,

(6) $$\frac{dw}{dz} = Ve^{i\alpha}(\zeta + l - 2ae^{-i\alpha} \cos \beta) \frac{\zeta^2}{(\zeta - s)^2(\zeta - l)},$$

the vanishing factor $\zeta + l$ having disappeared.

Putting $\zeta = -l$ we get from (6)

$$u_H - iv_H = - \left(\frac{dw}{dz} \right)_{\zeta = -l} = \frac{2aVl^2 \cos \beta}{(l + s)^2 (- 2l)} = - \frac{lV \cos \beta}{ae^{2i(\beta-\alpha)}}$$

from (5). Changing the sign of i we have

(7) $$u_H + iv_H = \frac{lV \cos \beta}{a} e^{i(\pi+2\beta-2\alpha)}.$$

Now from 6·4 the tangent at the cusp makes an angle $2\beta - 2\alpha$ with the real axis and therefore the wind streams smoothly past the cusp with speed $lV \cos \beta/a$.

The forward stagnation point A_1 on the circle transforms into the only stagnation point A on the profile, as is seen from (4).

We see from (3) that, at a great distance, when $|\zeta|$ is large, $dw/d\zeta = Ve^{i\alpha}$ nearly, and then from (4) that $dw/dz = dw/d\zeta = Ve^{i\alpha}$, so that there is the same uniform wind V at the same incidence to the real axis in both planes. The flow pattern in the ζ-plane (fig. 5·31 (a)) will transform into the flow pattern in the z-plane, the general form of some of the streamlines being shown in fig. 6·8.

Notes. (i) The problem of choosing a Joukowski profile is a purely geometrical one. Any circle will transform into some profile. The question of airflow does not enter at this stage. Naturally, however, the choice will be guided by the knowledge of shapes which have proved suitable.

(ii) The choice of wind speed and incidence at which the profile is to be used is perfectly free and has nothing to do with the transformation.

(iii) Having chosen (i) and (ii), in order to investigate mathematically the case of smooth flow past the cusp, it is necessary to arrange that the angle β or, what amounts to the same thing in view of 5·31 (2), the circulation is such that the point H_1 which transforms into the cusp is a stagnation point on the circle. See 7·11.

(iv) That there is the same uniform wind at infinity in both planes is really an inevitable consequence of the fact that when $|\zeta|$ is large (1) becomes $z = \zeta$, so that the complex potential is the same in both planes.

EXAMPLES VI

1. In the Joukowski transformation find the maps of the points $\zeta = \pm il$, $\zeta = \pm 3l \pm 4il$, and calculate SP and HP in each case.
Find also the angle SPH, and the scale of the mapping.

2. In the Joukowski transformation show that, if m is the scale of the mapping,

$$m^2 = \frac{SP \cdot HP}{OP_1{}^2}.$$

3. If the circle $|\zeta| = r$ is transformed into an ellipse by the transformation $z = \zeta + l^2/\zeta$, show that the semi-axes of the ellipse and its eccentricity are respectively

$$\frac{l^2 + r^2}{r}, \quad \frac{r^2 - l^2}{r}, \quad \frac{2lr}{l^2 + r^2}.$$

Prove also that the circle $|\zeta| = (a + b)/2$ is transformed into the ellipse of semi-axes a, b if $l^2 = (a^2 - b^2)/4$.

4. The circle $|\zeta| = a$ is transformed into a flat profile by

$$z = \zeta + \frac{a^2}{\zeta}.$$

Prove that near $z = 2a$

$$\frac{d\zeta}{dz} = \tfrac{1}{2} \frac{\sqrt{a}}{\sqrt{(z - 2a)}} + \cdots.$$

5. Map, numerically, the circle $\xi^2 + \eta^2 = 4$ by the transformation

$$z = \zeta + \frac{1}{\zeta}.$$

6. In the case of a circular arc profile show that the tangents to the profile at S and H make with the chord an angle equal to twice the angle which $\hat{C}H_1$ makes with the chord.

Prove also that, if the camber is small, the camber is equal to one quarter of the angle which the tangent at H to the profile makes with the chord.

7. In the general Joukowski profile (6·4) show that, if the centre of the circle is the point $se^{i\mu}$, the chord HA is

$$4l + \frac{4s^2 \cos^2 \mu}{l + 2s \cos \mu},$$

and that for slender profiles of small camber the chord is approximately $4a$.

8. Use the method 6·5 to map the points $\zeta = \pm 3l \pm 4il$, marking the corresponding points $P_1{}' \, P_1{}''$, and Q, and in each case calculate the scale of the mapping from $m = \dot{P}_1 P_1{}'/OP_1$.

9. Apply the construction of 6·51 to draw the profiles obtained from the circles through H_1 whose centres are at the points

$$\frac{l}{10} e^{i\pi/6}, \quad \frac{l}{10} e^{i\pi/3}$$

respectively.

Note the differences in the two profiles, and measure the camber and thickness ratio in each case.

10. In the Joukowski transformation, if $\zeta = re^{i\theta}$, prove that

$$x = l \cos \theta \left(\frac{r}{l} + \frac{l}{r} \right), \quad y = l \sin \theta \left(\frac{r}{l} - \frac{l}{r} \right).$$

11. A symmetrical Joukowski profile is got by transforming the circle

$$(\xi - kl)^2 + \eta^2 = l^2(1 + k)^2.$$

If $\zeta = re^{i\theta}$, prove that

$$\frac{r^2}{l^2} = 1 + 2k\left(1 + \frac{r \cos \theta}{l}\right).$$

If k^2 is negligible, show that

$$\frac{r}{l} = 1 + k(1 + \cos \theta), \quad \frac{l}{r} = 1 - k(1 + \cos \theta),$$

and hence that

$$x = 2l \cos \theta, \quad y = 2kl \sin \theta (1 + \cos \theta).$$

Hence show that the maximum thickness ratio occurs at the quarter point and that its value is $(3k \sqrt{3})/4$.

12. Construct tangents to the Joukowski profiles of Ex. 9.

13. Taking the chord of the Joukowski profile to be determined as the intercept of the profile on the x-axis, show that the tangent to the profile at the leading edge of the chord is parallel to the tangent to the circle at the point which transforms into the leading edge.

14. Supply the intermediate steps which lead to 6·8 (6).

15. If q_z, q_ζ are air speeds at the corresponding points P, P_1 prove that

$$q_z = q_\zeta \frac{OP_1}{P_1 P_1'}.$$

16. Show that the point $(a(h + \cos\theta), a(k + \sin\theta))$ describes a circle, a, h, k being constants.

If the Joukowski transformation is applied to this circle, show that

$$\frac{x}{a} = (h + \cos\theta)\left\{1 + \frac{l^2}{a^2(1 + h^2 + k^2 + 2h\cos\theta + 2k\sin\theta)}\right\},$$

$$\frac{y}{a} = (k + \sin\theta)\left\{1 - \frac{l^2}{a^2(1 + h^2 + k^2 + 2h\cos\theta + 2k\sin\theta)}\right\}.$$

Apply this transformation to trace the aerofoil for which $h = 0{\cdot}04$, $k = 0{\cdot}05$ and $l^2 = 0{\cdot}8a^2$, showing that it has a rounded trailing edge.

17. In a symmetrical Joukowski profile prove that the maximum thickness occurs in the forward quarter of the chord.

CHAPTER VII

THEORY OF TWO-DIMENSIONAL AEROFOILS

7·0. Types of profile. We have already seen that the flow past a Joukowski aerofoil is obtained by transforming the region outside a given circle into the region outside the profile by the transformation

(1)
$$z = \zeta + \frac{l^2}{\zeta}.$$

We shall now proceed to consider more general transformations. We regard both the circle and the transformation as given and we seek the properties of the resulting profile. The converse problem of finding the transformation which will map the region outside a given profile into the region outside a circle is more difficult and can only be solved in general by methods of successive approximation.

Good wing shapes are characterised by a blunt nose and a sharp trailing edge.

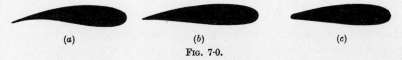

(a)   (b)   (c)

Fig. 7·0.

The trailing edge can be (see fig. 7·0) (a) a cusp, as in the Joukowski profile, (b) a point with distinct tangents (Kármán-Trefftz, von Mises), (c) rounded off (Carafoli). From the constructional point of view (a) is difficult to make.

7·01. Conditions to be satisfied. To obtain the flow past a wing due to a given wind at infinity it is desirable that a uniform wind in the plane of the circle (the ζ-plane) should correspond with a uniform wind in the plane of the profile (the z-plane).

The complex potential for a uniform wind directed as shown in fig. 3·7 (i) is given by

$$w = Ve^{i\alpha}\zeta, \quad \text{or} \quad w = Ve^{i\alpha}z,$$

according as we consider the ζ- or the z-plane. Thus our mapping function $z = F(\zeta)$ should be such that for large values of $|\zeta|$ and $|z|$ we shall have $z = \zeta$ approximately. This requirement is satisfied by a transformation of the form

(2)
$$z = \zeta + \frac{a_1}{\zeta} + \frac{a_2}{\zeta^2} + \dots = F(\zeta),$$

where the series converges for sufficiently large values of $|\zeta|$.

The Joukowski transformation (1) is a special case of this, where $a_1 = l^2$ and is purely real, while $a_2 = a_3 = \ldots = 0$. In what follows $a_1, a_2, \ldots$, will be, in general, complex numbers.

Since the mapping ceases to be conformal at a zero * of $dz/d\zeta$ and since we are mapping the region outside the circle on the region outside the profile, it follows that no zero of $dz/d\zeta = F'(\zeta)$ may lie *outside* the given circle. If $F'(\zeta)$ has a zero, say $\zeta = v$, *on* the given circle, the corresponding point, $z = z_0$, of the profile will be, in general, either a point with two distinct tangents, or a cusp.

Proof.

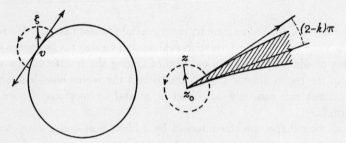

FIG. 7·01.

Let
$$F'(\zeta) = (\zeta - v)^{k-1} f(\zeta), \quad k > 1,$$

where $f(\zeta)$ has no zeros outside the given circle and where $f(v) = n \neq 0$. Then, if we restrict our consideration to small values of $|\zeta - v|$, we have

$$\frac{dz}{d\zeta} = F'(\zeta) = n(\zeta - v)^{k-1},$$

and by integration

(3)
$$z - z_0 = \frac{n}{k}(\zeta - v)^k.$$

Thus

$$\arg (z - z_0) = k \arg (\zeta - v) + \arg \frac{n}{k}.$$

Therefore, if as shown in fig. 7·01, we take ζ round the point v so that $\arg (\zeta - v)$ increases by π, then $\arg (z - z_0)$ will increase by $k\pi$ and the tangents to the profile will enclose the angle $(2 - k)\pi$. Therefore z_0 will be an ordinary point only if this is an odd multiple of π, which, in general, is not the case.

Q.E.D.

---

* A *zero* of $f(z)$ is a value of z for which $f(z) = 0$. Thus if z_0 is a zero, $z - z_0$ is a factor of $f(z)$, or more generally $f(z) = (z - z_0)^n g(z)$, where $n > 0$ and $g(z_0) \neq 0$; if $n = 1$, z_0 is said to be a *simple zero* (cf. simple pole 3·51). Thus near a simple zero $f(z)$ behaves like $A(z - z_0)$ where A is a constant, just as near a simple pole z_1, $f(z)$ behaves like $A/(z - z_1)$.

The only cases of aerodynamic interest are those in which the angle in question is zero or acute, i.e.

$$0 \leqslant (2 - k)\pi < \pi/2, \quad 1{\cdot}5 < k \leqslant 2,$$

the case $k = 2$ of course corresponding with a cusp (see 6·4). It should be observed that profiles usually present sharp points or cusps at the trailing edge, but that in extreme cases such points can occur elsewhere as well. Consider for instance the circular arc Joukowski profiles of 6·3.

In connection with (3) we note that near $\zeta = v$

$$\zeta - v = \left\{\frac{k}{n}(z - z_0)\right\}^{\frac{1}{k}},$$

so that

$$\frac{d\zeta}{dz} = \frac{\text{constant}}{(z - z_0)^{1 - \frac{1}{k}}}.$$

Thus as $z \to z_0$, $d\zeta/dz \to \infty$ and we write

(4)
$$\frac{d\zeta}{dz} = O\left[\frac{1}{(z - z_0)^{1 - \frac{1}{k}}}\right]$$

where the notation means that $d\zeta/dz$ is of the same order of magnitude as the number in the square brackets, or that the ratio of $d\zeta/dz$ to this number remains of finite modulus.

7·02. Origin at the centre of the circle.

In some cases it is convenient to take the origin of coordinates, in both the ζ-plane and the z-plane, at the centre C of the circle which is to be transformed. If C is the point $\zeta = s$, $z = s$ (see e.g. fig. 6·4), we effect the transformation by writing $\zeta + s$, $z + s$ in place of ζ and z. The transformation 7·01 (2) then becomes

$$z = \zeta + \frac{a_1}{\zeta + s} + \frac{a_2}{(\zeta + s)^2} + \dots.$$

If $|\zeta|$ is large, expansion of each term by the binomial theorem gives

(1)
$$z = \zeta + \frac{a_1}{\zeta} + \frac{a_2 - a_1 s}{\zeta^2} + \dots.$$

We also note that the transformation (1) can be *reversed* to give

(2)
$$\zeta = z - \frac{a_1}{z} - \dots = z\left(1 - \frac{a_1}{z^2} - \dots\right),$$

a result which can easily be verified to this degree of accuracy by substitution in (1).

7·03. Some properties of profiles obtained by transforming a circle.

If a is the radius of the circle, C its centre, and c the chord of the

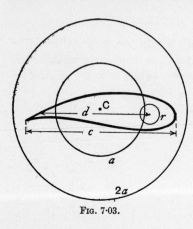

profile, defined as the longest line which can be drawn to join two points of the profile, we state, without proof, the following properties :

(i) $c \geqslant 2a \geqslant \frac{1}{2}c$, i.e. the diameter of the circle cannot exceed the chord nor be less than half the chord.

(ii) If we draw a circle centre C and radius $2a$, the whole profile lies entirely inside this circle.

The thinner and flatter the profile the closer it approaches this circle ; the extreme case is the flat aerofoil (6·11).

FIG. 7·03.

(iii) If r is the radius of a circle which lies entirely inside the profile and if d is the distance of its centre from *any* point of the profile, then

$$4a > \frac{(d + r)^2}{d}.$$

(iv) The centre C is clearly the centroid of the circumference of the circle, radius a, supposed uniformly weighted. If ds_1 is an arc of the circle and ds the *corresponding* arc of the profile, and if we suppose ds to carry the same load as ds_1, then C is the centroid of the profile thus weighted. It follows that C is interior to every convex curve which encloses the profile.

7·1. Aerodynamic force.

The aerodynamic force on the profile is due to the aerodynamic pressure thrusts on the elements of its periphery. It is

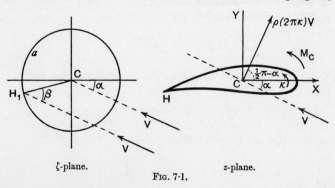

ζ-plane.   z-plane.

FIG. 7·1.

known that a system of forces acting on a rigid body (and we shall assume our aerofoil to be rigid) can be replaced, at any chosen *base point*, by a force acting at that base point, and a couple. Moreover, the magnitude and direction of the force are the same for all base points, whereas the moment of the couple depends upon the particular base point selected.

For the present investigation we shall take as base point the centre C of the circle. This point is called the *centre of the profile*; the actual position which it occupies with respect to the profile is shown when the points of the circle and the corresponding points of the profile are marked in the *same* Argand diagram (see for example fig. 6·51 (*c*)). In the present case we show the two Argand diagrams separately, fig. 7·1, and take the origin at the centre in both diagrams.

If α is the incidence, the complex potential for the flow past the circle is $Ve^{i\alpha}\zeta + a^2 Ve^{-i\alpha}/\zeta$ (see 5·2), and if in addition we have a circulation $2\pi\kappa$ of strength κ, we get (see 5·31)

$$(1) \qquad w = Ve^{i\alpha}\zeta + \frac{a^2 Ve^{-i\alpha}}{\zeta} + i\kappa \log \zeta.$$

For sufficiently large values of $|z|$ we may use 7·02 (2) to give, in the z-plane,

$$w = Ve^{i\alpha}\left(z - \frac{a_1}{z} - \dots\right) + \frac{Ve^{-i\alpha}a^2}{z}\left(1 - \frac{a_1}{z^2} - \dots\right)^{-1}$$
$$+ i\kappa\left[\log z + \log\left(1 - \frac{a_1}{z^2} - \dots\right)\right]$$

$$(2) \qquad = Ve^{i\alpha}z + i\kappa \log z + \frac{Ve^{-i\alpha}a^2 - Ve^{i\alpha}a_1}{z} + \dots ,$$

where the dots indicate omitted powers of $1/z$.

Comparison with 5·5 shows that here

$$(3) \qquad A = -i\kappa, \quad B = Ve^{-i\alpha}a^2 - Ve^{i\alpha}a_1,$$

and therefore from 5·5 (4) and (5)

$$(4) \qquad X + iY = 2\pi i\kappa\rho Ve^{-i\alpha} = 2\pi\kappa\rho Ve^{i\left(\frac{\pi}{2} - \alpha\right)},$$

$$(5) \qquad M_C = \text{real part of } (-2\pi\rho V^2\, ia_1 e^{2i\alpha}).$$

7·11. Joukowski's hypothesis.

Let H_1 (e.g. in fig. 7·1) be the point which transforms into the trailing edge H of the profile. If q_1 is the airspeed at H_1 in the plane of the circle (ζ-plane), and if q is the airspeed at H in the plane of the profile (z-plane), we have

$$q_1 = \left|\frac{dw}{d\zeta}\right| = \left|\frac{dw}{dz}\right| \times \left|\frac{dz}{d\zeta}\right| = q\left|\frac{dz}{d\zeta}\right|.$$

Now at a sharp trailing edge $dz/d\zeta = 0$. It follows that if q is to be finite we must have $q_1 = 0$, in other words H_1 must be a stagnation point for the flow round the circle. Joukowski's hypothesis is that the circulation in the case of a properly designed aerofoil, in its working range of incidence, always adjusts itself so that the airspeed at the trailing edge is finite.

Adopting this hypothesis we must choose the circulation so as to make H_1 a stagnation point. If β is the angle between the normal at H_1 to the circle and the direction of **V**, it follows from 5·311 that the strength of the circulation must be $\kappa = 2aV \sin \beta$.

7·12. The Lift.

From 7·1 (4) we have

$$X + iY = \rho \cdot 2\pi\kappa \cdot Ve^{i(\frac{1}{2}\pi - \alpha)}.$$

This shows that the force whose components are (X, Y) is of magnitude

$$(air\ density) \times (circulation) \times (speed)$$

and is perpendicular to the direction of motion of the aerofoil (or to the asymptotic wind velocity, if the aerofoil is taken to be at rest) and is therefore a *lift*. This is the theorem of Kutta and Joukowski (5·5). The state of affairs is illustrated in fig. 7·1.

The relation between the direction of the lift, the relative wind, and the sense of the circulation is given by the following rule:

To get the direction of the lift, rotate the *relative* wind velocity vector through a right angle in the sense *opposite* to that of the circulation.

Notes. (i) The magnitude of the lift is independent of the shape of the aerofoil but, for given V, decreases with increasing height, for then ρ decreases.

(ii) The magnitude of the lift is the force per *unit span*, for we are dealing with two-dimensional motion, and therefore we are concerned with that part of an infinitely long cylindrical aerofoil which lies between two parallel planes at unit distance apart. If the distance between the planes were h, the lift would be $\rho(2\pi\kappa)\,Vh$. This is actually the case when the aerofoil is in a "two-dimensional" wind tunnel, i.e. when the aerofoil is a cylinder extending right across the tunnel from one wall to a parallel wall.

7·13. Lift coefficient.

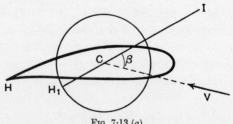

FIG. 7·13 (a).

If we draw the circle and the profile in the same Argand diagram as in fig. 7·13 (a) the line joining the centre of the profile to the rear stagnation point of the circle is called the *first axis* of the profile or Axis I. Comparing figs.

5·31 (b), 7·13 (a), we see that the strength of the circulation as calculated in 5·31 is

$$\kappa = 2aV \sin \beta,$$

where β is the angle between Axis I and the direction of motion.

Def. The angle β is called the *absolute incidence*. Thus absolute incidence is the incidence when the chord of the profile is considered to be along Axis I, and the lift is

$$L = 4\pi \rho \, a \, V^2 \sin \beta.$$

The dimensionless number

$$C_L = \frac{L}{\frac{1}{2}\rho V^2 c},$$

where c is the chord, is called the *lift coefficient* of the profile.

Thus

$$C_L = \frac{8\pi a}{c} \sin \beta.$$

The graph of C_L against absolute incidence is therefore a sine curve.

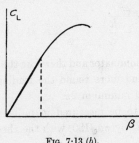

FIG. 7·13 (b).

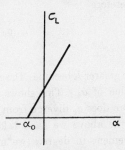

FIG. 7·13 (c).

In the practical range of incidence β is sufficiently small to allow the approximation $\sin \beta = \beta$ and the graph is therefore a straight line, whose gradient is conventionally denoted by a_0. Thus

$$C_L = a_0 \beta.$$

The theoretical value of a_0 is therefore $8\pi a/c$ and, in general, $c = 4a$ approximately so that

$$a_0 = 2\pi.$$

If, instead of absolute incidence we use *geometrical incidence* α, measured from the chord, we shall have $\beta = \alpha_0 + \alpha$ where α_0 is a constant. The (C_L, α) graph is still a straight line, of the same slope a_0, and clearly when the graph is drawn α_0 can be measured from it and thus the direction of Axis I can be inferred.

It also appears from the above considerations that the lift and C_L both vanish when $\beta = 0$, i.e. when the direction of motion is parallel to Axis I.

Thus we infer that Axis I gives the direction of motion which entails no lift. For this reason Axis I may also be called the *axis of zero lift*.

The theoretical value of a_0 can be delimited by the use of the theorems of 7·03.

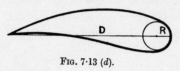

FIG. 7·13 (*d*).

From (i) we have

$$\frac{1}{2} \geqslant \frac{a}{c} \geqslant \frac{1}{4}, \text{ and therefore}$$

$$4\pi \geqslant a_0 \geqslant 2\pi.$$

Again, if R is the radius of the circle which osculates the profile at the leading edge of the chord we have

$$c = D + R,$$

where D is the remainder of the chord. Thus, from 7·03 (iii),

$$4a > \frac{c^2}{D},$$

and therefore

$$a_0 > \frac{2\pi c}{c - R} = \frac{2\pi}{1 - \dfrac{R}{c}}.$$

The greater R becomes, the smaller the denominator and therefore the greater the value of a_0. This shows the thicker and more round the leading edge, the more does a_0 diverge from its theoretical minimum 2π.

In the above we have taken incidence to be measured in radians. For measurement in degrees we must replace a_0 by $\pi a_0/180$, with the theoretical minimum $2\pi^2/180$, or about 1/9.

7·14. Pitching moment coefficient. If in the transformation we put $a_1 = l^2 e^{i\mu}$, we get from 7·1 (5) the pitching moment about the centre

(1) $\qquad M_C = \text{real part of } (-2\pi\rho V^2 i l^2 e^{i(2\alpha+\mu)}) = 2\pi\rho l^2 V^2 \sin(2\alpha + \mu).$

With the axes of reference used in fig. 7·1 we see that the pitching moment is positive when it tends to raise the nose of the profile, i.e. to increase the incidence.

The *pitching moment coefficient* about the centre is the dimensionless number

(2) $\qquad C_m = \dfrac{M_C}{\frac{1}{2}\rho V^2 c^2} = \dfrac{4\pi l^2}{c^2} \sin(2\alpha + \mu).$

Observe that the pitching moment, and therefore its coefficient, depends on the base point. In the above the chosen base point is the centre of the profile.

It also appears that the pitching moment with respect to the centre vanishes when $\alpha = -\mu/2$. This gives, for every profile, a perfectly definite direction of motion which entails zero pitching moment with respect to the centre of the profile. The line drawn through the centre in this direction is called the *second axis* of the profile or Axis II. This axis may be called the *axis of zero pitching moment with respect to the centre.*

It follows from the definition that when the direction of motion is that of Axis II, the lift passes through the centre of the profile and is perpendicular to Axis II. It

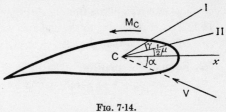

Fig. 7·14.

is also clear that the definition of Axis II implies the experimental means of determining its direction. We shall denote by γ the angle between Axes I and II.

In the case of a symmetrical profile it is obvious that $\gamma = 0$. We shall see later (7·7, 8·36) that there exist unsymmetrical profiles for which $\gamma = 0$.

In the range in which the incidence α is small so that $\sin 2\alpha = 2\alpha$, $\cos 2\alpha = 1$, we get from (2)

$$C_m = \frac{4\pi l^2}{c^2}\,(2\alpha \cos \mu + \sin \mu),$$

which shows that the (C_m, α) graph is also a straight line.

From fig. 7·14 it appears that $\alpha + \frac{1}{2}\mu + \gamma = \beta$, and therefore

(3) $$M_C = 2\pi\rho l^2 V^2 \sin (2\beta - 2\gamma).$$

We also observe that at zero absolute incidence the moment is

$$- 2\pi\rho l^2 V^2 \sin 2\gamma,$$

which is negative, i.e. tending to depress the nose, when γ is positive, and positive when γ is negative (Axis I " below " Axis II). The fact that, although the lift, at zero absolute incidence, vanishes, there is still a moment is explained by the observation that the resultant pressure thrust on the fore part of the profile is downwards while that on the after part is upwards, in fact the aerodynamic action is a couple.

In the case of a Joukowski profile got from the transformation

$$z = \zeta + l^2/\zeta, \quad \mu = 0,$$

and therefore Axis II is parallel to the real axis.

In the case of the general transformation

$$z = \zeta + \frac{l^2 e^{i\mu}}{\zeta} + \frac{a_2}{\zeta^2} + \dots;$$

if we turn *both* axes of reference through the angle ϵ we write $ze^{i\epsilon}$ for z and $\zeta e^{i\epsilon}$ for ζ, thus getting

$$z = \zeta + \frac{l^2 e^{i\mu}}{\zeta e^{2i\epsilon}} + \frac{a_2}{\zeta^2 e^{3i\epsilon}} + \dots.$$

If we take $\epsilon = \mu/2$ this becomes

$$z = \zeta + \frac{l^2}{\zeta} + \frac{a_2 e^{-3i\mu/2}}{\zeta^2} + \dots,$$

and the real axis as appears from fig. 7·14 is now parallel to Axis II.

Thus Axis II is parallel to the real axis when the mapping transformation is referred to axes such that the coefficient of $1/\zeta$ is real and positive.

7·2. Focus of a profile.

The *focus* or *aerodynamic centre* is the point such that the moment of aerodynamic force about it is independent of the incidence.

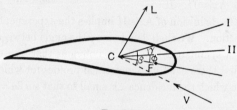

Fig. 7·2.

To establish the existence of the focus we note that if F is any point,

$$M_F = M_C - CF \cos (\beta - \gamma - \phi) \,.\, L,$$

where ϕ is the angle between CF and Axis II as shown in fig. 7·2.

Using the values (7·14, 7·13)

$$M_C = 2\pi\rho V^2 l^2 \sin 2(\beta - \gamma), \quad L = 4\pi\rho a V^2 \sin \beta,$$

we have

$$M_F = 2\pi\rho V^2 \{l^2 \sin (2\beta - 2\gamma) - 2a\,CF\,.\,\sin \beta \cos (\beta - \gamma - \phi)\}$$

$$= 2\pi\rho V^2 \{l^2 \sin (2\beta - 2\gamma) - a\,.\,CF\,.\,\sin (2\beta - \gamma - \phi) - a\,.\,CF\,.\,\sin (\gamma + \phi)\}.$$

This will be independent of β, the absolute incidence, if we take

$$l^2 = a\,.\,CF, \quad \phi = \gamma.$$

This proves the existence of the focus F and gives its position as distant l^2/a from the centre on a line which is the reflection of Axis I in Axis II.

The moment about the focus is

$$M_F = -\,2\pi\rho V^2 l^2 \sin 2\gamma.$$

Our diagrams have been drawn on the assumption that Axis I is above Axis II in the sense indicated in fig. 7·2. In this case the pitching moment *about the focus* is negative. If, however, Axis II were above Axis I, γ would change sign and the moment would become positive. The relative positions of Axes I and II therefore correspond with different dynamical properties of the profile.

Moreover, if $\gamma = 0$, we have $M_F = 0$ at all incidences and therefore the lift always passes through the focus. In this case the aerofoil is said to have a *centre of lift*.

In the case of the Joukowski profile obtained by transforming the circle shown in fig. 6·4, centre $\zeta = s$, the focus is the point $z = s + l^2 e^{-i\epsilon}/a$, which is on CA_1.

For a flat aerofoil the focus is the quarter point midway between the centre and the leading edge.

7·21. Metacentric parabola.

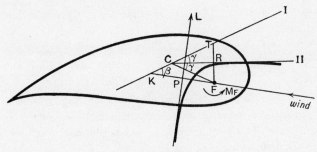

FIG. 7·21.

Let L be the actual line of action of the lift $4\pi\rho a V^2 \sin\beta$. The direction of L is perpendicular to the wind. Let the line L meet the line KF, which is drawn through the focus F parallel to the wind, at P, the point K being on Axis I. Taking moments about the focus F

$$M_F + FP \cdot L = 0,$$

$$FP = -\frac{M_F}{L} = \tfrac{1}{2}\frac{l^2}{a}\frac{\sin 2\gamma}{\sin\beta} = \tfrac{1}{2}CF \cdot \frac{FK}{FC} = \tfrac{1}{2}FK,$$

using the sine formula for the triangle FKC.

Thus the locus of P is a straight line parallel to Axis I and midway between F and Axis I. From a known property of the parabola that the foot of the perpendicular to a tangent from the focus lies on the tangent at the vertex, it follows that the line of action of the lift touches a parabola whose focus is F and whose directrix is Axis I. This is called the *metacentric parabola*.

To find the resultant lift we draw that tangent to this parabola which is perpendicular to the wind direction.

Axis II touches the metacentric parabola, for if FRT is perpendicular to Axis II, $FR = RT$ and hence R lies on the tangent at the vertex.

Since perpendicular tangents intersect on the directrix the corresponding lift passes through C (see 7·14).

7·3. Centre of pressure.

If AH is the chord of the aerofoil (taken as the double tangent in fig. 7·3 (a)), the point P where the line of action of the lift L meets the chord is

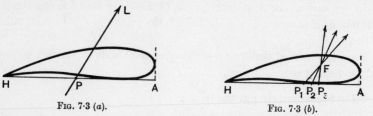

FIG. 7·3 (a).   FIG. 7·3 (b).

called the *centre of pressure*. The position of the centre of pressure thus depends on the particular choice of chord.

The *centre of pressure coefficient* is defined by

$$C_p = \frac{AP}{AH} = \frac{\text{distance of centre of pressure from leading edge of chord}}{\text{length of chord}}.$$

One of the desirable properties of an aerofoil is that the travel of the centre of pressure in the working range of incidence should not be large.

The positions of P for varying incidence can be obtained at once by drawing tangents to the metacentric parabola.

When Axes I and II coincide we have seen (7·2) that a *centre of lift* exists, namely the focus.

The existence of a centre of lift does not imply the existence of a fixed centre of pressure unless the chord is chosen to pass through the centre of lift (fig. 7·3 (b)).

7·31. Centre of pressure of a Joukowski rudder.

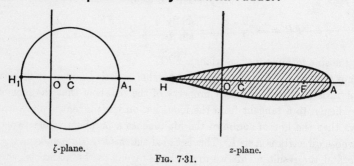

ζ-plane. z-plane.

FIG. 7·31.

Such a rudder is obtained by transforming a circle of radius a whose centre C is on the real axis. Let $OC = f$, then $OH_1 = a - f$ and the transformation is

$$z = \zeta + \frac{(a-f)^2}{\zeta}.$$

Clearly, from the symmetry of the profile, Axes I and II coincide, and so there is a centre of lift which is also a fixed centre of pressure if we take the axis of symmetry HA as chord. We then have

$$z_H = -OH, \quad z_A = OA, \quad \zeta_{H_1} = -(a-f), \quad \zeta_{A_1} = a+f,$$

and so

$$c = AH = z_A - z_H = \zeta_{A_1} + \frac{(a-f)^2}{\zeta_{A_1}} - \zeta_{H_1} - \frac{(a-f)^2}{\zeta_{H_1}}$$

$$= \frac{4a^2}{a+f} = 4a\left(1 - \frac{f}{a} + \frac{f^2}{a^2} - \ldots\right).$$

Thus if f/a is small we have $c = 4a$ nearly. Again, if F is the focus, we have from 7·2,

$$AF = OA - (OC + CF) = \frac{a^3 + 2af^2 - f^3}{a(a+f)}.$$

Hence
$$C_p = \frac{AF}{c} = \frac{a^3 + 2af^2 - f^3}{4a^3} = \tfrac{1}{4} + \tfrac{1}{2}\left(\frac{f}{a}\right)^2 - \tfrac{1}{4}\left(\frac{f}{a}\right)^3.$$

This is an exact result, but if f^2/a^2 is small we see that $C_p = \tfrac{1}{4}$.

The point whose distance from the leading edge of the chord is $c/4$ is called the *quarter-point* of the aerofoil.

In the case of a symmetrical Joukowski aerofoil the quarter-point is, to a good approximation, the centre of pressure.

The centre of pressure of a flat aerofoil is at the quarter-point (cf. 7·2), as is seen by putting $f = 0$, and therefore coincides with the focus.

In the case of unsymmetrical aerofoils the quarter-point Q of the chord may be used as a convenient reference point.

If P is the centre of pressure, taking moments about P gives

$$PQ \cdot L + M_Q = 0,$$

where M_Q is the moment about Q and PQ is positive when P is aft of Q. Thus dividing by $\tfrac{1}{2}\rho V^2 c^2$ we get

$$(C_p - \tfrac{1}{4}) C_L = -C_{m_Q}.$$

7·4. Centroid of the circulation.

The contribution, to the circulation round a profile, of the arc ds is

$$q\, ds = -\frac{\partial \phi}{\partial s}\, ds = -dw,$$

since ψ is constant, for the profile is a streamline. The centroid of the circulation is the point whose coordinates are (x_c, y_c) given by

$$x_c = \int xq\, ds \Big/ \int q\, ds, \quad y_c = \int yq\, ds \Big/ \int q\, ds,$$

so that

$$z_c = \int z\, dw \Big/ \int dw = \int z\frac{dw}{dz}\, dz \Big/ \int \frac{dw}{dz}\, dz.$$

To calculate these integrals we take the origin at the centre of the profile and use 7·1 (2), whence

(1) $$z_c = \frac{V}{i\kappa}(a_1 e^{i\alpha} - a^2 e^{-i\alpha}).$$

7·41. The third axis of the profile.

The locus of the centroid of the circulation is called the *third axis* of the profile or Axis III.

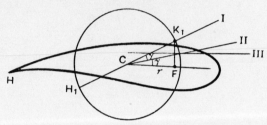

FIG. 7·41.

Let F be the focus, K_1 the second point in which Axis I cuts the circle which transforms into the profile as shown in fig. 7·41. Then Axis III is the perpendicular bisector of the line FK_1.

Proof. When the absolute incidence β is given and the origin is at the centre of the profile, the position of the centroid of the circulation is given by 7·4 (1). It is convenient to take Axis II for x-axis, in which case $a_1 = l^2$, $\alpha = \beta - \gamma$, and $\kappa = 2aV \sin \beta$, so that

$$z = \frac{l^2 e^{i(\beta-\gamma)} - a^2 e^{-i(\beta-\gamma)}}{2ai \sin \beta}$$

referred to Axis II as x-axis.

Let $CF = l^2/a = r$. Then equating real and imaginary parts we get

$$2x = \frac{a+r}{\sin \beta} \sin(\beta - \gamma),$$

$$2y = \frac{a-r}{\sin \beta} \cos(\beta - \gamma).$$

Multiplying these equations by $\cos \gamma/(a+r)$ and $\sin \gamma/(a-r)$ respectively, and then adding, we get

$$\frac{2x}{a+r} \cos \gamma + \frac{2y}{a-r} \sin \gamma = 1,$$

which proves that the locus is a straight line whose gradient is

$$- (a-r) \cot \gamma/(a+r).$$

Since II is the x-axis, K_1 and F are respectively the points $(a \cos \gamma, a \sin \gamma)$ $(r \cos \gamma, -r \sin \gamma)$.

The middle point of FK_1 is $(\frac{1}{2}(a+r) \cos \gamma, \frac{1}{2}(a-r) \sin \gamma)$ which clearly lies on the above line and the gradient of FK_1 is $(a + r) \tan \gamma/(a - r)$ so that Axis III is perpendicular to FK_1.   Q.E.D.

Corollary. Axis III is a tangent to the metacentric parabola, for it bisects FK_1 at right angles.

It can be proved that the line of action of the aerodynamic force passes through the centroid of the circulation. Hence at any incidence this centroid is determined by the intersection of the line of action of the aerodynamic force with Axis III.

7·5. Force at a sharp point of a profile.

When a flat plate aerofoil is presented at absolute incidence different from zero it will experience a force. Since the air pressure thrusts act perpendicularly to the plate, this force should be perpendicular to the plate. If we calculate the force by means of the Blasius formula, we get a force perpendicular to the asymptotic wind. This apparent paradox is explained by the action of the air at the sharp edges of the plate. It will appear that the sharpness of the trailing edge causes no anomaly but that the sharp leading edge entails a particular behaviour.

In fig. 7·5 (a) there is shown a profile having a sharp point at B. That this point also happens to be the trailing edge in the diagram has nothing to do with the argument which follows.

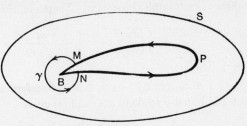

Fig. 7·5 (a).

Let us draw a circular arc γ whose centre is B to meet the profile at M and N. Let P denote the contour consisting of the part of the profile outside γ, and let S be a contour which surrounds both P and γ.

Then by Cauchy's theorem (3·5)

$$(1) \qquad \tfrac{1}{2} i\rho \int_{(S)} \left(\frac{dw}{dz}\right)^2 dz = \tfrac{1}{2} i\rho \int_{(\gamma)} \left(\frac{dw}{dz}\right)^2 dz + \tfrac{1}{2} i\rho \int_{(P)} \left(\frac{dw}{dz}\right)^2 dz.$$

We shall, as in the theorem of Blasius, write (1) in the form

$$(2) \qquad X_S - iY_S = (X_\gamma - iY_\gamma) + (X_P - iY_P).$$

In (2) (X_S, Y_S) is the force calculated by direct application of the theorem of Blasius, and the result of this calculation does not depend on γ or P. Also when $\gamma \to 0$ (i.e. when the radius of the arc $\gamma \to 0$) the force (X_P, Y_P) will tend to the force (X, Y) on the whole profile. Therefore

$$(3) \qquad X - iY = (X_S - iY_S) - (X_0 - iY_0), \text{ where}$$

$$(4) \qquad X_0 - iY_0 = \lim_{\gamma \to 0} \tfrac{1}{2} i\rho \int_{(\gamma)} \left(\frac{dw}{dz}\right)^2 dz.$$

Let us calculate $X_0 - iY_0$.

If B is the point z_0 and the profile is obtained by transforming a circle in the ζ-plane we have from 7·01 (4)

$$\frac{d\zeta}{dz} = O\left(\frac{1}{(z-z_0)^{1-\frac{1}{k}}}\right), \quad 1 < k \leqslant 2,$$

in the neighbourhood of B. Here $(2-k)\pi$ measures the angle between the tangents at B which is by hypothesis a sharp point, but not necessarily the trailing edge.

Again

$$\frac{dw}{dz} = \frac{dw}{d\zeta}\frac{d\zeta}{dz},$$

and in the ζ-plane $dw/d\zeta$ is finite everywhere. Therefore dw/dz is of the same order of magnitude as $d\zeta/dz$ and we can write near B

(5)
$$\frac{dw}{dz} = \frac{c}{(z-z_0)^{1-\frac{1}{k}}},$$

where c is a (complex) constant. Therefore, if $k \neq 2$,

$$\int\left(\frac{dw}{dz}\right)^2 dz = \int\frac{c^2\,dz}{(z-z_0)^{2-\frac{2}{k}}} = \frac{c^2(z-z_0)^{\frac{2}{k}-1}}{\frac{2}{k}-1}.$$

If $1 < k < 2$, $2/k - 1 > 0$ and therefore when $\gamma \to 0$, i.e. as $z \to z_0$, the integral round γ tends to zero and (4) gives

$$X_0 - iY_0 = 0, \quad \text{if } 1 < k < 2,$$

i.e. when the sharp point has two distinct tangents it makes no difference and the force on the profile as given in (3) is the same as the Blasius force.

If, however, $k = 2$, the sharp point becomes a *cusp* and things are different. In this case (4) gives (3·52)

$$X_0 - iY_0 = \lim_{\gamma \to 0} \tfrac{1}{2}i\rho\int_{(\gamma)}\frac{c^2}{z-z_0}\,dz = -\pi\rho c^2.$$

Let $c = c_0 e^{i\lambda}$ where c_0 is real. Then

(6) $$X_0 + iY_0 = -\pi\rho c_0{}^2 e^{-2i\lambda} = \pi\rho c_0{}^2 e^{i(\pi-2\lambda)}.$$

We now use the fact that at M the air velocity is tangential to the profile. If (u_M, v_M) is the velocity at M and if $z - z_0 = re^{i\theta}$, we have from (5)

$$u_M - iv_M = -\left(\frac{dw}{dz}\right)_M = \frac{-c}{(z-z_0)^{\frac{1}{2}}} = \frac{-c_0 e^{i\lambda}}{r^{\frac{1}{2}}e^{i\theta/2}},$$

and therefore, changing the sign of i,

$$u_M + iv_M = \frac{c_0}{r^{\frac{1}{2}}}e^{i\left(\frac{\theta}{2}-\lambda+\pi\right)}.$$

But if the radius of γ is sufficiently small, the tangent to the profile at B coincides with MB, and with the tangent to the cusp at B, and the directed line MB makes the angle $\pi + \theta$ with the x-axis.

Thus $\arg (u_M + iv_M) = \pi + \theta$ and therefore

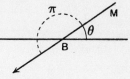

(7) $\qquad \dfrac{\theta}{2} - \lambda + \pi = \pi + \theta,$*

so that $\lambda = -\,\theta/2$ and therefore from (6)

(8) $\qquad X_0 + iY_0 = \pi\rho c_0{}^2 e^{i(\pi+\theta)}$

Fig. 7·5 (b).

which is a force directed along the tangent at the cusp in the sense MB, i.e. outwards from the profile, a force which might be described as a *suction* force at the cusp.

As may be seen from (5) the presence of this force (X_0, Y_0) depends on dw/dz becoming infinite at the cusp. In the case of a Joukowski profile the value of dw/dz has been shown to be finite at the trailing edge (6·8) and therefore the suction there is zero.

The above reasoning also shows that if a profile presents a second cusp there will also be a suction force at that point.

7·51. The flat aerofoil. This is obtained by transforming the circle $|\zeta| = a$ by the transformation

(1) $\qquad\qquad\qquad z = \zeta + \dfrac{a^2}{\zeta}.$

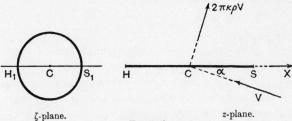

ζ-plane.   z-plane.

Fig. 7·51.

As we have seen there will be no suction at H, the trailing edge, but there will be a suction X_0 at S the leading edge, $z = 2a$. To apply the method of the preceding section we have from (1) solved as a quadratic in ζ

$$2\zeta = z + \sqrt{(z^2 - 4a^2)},$$

the positive sign being taken because at distant parts we must have $\zeta = z$. This gives

$$2\frac{d\zeta}{dz} = 1 + \frac{z}{\sqrt{[(z-2a)(z+2a)]}}.$$

---

* It may be observed that in deducing (7) we have ignored a possible integral multiple of 2π which does not affect the value of $e^{i\lambda}$. Observe also that (8) follows whether we take θ or $\pi + \theta$ for the argument of the velocity, i.e. whether the velocity is in the sense MB or BM.

Near $z = 2a$ this becomes very large, the dominant term being obtained by putting $z = 2a$ except in the factor $(z - 2a)$ which is responsible for the largeness. Thus near $z = 2a$ we have, very nearly,

(2)
$$\frac{d\zeta}{dz} = \frac{\sqrt{a}}{2\sqrt{(z - 2a)}}.$$

Again,

$$w = Ve^{i\alpha}\zeta + \frac{Ve^{-i\alpha}a^2}{\zeta} + i\kappa \log \zeta, \quad \kappa = 2aV \sin \alpha.$$

Thus

$$\frac{dw}{d\zeta} = Ve^{i\alpha} + \frac{2iaV \sin \alpha}{\zeta} - \frac{a^2 Ve^{-i\alpha}}{\zeta^2},$$

and near $\zeta = a$, which corresponds with $z = 2a$, we have

(3)
$$\frac{dw}{d\zeta} = 4iV \sin \alpha.$$

Combining (2) and (3) we have near S

$$\frac{dw}{dz} = \frac{2iV \sin \alpha \sqrt{a}}{\sqrt{(z - 2a)}}.$$

Thus from 7·5 (5) we have $c = 2iV \sin \alpha \sqrt{a}$, and therefore from 7·5 (8)

$$X_0 + iY_0 = 4\pi\rho V^2 a \sin^2 \alpha.$$

So that $Y_0 = 0$, $X_0 = 4\pi\rho V^2 a \sin^2 \alpha$. The force given by the Blasius integral is of course the Joukowski lift.

$$2\pi\kappa\rho Ve^{i\left(\frac{\pi}{2} - \alpha\right)} = 4\pi\rho a V^2 \sin \alpha(\sin \alpha + i \cos \alpha).$$

Subtracting X_0 from this we have the resultant force $4\pi i \rho a V^2 \sin \alpha \cos \alpha$ which is perpendicular to the aerofoil.

The foregoing investigation is intended to show that the apparent paradox in the case of the flat plate has no theoretical substance. On the other hand, the further inference is that cusps on profiles must be avoided except at the trailing edge. When, therefore, we treat an aerofoil as a rectangular plate, or a part of the surface of a cylinder, or other limiting cases of extreme thinness, it must be remembered that this is only a convenient mathematical simplification, and that the skeleton thus used must, in practice, be clothed and the cuspidal edge rounded off. To such an aerofoil we can apply the Joukowski theory.

7·6. Kármán-Trefftz profiles.
These may be regarded as a generalisation of the Joukowski profiles. They have the constructional advantage that the cusp is replaced by a sharp point at which there are two distinct tangents. The Joukowski transformation (6·1) can be written, from 6·1 (2) and (4), as

$$\frac{z - 2l}{z + 2l} = \frac{(\zeta - l)^2}{(\zeta + l)^2}.$$

The required generalisation is obtained by replacing 2 by k, where $k < 2$, so that

$$(1) \qquad \frac{z - kl}{z + kl} = \frac{(\zeta - l)^k}{(\zeta + l)^k};$$

when $k = 2$ we get the Joukowski profiles.

Referring to fig. 7·6 (a), let us put

$$\arg(\zeta - l) = \chi_1, \quad \arg(\zeta + l) = \omega_1, \quad \arg(z - kl) = \chi, \quad \arg(z + kl) = \omega,$$

the points S_1, H_1, S, H being $(\pm l, 0)$ and $(\pm kl, 0)$. It follows from (1) that

$$\arg(z - kl) - \arg(z + kl) = k \arg(\zeta - l) - k \arg(\zeta + l)$$

or

$$(2) \qquad \chi - \omega = k(\chi_1 - \omega_1).$$

Note that when P_1 is below the real axis χ_1, ω_1 are greater than π and $\chi_1 - \omega_1$ is negative.

Let us transform the circle $|\zeta| = l$ whose centre C is therefore at the origin. Using one Argand diagram for ζ and z, if P_1 and P correspond we have from (2) applied to fig. 7·6 (a), $S\hat{P}H = k\, S_1\hat{P}_1H_1$.

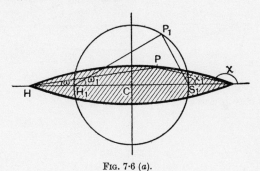

FIG. 7·6 (a).

As P_1 describes the upper semicircle $S_1\hat{P}_1H_1$ remains constant and equal to $\pi/2$ so that $S\hat{P}H$ remains constant and equal to $k\pi/2$. Thus P describes a circular arc also above the real axis and passing through S, H.

When P_1 describes the lower semicircle, $\chi_1 - \omega_1 = -\pi/2$ and so $\chi - \omega = -k\pi/2$. Therefore P describes a circular arc through S and H which is the reflection in the real axis of the first circular arc.

Observe that the nearer k is to 2 the nearer the angle $k\pi/2$ to π and the flatter the two arcs just obtained.

The angle between the tangents at the point H is $2\pi - k\pi$, as is easily seen by considering what happens to $S\hat{P}H$ when P_1 moves along the upper semi-circle to coincide with H_1.

If we transform a circle centre the origin but whose radius is larger than l *
the sharp points disappear and we get a profile of elliptical appearance sur-

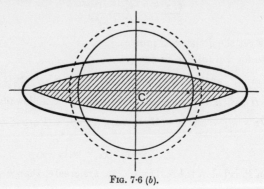

FIG. 7·6 (b).

rounding the two circular arcs of fig. 7·6 (a) which form, as it were, the *core of the profile.*

The transformation of a circle whose centre is on the real axis and which

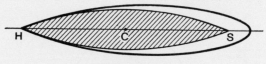

FIG. 7·6 (c)

passes through H_1 but encloses S_1 (see fig. 7·31) leads to a symmetrical profile
with the same core suitable for a rudder, strut or fin.

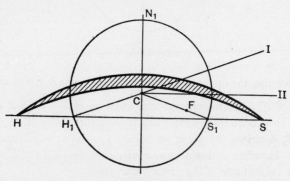

FIG. 7·6 (d).

If the circle to be transformed passes through S_1 and H_1 but has its centre
on the imaginary axis the reasoning given above still applies to show that as
P_1 describes the major arc $S_1N_1H_1$, the angle in which is, say, ϵ, then P describes

---

* The dotted circle in fig. 7·6 (b).

an arc through S, H the angle in which is $k\epsilon$. Likewise when P_1 describes the minor arc, P will describe an arc through S and H the angle in which is $2\pi - k(\pi - \epsilon)$. The angle at the point H enclosed by the tangents is, as

before, $(2 - k)\pi$. Thus the transformed profile is bounded by two circular arcs, crescent shaped as in fig. 7·6 (d), or biconvex as in fig. 7·6 (a).

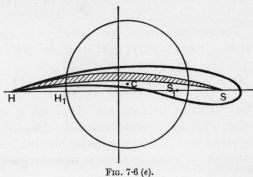

If we displace the centre of the given circle to lie on neither axis we get a profile of which the crescent is the core.

Since the angle be-

Fig. 7·6 (e).

tween the tangents at H is $(2 - k)\pi$, it follows that $(2 - k)\pi < \pi$ for H to be a sharp point, i.e. we must have $k > 1$.

From (1) we get

$$\frac{z}{kl} = \frac{\left(1 + \dfrac{l}{\zeta}\right)^k + \left(1 - \dfrac{l}{\zeta}\right)^k}{\left(1 + \dfrac{l}{\zeta}\right)^k - \left(1 - \dfrac{l}{\zeta}\right)^k}.$$

Expanding by the binomial theorem we get

$$(2) \qquad z = \zeta + \frac{(k^2 - 1)}{3}\frac{l^2}{\zeta} + \dots.$$

Since $1 < k \leqslant 2$ the coefficient of $1/\zeta$ is real and positive and therefore (cf. 7·14) Axis II is parallel to the real axis. Axis I is, of course, the line H_1C. It follows from the construction given in 7·2 that the focus F is on the line CA_1. This result is true also when $k = 2$, i.e. for the Joukowski profile (see fig. 6·4).

7·7. Von Mises profiles.

Fig. 7·7 (a) shows a profile whose skeleton presents a point of inflexion. Such a skeleton is sometimes described as **S**-shaped. The skeletons of Joukowski and Kármán-Trefftz profiles are based on circular arcs, and therefore such profiles cannot include the **S**-shaped variety. The French term *relevé* is commonly employed to describe the

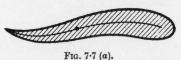

Fig. 7·7 (a).

turning up of the tail illustrated in fig. 7·7 (a). In unsymmetrical profiles *relevé* is essential for the existence of a centre of lift (7·2).

In the Joukowski and Kármán-Trefftz type of transformation $dz/d\zeta$ has zeros at $\zeta = \pm l$ and nowhere else. Thus, for example, we have from 6·1 (1)

(1)
$$\frac{dz}{d\zeta} = \left(1 - \frac{l^2}{\zeta^2}\right) = \left(1 - \frac{l}{\zeta}\right)\left(1 + \frac{l}{\zeta}\right).$$

The generalisation of this, made by von Mises, is to postulate a certain number of zeros, say n, of $dz/d\zeta$ in addition to the zero at $\zeta = -l$. Thus

(2)
$$\frac{dz}{d\zeta} = \left(1 + \frac{l}{\zeta}\right)\left(1 - \frac{v_1}{\zeta}\right)\left(1 - \frac{v_2}{\zeta}\right)\ldots\left(1 - \frac{v_n}{\zeta}\right).$$

This reduces to the Joukowski case when $v_1 = l$, $v_2 = v_3 = \ldots = v_n = 0$. As $\zeta \to -l$ so $dz/d\zeta$ tends to zero like $l + \zeta$, and, therefore, as was shown in 6·4, there is a cusp at the point corresponding with $\zeta = -l$. If we wish to have a sharp point at which the tangents enclose the angle $(2 - k)\pi$ we put, instead of (2),

(3)
$$\frac{dz}{d\zeta} = \left(1 + \frac{l}{\zeta}\right)^{k-1}\left(1 - \frac{v_1}{\zeta}\right)\left(1 - \frac{v_2}{\zeta}\right)\ldots\left(1 - \frac{v_n}{\zeta}\right),$$

and for large values of $|\zeta|$ we have

$$\frac{dz}{d\zeta} = 1 + \frac{A}{\zeta} + \frac{B}{\zeta^2} + \ldots, \text{ where}$$

(4)
$$A = (k - 1)\, l - v_1 - v_2 - \ldots - v_n.$$

Integrating this we get

$$z = \zeta + A \log \zeta - \frac{B}{\zeta} - \ldots,$$

and this is unsuitable for a transformation of the type for which $z \to \zeta$ at infinity on account of the presence of the logarithm. Therefore we must have $A = 0$.

The condition $A = 0$ means that the origin is the centroid of unit masses placed at $v_1, v_2, \ldots, v_n$ and a mass $(k - 1)$ units at the point $-l$. Since to get an aerofoil profile none of the zeros of $dz/d\zeta$ may be outside the circle it follows that the origin must be *inside the circle*, and indeed inside every convex contour which encloses the zeros. In the Joukowski case, $n = 1$, $k = 2$, the origin is the centroid of unit masses placed at $-l$ and l.

As a simple case consider $n = 2$, $k = 2$. Then

$$\frac{dz}{d\zeta} = \left(1 + \frac{l}{\zeta}\right)\left(1 - \frac{v_1}{\zeta}\right)\left(1 - \frac{v_2}{\zeta}\right), \quad l = v_1 + v_2,$$

and therefore integrating and omitting the irrelevant constant

$$z = \zeta + \frac{l^2 - v_1 v_2}{\zeta} - \frac{l v_1 v_2}{2\zeta^2}.$$

To satisfy the condition $v_1 + v_2 = l$ we may put

$$v_1 = \tfrac{1}{2}l(1 + \lambda e^{i\nu}), \quad v_2 = \tfrac{1}{2}l(1 - \lambda e^{i\nu}),$$

where λ and ν are real. We then get

(5) $$z = \zeta + \frac{l^2(3 + \lambda^2 e^{2i\nu})}{4\zeta} - \frac{l^3(1 - \lambda^2 e^{2i\nu})}{8\zeta^2}.$$

To get the direction of the second axis, as in 7·14, we put

$$L^2 e^{i\mu} = \tfrac{1}{4}l^2(3 + \lambda^2 e^{2i\nu}),$$

so that, equating real and imaginary parts and reducing,

(6) $\cot \mu = \cot 2\nu + \dfrac{3}{\lambda^2} \operatorname{cosec} 2\nu, \quad L^2 = \tfrac{1}{4}l^2[\lambda^4 + 9 + 6\lambda^2 \cos 2\nu]^{\frac{1}{2}}.$

The inclination of Axis II to the real axis (7·14) is $\tfrac{1}{2}\mu$, and the distance of the focus from the centre is L^2/a.

If ϵ is the inclination of Axis I to the real axis, the pitching moment with respect to the focus will be positive or negative according as Axis II is above or below Axis I in the sense explained in 7·2, i.e. according as $\mu/2 > \epsilon$ or $< \epsilon$.

If we take $\mu/2 = \epsilon$, Axes I and II coincide and there is then a centre of lift. Thus, to get unsymmetrical profiles with a centre of lift we take the centre C of the circle to be transformed anywhere on the line through $\zeta = -l$ inclined at $\tfrac{1}{2}\mu$ to the real axis, subject of course to the conditions that the circle passes through $\zeta = -l$ and encloses v_1 and v_2.

As a numerical example take $\lambda = 1$, $\nu = \pi/8$. Then from (6) we get $\tfrac{1}{2}\mu = 5° 24'$, $L^2 = 0.943l^2$.

Fig. 7·7 (b) shows a sketch, not to scale, of the type of profile and the zeros v_1, v_2.

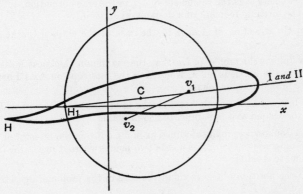

Fig. 7·7 (b).

Any number of profiles may be devised on the above basis. We note, however, that the simple geometrical construction of z which applied in the case of the Joukowski profile is no longer available and the values of z must be worked out individually from the mapping function (e.g. (5)). There are however, link mechanisms which allow the direct tracing of such profiles.

7·8. Carafoli profiles. All the profiles hitherto considered have a sharp point or cusp at the trailing edge. This circumstance arises from the fact that the circle from which they are obtained passes through H_1 which is a zero of $dz/d\zeta$. If we apply any of the mappings already described to a circle which is concentric with, but slightly larger than, the circle mentioned, the point H_1 will lie inside and the mapping will be conformal over the whole boundary so that the sharp point will be rounded off and a Carafoli type of profile will result. See figs. 7·0 (c), 7·6 (b), 6·11. The constructional advantages are obvious, but the theory calls for no special amplification.

EXAMPLES VII

1. The Joukowski transformation $z = \zeta + l^2/\zeta$ is applied to the circle $|\zeta - s| = a$. Examine the forms of $dz/d\zeta$ and $d\zeta/dz$ when the origin is taken at the centre of the circle.

2. Show that the Joukowski transformation is reversed by
$$\zeta = \tfrac{1}{2}z + \tfrac{1}{2}\sqrt{(z^2 - 4l^2)},$$
and hence expand ζ in ascending powers of $1/z$.

3. Prove that, in the case of a circular arc profile obtained by transforming a circle centre C by the Joukowski transformation, the centre C is also the centroid of the profile, assuming corresponding arcs of circle and profile to carry equal loads, and the circle to be uniformly loaded.

4. In the case of a flat aerofoil, prove that $C_L = 2\pi \sin \beta$, and calculate C_m with respect to the leading edge.

5. Prove that the pitching moment with respect to the centre is the real part of $2\pi\rho V^2 i\bar{a}_1 e^{-2i\alpha}$, where a_1 is the coefficient of $1/\zeta$ in the transformation. Deduce the pitching moment of a Joukowski aerofoil.

6. Find the direction of Axis I in the case of the aerofoil Clarke YH. See Ex. I, 10, 21.

7. In a Joukowski profile prove that the maximum thickness measured perpendicularly to the x-axis is $5·2 h$, where h is the intercept on Axis I between the centre of the profile and the y-axis.

8. The bisector of the angle between the tangent at the cusp of a Joukowski profile and the second axis is parallel to the first axis.

9. For a Joukowski profile, obtained as in fig. 6·51 (c), the bisector of the angle between the tangents to the profile at the two points where it meets the imaginary axis is parallel to Axis II.

10. Find the position of the focus in the case of the profile of fig. 6·51 (c), and the profiles of Ex. VI, 9.

11. if the transformation is
$$z = \zeta + \frac{l^2 e^{2i\mu}}{\zeta} + \dots,$$

prove that the moment about the centre of the profile is
$$2\pi\rho V^2 l^2 \sin 2(\mu + \alpha),$$
and hence show that when the wind is along the second axis the lift passes through the centre of the profile.

Deduce that the second axis is a tangent to the metacentric parabola.

Prove that the lift is greatest when the wind is along the axis of the metacentric parabola.

12. Calculate the position of the centroid of the circulation in the case of flow past a circular cylinder, at incidence α to the x-axis, with circulation.

13. Calculate the position of the centroid of the circulation for a circular arc aerofoil of chord c and camber m in a wind at incidence α.

14. Find the position of the third axis of the profile of fig. 6·51 (c), and the profile of Ex. VI, 9, showing in each case the corresponding metacentric parabola.

15. Prove that for any given incidence β the line of action of the lift passes through the centroid of the circulation.

16. Calculate the suction at the leading edge of a circular arc aerofoil.

Find the aerodynamic force on the aerofoil.

17. Show that the transformation
$$\frac{z - nc}{z + nc} = \left(\frac{\zeta - c}{\zeta + c}\right)^n, \quad 1 < n \leqslant 2$$
transforms the circle $\xi^2 + (\eta - h)^2 = c^2 + h^2$ into two circular arcs and that the chord of the aerofoil is $2nc$.

Show that the transformation can be written
$$z = \zeta + \frac{(n^2 - 1)c^2}{3\zeta} + \dots.$$

18. In a symmetrical biconvex Kármán-Trefftz profile, calculate the upper camber in terms of the angle between the tangents at the trailing edge.

19. Calculate the thickness ratio and the camber of a crescent-shaped aerofoil of the type shown in fig. 7·6 (d) obtained from the circle $x^2 + (y - h)^2 = a^2$.

Obtain approximate formulae applicable when h/a is sufficiently small.

20. Plot point by point the von Mises aerofoil of fig. 7·7 (b), marking the Axes and the centre of lift.

Discuss the position of the third Axis.

21. Use the method of 7·5 to find the moment of the suction force at a cusp, and hence verify that the force is localised at the cusp.

CHAPTER VIII

THIN AEROFOILS

8·0. Geometry of profiles. Consider a general type of profile rounded at both ends (Carafoli type, 7·8). We shall assume that the profile has a unique

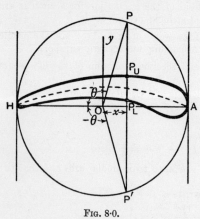

double normal AH, that is to say a line which is normal to the profile at A and at H, and that the profile lies entirely between the tangents at A and H. This double normal will be taken as the chord of the profile, and, as hitherto, A will be the anterior or leading edge and H the hinder or trailing edge.

If H is a sharp point or a cusp AH will be the line through H which is normal to the profile at the leading edge.

We take the mid-point of AH as origin and the x-axis along the chord.

FIG. 8·0.

Draw the circle on AH as diameter. A double ordinate PP' of this circle will meet the upper curve of the profile in a point P_U and the lower curve in a point P_L. The angle $POH = \theta$ will be called the *eccentric angle* of P_U and the angle $P'OH = -\theta$ will then be the eccentric angle of P_L. Thus as θ increases from 0 to π, the point P_U will trace the upper line of the profile and the point P_L will trace the lower line.

We suppose the equation of the profile to be expressed by

$$(1) \qquad x = -\tfrac{1}{2}c \cos \theta, \quad y = cf(\theta)$$

in terms of θ as a parameter. Thus for any value of θ, $0 < \theta < \pi$ we shall have the same value of x as given by (1) but

$$(2) \qquad y_U = cf(\theta), \quad y_L = cf(-\theta),$$

according as we refer to the upper or lower line of the profile.

The camber line (1·14) is then defined by

$$(3) \qquad x = -\tfrac{1}{2}c \cos \theta, \quad y_C = \tfrac{1}{2}(y_U + y_L) = \tfrac{1}{2}c\{f(\theta) + f(-\theta)\}.$$

This function $y_C = y_C(\theta)$ is an even function of θ, since from (3)

$$y_C(\theta) = y_C(-\theta).$$

The function $y_C(\theta)$ is the *camber line function*, and the curve described by (x, y_C), dotted in fig. 8·0, is the camber line.

The *thickness function* $y_T = y_T(\theta)$ is defined by

$$(4) \qquad y_T = \tfrac{1}{2}(y_U - y_L) = \tfrac{1}{2}c\{f(\theta) - f(-\theta)\},$$

where $0 \leqslant \theta \leqslant \pi$. The thickness function is an odd function of θ, for clearly $y_T(-\theta) = -y_T(\theta)$. A knowledge of y_C and y_T allows us to build up the profile from the equations

$$(5) \qquad y_U = y_C + y_T, \quad y_L = y_C - y_T, \quad 0 \leqslant \theta \leqslant \pi.$$

Since dy/dx is, in general, infinite at $H(\theta = 0)$ and at $A(\theta = \pi)$ we may reasonably write the slope of the profile * in the form

$$(6) \qquad \frac{dy}{dx} = -\tfrac{1}{2}\lambda \tan \tfrac{1}{2}\theta + \tfrac{1}{2}\tau \cot \tfrac{1}{2}\theta + F(\theta),$$

where it will be assumed that $F(\theta)$ can be expanded in a convergent Fourier series

$$(7) \qquad F(\theta) = \tfrac{1}{2}c_0 + \sum_{n=1}^{\infty} (a_n \cos n\theta + b_n \sin n\theta).$$

To determine a_n we multiply both sides of (7) by $\cos n\theta$ and then integrate from $-\pi$ to π. Since, as is readily proved,

$$\int_{-\pi}^{\pi} \sin m\theta \cos n\theta \, d\theta = 0, \quad \int_{-\pi}^{\pi} \cos^2 n\theta \, d\theta = \pi, \quad \int_{-\pi}^{\pi} \cos m\theta \cos n\theta \, d\theta = 0, \quad m \neq n,$$

we get

$$(8) \quad c_0 = \frac{1}{\pi}\int_{-\pi}^{\pi} F(\theta)\,d\theta, \quad a_n = \frac{1}{\pi}\int_{-\pi}^{\pi} F(\theta) \cos n\theta \, d\theta = \frac{2}{\pi}\int_{0}^{\pi} \frac{dy_C}{dx} \cos n\theta \, d\theta$$

$$b_n = \frac{1}{\pi}\int_{-\pi}^{\pi} F(\theta) \sin n\theta \, d\theta = \frac{2}{\pi}\int_{0}^{\pi} \frac{dy_T}{dx} \sin n\theta \, d\theta$$

Since $dy/d\theta = (dy/dx)(dx/d\theta)$ we get from (6)

$$(9) \qquad \frac{1}{c}\frac{dy}{d\theta} = -\tfrac{1}{4}\lambda(1 - \cos \theta) + \tfrac{1}{4}\tau(1 + \cos \theta) + \tfrac{1}{2} \sin \theta F(\theta).$$

The radii of curvature ρ_A, ρ_H at the leading and trailing edges may be calculated from the formula

$$\rho = \pm (\dot{x}^2 + \dot{y}^2)^{3/2}/(\dot{x}\ddot{y} - \ddot{x}\dot{y}),$$

where the dot refers to differentiation with respect to θ. The reader may readily verify that

$$(10) \qquad\qquad c\lambda^2 = 2\rho_A, \quad c\tau^2 = 2\rho_H.$$

Again, $\displaystyle\int_{-\pi}^{\pi} \frac{dy}{d\theta} \, d\theta = 0$ and therefore from (9)

$$(11) \qquad\qquad b_1 = \lambda - \tau.$$

---

* M. J. Brennan and A. C. Stevenson, *Simplified Two-Dimensional Aerofoil Theory*, Aircraft Engineering, xviii (1946) 182.

8·01. Thin aerofoils of small camber. An aerofoil is said to be *thin* if the upper and lower lines of the profiles differ but slightly from the camber line, so that to a first approximation the profile may be replaced by its camber line. The theory is further simplified by the assumption, which will be made in what follows, that the slope of the camber line is so small that $(dy/dx)^2$ can be neglected. We then have for an arc ds of the camber line

$$\left(\frac{ds}{dx}\right)^2 = 1 + \left(\frac{dy}{dx}\right)^2 = 1,$$

so that $ds = dx = \frac{1}{2}c \sin \theta \, d\theta$.

In fig. 8·01 to this order of approximation the arcs $P_U Q_U$ and $P_L Q_L$ of the profile are to be regarded as equal to the arc PQ of the camber line and of length dx.

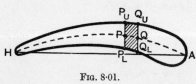

FIG. 8·01.

If we imagine the profile to be divided into elementary sections such as $P_U Q_U Q_L P_L$, the circulation round the whole profile is the sum of circulations round the elementary sections. In evaluating such a circulation we shall suppose that the velocity at points of $P_U P_L$ and $Q_U Q_L$ is zero, while the velocity along $Q_U P_U$ and $Q_L P_L$ is the actual velocity with which the air flows along these arcs.

Each elementary section will experience the lift appropriate to its circulation and the aerodynamic force on the whole profile will be the resultant of the aerodynamic forces on the component sections.

This is the circulation (or vortex sheet) method of approach which we shall proceed to develop.

A second method will be explained in 8·4.

8·1. The flat aerofoil. Here the profiles are straight lines which coincide with the chord c. Such profiles arise from transforming the circle $|\zeta| = \frac{1}{4}c$ by the Joukowski transformation

$$z = \zeta + \frac{c^2}{16\zeta},$$

and the complex potential is

$$w = Ve^{i\alpha}\zeta + \frac{Ve^{-i\alpha}c^2}{16\zeta} + 2iV\frac{c}{4}\sin \alpha \log \zeta.$$

If ζ is the point in the circle which corresponds with the point P of the aerofoil, with the notation of fig. 8·1 we can write

$$\zeta = \tfrac{1}{4}ce^{i(\pi-\theta)} = -\tfrac{1}{4}ce^{-i\theta},$$

and this value of ζ gives

$$z = \tfrac{1}{4}ce^{i(\pi-\theta)} + \tfrac{1}{4}ce^{-i(\pi-\theta)} = \tfrac{1}{2}c \cos (\pi - \theta),$$

which shows that P and ζ are related as in fig. 8·1.

Substituting the above value of ζ in the complex potential, a simple reduction shows that on the aerofoil

$$w = \tfrac{1}{2}cV[-\cos(\theta - \alpha) - (\pi - \theta)\sin\alpha], \quad dw = \tfrac{1}{2}cV[\sin(\theta - \alpha) + \sin\alpha]d\theta.$$

Now the circulation round the element PQ is the decrease in dw as we go from Q to P along the upper surface and from P to Q along the lower

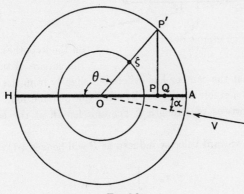

FIG. 8·1.

surface. Also if θ is the eccentric angle of P regarded as belonging to the upper surface, the eccentric angle of P regarded as belonging to the lower surface is $-\theta$ so that the required circulation is $[w'(\theta) + w'(-\theta)]\,d\theta$ or

$$\tfrac{1}{2}cV(2\sin\alpha + \sin\overline{\theta - \alpha} - \sin\overline{\theta + \alpha})\,d\theta = cV\sin\alpha(1 - \cos\theta)\,d\theta$$
$$= 2V\sin\alpha\tan\tfrac{1}{2}\theta\,dx,$$

since $dx = \tfrac{1}{2}c\sin\theta\,d\theta$.

Thus the circulation can be regarded as distributed at the rate

$$2V\sin\alpha\tan\tfrac{1}{2}\theta$$

per unit length at the point whose eccentric angle is θ. If we write

(1) $$2\pi k = 2V\sin\alpha\tan\tfrac{1}{2}\theta,$$

the circulation round the whole profile is

(2) $$2\pi\kappa = \int_{-c/2}^{c/2} 2\pi k \cdot dx.$$

When the lift on the aerofoil is given, κ is given and we may regard (1) as the appropriate solution * of the equation (2), for k as unknown, when κ is known. Observe that $k = 0$ at the trailing edge and $k = \infty$ at the leading edge.

---

* Equation (2) has of course many solutions. That represented by (1) fits our problem.

8·2. The general problem. Consider a thin aerofoil. Take the x-axis along the chord and the origin at the centre of the chord.

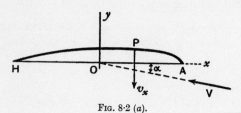

FIG. 8·2 (a).

If $2\pi\kappa$ is the circulation round the aerofoil, we write

$$(1) \quad 2\pi\kappa = \int_{(HA)} 2\pi k \, ds$$

$$= \int_{-c/2}^{c/2} 2\pi k \, dx$$

on the hypothesis (see 8·01) that the square of the slope dy/dx is negligible, so that the camber line differs but slightly from the chord. We can then regard the aerofoil as replaced by a suitable distribution, along the chord, of rectilinear vortices of strength k per unit length at the position x of the chord.

The total downward velocity induced at P will be (see 4·7)

$$(2) \qquad v_x = \int_{-c/2}^{c/2} \frac{k \, d\xi}{\xi - x}.$$

There will also be a small component u parallel to the x-axis.

Referring to fig. 8·2 (b) we see that the components of velocity at P parallel to the x- and y-axes are got by compounding the wind velocity with the induced velocity and are respectively

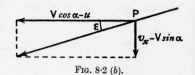

FIG. 8·2 (b).

$$- V \cos \alpha + u, \qquad V \sin \alpha - v_x.$$

The resultant velocity at P therefore makes with the chord the angle

$$\epsilon = \tan^{-1} \frac{v_x - V \sin \alpha}{V \cos \alpha - u}.$$

Since u and α are both small we see that ϵ is also small and therefore

$$\epsilon = - \alpha + v_x/V.$$

Now the air must flow along the aerofoil and therefore we must have at P

$$(3) \qquad \frac{dy}{dx} = \frac{v_x}{V} - \alpha.$$

Here dy/dx is known at every point, for the form of the aerofoil is given, and therefore (3) determines v_x. It remains to find k from (2) subject to the condition (1).

8·3. Glauert's method of solution.

We have seen that in the case of the flat aerofoil, in terms of the eccentric angle,

$$k = \frac{V}{\pi} \alpha \tan \tfrac{1}{2}\theta.$$

In the general case we therefore assume

(1) $$k = \frac{V}{\pi} \{A_0 \tan \tfrac{1}{2}\theta + A_1 \sin \theta + A_2 \sin 2\theta + A_3 \sin 3\theta + \ldots\},$$

where $A_0, A_1, \ldots$, are dimensionless constants to be evaluated, the first term being suggested by the case of the flat aerofoil and the remaining terms being perturbations produced by the camber. The form assumed for k is an odd function of θ, changing sign when θ is replaced by $-\theta$.

The method consists of the application of 8·2 (2) to evaluating v_x in terms of the A_n, which are then determined by 8·2 (3). The lift and moment are then calculated in terms of the now known constants A_0, A_1, A_2.

Glauert's method has its proper application to a thin aerofoil reduced to its camber line. No account is taken of thickness, so that here $y = y_C$.

8·31. Calculation of the induced velocity.

In 8·2 (2) put

$$x = -\tfrac{1}{2}c \cos \theta, \quad \xi = -\tfrac{1}{2}c \cos \phi \quad \text{so that}$$
$$\xi - x = \tfrac{1}{2}c(\cos \theta - \cos \phi), \quad d\xi = \tfrac{1}{2}c \sin \phi \, d\phi. \quad \text{Then}$$

$$k \, d\xi = \frac{V}{\pi}\left\{A_0 \tan \tfrac{1}{2}\phi + \sum_{n=1}^{\infty} A_n \sin n\phi\right\} \tfrac{1}{2}c \sin \phi \, d\phi$$
$$= \frac{V}{\pi}\left\{A_0(1 - \cos \phi) + \tfrac{1}{2}\sum_{n=1}^{\infty} A_n(\cos (n - 1)\phi - \cos (n + 1)\phi)\right\} \tfrac{1}{2}c \, d\phi$$

and therefore

$$v_x = \frac{V}{\pi}\left\{A_0(I_1 - I_0) + \tfrac{1}{2}\sum_{n=1}^{\infty} A_n(I_{n+1} - I_{n-1})\right\},$$

where (see 4·71) $I_n = \int_0^{\pi} \frac{\cos n\phi}{\cos \phi - \cos \theta} \, d\phi = \pi \frac{\sin n\theta}{\sin \theta}.$ Thus

$$v_x = \frac{V}{\pi}\left\{\pi A_0 + \tfrac{1}{2}\sum_{n=1}^{\infty} \frac{\pi A_n}{\sin \theta}(\sin (n+1)\theta - \sin (n - 1)\theta)\right\}$$
$$= V\left\{A_0 + \sum_{n=1}^{\infty} A_n \cos n\theta\right\}.$$

8·32. Determination of the coefficients A_n.

Substituting the value of v_x in 8·2 (3) we have, noting that here $y = y_C$,

(1) $$\frac{dy}{dx} = -\alpha + A_0 + \sum_{m=1}^{\infty} A_m \cos m\theta.$$

Integrating from 0 to π we get

(2) $$A_0 = \alpha + \frac{1}{\pi}\int_0^{\pi} \frac{dy}{dx} \, d\theta.$$

To determine A_n when $n \neq 0$ multiply (1) by $\cos n\theta$ and integrate from 0 to π. Since

$$\int_0^\pi \cos m\theta \cos n\theta \, d\theta = \tfrac{1}{2}\pi \text{ or } 0,$$

according as $m = n$ or $m \neq n$, we have

$$(3) \qquad A_n = \frac{2}{\pi} \int_0^\pi \frac{dy}{dx} \cos n\theta \, d\theta.$$

In particular

$$(4) \qquad A_1 = \frac{2}{\pi} \int_0^\pi \frac{dy}{dx} \cos \theta \, d\theta, \quad A_2 = \frac{2}{\pi} \int_0^\pi \frac{dy}{dx} \cos 2\theta \, d\theta.$$

Note that only A_0 depends on incidence. Observe also that if the incidence is determined by

$$(5) \qquad \alpha_i = -\frac{1}{\pi} \int_0^\pi \frac{dy}{dx} \, d\theta,$$

then $A_0 = 0$, and the circulation as given in 8·3 (1) is nowhere infinite. The angle α_i defined by (5) is called * the *ideal angle of attack*, and $A_0 = \alpha - \alpha_i$.

We can also show that the coefficients A_n are identical with the coefficients a_n of 8·0 (7). For

$$a_n = \frac{1}{\pi} \int_0^\pi \left(\frac{dy_U}{dx} + \tfrac{1}{2}\lambda \tan \tfrac{1}{2}\theta - \tfrac{1}{2}\tau \cot \tfrac{1}{2}\theta \right) \cos n\theta \, d\theta$$
$$+ \frac{1}{\pi} \int_{-\pi}^0 \left(\frac{dy_L}{dx} + \tfrac{1}{2}\lambda \tan \tfrac{1}{2}\theta - \tfrac{1}{2}\tau \cot \tfrac{1}{2}\theta \right) \cos n\theta \, d\theta.$$

Replacing θ by $-\theta$ in the second integral, the terms in λ, τ cancel, and 8·0 (3) then gives

$$(6) \qquad a_n = \frac{2}{\pi} \int_0^\pi \frac{dy_C}{dx} \cos n\theta \, d\theta = A_n.$$

Observe also that

$$(7) \qquad c_0 = \frac{2}{\pi} \int_0^\pi \frac{dy_C}{dx} \, d\theta = 2(A_0 - \alpha) = -2\alpha_i.$$

8·33. The Lift. By the Kutta-Joukowski theorem we have

$$L = 2\pi\kappa\rho V = 2\pi\rho V \int_{-c/2}^{c/2} k \, dx$$
$$= 2\rho V^2 \int_0^\pi \left(A_0 \tan \tfrac{1}{2}\theta + \sum_1^\infty A_n \sin n\theta \right) \tfrac{1}{2}c \sin \theta \, d\theta.$$

Now, if $n > 1$, $\int_0^\pi \sin n\theta \sin \theta \, d\theta = 0$. Therefore

$$L = c\rho V^2 \int_0^\pi (2A_0 \sin^2 \tfrac{1}{2}\theta + A_1 \sin^2 \theta) \, d\theta = \pi c \rho V^2 (A_0 + \tfrac{1}{2}A_1),$$

---

* Theodorsen, *N.A.C.A.*, Technical Report No. 383.

and therefore the lift coefficient is

(1) $C_L = \pi(2A_0 + A_1) = \pi\left(2\alpha + \dfrac{2}{\pi}\displaystyle\int_0^\pi \dfrac{dy}{dx}(1 + \cos\theta)\,d\theta\right) = 2\pi(\lambda_1 + \alpha)$

where

(2) $$\lambda_1 = \dfrac{1}{\pi}\int_0^\pi \dfrac{dy}{dx}(1 + \cos\theta)\,d\theta.$$

This result shows that $\lambda_1 + \alpha$ is the absolute incidence, and that the slope of the $(C_L,\ \alpha)$ graph has its theoretical value 2π (see 7·13), and therefore the angle λ_1 defines the *direction* of the axis of zero lift, see fig. 8·33.

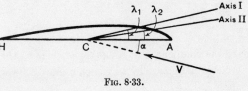

Fig. 8·33.

Since our aerofoil by hypothesis differs but little from the flat aerofoil, we may take the centre of the chord to be the centre of the profile so that λ_1 determines Axis I completely.

We can transform λ_1 by integrating by parts. We have

$$\dfrac{dy}{dx}(1 + \cos\theta) = 2\dfrac{dy}{d\theta}\cdot\dfrac{1 + \cos\theta}{c\sin\theta} = \dfrac{2}{c}\dfrac{dy}{d\theta}\cot\tfrac{1}{2}\theta,$$

and therefore

$$\lambda_1 = \dfrac{2}{\pi c}\int_0^\pi \dfrac{dy}{d\theta}\cot\tfrac{1}{2}\theta\,d\theta = \dfrac{2}{\pi c}\left[y\cot\tfrac{1}{2}\theta\right]_0^\pi + \dfrac{2}{\pi c}\int_0^\pi y\cdot\tfrac{1}{2}\operatorname{cosec}^2\tfrac{1}{2}\theta\,d\theta.$$

The integrated part vanishes at the leading edge, $\theta = \pi$, and at the trailing edge it assumes the indeterminate form $0 \times \infty$. Now

$$\lim_{\theta\to0} y\cot\tfrac{1}{2}\theta = \lim_{\theta\to0}\dfrac{y}{\sin\tfrac{1}{2}\theta} = \lim_{\theta\to0}\dfrac{dy/d\theta}{d(\sin\tfrac{1}{2}\theta)/d\theta},$$

by l'Hospital's theorem (1·9). Also

$$\dfrac{dy}{d\theta} = \dfrac{dy}{dx}\times\dfrac{dx}{d\theta} = \tfrac{1}{2}c\sin\theta\,\dfrac{dy}{dx}.$$

Hence

$$\lim_{\theta\to0} y\cot\tfrac{1}{2}\theta = \lim_{\theta\to0}\left(\dfrac{dy}{dx}\times 2c\sin\tfrac{1}{2}\theta\right) = 0,$$

unless dy/dx is infinite at the trailing edge.* Thus

$$\lambda_1 = \dfrac{2}{\pi}\int_0^\pi \dfrac{y}{c}\cdot\dfrac{1}{1 - \cos\theta}\,d\theta,$$

which can be evaluated graphically or by approximate formulae which avoid the end points.†

---

* When the profile is replaced by its camber line, as is here the case, dy/dx will not become infinite.

† Milne-Thomson, *Calculus of Finite Differences*, 7·33.

8·34. The pitching moment. We take moments about the centre O. The quarter point Q of the chord is distant $\frac{1}{4}c$ from the leading edge.

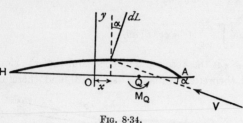

FIG. 8·34.

The lift on the element, whose abscissa is x, is

$$dL = 2\pi k \, dx \, \rho V$$

at right angles to the wind. Thus the pitching moment, positive when it tends to raise the nose, is

$$M_O = \int x \, dL,$$

since the component of lift perpendicular to the chord is $dL \cos \alpha = dL$, and the component parallel to the chord is $dL \sin \alpha = dL \, . \, \alpha$ and is of the second order.

Substituting for k and x we get

$$M_O = - \tfrac{1}{4}\rho V^2 c^2 \int_0^\pi \left\{ A_0 \tan \tfrac{1}{2}\theta + \sum_1^\infty A_n \sin n\theta \right\} 2 \sin \theta \cos \theta \, d\theta.$$

Now
$$\int_0^\pi \tan \tfrac{1}{2}\theta \, 2 \sin \theta \cos \theta \, d\theta = 2 \int_0^\pi (\cos \theta - \cos^2 \theta) \, d\theta = - \pi.$$

$$\int_0^\pi \sin n\theta \, 2 \sin \theta \cos \theta \, d\theta = \int_0^\pi \sin n\theta \sin 2\theta \, d\theta.$$

If $n = 2$ this is $\tfrac{1}{2}\pi$, for any other integral value of n it is zero.

Therefore

(1) $M_O = \tfrac{1}{4}\pi\rho V^2 c^2 (A_0 - \tfrac{1}{2}A_2) = \tfrac{1}{4}\pi\rho V^2 c^2 (\alpha + \lambda_2),$

where from 8·32 (2) and (4)

(2) $\lambda_2 = \dfrac{1}{\pi} \displaystyle\int_0^\pi \dfrac{dy}{dx} (1 - \cos 2\theta) \, d\theta = \dfrac{4}{\pi c} \int_0^\pi \dfrac{dy}{d\theta} \sin \theta \, d\theta = - \dfrac{4}{\pi} \int_0^\pi \dfrac{y}{c} \cos \theta \, d\theta,$

on integrating by parts. This form is suited to graphical integration. It follows from (1) that M_O is zero when $\alpha = - \lambda_2$ so that (7.14) λ_2 gives the direction of Axis II, see fig. 8·33.

The pitching moment coefficients about the centre and the quarter point are related by

(3) $C_{m_Q} = C_{m_O} - \tfrac{1}{4}C_L = \tfrac{1}{2}\pi(\lambda_2 - \lambda_1)$

which is independent of incidence and we have the important result that for a thin aerofoil of small camber *the quarter point of the chord is the focus of the profile* (7·2).

If the pitching moment about the focus is zero, the aerofoil has a centre of lift, and, since here the focus is on the chord, a fixed centre of pressure. The condition for this is $\lambda_1 = \lambda_2$.

8·35. Travel of the centre of pressure. The position of the centre of pressure (from 7·31) aft of the quarter point is given by

(1)
$$(C_p - \tfrac{1}{4})(\lambda_1 + \alpha) = \tfrac{1}{4}(\lambda_1 - \lambda_2),$$

where $\lambda_1 > \lambda_2$ if Axis II is below Axis I (7·2). The graph of $C_p - \tfrac{1}{4}$ plotted against incidence (note that $\lambda_1 + \alpha$ is absolute incidence) is therefore a rectangular hyperbola.

Thus as incidence increases the centre of pressure moves towards the quarter point. A similar result

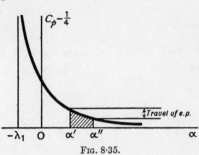

Fig. 8·35.

and graph follow if $\lambda_1 < \lambda_2$. The travel in a working range α' to α'' is therefore

(2)
$$\tfrac{1}{4}c(\lambda_1 - \lambda_2)\left(\frac{1}{\lambda_1 + \alpha'} - \frac{1}{\lambda_1 + \alpha''}\right).$$

8·36. Unsymmetrical aerofoil with a fixed centre of pressure. We have just seen that the thin aerofoil has a fixed centre of pressure if $\lambda_1 = \lambda_2$, i.e. if

(1)
$$\int_0^\pi \frac{y}{c}\left(\frac{1}{1 - \cos\theta} + 2\cos\theta\right)d\theta = 0.$$

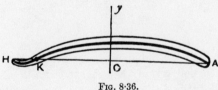

Fig. 8·36.

Let us take as camber line a cubic curve passing through the points $\tfrac{1}{2}c$, $-\tfrac{1}{2}\epsilon c$, $-\tfrac{1}{2}c$, the points A, K, H in fig. 8·36. The equation of such a line is

$$\frac{y}{c} = \mu\left(\frac{x}{c} - \frac{1}{2}\right)\left(\frac{x}{c} + \frac{1}{2}\right)\left(\frac{x}{c} + \frac{1}{2}\epsilon\right),$$

where μ is an arbitrary number which can be so chosen as to give the mean camber any suitable value. In terms of the eccentric angle we have

$$\frac{y}{c} = \tfrac{1}{8}\mu(1 - \cos^2\theta)(\cos\theta - \epsilon).$$

Substituting in (1) there will be a fixed centre of pressure if

$$\int_0^\pi [(1 + \cos\theta)(\cos\theta - \epsilon) + 2\sin^2\theta(\cos^2\theta - \epsilon\cos\theta)]d\theta = 0.$$

Performing the simple integrations we get $\epsilon = 3/4$, which means that K is distant $c/8$ from the trailing edge.

We may compare this result with the remarks concerning the necessity of the **S**-shape in connection with von Mises profiles (7·7).

8·37. Effect of operating a flap.

Fig. 8·37 (a) shows an aerofoil profile whose after part is movable about a hinge at P on the camber line so

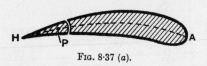

FIG. 8·37 (a).

that the part PH, the *flap*, can be raised or lowered from the neutral position shown in the diagram. In an aerofoil of finite aspect ratio the flap movement usually affects only part of a wing. Ailerons are flaps near the wing tips and are arranged so that the port and starboard ailerons move in opposite senses, one up and one down. The investigation which follows will throw light on the general effect. Here

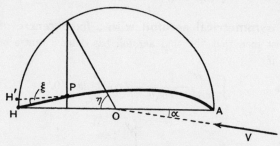

FIG. 8·37 (b).

we are concerned with the two-dimensional problem and for simplicity we assume that the aerofoil is *thin*, that the portion PH of the camber line is *straight*, and that the angle ξ through which the flap is rotated is *small*.

For the thin aerofoil shown in fig. 8·37 (b) the eccentric angle which defines the position of the hinge P is η. In the neutral position PH of the flap we have from 8·33, y denoting y_C,

$$C_L = 2\pi(\alpha + \lambda_1) = 2\pi\alpha + 2\int_0^\pi \frac{dy}{dx}(1 + \cos\theta)\,d\theta$$

$$= 2\pi\alpha + 2\int_0^\eta \frac{dy}{dx}(1 + \cos\theta)\,d\theta + 2\int_\eta^\pi \frac{dy}{dx}(1 + \cos\theta)\,d\theta$$

The effect of raising the flap is to decrease dy/dx to $dy/dx - \xi$ on the raised part PH' and to leave it unaltered on the part PA. The lift coefficient is thereby altered to C_L' where

$$(1) \quad C_L' = 2\pi\alpha + 2\int_0^\eta \left(\frac{dy}{dx} - \xi\right)(1 + \cos\theta)\,d\theta + 2\int_\eta^\pi \frac{dy}{dx}(1 + \cos\theta)\,d\theta.$$

Thus

$$(2) \qquad C_L' - C_L = -2\xi \int_0^\eta (1 + \cos \theta) d\theta = - 2\xi(\eta + \sin \eta).$$

Thus the effect of raising the flap is to decrease the lift coefficient, the effect of lowering the flap is to increase the lift coefficient. Therefore, in particular, when the flaps are lowered just before landing increased lift is obtained (and also increased drag). We also observe that in the case of ailerons, if the port aileron is raised and the starboard aileron depressed, the lift on the port wing is decreased and that on the starboard wing is increased, causing a rolling moment which tends to raise the starboard wing tip.

Another way of looking at (1) is to write it in the form

$$C_L' = 2\pi(\alpha' + \lambda_1), \quad \alpha' = \alpha - \frac{\xi}{\pi}(\eta + \sin \eta),$$

so that C_L' is the lift coefficient of the original aerofoil but at a decreased incidence α' when the flap is raised, in other words, raising the flap effectively decreases the incidence by the amount $\alpha - \alpha'$.

The effect on the moment about the focus is obtained in a similar manner from 8·34 (3), which gives

$$C_{mQ}' - C_{mQ} = \tfrac{1}{2}\xi \int_0^\eta (\cos \theta + \cos 2\theta) d\theta = \tfrac{1}{2}\xi \sin \eta \, (1 + \cos \eta).$$

Thus raising the flap increases the tendency of the nose to lift, and lowering the flap tends to put the nose down.

8·4. The pressure method.

This method * is based on the geometrical considerations of 8·0 supplemented by a linearisation which reduces the theory to a boundary condition problem.

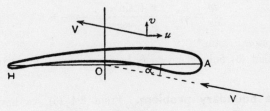

Fig. 8·4.

Suppose the aerofoil in a wind V at incidence α. If (U_x, U_y) is the velocity at any point, we may write

$$(U_x, U_y) = (- V \cos \alpha + u, \ V \sin \alpha + v)$$

---

* *Op. cit.*, p. 134, called here the pressure method because of its use in deriving aerofoil shapes given the pressure distribution.

where (u, v) is the *incremental* or *perturbation velocity* imposed on the steady flow by the presence of the aerofoil. Thus we may write

(1) $U_x - iU_y = -V \cos \alpha - iV \sin \alpha + u - iv.$

The complex velocity $U_x - iU_y$ and therefore the perturbation velocity $u - iv$ is a function of z only.

Assumption I. u, v, α are so small that their squares and products may be neglected.

This assumption will fail at a stagnation point for then $u = V \cos \alpha = V$. This will subsequently be found to entail infinite pressure at the forward stagnation point, except for the ideal angle of attack.

If P is the pressure at any point and Π is the pressure in the unperturbed wind, Bernoulli's theorem gives

$$P + \tfrac{1}{2}\rho[(-V \cos \alpha + u)^2 + (V \sin \alpha + v)^2] = \Pi + \tfrac{1}{2}\rho V^2,$$

and therefore from Assumption I

$$p = P - \Pi = \tfrac{1}{2}\rho V^2 \cdot \frac{2u}{V},$$

where p is the *pressure excess*.

It is convenient to use the following units : unit of length the chord c ; unit of velocity V ; unit of pressure $\tfrac{1}{2}\rho V^2$.

With this system of units Bernoulli's theorem assumes the (linearised) form

(2) $p = 2u,$

and the chord of the aerofoil stretches from $(-\tfrac{1}{2}, 0)$ to $(\tfrac{1}{2}, 0)$. Observe that (2) fails at a stagnation point (where $p = 1$).

The condition to be satisfied is that at the surface of the aerofoil the flow must be tangential, therefore

(3) $\dfrac{dy}{dx} = \dfrac{U_y}{U_x} = \dfrac{v + \sin \alpha}{u - \cos \alpha} = -v - \alpha,$

using Assumption I again.*

Observe that (3) fails where dy/dx becomes infinite, in particular at the leading edge.

8·41. The boundary problem. From 8·4 (3) we have to choose $u - iv$ so that, at the surface of the aerofoil,

$$2iv = -2i\frac{dy}{dx} - 2i\alpha = i\lambda \tan \tfrac{1}{2}\theta - i\tau \cot \tfrac{1}{2}\theta - 2iF(\theta) - 2i\alpha$$

from 8·0 (6), so that using 8·0 (7), and writing A_n, B_n instead of a_n, b_n, we must have

---

* In 8·2 (3) the sign of v_x is opposite to the sign of v here.

(1) $2iv = i\lambda \tan \tfrac{1}{2}\theta - i\tau \cot \tfrac{1}{2}\theta - i(2\alpha + c_0) - 2i \sum\limits_{n=1}^{\infty} (A_n \cos n\theta + B_n \sin n\theta).$

Now the transformation

(2) $$z = \tfrac{1}{4}\left(\zeta + \frac{1}{\zeta}\right)$$

transforms the region exterior to the circle $\zeta\overline{\zeta} = 1$ into the region exterior to the line joining $(-\tfrac{1}{2}, 0)$ to $(\tfrac{1}{2}, 0)$. On this circle we have (cf. fig. 8·1)

$$\zeta = -e^{-i\theta}, \;\; \overline{\zeta} = -e^{i\theta} = \frac{1}{\zeta}.$$

Also if $\qquad u - iv = f(\zeta),$ then $2iv = \bar{f}(\bar{\zeta}) - f(\zeta).$

Now $2\cos\theta = e^{i\theta} + e^{-i\theta}$ and $2i\sin\theta = e^{i\theta} - e^{-i\theta}$, and therefore

$$i \tan \tfrac{1}{2}\theta = \frac{e^{\frac{1}{2}i\theta}}{e^{\frac{1}{2}i\theta} + e^{-\frac{1}{2}i\theta}} - \frac{e^{-\frac{1}{2}i\theta}}{e^{\frac{1}{2}i\theta} + e^{-\frac{1}{2}i\theta}} = \frac{1}{1 + e^{-i\theta}} - \frac{1}{e^{i\theta} + 1} = \frac{1}{\overline{\zeta} - 1} - \frac{1}{\zeta - 1},$$

on the boundary of the circle. Similarly, we show that on the boundary of the circle

$$- i \cot \tfrac{1}{2}\theta = -\frac{1}{\overline{\zeta} + 1} + \frac{1}{\zeta + 1}.$$

Again,

$$1 = \frac{e^{\frac{1}{2}i\theta}}{e^{\frac{1}{2}i\theta} + e^{-\frac{1}{2}i\theta}} + \frac{e^{-\frac{1}{2}i\theta}}{e^{\frac{1}{2}i\theta} + e^{-\frac{1}{2}i\theta}} = \frac{1}{e^{-i\theta} + 1} + \frac{1}{e^{i\theta} + 1},$$

and therefore on the boundary of circle

$$-i = \frac{i}{\overline{\zeta} - 1} - \left(\frac{-i}{\zeta - 1}\right).$$

Moreover, $\quad 2\cos n\theta = (-\overline{\zeta})^{-n} + (-\zeta)^{-n}, \quad 2i \sin n\theta = (-\zeta)^{-n} - (-\overline{\zeta})^{-n},$

and so, substituting in (1), we can pick out $f(\zeta)$, in fact

(3) $\qquad f(\zeta) = \dfrac{\lambda - i(2\alpha + c_0)}{\zeta - 1} - \dfrac{\tau}{\zeta + 1} + \sum\limits_{n=1}^{\infty} (iA_n + B_n)(-\zeta)^{-n}.$

Since $f(\zeta) \to 0$ when $|\zeta| \to \infty$, this is the solution of (1) in the case of the flat aerofoil obtained by applying (2) to the circle $\zeta\overline{\zeta} = 1$.

Assumption II. The boundary problem is solved by $u - iv = f(\zeta)$, where $f(\zeta)$ is given by (3), provided that the aerofoil approximates sufficiently closely to a flat plate.

Separating the real and imaginary parts in $u - iv = f(\zeta)$ and using 8·4 (2) we get

(4) $p = 2u = -\lambda - \tau - (2\alpha + c_0) \tan \tfrac{1}{2}\theta + 2 \sum\limits_{n=1}^{\infty} (-A_n \sin n\theta + B_n \cos n\theta).$

(5) $v = \tfrac{1}{2}\lambda \tan \tfrac{1}{2}\theta - \tfrac{1}{2}\tau \cot \tfrac{1}{2}\theta - \tfrac{1}{2}(2\alpha + c_0) - \sum\limits_{n=1}^{\infty} (A_n \cos n\theta + B_n \sin n\theta).$

Notes. (i) The circulation round the element dx is

$$2\pi k\, dx = [u(\,-\theta) - u(\theta)]\, dx.$$

Since $\cos n\theta$ is an even function of θ, we get

$$k = \frac{1}{\pi}\left[(\alpha + \tfrac{1}{2}c_0)\tan\tfrac{1}{2}\theta + \sum_{n=1}^{\infty} A_n \sin n\theta\right].$$

Comparison with 8·3 (1) shows that (see also 8·32 (7))

$$\tfrac{1}{2}c_0 = A_0 - \alpha.$$

(ii) From 8·32 we see that the *ideal angle of attack* is given by $\alpha_i = -\tfrac{1}{2}c_0$.

(iii) From (4) it appears that the pressure is infinite at the leading edge ($\theta = \pi$) except when the incidence is the ideal angle of attack. This infinite pressure is a consequence of applying Assumption I to the stagnation point at the leading edge. It would not appear in an exact solution.

(iv) The pressure excess p depends only on the incremental component velocity along the chord and has the same sign.

(v) Since the aerodynamic force depends only on the pressure distribution given by (4), the component v contributes nothing to the force calculated from the pressure distribution.

(vi) The component v becomes infinite at the leading and trailing edges. This result has no further physical implications on account of (v).

(vii) Referring to 8·31 we see that

$$v + v_x = \tfrac{1}{2}\lambda\tan\tfrac{1}{2}\theta - \tfrac{1}{2}\tau\cot\tfrac{1}{2}\theta - \sum_{n=1}^{\infty} B_n \sin n\theta,$$

which is an odd function of θ but is not zero in general. For an infinitely thin aerofoil the contributions to v_x of that part of v which is an odd function of θ will cancel.

(viii) On account of the choice of units the lift and pitching moment will appear as C_L and C_m, for $\tfrac{1}{2}\rho V^2 = 1$ and $c = 1$.

To calculate these we have

$$C_L = \int_{-\frac{1}{2}}^{\frac{1}{2}} \{p(-\theta) - p(\theta)\}\, dx, \quad C_m = \int_{-\frac{1}{2}}^{\frac{1}{2}} x\{p(-\theta) - p(\theta)\}\, dx,$$

which yield the results already obtained in 8·33 and 8·34. This agreement is due to the fact that only those parts of p which are odd functions of θ contribute to C_L, so that the integrals are the same in the two methods.

(ix) For *zero lift* we have from 8·33 (1), $2A_0 + A_1 = 0$, and therefore from (i) the incidence α_0 for zero lift is

$$\alpha_0 = -\tfrac{1}{2}(c_0 + A_1) = \alpha_i - \tfrac{1}{2}A_1 = -\lambda_i$$

in the notation of 8·33.

8·42. The inverse problem. Given the distribution of pressure over the whole surface of the aerofoil to find the form of the profile.

Let $p(\theta)$ be the given pressure at the point whose eccentric angle is θ. Then we can write

(1) $$p(\theta) = p_C(\theta) + p_T(\theta)$$

where $p_C(\theta)$ is an odd function of θ, i.e. $p_C(-\theta) = -p_C(\theta)$, and $p_T(\theta)$ is an even function of θ, i.e. $p_T(\theta) = p_T(-\theta)$. When $p(\theta)$ is known we can determine p_C, p_T from

(2) $$p_C(\theta) = \tfrac{1}{2}[p(\theta) - p(-\theta)], \quad p_T(\theta) = \tfrac{1}{2}[p(\theta) + p(-\theta)].$$

Now from 8·41 (4), separating the odd and even parts,

(3) $$p_C(\theta) = -(2\alpha - 2\alpha_i)\tan\tfrac{1}{2}\theta - 2\sum_{n=1}^{\infty} A_n \sin n\theta.$$

(4) $$p_T(\theta) = -\lambda - \tau + 2\sum_{n=1}^{\infty} B_n \cos n\theta.$$

It follows that when $p_C(\theta)$ is given, we can determine the coefficients A_n, and when $p_T(\theta)$ is given, we can determine the coefficients B_n.

Now from 8·0 (5) we have

$$\frac{dy}{dx} = \frac{dy_C}{dx} + \frac{dy_T}{dx},$$

and therefore from 8·0 (6), since dy_C/dx is an even function and dy_T/dx is an odd function of θ,

(5) $$\frac{dy_C}{dx} = -\alpha_i + \sum_{n=1}^{\infty} A_n \cos n\theta,$$

$$\frac{dy_T}{dx} = -\tfrac{1}{2}\lambda\tan\tfrac{1}{2}\theta + \tfrac{1}{2}\tau\cot\tfrac{1}{2}\theta + \sum_{n=1}^{\infty} B_n \sin n\theta.$$

Thus p_C determines the coefficients A_n and therefore from (5) the camber line function y_C, while p_T determines the coefficients B_n and therefore from (5) the thickness function y_T. The profile can then be constructed by using 8·0 (5).

The problem is thus reduced to finding the camber line function and the thickness function. It will now be clear that we can assign to p_C any arbitrary continuous pressure distribution which is an odd function of θ, and to p_T any arbitrary continuous pressure distribution which is an even function of θ.

8·43. Determination of a camber line. To illustrate the method by a specific example, let the pressure on the upper surface of the aerofoil be given by

(1) $$p_C = -p_0(1 - \cos\theta), \quad 0 \leqslant \theta \leqslant \tfrac{1}{2}\pi,$$
$$p_C = -p_0, \qquad\qquad \tfrac{1}{2}\pi \leqslant \theta < \pi,$$

and let $p_C = 0$ when $\theta = \pi$. Since p_C is to be an odd function, the pressure on the lower surface will be $p_0(1 - \cos\theta)$ and p_0 for the same ranges of values of θ.

Fig. 8·43 (i) shows the form of pressure distribution. The thrust C_L on the aerofoil is therefore

$$(2) \qquad C_L = 2[\tfrac{1}{2}p_0 + \tfrac{1}{2}p_0 \times \tfrac{1}{2}], \quad p_0 = \tfrac{2}{3}C_L.$$

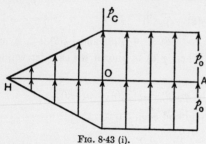

FIG. 8·43 (i).

We suppose that the aerofoil will be designed for the ideal angle of attack, i.e. $\alpha = \alpha_i$.

Then 8·42 (3) gives

$$(3) \quad p_C = -2 \sum_{n=1}^{\infty} A_n \sin n\theta.$$

To determine A_n multiply by $\sin n\theta$ and integrate from 0 to π. Then all the trigonometrical integrals on the right vanish except that which contains A_n and so

$$A_n = -\frac{1}{\pi}\int_0^{\pi} p_C \sin n\theta \, d\theta = \frac{p_0}{\pi}\int_0^{\frac{1}{2}\pi} (1 - \cos\theta)\sin n\theta \, d\theta + \frac{p_0}{\pi}\int_{\frac{1}{2}\pi}^{\pi} \sin n\theta \, d\theta.$$

Performing the simple integrations and using (2) we get

$$(4) \quad A_1 = \frac{C_L}{\pi}, \quad A_n = -\frac{2C_L}{3\pi}\left\{\frac{(-1)^n - 1}{n} + \tfrac{1}{2}\frac{\sin\frac{1}{2}n\pi + 1}{n+1} - \tfrac{1}{2}\frac{\sin\frac{1}{2}n\pi - 1}{n-1}\right\}.$$

To find dy_C/dx, we have to evaluate 8·42 (5). Now

$$(5) \qquad \sum_{n=1}^{\infty} A_n \cos n\theta = \text{real part of } \Sigma A_n t^n,$$

where $t = e^{i\theta}$. We use the logarithmic series

$$\log(1+z) = z - \tfrac{1}{2}z^2 + \tfrac{1}{3}z^3 - \dots, \quad -\log(1-z) = z + \tfrac{1}{2}z^2 + \tfrac{1}{3}z^3 + \dots$$

Having regard to the value of A_n given by (4), we want the following sums:

$$\sum_{n=1}^{\infty} \frac{(-1)^n t^n}{n} = -\log(1+t), \quad \sum_{n=1}^{\infty} \frac{t^n}{n} = -\log(1-t),$$

$$\sum_{n=1}^{\infty} \frac{t^n}{n+1} = \tfrac{1}{2}t + \tfrac{1}{3}t^2 + \tfrac{1}{4}t^3 + \dots = -\frac{1}{t}\log(1-t) - 1,$$

$$\sum_{n=2}^{\infty} \frac{t^n}{n-1} = t^2 + \tfrac{1}{2}t^3 + \tfrac{1}{3}t^4 + \dots = -t\log(1-t),$$

$$\sum_{n=1}^{\infty} \frac{\sin\frac{1}{2}n\pi \, t^n}{n+1} = \tfrac{1}{2}t - \tfrac{1}{4}t^3 + \tfrac{1}{6}t^5 - \dots = \frac{1}{2t}\log(1+t^2)$$

$$\sum_{n=2}^{\infty} \frac{\sin\frac{1}{2}n\pi \, t^n}{n-1} = -\tfrac{1}{2}t^3 + \tfrac{1}{4}t^5 - \tfrac{1}{6}t^7 = -\tfrac{1}{2}t\log(1+t^2).$$

Therefore from (4),

$$(6) \quad -\frac{3\pi}{2C_L}\sum_{n=1}^{\infty} A_n t^n$$

$$= \left(1 - \tfrac{1}{2}t - \frac{1}{2t}\right)\log(1-t) - \log(1+t) - \tfrac{1}{2} + \tfrac{1}{4}\left(t + \frac{1}{t}\right)\log(1+t^2).$$

Now since $t = e^{i\theta}$ and $x = -\frac{1}{2}\cos\theta$ we have

$$1 - t = e^{\frac{1}{2}i\theta}(e^{-\frac{1}{2}i\theta} - e^{\frac{1}{2}i\theta}) = -2ie^{\frac{1}{2}i\theta}\sin\tfrac{1}{2}\theta = -ie^{\frac{1}{2}i\theta}\sqrt{(2 + 4x)},$$

and similarly

$$1 + t = e^{\frac{1}{2}i\theta}\sqrt{(2 - 4x)}, \quad 1 + t^2 = -e^{i\theta}\,4x.$$

Therefore from (5), (6) and 8·42 (5)

$$\frac{dy_C}{dx} = -\alpha_i - \frac{C_L}{3\pi}\{(1 + 2x)\log(2 + 4x) - \log(2 - 4x) - 1 - x\log 16x^2\}.$$

Integrating, we get

$$y_C = k - \alpha_i x + \frac{C_L}{3\pi}\{\tfrac{1}{2}x^2\log 4x^2 + \tfrac{1}{2}x - \tfrac{1}{4}(1 + 2x)^2\log(1 + 2x)$$
$$- \tfrac{1}{2}(1 - 2x)\log(1 - 2x)\},$$

a result which can be easily verified by direct differentiation. Since $y_C = 0$ when $x = \frac{1}{2}$ and when $x = -\frac{1}{2}$, we can determine the unknown constants k and α_i. Remembering that $z\log z \to 0$ when $z \to 0$, we get

(7) $$k = \frac{C_L}{3\pi}\log 2, \quad \alpha_i = \frac{C_L}{6\pi} = 0\!\cdot\!0530517C_L,$$

and the equation of the camber line is therefore

(8) $$y_C = \frac{C_L}{3\pi}\{\log 2 + \tfrac{1}{2}x^2\log 4x^2 - \tfrac{1}{4}(1 + 2x)^2\log(1 + 2x)$$
$$- \tfrac{1}{2}(1 - 2x)\log(1 - 2x)\},$$

where the logarithms are all to base e.

The form of the camber line is shown in fig. 8·43 (ii), the ordinates being y_C/C_L.

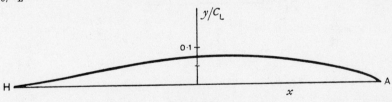

y/C_L

$0\cdot1$

H x A

FIG. 8·43 (ii).

Since $\alpha_i = C_L/6\pi$, $A_1 = C_L/\pi$, $A_2 = -4C_L/9\pi$, we get from 8·41 (ix) the incidence for zero lift $-\lambda_1 = -2\alpha_i$.

From 8·34, $\quad C_{mQ} = -\tfrac{1}{2}\pi(\lambda_1 - \lambda_2) = -\tfrac{1}{4}\pi(A_1 + A_2) = -\dfrac{5C_L}{36}.$

Thus from 8·35 (2), assuming a working range of the incidence from $\alpha' = 0$ to $\alpha'' = 3\alpha_i$, we find the travel of the centre of pressure to be $1/8$ of the chord.

As a numerical example, suppose it is required to design a high lift aerofoil, with $C_L = 0\!\cdot\!8$ at the ideal angle of attack, and with the assumed pressure distribution. Then

$$\alpha_i = 0\!\cdot\!0424413 = 2°\,26', \quad \lambda_1 = 4°\,52',$$

and the lift coefficient is then given for any other incidence by the graph.

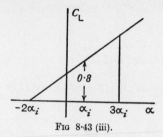

FIG 8·43 (iii).

Thus in the assumed working range $0 < \alpha < 3\alpha_i$, the lift coefficient varies nearly from 0·53 to 1·33.

8·44. Determination of a thickness function.

Again we take a specific distribution to illustrate the method. Consider then

$$(1) \qquad \begin{aligned} p_T &= -\{\mu - \delta \cos \theta\} \quad \text{for} \quad 0 \leqslant |\theta| \leqslant \tfrac{1}{2}\pi, \\ p_T &= -\{\mu + \epsilon \cos \theta\} \quad \text{for} \quad \tfrac{1}{2}\pi \leqslant |\theta| \leqslant \pi, \end{aligned}$$

which is a linear distribution of pressure defect symmetrical on both faces and which therefore causes no lift or moment.

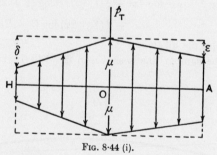

FIG. 8·44 (i).

From 8·42 (4) we have

$$(2) \qquad p_T = -\lambda - \tau + 2 \sum_{n=1}^{\infty} B_n \cos n\theta,$$

and therefore multiplying by $\cos n\theta$ and integrating from 0 to π we get, if $n \geqslant 2$,

$$\pi B_n = \int_0^{\pi} p_T \cos n\theta \, d\theta = - \int_0^{\frac{1}{2}\pi} (\mu - \delta \cos \theta) \cos n\theta \, d\theta - \int_{\frac{1}{2}\pi}^{\pi} (\mu + \epsilon \cos \theta) \cos n\theta \, d\theta,$$

$$(3) \quad B_n = \frac{1}{2\pi} (\delta + \epsilon) \left\{ \frac{\sin \frac{1}{2}(n + 1)\pi}{n + 1} + \frac{\sin \frac{1}{2}(n - 1)\pi}{n - 1} \right\}, \quad B_1 = \tfrac{1}{4}(\delta - \epsilon),$$

where B_1 is calculated separately but by the same method.

Also integrating (2) from 0 to π we get

$$(4) \qquad \lambda + \tau = \mu - \frac{1}{\pi} (\delta + \epsilon).$$

It is convenient to write

(5) $\beta = \delta - \epsilon$ = total rise in pressure along the chord;

$\gamma = \frac{1}{2}(\delta + \epsilon)$ = mean of the leading and trailing edge pressures above the minimum pressure.

Then

(6) $$\lambda + \tau = \mu - 2\gamma/\pi, \quad B_1 = \beta/4 = \lambda - \tau$$

from 8·0 (11), and therefore from 8·0 (10) the radii of curvature ρ_A, ρ_H at the leading and trailing edges of the aerofoil are given by

(7) $$(2\rho_A)^{\frac{1}{2}} = \lambda = \frac{1}{2}\mu - \frac{\gamma}{\pi} + \frac{\beta}{8},$$

(8) $$(2\rho_H)^{\frac{1}{2}} = \tau = \frac{1}{2}\mu - \frac{\gamma}{\pi} - \frac{\beta}{8}.$$

To find the thickness function y_T we use 8·42 (5) which entails evaluating

(9) $$\sum_{n=1}^{\infty} B_n \sin n\theta = \text{imaginary part of } \sum_{n=1}^{\infty} B_n t^n, \text{ where } t = e^{i\theta}.$$

Now

$$\sum_{n=2}^{\infty} \frac{\sin \frac{1}{2}(n+1)\pi}{n+1} t^n = -\frac{t^2}{3} + \frac{t^4}{5} - \frac{t^6}{7} + \dots,$$

$$\sum_{n=2}^{\infty} \frac{\sin \frac{1}{2}(n-1)\pi}{n-1} t^n = t^2 - \frac{t^4}{3} + \frac{t^6}{5} - \dots.$$

It is readily verified from the logarithmic series (8·43) that

$$t - \frac{t^3}{3} + \frac{t^5}{5} - \dots = -\frac{1}{2}i\{\log(1+it) - \log(1-it)\} = \frac{1}{2}i \log \frac{1-it}{1+it}.$$

Therefore

$$1 + \frac{\pi}{\gamma} \sum_{n=2}^{\infty} B_n t^n = \left(\frac{i}{2t} + \frac{it}{2}\right) \log \frac{1-it}{1+it} = i \cos\theta \log \frac{1+\sin\theta - i\cos\theta}{1-\sin\theta + i\cos\theta}.$$

Also the real part of $\log(a + ib) = \frac{1}{2} \log(a^2 + b^2)$.

Therefore

$$\frac{\pi}{\gamma} \sum_{n=2}^{\infty} B_n \sin n\theta = \frac{1}{2} \cos\theta \log \frac{1+\sin\theta}{1-\sin\theta},$$

and therefore from 8·42 (5)

$$\frac{dy_T}{dx} = -\frac{1}{2}\lambda \tan \frac{1}{2}\theta + \frac{1}{2}\tau \cot \frac{1}{2}\theta + \frac{1}{4}\beta \sin\theta + \frac{\gamma}{2\pi} \cos\theta \log \frac{1+\sin\theta}{1-\sin\theta}.$$

But

$$\frac{dy_T}{d\theta} = \frac{dy_T}{dx} \frac{dx}{d\theta} = \frac{1}{2} \sin\theta \frac{dy_T}{dx}.$$

Therefore

$$\frac{dy_T}{d\theta} = -\frac{1}{4}\lambda(1 - \cos\theta) + \frac{1}{4}\tau(1 + \cos\theta) + \frac{\beta}{16}(1 - \cos 2\theta)$$

$$+ \frac{\gamma}{4\pi} \sin\theta \cos\theta \log \frac{1+\sin\theta}{1-\sin\theta}.$$

$$= \left(\frac{1}{4}\mu - \frac{\gamma}{2\pi}\right) \cos\theta - \frac{\beta}{16} \cos 2\theta + \frac{\gamma}{4\pi} \sin\theta \cos\theta \log \frac{1+\sin\theta}{1-\sin\theta}.$$

Integrating and observing that $y = 0$ when $\theta = 0$ and $\theta = \pi$ so that the constant of integration is zero, we get

$$(10) \quad y_T = \tfrac{1}{4}\mu \sin\theta - \tfrac{1}{32}\beta \sin 2\theta - \frac{\gamma}{4\pi}\left\{\sin\theta + \tfrac{1}{2}\cos^2\theta \log\frac{1+\sin\theta}{1-\sin\theta}\right\}.$$

The correctness of the integration can be verified by differentiation. This is the required thickness function.

Note that the coefficients of μ and γ are unaltered when θ is replaced by $\pi - \theta$, so that a thickness function for which $\beta = 0$ would give an oval with symmetry about both the x- and y-axes, and it is clear that the β term will give the oval its characteristic aerofoil shape. If the pressure distribution is $p = \text{constant} = \mu$ all round the aerofoil, then $\delta = \epsilon = 0$ and therefore $\beta = \gamma = 0$, so that a constant pressure round the aerofoil corresponds with the elliptic thickness function

$$x = -\tfrac{1}{2}\cos\theta, \quad y = \tfrac{1}{4}\mu\sin\theta.$$

As an example Brennan and Stevenson give the symmetrical aerofoil ($y_C = 0$) for the values $\mu = 0\cdot40$, $\beta = 0\cdot55$, $\gamma = 0\cdot35$ or $\delta = 0\cdot625$, $\epsilon = 0\cdot075$. For these (7) and (8) give

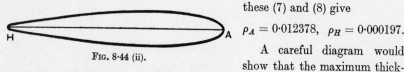

$$\rho_A = 0\cdot012378, \quad \rho_H = 0\cdot000197.$$

Fig. 8·44 (ii).

A careful diagram would show that the maximum thickness occurs at about $x = 0\cdot05$ and is then about 15 per cent. of the chord.

If we combine the distributions p_C (with $C_L = 0\cdot8$) of 8·43 and the present p_T we get, see Ex. VIII, 6, the characteristic aerofoil shape.

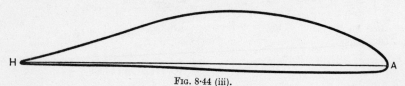

Fig. 8·44 (iii).

The foregoing simple illustrations sufficiently exemplify the general procedure which can always be carried out by methods of numerical integration for any given pressure distribution.

8·5. Substitution vortex. If we transform the circle $|\zeta| = a$ into an aerofoil by a transformation of the type

$$(1) \qquad z = \zeta + \frac{a_1}{\zeta} + \frac{a_2}{\zeta^2} + \dots,$$

the streamlines in the plane of the circle C will transform into the streamlines in the plane of the aerofoil. If the motion in the plane of the circle (ζ-plane) consists of circulation $2\pi\kappa$ only, the streamlines are circles concentric with C. Now the transformation (1) is such that for sufficiently large values of $|\zeta|$ the plane is undistorted, i.e. $z = \zeta$. Thus at sufficiently great distances from the aerofoil the streamlines will be also concentric circles, in fact nearly the same circles as those in the ζ-plane, and the circulation will of course be rigorously $2\pi\kappa$.

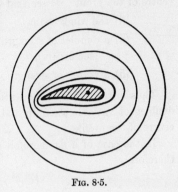

It follows that at sufficiently great distances from the aerofoil, the flow is that due to a vortex of strength κ suitably placed. This vortex is called the *substitution vortex* of the aerofoil because the perturbation in the flow due to the aerofoil can be equally well calculated at sufficiently great distances by supposing the aerofoil to be removed and the vortex substituted for it.

Fig. 8·5.

To obtain some idea of what constitutes a " sufficiently great distance ", if $a_1 = l^2 e^{i\mu}$ and $c = 4a$, we have $l < a$ and therefore $l^2 < c^2/16$. Thus, if $|z| = r$,

$$\left|\frac{a_1}{z}\right| < \frac{c^2}{16r},$$

so that if $r > c$, we may take $z = \zeta$ in (1) with an error of less than $1/16$. Thus as a reasonable working rule we can use the substitution vortex at distances from the centre of the profile greater than the length of the chord.

To find the position of the substitution vortex, observe that for a vortex of strength κ at the point z_0 and a uniform wind $Ve^{i\alpha}$ we have

$$w = Ve^{i\alpha}z + i\kappa \log (z - z_0) = Ve^{i\alpha}z + i\kappa \log z + i\kappa \log \left(1 - \frac{z_0}{z}\right),$$

$$(2) \qquad w = Ve^{i\alpha}z + i\kappa \log z - \frac{i\kappa z_0}{z} + O\left(\frac{z_0^2}{z^2}\right).$$

Again, for the aerofoil we have

$$w = Ve^{i\alpha}\zeta + \frac{Ve^{-i\alpha}a^2}{\zeta} + i\kappa \log \zeta,$$

together with (1). Reversing (1) as in 7·02 (2), and substituting for ζ in terms of z, we get, to the same order of approximation,

$$(3) \qquad w = Ve^{i\alpha}z + \frac{Va^2e^{-i\alpha} - a_1 Ve^{i\alpha}}{z} + i\kappa \log z + O\left(\frac{a_1}{z^2}\right).$$

Comparing (2) and (3) and neglecting the O-terms we have

$$(4) \qquad z_0 = \frac{iV}{\kappa}(a^2e^{-i\alpha} - a_1e^{i\alpha}).$$

This point is the centroid of the circulation found in 7·4. It should be observed that the position of the substitution vortex depends, in general, on incidence and on the ratio V/κ.

Equation (3) above, with the O-term omitted, allows us to adopt a somewhat different point of view. Remembering that the origin has been taken at the centre of the profile, we see that w is composed of three terms due to

    (i) the windstream,

    (ii) a doublet of strength $V(a^2 e^{-i\alpha} - a_1 e^{i\alpha})$ at the centre of the profile (see 3·7 (iv)),

    (iii) a vortex of strength κ also at the centre of the profile.

Thus instead of the substitution vortex at a variable point, we can use a doublet of variable strength and a vortex both at a fixed point, namely the centre of the profile.

In the case of a flat aerofoil of chord c we have $a = c/4$, $a_1 = c^2/16$ and therefore

$$z_0 = \frac{iV}{\kappa} \cdot \frac{c^2}{16}(e^{-i\alpha} - e^{i\alpha}) = \frac{V}{8\kappa}c^2 \sin\alpha.$$

The value of κ which gives zero velocity at the trailing edge is $\frac{1}{2}cV \sin\alpha$ and therefore, in this case, $z_0 = \frac{1}{4}c$; the substitution vortex is at the quarter point, i.e. at the focus.

We have seen (8·34) that for thin profiles of small camber the focus and quarter point coincide. Thus here also we may suppose the substitution vortex placed at the quarter point.

8·6. The two-dimensional biplane.

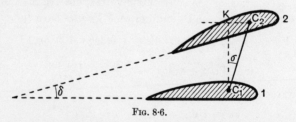

Fig. 8·6.

Fig. 8·6 shows the section of a two-dimensional biplane by a plane perpendicular to the span, giving two profiles, 1 and 2, whose chords are inclined at the angle δ called the *décalage*, which is reckoned positive when the upper wing is at greater incidence than the lower, as in the diagram. Let C_1, C_2 be the centres of the profiles. If we project the line $C_1 C_2$ on to the normal to the chord of the lower profile, the length $C_1 K$ so obtained is called the *gap* and will be denoted by h. The distance $C_2 K = s$, by which the centre of the upper profile is in advance of the centre of the lower profile, is called the *stagger*. In the diagram we have *forward stagger*, i.e. s is positive. When s is negative we have *backward stagger*. The *angle of stagger*, σ, is the angle between $C_1 C_2$ and the normal to the chord of the lower profile. reckoned positive in the case of forward stagger. The above definitions are purely geometrical and largely arbitrary, in the sense that reference points other than C_1 and C_2 could be taken.

If, instead of the chord of the lower wing, the direction of motion is taken as fundamental, the corresponding projections of C_1C_2 give what is known as the *aerodynamic gap* and *aerodynamic stagger* respectively, which are functions of incidence.

Exact theories of the two-dimensional general biplane have been founded on conformal transformations which map two circles into two aerofoil profiles. Such theories are of great mathematical interest, but inasmuch as the interference due to the trailing vortices in a biplane of finite span is of the same order of magnitude as the interaction between the two planes in the two-dimensional theory, there is nothing to be gained by entering into the necessarily complicated analytical details. We shall, therefore, be content with outlining an approximate theory which gives, at least to the second approximation, the same results as the exact analysis.

8·7. Approximate theory of the biplane.
We suppose the biplane wings to consist of thin aerofoils of small camber. To a first approximation the effect of one wing on the other can be calculated by means of a substitution vortex, placed at the quarter point of the chord. The condition that the velocity at each trailing edge shall be finite leads to two equations to determine the circulations. We can then proceed, if desired, to a second approximation using 8·5 (4), and the circulation first calculated.

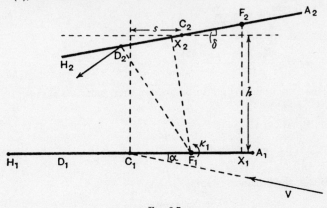

FIG. 8·7.

Fig. 8·7 shows two thin aerofoils represented by their chords c_1, c_2. The points C, F, D are points of quadrisection, so that we may take F_1, F_2 as the foci, C_1, C_2 as the centres of the profiles and D_1, D_2 as the rear stagnation points of the circles which transform into the aerofoils ; X_1, X_2 are the projections of

F_2, F_1 on the respective chords. The *décalage* δ and the incidence α are small angles whose cosines may be taken as unity. Observe that α is the incidence of the lower aerofoil and therefore $\alpha + \delta$ is the incidence of the upper.

To find the strength of the circulation κ_2 about the upper aerofoil we replace the lower aerofoil by a substitution vortex of strength κ_1 at its focus F_1 and then calculate the velocity at D_2 tangential to the circle which transforms into the upper aerofoil, as if the aerofoil were absent (see 5·311). This velocity is the component perpendicular to the chord c_2 of the velocity due to the stream and vortex. Since the radius of the circle is $c_2/4$ we get from 5·311

$$\kappa_2 = \tfrac{1}{2}c_2 \left(V(\alpha + \delta) - \kappa_1 \frac{X_2 D_2}{F_1 D_2{}^2} \right).$$

Similarly considering the lower aerofoil we get

$$\kappa_1 = \tfrac{1}{2}c_1 \left(V\alpha - \kappa_2 \frac{X_1 D_1}{F_2 D_1{}^2} \right).$$

If we take C_1 as origin and the chord c_1 as x-axis, we can write down the coordinates of the points as follows:

$$F_1(\tfrac{1}{4}c_1, 0), \quad D_1(-\tfrac{1}{4}c_1, 0), \quad X_1(s + \tfrac{1}{4}c_2, 0),$$
$$F_2(s + \tfrac{1}{4}c_2, h + \tfrac{1}{4}c_2\delta), \quad D_2(s - \tfrac{1}{4}c_2, h - \tfrac{1}{4}c_2\delta), \quad X_2(\tfrac{1}{4}c_1 - h\delta, h - s\delta + \tfrac{1}{4}c_1\delta),$$

where squares and higher powers of δ are neglected.

If we introduce the mean chord $c = \tfrac{1}{2}(c_1 + c_2)$, the equations for the circulations can be written

$$(1) \qquad \frac{2\kappa_1}{Vc_1} = \alpha - \frac{2\kappa_2}{Vc_2} \cdot \mu_2, \quad \frac{2\kappa_2}{Vc_2} = \alpha + \delta - \frac{2\kappa_1}{Vc_1} \cdot \mu_1,$$

where

$$(2) \qquad \mu_1 = \frac{\tfrac{1}{2}c_1(\tfrac{1}{2}c - s - h\delta)}{h^2 + (\tfrac{1}{2}c - s)^2 - \tfrac{1}{2}c_2 h\delta}, \quad \mu_2 = \frac{\tfrac{1}{2}c_2(\tfrac{1}{2}c + s)}{h^2 + (\tfrac{1}{2}c + s)^2 + \tfrac{1}{2}c_2 h\delta}.$$

In a biplane it is usual for h/c to be greater than unity, so that μ_1, μ_2 are of order $c^2/4h^2$ at most.

Solving equations (1) we get

$$(3) \qquad \kappa_1 = \tfrac{1}{2}c_1 V \left(\frac{\alpha - \mu_2(\alpha + \delta)}{1 - \mu_1\mu_2} \right), \quad \kappa_2 = \tfrac{1}{2}c_2 V \left(\frac{\alpha + \delta - \mu_1\alpha}{1 - \mu_1\mu_2} \right),$$

and the lift on the biplane system is

$$L = 2\pi(\kappa_1 + \kappa_2)\rho V.$$

8·8. Equal biplane.

In the case of an equal unstaggered biplane without *décalage* we have

$$c_1 = c_2 = c, \quad s = 0, \quad \delta = 0, \quad \mu_1 = \mu_2 = \frac{c^2}{4h^2 + c^2},$$

and therefore from 8·7 (3),

$$(1) \qquad \kappa_1 = \kappa_2 = \tfrac{1}{2}cV\alpha \, \frac{1}{1 + \mu_1} = \tfrac{1}{2}cV\alpha \, \frac{1 + \dfrac{c^2}{4h^2}}{1 + \dfrac{c^2}{2h^2}},$$

so that the lift on the system is

$$(2) \qquad 2\pi\rho V(\kappa_1 + \kappa_2) = 2\pi\rho cV^2\alpha \, \frac{1 + \dfrac{c^2}{4h^2}}{1 + \dfrac{c^2}{2h^2}},$$

which is less than the lift would be if each plane were isolated ($h = \infty$), when there would be no interference effect, in other words the biplane has to fly at a greater incidence to produce the same lift as a monoplane. To obtain some idea of the nature of the interference effect which operates when the planes are not isolated, let us suppose that c/h is so small that we can neglect powers higher than the square. We can then replace both planes simultaneously by the substitution vortices (1) placed at the foci.

These vortices then repel one another with the force, calculated in 4·5,

$$F = 2\pi\rho\kappa_1\kappa_2/h = \tfrac{1}{2}\pi\rho c^2 V^2\alpha^2/h$$

to the order of approximation assumed. Each vortex also experiences the ordinary lift

$$L = 2\pi\kappa\rho V,$$

where L has half the value given by (2), so that we can write

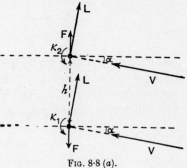

Fig. 8·8 (a).

$$\frac{F}{L} = \frac{1}{2}\frac{c\alpha}{hk}, \quad k = \frac{1 + \dfrac{c^2}{4h^2}}{1 + \dfrac{c^2}{2h^2}} = 1 - \frac{c^2}{4h^2}.$$

The upper plane therefore experiences the lift

$$L_2 = L + F\cos\alpha = L\left(1 + \frac{1}{2}\frac{c\alpha}{hk}\right),$$

while the lower plane experiences the lift

$$L_1 = L - F\cos\alpha = L\left(1 - \frac{1}{2}\frac{c\alpha}{hk}\right).$$

Thus the interference causes an increase of lift on the upper plane and an equal decrease on the lower plane.

We also notice that the upper plane experiences a drag $F \sin \alpha = F\alpha$, which is counterbalanced by an equal *negative* drag on the lower plane.

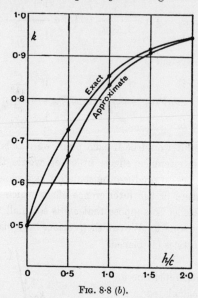

FIG. 8·8 (b).

Lastly, we give in fig. 8·8 (b) graphs to show the small divergence which exists between the approximate formula (2) and the result of exact two-dimensional theory applied to the biplane of this section.

The ordinates give the factor k mentioned above.

EXAMPLES VIII

1. An aerofoil profile is in the form of the segment of the parabola $4hx^2 = c^2(h - y)$ cut off by the x-axis. Determine the camber line function and the thickness function.

2. In the case of the elliptic aerofoil $x = -\frac{1}{2}c \cos \theta$, $y = \frac{1}{2}hc \sin \theta$, express dy/dx in the form

$$-\tfrac{1}{2}\lambda \tan \tfrac{1}{2}\theta + \tfrac{1}{2}\tau \cot \tfrac{1}{2}\theta + F(\theta)$$

and determine λ and τ.

Verify the relations $\lambda^2 = 2\rho_A/c$, $\tau^2 = 2\rho_H/c$. Show that by taking h small enough the aerofoil approximates to a flat plate.

3. If the function $F(\theta)$ can be expanded in the Fourier series,

$$\tfrac{1}{2}A_0 + \sum_{n=1}^{\infty} (A_n \cos n\theta + B_n \sin n\theta),$$

show that

$$A_n + iB_n = \frac{1}{\pi} \int_{-\pi}^{\pi} F(\theta) e^{in\theta}\, d\theta.$$

4. Prove that $z = \frac{1}{4}(\zeta + 1/\zeta)$ transforms the region exterior to the circle $\zeta\bar{\zeta} = 1$ into the region exterior to the line joining $(-\frac{1}{2}, 0)$ to $(\frac{1}{2}, 0)$.

5. Show that the circulation for a flat aerofoil may be considered distributed at the rate

$$2\pi k = 2V \sin \alpha \sqrt{\frac{c + 2x}{c - 2x}}$$

at the point $(x, 0)$, with the usual axes of reference.

6. Draw a graph to show the pressure distribution arising from the combination of the pressures given in 8·43 (1) and 8·44 (1).

7. Calculate λ_1, which gives the incidence for zero lift, for the parabolic aerofoil of Ex. 1.

8. Calculate λ_1 and λ_2 for a circular arc aerofoil of small camber.

9. Show that the moment coefficient about the leading edge of a thin aerofoil is

$$C_m = -\tfrac{1}{4}C_L - \tfrac{1}{4}\pi(A_2 + A_1),$$

and deduce the relation between the (C_m, α) and (C_L, α) graphs.

10. Show that the condition for the existence of a fixed centre of pressure can be deduced from consideration of the travel of the centre of pressure.

11. Calculate the mean camber of the cubic profile of 8·36.

12. Draw the travel of centre of pressure graph when Axis II is above Axis I. Prove that as incidence increases the centre of pressure moves towards the quarter point, whether Axis II is above or below Axis I.

13. If γ is the angle between Axis I and Axis II, show that if $\gamma < 0$, the pitching moment about the focus tends to increase the incidence, whereas if $\gamma > 0$, it tends to decrease the incidence.

14. If γ is the angle between Axis I and Axis II of a thin aerofoil, show that $\sin 2\gamma$ is proportional to $\lambda_1 - \lambda_2$.

15. If C_p is the centre of pressure coefficient and the origin is taken at the quarter point, prove that the equation of the line of action of the resultant aerodynamic force is

$$x + c(C_p - \tfrac{1}{4}) = \alpha y,$$

where α is the incidence to the chord, considered as small.

16. In Ex. 15 use 8·35 to show that the equation of the line of action of the lift may be written in the form

$$\alpha^2 y + \alpha(\lambda_1 y - x) - x\lambda_1 - \tfrac{1}{4}c(\lambda_1 - \lambda_2) = 0.$$

By expressing the condition that the above quadratic in α has equal roots, or otherwise, prove that the line of action of the lift envelops the metacentric parabola

$$(\lambda_1 y + x)^2 = -cy(\lambda_1 - \lambda_2).$$

Trace this parabola when $\lambda_1 > \lambda_2$ and when $\lambda_1 < \lambda_2$.

17. Draw graphs to show the change in C_L and C_m as functions of the eccentric angle of the hinge for a given flap angle ξ.

18. Show that the elliptic pressure distribution (at the ideal angle of attack)

$$x = -\tfrac{1}{2}\cos\theta, \quad p = -\frac{2C_L}{\pi}\sin\theta,$$

leads to the parabolic camber line function

$$y_C = \frac{C_L}{4\pi}(1 - 4x^2) \qquad \text{(Brennan and Stevenson.)}$$

19. Show that the pressure distribution

$$p = -\frac{2C_L}{\pi} \sin \theta \{1 - \tfrac{3}{2}k \cos \theta\}$$

leads to the cubic camber line

$$y = \frac{C_L}{4\pi} (1 - 4x^2)(1 + 2kx),$$

and that there is a fixed centre of pressure if $k = 4/3$. (B. and S.)

20. Find the pressure distribution which yields the quartic camber line

$$y_C = \frac{C_L}{\pi} \{1 - 4x^2\} \{a + bx + cx^2\}.$$

21. Show that constant pressure all round the profile leads to the elliptic thickness function

$$y_T = \tfrac{1}{2}(2\rho)^{\frac{1}{2}} \sin \theta,$$

where ρ is the radius of curvature at the leading and trailing edges.

(Brennan and Stevenson.)

22. Show that the pressure distribution

$$p = -(\lambda + \tau) + 2(\lambda - \tau) \cos \theta$$

leads to the thickness function

$$y_T = \tfrac{1}{4} \sin \theta \{(\lambda + \tau) - (\lambda - \tau) \cos \theta\}.$$ (B. and S.)

23. For the pressure distribution

$$p = -0.16\{1 - \tfrac{3}{2} \cos \theta\},$$

show that $\rho_A = 0.0098$, $\rho_H = 0.0002$.

Determine the thickness function and draw the profile. (B. and S.)

24. Prove that the substitution vortex of an aerofoil should be placed at the centroid of the circulation.

Hence show that the locus of possible positions of the substitution vortex is a straight line.

25. Show that the unstaggered biplane arrangement of two equal aerofoils reduces the circulation round each in the ratio $(1 - c^2/4h^2) : 1$ compared with the monoplane arrangement at the same incidence.

CHAPTER IX

INDUCED VELOCITY

9·1. Vector notation. In Chapter IV we considered the properties of two-dimensional vortex motion due to rectilinear vortices. It there appeared that the vorticity vector is perpendicular to the plane of the motion and that the existence of a rectilinear vortex implies the co-existence of a certain velocity distribution called the induced velocity. In the present chapter we investigate the properties of vorticity when the motion is no longer necessarily two-dimensional, and, in order to do this in the most physically intuitive manner it is desirable to employ the language and notation of vectors. Those readers to whom the subject of vector analysis is unfamiliar will find the necessary details in Chapter XXI. As there is some diversity in the notations used by different writers, we make a few preliminary observations.

If **a** and **b** are two vectors, we shall denote by **ab** their scalar product and by $\mathbf{a}_\wedge \mathbf{b}$ their vector product (see 21·12, 21·13).

The vector differentiation operator ∇ operating (21·3) on a scalar ϕ or a vector **a** yields

$$\nabla \phi = \operatorname{grad} \phi, \quad \nabla \mathbf{a} = \operatorname{div} \mathbf{a}, \quad \nabla_\wedge \mathbf{a} = \operatorname{curl} \mathbf{a}.$$

In cartesian coordinates x, y, z, with corresponding unit vectors **i**, **j**, **k**,

$$(1) \qquad \nabla = \mathbf{i}\frac{\partial}{\partial x} + \mathbf{j}\frac{\partial}{\partial y} + \mathbf{k}\frac{\partial}{\partial z}.$$

The fluid velocity **q** is

$$(2) \qquad \mathbf{q} = \mathbf{i}u + \mathbf{j}v + \mathbf{k}w.$$

The position vector of the point (x, y, z) is

$$(3) \qquad \mathbf{r} = \mathbf{i}x + \mathbf{j}y + \mathbf{k}z,$$

and therefore

$$(4) \qquad d\mathbf{r} = \mathbf{i}\,dx + \mathbf{j}\,dy + \mathbf{k}\,dz.$$

It follows that the change in a scalar function ϕ in going from **r** to $\mathbf{r} + d\mathbf{r}$ is

$$(5) \qquad d\phi = \frac{\partial \phi}{\partial x}\,dx + \frac{\partial \phi}{\partial y}\,dy + \frac{\partial \phi}{\partial z}\,dz = (d\mathbf{r}\,\nabla)\phi = d\mathbf{r} \cdot \nabla \phi,$$

where $d\mathbf{r}\,\nabla$ denotes the scalar product of (1) and (4).

9·2. The equation of motion.

In 1·41 it was proved that the component force due to pressure thrust on an infinitesimal volume of fluid $\delta\tau$, in the direction of the line element δs, is $-(\partial p/\partial s)\,\delta\tau$. Another way of stating

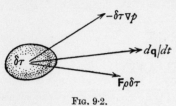

FIG. 9·2.

this fact is to say that the resultant thrust is the vector $-(\nabla p)\,\delta\tau$, and the above-mentioned force is the component of this vector in the direction of δs. Now consider a small volume $\delta\tau$ of air considered as an inviscid fluid.

If $\mathbf{q}$ is the velocity, $d\mathbf{q}/dt$ is the acceleration, and if $\mathbf{F}$ is the external force per unit *mass*, the external force on the volume is $\mathbf{F}\rho\,\delta\tau$ where ρ is the density. By Newton's second law of motion we have therefore

$$\mathbf{F}\rho\,\delta\tau - \delta\tau\,\nabla p = (\rho\,\delta\tau)\frac{d\mathbf{q}}{dt},$$

whence the equation of motion

$$\frac{d\mathbf{q}}{dt} = -\frac{1}{\rho}\nabla p + \mathbf{F}.$$

In the case of natural air the external force is due to gravity and is therefore the negative gradient of a potential * function Ω say, so that

$$\mathbf{F} = -\nabla\Omega,$$

and therefore the equation of motion is

(1)
$$\frac{d\mathbf{q}}{dt} = -\frac{1}{\rho}\nabla p - \nabla\Omega.$$

This equation holds whether the air is regarded as compressible or incompressible.

Another method of deriving this equation is given in 21·6.

9·21. Kelvin's circulation theorem.

When the external forces are conservative and derived from a one-valued potential, the circulation in any circuit which moves with the fluid (i.e. which always consists of the same fluid particles) is independent of the time.

Proof.

If C is the circuit, then from 21·12,

$$\operatorname{circ} C = \int_{(C)} \mathbf{q}\,d\mathbf{r}.$$

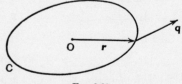

FIG. 9·21.

---

* If z is the height above a fixed level, $\Omega = gz$, where g is the acceleration due to gravity, and the force per unit mass is $-\nabla(gz) = -\mathbf{k}g$ using 9·1 (1), i.e. g vertically downwards, a well-known result.

Since $\mathbf{q} = d\mathbf{r}/dt$, we have

$$\frac{d}{dt}(\text{circ } C) = \int_{(C)} \left(\frac{d\mathbf{q}}{dt} \, d\mathbf{r} + \mathbf{q} \, d\mathbf{q} \right).$$

From the equation of motion, 9·2 (1),

$$\frac{d\mathbf{q}}{dt} = -\frac{1}{\rho} \nabla p - \nabla \Omega,$$

and therefore taking the scalar product by $d\mathbf{r}$ and using 9·1 (5)

$$\frac{d\mathbf{q}}{dt} \, d\mathbf{r} = -\frac{dp}{\rho} - d\Omega$$

and so

$$\frac{d}{dt}(\text{circ } C) = \text{change in} \left(-\int_{p_0}^{p} \frac{dp}{\rho} - \Omega + \tfrac{1}{2}q^2 \right)$$

on going once round the circuit, and this change is zero, for all the quantities on the right return to their initial values. Thus circ C is independent of t.

Q.E.D.

9·3. Vorticity. Let $\mathbf{q}$ be the velocity of the fluid at any point. The *vorticity vector* is defined by

(1)　　　　　　　　$\boldsymbol{\zeta} = \text{curl } \mathbf{q} = \nabla \wedge \mathbf{q}.$

In cartesian coordinates the components of vorticity are

(2)　　　　$\dfrac{\partial w}{\partial y} - \dfrac{\partial v}{\partial z}, \quad \dfrac{\partial u}{\partial z} - \dfrac{\partial w}{\partial x}, \quad \dfrac{\partial v}{\partial x} - \dfrac{\partial u}{\partial y}$

parallel to the axes of reference, where u, v, w are the corresponding components of velocity (see 9·1 (2)). Thus it appears that, when the velocity field is known, the corresponding vorticity can be calculated.

Conversely, given the vorticity field $\boldsymbol{\zeta}$ of a flow in which there are no singularities other than vortices, there is a corresponding velocity field $\mathbf{q}$ which (see 9·4) is uniquely determined by the given vorticity field.

Definition. The velocity field which coexists with a given distribution of vorticity and vanishes with it is called the *induced velocity field* (cf. 4·1).

Thus if $\boldsymbol{\zeta}$, $\mathbf{q}_i$ are the vorticity and induced velocity vectors at a point, we have

(3)　　　　　　　　$\boldsymbol{\zeta} = \nabla \wedge \mathbf{q}_i.$

Notice that $\boldsymbol{\zeta}$ and $\mathbf{q}_i$ occur *together* but neither can properly be said to *cause* the other.

Put in another way, the induced velocity field is the velocity field consistent with the existing vorticity distribution and without other singularities.

It should be observed that the induced velocity field may be, and often is, irrotational. A very simple example of this is the field round a rectilinear vortex (4·2). It is the circulation in any circuit embracing the vortex which gives the required intensity.

Returning to (3), if $\mathbf{q}_0$ is a velocity such that $\nabla_{\wedge} \mathbf{q}_0 = 0$, i.e. if $\mathbf{q}_0$ is the velocity of an irrotational flow, we still have

$$\boldsymbol{\zeta} = \nabla_{\wedge} (\mathbf{q}_i + \mathbf{q}_0) = \nabla_{\wedge} \mathbf{q}_i.$$

It follows that any existing velocity field can be regarded as composed of two parts : (i) the induced velocity field, (ii) an irrotational field ; and the velocity at any point is the sum of the velocities due to (i) and (ii).

It can be immediately verified from (2) and the cartesian form of ∇ (see 19·3 (10)) that

(4) $\nabla \boldsymbol{\zeta} = 0.$

9·31. Vortex lines, tubes and filaments.   A *vortex line* is a line whose tangent at each point is in the direction of the vorticity vector at that point.

FIG. 9·31 (a).

If through each point of the boundary of a closed curve we draw the vortex line which passes through that point, the lines so drawn constitute the surface of a *vortex tube* of which the curve is a cross-section.

A *vortex filament* is a vortex tube whose cross-section is of infinitesimal maximum dimensions.

Consider a portion AB of a vortex filament, the cross-sections of the tube at A and B being of areas σ_1, σ_2 respectively. By Gauss's theorem (19·4) applied to the surface and volume of the tube AB, we have

FIG. 9·31 (b).

$$\int \boldsymbol{\zeta} \mathbf{n} \, dS = \int \nabla \boldsymbol{\zeta} \, d\tau = 0,$$

where $\mathbf{n}$ is the unit normal vector (drawn outwards) at the element dS, since, from 9·3 (4), $\nabla \boldsymbol{\zeta} = 0$. Also $\boldsymbol{\zeta} \mathbf{n} = 0$ on the curved surface of the tube, for $\boldsymbol{\zeta}$ is tangential to this surface. Therefore the only contributions arise from the ends, i.e.

$$\boldsymbol{\zeta}_1 \mathbf{n}_1 \sigma_1 + \boldsymbol{\zeta}_2 \mathbf{n}_2 \sigma_2 = 0,$$

and so

$$\zeta_1 \sigma_1 = \zeta_2 \sigma_2.$$

This result may be called the "equation of continuity" for a vortex filament (cf. 1·23). It means that the product of the magnitude of the vorticity and the cross-sectional area, which may be called the *intensity*, is constant along the filament. This is Helmholtz's *second theorem*. It follows from this theorem that a vortex filament cannot terminate in the interior of the fluid for if it did, σ would have to vanish and therefore ζ would have to be infinite. Thus we see that a vortex filament, and therefore a vortex line, must either be closed (vortex ring) or must terminate on the boundary. Fig. 9·31 (c) shows a somewhat fanciful picture of the possible arrangements.

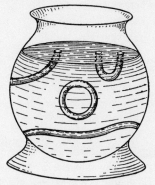

A vortex tube may be regarded as a bundle of vortex filaments, so that we may use the term *intensity of a vortex tube* in the sense that it is the sum of the intensities of the filaments which compose the tube. It follows that the intensity of a vortex tube remains constant along the tube.

Notice that steady motion is not assumed.

We now prove the *third* and *fourth* theorems of Helmholtz, which are as follows :

III. The fluid which forms a vortex tube continues to form a vortex tube.

Fig. 9·31 (c).

IV. The intensity of a vortex tube remains constant as the tube moves about.

Proof. Let A, B be a portion of a vortex tube. Draw a closed curve C on

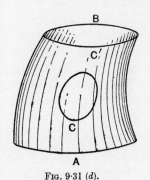

the curved surface of the tube. The normal component of the vorticity over the surface enclosed by C is zero, for all the vortex vectors are tangential to this surface. Therefore (21·7) by Stokes's theorem circ $C = 0$. As the tube moves about, by Kelvin's circulation theorem circ C remains zero. Thus the curved surface enclosed by C, although it may deform, remains on the surface of the vortex tube, for no vortex lines can pass through it. This proves III.

Fig. 9·31 (d).

Again, if we take a section of the tube whose bounding curve is a circuit C', by Kelvin's theorem circ C' remains constant as the tube moves about. But circ C' is the sum of the intensities of the vortex filaments which compose the tube. This proves IV.

9·4. The law of induced velocity.
Consider the *whole of space* * to be filled with fluid which is at rest at infinity in the sense that at great distances r the speed q is of order $1/r^2$ at least. Then it can be proved † that if ζ_Q is the

---

* This implies that the fluid has no rigid boundaries either internal, such as an aerofoil, or external, such as containing walls.

† Milne-Thomson, *Theoretical Hydrodynamics*, 18·22.

vorticity vector at Q, and if $d\tau_Q$ is an element of volume at Q, the induced velocity $\mathbf{q}_P$ at P is given by

(1)
$$\mathbf{q}_P = \frac{1}{4\pi}\int \zeta_Q \wedge \frac{\overrightarrow{QP}}{QP^3}\,d\tau_Q,$$

where the integral is taken throughout the fluid, assumed incompressible. This serves to define uniquely the induced velocity coexistent with the given vorticity field.

This result admits of a striking and simple formulation, which must, however, be regarded as entirely conventional.

Let us regard the velocity vector ζ as a (vector) density of vorticity per unit volume and call $\zeta_Q\,d\tau_Q$ the *amount* of vorticity in the volume $d\tau_Q$.

If we write

(2)
$$d\mathbf{q}_P = \zeta_Q\,d\tau_Q \wedge \frac{\overrightarrow{QP}}{4\pi QP^3},$$

we can say that the fluid element $d\tau_Q$ behaves as an amount of vorticity $\zeta_Q\,d\tau_Q$ which induces at P the element of velocity $d\mathbf{q}_P$ given by (2), and that by summing all the elementary velocities so determined we arrive at the complete velocity $\mathbf{q}_P$. If in fig. 9·4 we imagine ζ_Q in the plane of the paper, $d\mathbf{q}_P$ is directed into the paper.

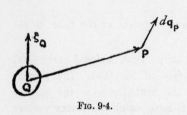

FIG. 9·4.

This way of looking at (1) is entirely analogous to the law of Biot and Savart concerning the magnetic field due to an electric current and is therefore sometimes called by their names. The analogy is no mere accident, but has its root in the fact that the mathematical formulation of the two cases is the same, but with a different physical meaning assigned to the symbols.

The formula (2) recalls also (21·13) the formula $\boldsymbol{\omega}_Q \wedge \overrightarrow{QP}$ which gives the velocity at the point P of a rigid body which has angular velocity $\boldsymbol{\omega}_Q$ about an axis through Q. In this comparison we could write

$$\boldsymbol{\omega}_Q = \frac{\zeta_Q\,d\tau_Q}{4\pi QP^3},$$

and in this connection the reader is invited to compare the interpretation of vorticity in terms of angular velocity as given in Helmholtz's first theorem (see 3·22).

It is important to get the law of induced velocity as formulated in (2) in the right perspective. If ten men together contribute a total of fifty shillings, no inference can be drawn as to their individual contributions, but it would be true to say, for example, that the result is the same as if five of the men each contributed seven shillings and the remaining five each contributed three shillings. In the same way all we know is that the resultant velocity is $\mathbf{q}_P$ as given by (1), and the induction law simply states that this final result would be attained if we supposed each fluid element $d\tau_Q$ to contribute the amount (2), and then added all the results. The statement certainly does not imply that $d\tau_Q$ does, in fact, contribute the amount (2), although there is often a tendency to think that it does. This impression must be guarded against ; otherwise an improper physical picture may be imagined. It is perfectly justifiable to calculate the partial assumed contributions of the elements of the fluid, but it is only their total (vector) sum which can be asserted to have physical reality. Any subsequent statements which may have the appearance of implying the contrary must be interpreted in the light of this explanation.*

Finally, we recall the statement in 9·3 that vorticity cannot properly be said to " cause " induced velocity.

9·5. Velocity induced by a vortex filament.

Consider an element ds of a vortex filament at the point Q. Let σ be the cross-sectional area at Q ζ the magnitude of the vorticity and $\mathbf{t}$ a unit vector along the tangent to the filament. Then $\boldsymbol{\zeta} = \zeta\mathbf{t}$, and the law of induced velocity gives for the velocity induced at P by the element

$$d\mathbf{q}_P = \frac{\zeta\mathbf{t}\sigma\,ds}{4\pi QP^3} \wedge \overrightarrow{QP}.$$

Now by Helmholtz's second theorem, $\zeta\sigma = K$ is the constant intensity of the filament. Thus

$$d\mathbf{q}_P = \frac{K}{4\pi}\,ds\,\mathbf{t} \wedge \frac{\overrightarrow{QP}}{QP^3},$$

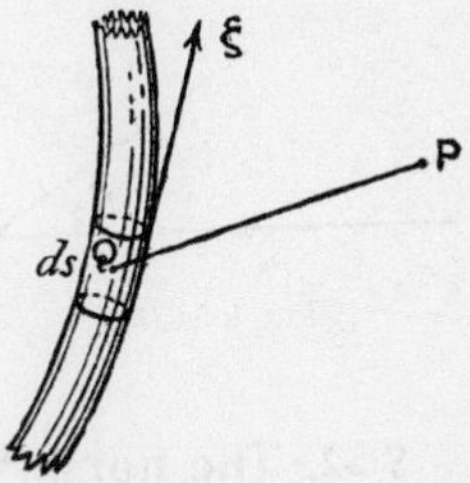

Fig. 9·5.

and $\mathbf{q}_P$ is obtained by integrating this expression along the filament. Here we can regard $K\mathbf{t}$ as the (vector) density of vorticity per unit length of filament.

9·51. Straight filament.

If a vortex filament contains a straight portion AB, we can calculate the contribution of that portion to the induced velocity integral.

If K is the intensity of the filament and AB the direction of the vorticity vector, the velocity induced at P by the element $MM' = ds$ is directed out of the plane of the paper and is of magnitude

$$dq_P = \frac{K}{4\pi}\,ds\,\frac{PM\sin\alpha}{PM^3} = \frac{K}{4\pi h^2}\sin^3\alpha\,ds,$$

---

* *There was a strong man on a Syndicate,*
  Who loved the exact truth to vindicate ;
    He rose to deny
    That his words could imply
  What their sense seemed intended to indicate.

T. R. GLOVER, *Cambridge Retrospect.*

where $h = PN$ is the distance of P from AB and α is the angle PMB. Now

$$ds = dAM = d(AN - MN) = d(-h\cot\alpha) = h\operatorname{cosec}^2\alpha\,d\alpha.$$

Thus $\qquad q_P = \dfrac{K}{4\pi h}\displaystyle\int_{\theta_1}^{\pi-\theta_2}\sin\alpha\,d\alpha = \dfrac{K}{4\pi h}(\cos\theta_1 + \cos\theta_2),$

where $\theta_1 = P\hat{A}B$ and $\theta_2 = P\hat{B}A$.

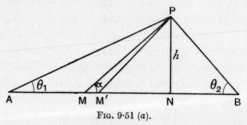

FIG. 9·51 (a).

Special Cases.

(i) Infinite filament $q_P = 2K/4\pi h = K/2\pi h$ as already obtained for an ordinary rectilinear vortex.

(ii) Semi-infinite filament

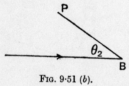

$$q_P = \frac{K}{4\pi h}(1 + \cos\theta_2).$$

FIG. 9·51 (b).

(iii) If PB is perpendicular to AB, $q_P = K/4\pi h$, which is half the result of (i), as is obvious from first principles.

9·52. The horseshoe vortex. This name is given to a vortex system consisting of three sides of a rectangle, two sides being of infinite length.

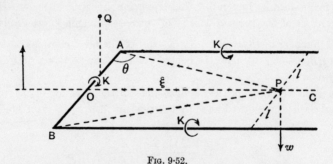

FIG. 9·52.

The induced velocity at any point can be obtained by applying the results of 9·51. Referring to fig. 9·52, let the finite side of the rectangle be $AB = 2l$, and let there be constant circulation K round all three arms of the "horse-shoe". Consider a point P in the plane of the system and on the perpendicular bisector OC of AB. Then if $OP = \xi$ we have for the velocity induced at P the value

$$w = \frac{K}{4\pi\xi}(\cos\theta + \cos\theta) + 2 \times \frac{K}{4\pi l}(1 + \sin\theta),$$

in a direction normal to the plane of the vortex and downwards.

Since $\cos\theta = l/\sqrt{(l^2 + \xi^2)}$, $\sin\theta = \xi/\sqrt{(l^2 + \xi^2)}$, we get at once

$$w = \frac{K}{2\pi l}\left\{1 + \frac{\sqrt{(l^2 + \xi^2)}}{\xi}\right\}.$$

This decreases from large values when ξ is small to the constant value $K/(\pi l)$ when $\xi \to \infty$. If P is not, as shown in fig. 9·52, within the arms but on the other side of AB, ξ is negative and w changes sign, being directed upwards, and as we go further from AB w decreases; finally to zero. To anticipate, the sign of w distinguishes downwash (11·21) from upwash.

If we consider points in the plane through AB perpendicular to the plane of the vortex it is clear from the induction law that at a point Q the only component of velocity parallel to OC is contributed by AB and is directed in the same sense as OC above the plane of the vortex and in the opposite sense below that plane. On the other hand the components of velocity in the plane QAB are due only to the infinite arms of the horseshoe. If we consider points on AB itself, AB makes no contribution and the velocity distribution along AB is just half that due to rectilinear vortices of strengths $K/2\pi$ at A and B. The distribution is shown in fig. 4·21 (b). The calculation of the components at other points offers no special difficulties but is somewhat complicated (see Ex. IX, 13, 14).

9·6. Vortex sheet. In 4·2 we defined a point rectilinear vortex as the idealised limit of a cylindrical region of vorticity whose cross-section shrinks to a point while the amount of vorticity remains unaltered. We use an analogous process in defining a vortex sheet.

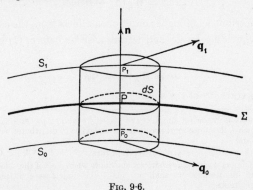

Fig. 9·6.

In fig. 9·6 **n** is the unit normal vector at the point P of a surface Σ. Let ϵ be an infinitesimal positive scalar and consider the points P_1, P_0 whose position vectors referred to P are $\frac{1}{2}\epsilon\mathbf{n}$, $-\frac{1}{2}\epsilon\mathbf{n}$ respectively. As P describes the surface Σ the points P_1, P_0 describe surfaces S_1, S_0 parallel to Σ which is halfway between them. Take an infinitesimal element of area of Σ, say dS, whose centroid is P. The normals to Σ at the boundary of dS, together with the surfaces S_1, S_0 will delimit a cylindrical element of volume $d\tau = \epsilon\, dS$.

Now imagine the above surfaces to be drawn in fluid which is moving irrotationally* everywhere except in that part which lies between S_1 and S_0. Let $\boldsymbol{\zeta}$ be the vorticity vector at P. Then we can write $\boldsymbol{\zeta}\,d\tau = \boldsymbol{\zeta}\epsilon\,dS = \boldsymbol{\omega}\,dS$, where

$$(1) \qquad\qquad \boldsymbol{\omega} = \boldsymbol{\zeta}\epsilon.$$

If we now let $\epsilon \to 0$, $\boldsymbol{\zeta} \to \infty$, in such a way that $\boldsymbol{\omega}$ remains unaltered, the surface Σ is called a *vortex sheet* of vorticity $\boldsymbol{\omega}$ per unit area.

Notice that in the above process the amount (9·4) of vorticity in the layer between S_1 and S_0 remains unaltered as the layer shrinks in thickness ($\epsilon \to 0$), and we can regard $\boldsymbol{\omega}$ as a (vector) density of vorticity per unit area.

The dimensions of $\boldsymbol{\omega}$ are LT^{-1}.

Before the passage to the limit, the velocity will be continuous throughout the fluid and if $\mathbf{q}$, $\mathbf{q}_1$, $\mathbf{q}_0$ are the velocities at P, P_1, P_0 we have †

$$(2) \qquad\qquad \mathbf{q}_1 = \mathbf{q} + \tfrac{1}{2}\epsilon(\mathbf{n}\,\nabla)\mathbf{q}, \qquad \mathbf{q}_0 = \mathbf{q} - \tfrac{1}{2}\epsilon(\mathbf{n}\,\nabla)\mathbf{q},$$

whence by addition

$$(3) \qquad\qquad \mathbf{q} = \tfrac{1}{2}(\mathbf{q}_0 + \mathbf{q}_1).$$

This result is true however small ϵ may be. Thus the velocity of a point P of a vortex sheet is the arithmetic mean of the velocities of the points just above and just below P on the normal at P.

If we apply the particular form 21·4 (4) of Gauss's theorem to the elementary cylinder of volume $d\tau$ in fig. 9·6, we get, approximately,

$$\boldsymbol{\zeta}\epsilon\,dS = \mathbf{n}_\wedge(\mathbf{q}_1 - \mathbf{q}_0)dS,$$

neglecting a contribution of higher order of smallness from the curved surface of the cylinder. Dividing by dS and letting $\epsilon \to 0$ as before, (1) gives the exact result

$$(4) \qquad\qquad \boldsymbol{\omega} = \mathbf{n}_\wedge(\mathbf{q}_1 - \mathbf{q}_0)$$

for the surface vorticity of the sheet.‡

---

* This hypothesis is made for simplicity of statement and to preserve the analogy with 4·2. The definition which follows need not preclude the existence of distributed vorticity outside the layer between S_1 and S_0.

† From 21·3 (13) putting $d\mathbf{r} = \frac{1}{2}\epsilon\mathbf{n}$ and $-\frac{1}{2}\epsilon\mathbf{n}$ in turn.

‡ Observe that in (4) $\mathbf{q}_1$ and $\mathbf{q}_0$ are strictly the limits when $\epsilon \to 0$ of the velocities at P_1 and P_0. Since ϵ is already infinitesimal we shall not make the unnecessary distinction between the velocities and their limits.

It is clear from this that a non-zero value of $\boldsymbol{\omega}$ is associated with a discontinuity of the components of $\mathbf{q}_0$, $\mathbf{q}_1$ perpendicular to $\mathbf{n}$. It follows that *a surface across which the tangential velocity changes abruptly is a vortex sheet.*

It also appears from (4) that $\boldsymbol{\omega}$ is perpendicular to $\mathbf{n}$ and is therefore tangential to the vortex sheet.

Note also that we can apply the induction law 9·4 (1) to find uniquely the velocity induced by a vortex sheet if we replace $\boldsymbol{\zeta}_Q \, d\tau_Q$ by $\boldsymbol{\omega}_Q \, dS_Q$, thereby converting 9·4 (1) into an integral over the surface of the vortex sheet.

9·61. Velocity of a point of a vortex sheet.

Referring to the vortex sheet Σ of fig. 9·6, since only the tangential components of $\mathbf{q}_1$, $\mathbf{q}_0$ contribute to the surface vorticity $\boldsymbol{\omega}$, and since the normal components must be continuous, for this is the condition that the layer or sheet shall subsist, we have

$$(1) \qquad\qquad \mathbf{n}(\mathbf{q}_1 - \mathbf{q}_0) = 0.$$

Now from the triple vector product rule (21·2) we have

$$\mathbf{n}_\wedge \boldsymbol{\omega} = \mathbf{n}_\wedge [\mathbf{n}_\wedge (\mathbf{q}_1 - \mathbf{q}_0)] = -(\mathbf{q}_1 - \mathbf{q}_0)\mathbf{n}\mathbf{n} + \mathbf{n}[\mathbf{n}(\mathbf{q}_1 - \mathbf{q}_0)],$$

so that, using (1),

$$(2) \qquad\qquad \boldsymbol{\omega}_\wedge \mathbf{n} = \mathbf{q}_1 - \mathbf{q}_0$$

Therefore from 9·6 (3)

$$(3) \qquad\qquad \mathbf{q}_1 = \mathbf{q} + \tfrac{1}{2}\boldsymbol{\omega}_\wedge \mathbf{n}, \qquad \mathbf{q}_0 = \mathbf{q} - \tfrac{1}{2}\boldsymbol{\omega}_\wedge \mathbf{n}.$$

Hence for any vortex sheet *at rest* ($\mathbf{q}=0$ at every point of the sheet) the induced velocities at adjacent points on opposite sides of the sheet are equal but opposite vectors.

The element dS of the sheet cannot induce any velocity at its centroid P (cf. 4·1), and so $\mathbf{q}$, the velocity of P, must be the velocity induced at P by the rest of the sheet excluding the infinitesimal element dS, while $\tfrac{1}{2}\boldsymbol{\omega}_\wedge \mathbf{n}$ and $-\tfrac{1}{2}\boldsymbol{\omega}_\wedge \mathbf{n}$ are the velocities induced at P_1 and P_0 by dS.

Consider the special case in which the fluid *speeds* are the same at P_1 and P_0, i.e. $q_1 = q_0$, so that $(\mathbf{q}_1 - \mathbf{q}_0)(\mathbf{q}_1 + \mathbf{q}_0) = 0$. Then either (i) $\mathbf{q}_1 - \mathbf{q}_0 = 0$ which gives no vortex sheet, a trivial case which we ignore, or (ii) $\mathbf{q}_0 + \mathbf{q}_1 = 0$, i.e. the vortex sheet is at rest by 9·6 (3), or (iii) $\mathbf{q}_1 - \mathbf{q}_0$ and $\mathbf{q}_1 + \mathbf{q}_0$ are perpendicular vectors.

In this last case $\mathbf{q} = \tfrac{1}{2}(\mathbf{q}_1 + \mathbf{q}_0)$ and $\boldsymbol{\omega} = \mathbf{n}_\wedge (\mathbf{q}_1 - \mathbf{q}_0)$ are both perpendicular to $\mathbf{q}_1 - \mathbf{q}_0$, and therefore the tangential component of $\mathbf{q}$ must be parallel to $\boldsymbol{\omega}$.

From this we conclude that if the fluid speed is unaltered on passing through the vortex sheet at all points of it, then either the vortex sheet is at rest, or else the streamlines of the tangential motion in the vortex sheet and the vortex lines in the vortex sheet coincide.

This state of affairs must hold in the case of the vortex sheet wake (9·63) behind an aerofoil, for outside the wake the motion is irrotational and therefore the constant in Bernoulli's equation must be the same everywhere. Since the pressure must be continuous in passing through the sheet, q^2 has the same value on either side of the sheet at adjacent points separated by it.

9·63. Application to aerofoils.   The principal aerodynamic application of vortex sheets is to the surface and wake of an aerofoil.

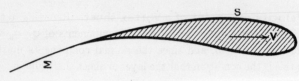

Fig. 9·63.

Consider an aerofoil advancing with velocity **V** in air otherwise at rest. As it advances the aerofoil divides the air into two streams, one of which passes over its upper surface, the other over its lower surface, and both streams reunite, if the aerofoil is properly shaped, above and below an interface Σ which forms the wake. This process will be considered in detail in the next chapter, but we have here to observe that the two streams arrive at the trailing edge with different velocities so that Σ is a vortex sheet in the sense already described.

Further, if we are to apply the induction law, it is necessary that the fluid should be unbounded (9·4); the presence of the surface S of the aerofoil cannot be tolerated, for it forms a boundary. We therefore adopt the following artifice for the purpose of calculation. We suppose the aerofoil to be removed and its place taken by air of the same density as the surrounding atmosphere but moving with the same velocity **V** as the aerofoil. This supposes, of course, a suitable pressure distribution to be supplied. The surface S of the aerofoil now becomes a vortex sheet across which the velocity changes abruptly and we can apply the induction law, for the whole region is occupied by air, at rest at infinity. Moreover, the application of the law will give the actual velocity $\mathbf{q}_P$ of the air at any point P.

If now we superpose on the whole system a velocity $-\mathbf{V}$, the air will stream past the aerofoil (which will now be replaced by air at rest) and the velocity at the point P will now be $\mathbf{q}_P - \mathbf{V}$. At infinity the air is no longer at rest but has the velocity $-\mathbf{V}$, and one of the conditions for the application of the induction law (rest at infinity) is violated. On the other hand superposition of the same velocity $-\mathbf{V}$ on the system clearly does not alter the surface vorticity associated with the sheets Σ and S, for the tangential components are equally changed on either side of a sheet and the *difference* of velocities is therefore unaltered. The use of the induction law, without regard to the validity of such a process, must therefore lead mathematically at the point P to the same velocity $\mathbf{q}_P$ as before. But we have seen that the actual velocity at P is $\mathbf{q}_P - \mathbf{V}$. It follows that we may in this case also apply the induction law provided that we interpret the result as giving, not the actual velocity at P, but the *perturbation velocity*, that is to say, the velocity by which the actual velocity at P differs from the general velocity $-\mathbf{V}$ of the stream. The actual air velocity at P will then be perturbation velocity + velocity of stream, and the perturbation velocity is calculated by the direct application of the induction law.

There is another point of view which is useful in this connection. When the air streams past an aerofoil at rest and when, with a view to applying the induction law, we replace the aerofoil by air at rest, the distribution of surface vorticity must be such that the component of the perturbation velocity normal to S at any point must just cancel the component of the stream normal to S at that point, so that there is no flux through the boundary and the aerofoil shape is preserved. This principle was used in the theory of thin aerofoils, in particular in deriving 8·2 (3), and the reader is invited to read that section in the light of the present explanation.

EXAMPLES IX

1. Write out the equation of motion in the equivalent form of three cartesian equations.

2. If $\boldsymbol{\zeta}$ is the vorticity vector, show that the equation of motion can be written

$$\frac{\partial \mathbf{q}}{\partial t} - \mathbf{q}_\wedge \boldsymbol{\zeta} = -\frac{1}{\rho} \nabla p - \nabla(\tfrac{1}{2}q^2 + \Omega).$$

3. Prove that in the steady motion of incompressible air

$$\mathbf{q}_\wedge \boldsymbol{\zeta} = \nabla \left(\frac{p}{\rho} + \tfrac{1}{2}q^2 + \Omega \right).$$

4. Show that the aerodynamic pressure p satisfies the equation of motion

$$\frac{d\mathbf{q}}{dt} = -\frac{1}{\rho} \nabla p,$$

and hence show that for incompressible air the acceleration is derivable from a potential function.

5. Write out the cartesian components (ξ, η, ζ) of the vorticity vector $\boldsymbol{\zeta}$, and then verify that

$$\nabla \boldsymbol{\zeta} = \frac{\partial \xi}{\partial x} + \frac{\partial \eta}{\partial y} + \frac{\partial \zeta}{\partial z} = 0.$$

6. If C and C' are two sections of a vortex tube and P is a point on C and Q is a point on C', prove Helmholtz's fourth theorem by considering the circulation in the circuit formed of C, PQ, $-C'$, QP, where $-C'$ denotes that C' is described in the sense opposite to the sense of description of C.

7. Show that the velocity due to a rectilinear segment AB of a vortex filament is perpendicular to the plane PAB and equal to

$$\frac{K}{4\pi p}(\cos PAB + \cos PBA),$$

where p is the perpendicular from P to AB.

Calculate the velocity at any point due to a rectangular vortex " ring " the sides of the rectangle being given by

$$z = 0, \quad x = \pm a, \quad y = \pm b.$$

8. Use the induction law to prove that a circular vortex ring of intensity Γ induces an axial velocity $\Gamma/2R$ at the centre of the ring, where R is the radius.

9. A circular vortex ring of radius R has its centre at O, and P is any point on the normal at O to the plane of the ring. Any point Q is taken on the arc of length s cut from the ring by a chord AB. Prove that the contribution of the arc s to the velocity induced at P is

$$\mathbf{v} = \frac{\Gamma}{4\pi PQ^3}\left\{\frac{sR \cdot \overrightarrow{OP}}{OP} + \overrightarrow{AB}_\wedge \overrightarrow{OP}\right\}.$$

10. With the notation of the previous example prove that

$$\mathbf{v} = \frac{\Gamma}{4\pi R^2 \cdot OP}\{s\sin^3\phi \cdot \overrightarrow{OP} + \overrightarrow{AB}_\wedge \overrightarrow{OP}\sin^2\phi\cos\phi\},$$

where ϕ is the angle OPQ.

Deduce that the whole ring induces at P the axial velocity of magnitude $\Gamma\sin^3\phi/2R$, and that if P coincides with O, the velocity is $\Gamma/2R$.

11. Prove that a circular vortex ring of radius R induces momentum parallel to the axis of the ring of amount $\rho\pi R^2\Gamma$, where Γ is the intensity of the ring.

Deduce that a vortex pair of two rectilinear vortices at distance l apart induces momentum of amount $\rho\Gamma l$ per unit length of vortex.

12. A point P in the plane of a horseshoe vortex is between the arms and equidistant from all three filaments. Prove that induced velocity at P is

$$\frac{K(1 + \sqrt{2})}{\pi AB},$$

where K is the intensity and AB is the length of the finite side of the horseshoe.

13. In the horseshoe vortex of fig. 9·52, the origin is taken at O the x-axis along PO in the sense P to O, the y-axis along OA and the z-axis downwards to form a right-handed triplet. Prove that the components (u, v, w) of the induced velocity at (x, y, z) are given by

$$u = \frac{Kz}{4\pi(x^2 + z^2)}\left\{\frac{y + l}{\sqrt{(x^2 + (y + l)^2 + z^2)}} - \frac{y - l}{\sqrt{(x^2 + (y - l)^2 + z^2)}}\right\},$$

$$v = \frac{Kz}{4\pi(z^2 + (y - l)^2)}\left\{1 - \frac{x}{\sqrt{(x^2 + (y - l)^2 + z^2)}}\right\}$$

$$- \frac{Kz}{4\pi(z^2 + (y + l)^2)}\left\{1 - \frac{x}{\sqrt{(x^2 + (y + l)^2 + z^2)}}\right\},$$

$$w = \frac{-Kx}{4\pi(x^2 + z^2)}\left\{\frac{y + l}{\sqrt{(x^2 + (y + l)^2 + z^2)}} - \frac{y - l}{\sqrt{(x^2 + (y - l)^2 + z^2)}}\right\}$$

$$- \frac{K(y - l)}{4\pi(z^2 + (y - l)^2)}\left\{1 - \frac{x}{\sqrt{(x^2 + (y - l)^2 + z^2)}}\right\}$$

$$+ \frac{K(y + l)}{4\pi(z^2 + (y + l)^2)}\left\{1 - \frac{x}{\sqrt{(z^2 + (y + l)^2 + z^2)}}\right\}$$

14. With the notation of the preceding example, show that, when $x = 0$, u is negative above the horseshoe and positive below it.

Prove also when $y^2 + z^2$ is large compared with l^2

$$v = \frac{2Klyz}{2\pi(y^2 + z^2)^2}, \quad w = \frac{-Kl(y^2 - z^2)}{2\pi(y^2 + z^2)^2},$$

approximately.

CHAPTER X

AEROFOIL OF FINITE ASPECT RATIO

10·1. Steady motion. Reduced to its essential mechanical principles
a monoplane aircraft may be regarded as an aerofoil of finite span pulled
through the air by a tractive force. To fix our ideas, let us consider an aerofoil

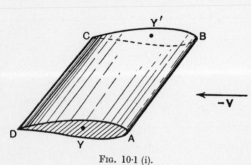

FIG. 10·1 (i).

of rectangular plan moving
horizontally in the plane of
symmetry, and *without rotation*, with *constant* velocity **V**,
in air otherwise at rest. By
the first law of motion the
tractive force, the weight, and
the aerodynamic force due to
pressure must form a system
in equilibrium. The pressure
distribution will therefore be
exactly the same if we imagine the constant velocity − **V** to be impressed on
the whole system of air and aerofoil. The aerofoil will then be at rest and the
air will stream past it with the general velocity − **V**, that is to say, the air will
be moving with velocity − **V** at a great distance upstream of the aerofoil;
near the aerofoil, and in the wake, the velocity distribution of the general
stream − **V** will be modified by the presence of the aerofoil. Our purpose is to
examine the general character of this perturbed airflow, which, on the present
hypothesis, is a steady motion.

Since the aerofoil is known in these circumstances to experience a lift
perpendicular to the wind and, in this case, vertically upwards, we conclude
that the upward vertical com-
ponent of the aerodynamic force
on the lower surface of the aerofoil must exceed the downward

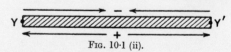

FIG. 10·1 (ii).

vertical component on the upper surface of the aerofoil. This shows that the
average pressure over the lower surface must exceed the average pressure
over the upper surface. Now consider fig. 10·1 (ii) which represents a section
of the aerofoil of fig. (i) by a vertical plane parallel to the span.

As we go from the median plane to a tip point such as Y whether along the
upper or the lower surface we must arrive at the same pressure at Y. Thus
there is a drop in pressure as we move outwards along the lower surface towards

Y and a further drop as we move inwards from Y along the upper surface towards the median plane of symmetry. Since the air is urged in the direction of decreasing pressure, it follows that an air particle which impinges on the leading edge AB and passes over the upper surface, which we shall call the *suction side* of the aerofoil, acquires a velocity component parallel to the span and towards the median plane. A corresponding particle which passes over the lower surface, which we shall call the *pressure side* of the aerofoil, acquires a velocity component parallel to the span but away from the median plane. The paths of such particles are shown schematically in fig. 10·1 (iii).

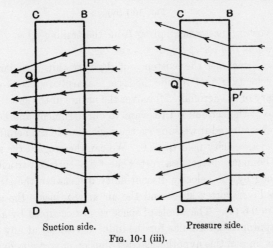

Suction side.   Pressure side.

FIG. 10·1 (iii).

The particles which leave the trailing edge CD at a given point Q must therefore have impinged on the leading edge AB at different points P and P' according as they have arrived at Q via the suction or pressure sides, moreover these particles will arrive at Q with different velocities, as is clearly seen from fig. (iii). On the other hand the particles in question must have the same speed q, for the pressure is continuous and since, by Bernoulli's theorem,

$$\frac{p}{\rho} + \tfrac{1}{2}q^2$$

is constant throughout the fluid, it follows that the value of q must be the same at Q whether the particle arrives there by the suction or the pressure side.

This discontinuity of direction of the velocity at the trailing edge means that the interface between the two streams of air from the suction and pressure sides of the aerofoil is a vortex sheet, and the equality of the speeds shows (see 9·61) that the vortex lines and the streamlines, with which they then coincide, bisect the angle between the two velocities at a point such as Q.

We now arrive at a preliminary picture of the flow past an aerofoil. The

oncoming air stream divides at the leading edge, passes over the suction and
the pressure sides to reunite in a vortex sheet Σ which forms the wake, Σ being

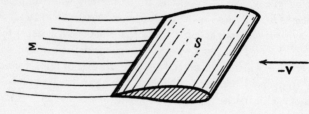

FIG. 10·1 (iv).

the locus of vortex lines which spring from the trailing edge. These vortex
lines are known as *free vortex lines*.

Further we remember the existence of the very thin boundary layer (1·8)
in which the velocity $\mathbf{q}_1$ just outside the layer is sharply reduced to $\mathbf{q}_0 = 0$ at the
actual boundary of the aerofoil. Moreover the nature of the substance enclosed
by the surface of the aerofoil is irrelevant to our argument and nothing in our
line of thought is altered if we suppose the region inside the aerofoil to contain
air at rest, i.e. a velocity field $\mathbf{q}_0 = 0$. We can then regard the surface of the
aerofoil (plus boundary layer) as a vortex sheet S of surface vorticity $\boldsymbol{\omega}$ per unit
area determined by the velocity discontinuity $\mathbf{q}_1$ between the air just outside
the boundary layer vortex sheet and the air at rest inside the aerofoil, with
$\boldsymbol{\omega} = \mathbf{n} \wedge \mathbf{q}_1$ as in 9·6 (4). The whole of space is now occupied by air so that we
can use the appropriate formulae for the induced velocity at any point.

The vortex lines of this hypothetical distribution of vorticity $\boldsymbol{\omega}$ are known as
bound vortex lines since they remain on the surface. The adjective " bound "
must not be taken to imply that points of the vortex sheet are bound in the
sense of having no velocity along the sheet. The velocity of a point of this
vortex sheet will be $\frac{1}{2}\mathbf{q}_1$, where $\mathbf{q}_1 = \boldsymbol{\omega} \wedge \mathbf{n}$ from 9·61 (2) remembering that
$\mathbf{q}_0 = 0$.

Since $\boldsymbol{\omega}$ is perpendicular to $\mathbf{n}$, we have $\mathbf{n}\boldsymbol{\omega} = 0$ and therefore from the
triple vector product rule (19·2)

$$\mathbf{q}_1 \wedge \boldsymbol{\omega} = (\boldsymbol{\omega} \wedge \mathbf{n}) \wedge \boldsymbol{\omega} = \mathbf{n}\omega^2 - \boldsymbol{\omega}(\mathbf{n}\boldsymbol{\omega}) = \mathbf{n}\omega^2,$$

whence it appears that the streamlines of the motion just outside the boundary
layer (determined from the velocity field $\mathbf{q}_1$) are identical with the streamlines
of the idealised vortex sheet S (determined from the velocity field $\frac{1}{2}\mathbf{q}_1$) and that
the vortex lines and streamlines cannot coincide since $\boldsymbol{\omega} \neq 0$.

10·2. The generation of the vortex system. In discussing the
nature of the flow past an aerofoil we considered the flow of a steady stream
past a fixed aerofoil, and, further, we supposed that the steady motion was
established. In order to see how the vortex system in the established steady

motion might be generated, it is more physically intuitive to suppose the aerofoil to be started into motion suddenly, in still air, with the velocity **V** originated by a suitably applied impulse, the velocity **V** of the aerofoil being subsequently maintained by a suitable tractive force.

It is evident that after a finite time t has elapsed from the start of the motion, the vortex sheet formed at the interface between the two sheets into which the advancing aerofoil separates the fluid will be of finite extent.

Fig. 10·2 (i) is intended to illustrate the general state of affairs, the wake Σ being represented by $CDEF$. Since vortex lines cannot terminate in the fluid

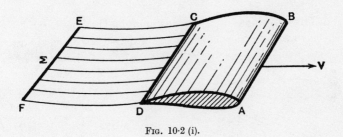

Fig. 10·2 (i).

(9·31) the sheet Σ will be bounded at its after end by a vortex line EF known as the *starting vortex*. As t increases so the wake Σ lengthens.

Fig. 10·2 (ii) shows schematically the vortex lines viewed from the suction side of the aerofoil. If, as described in 9·63, we regard the aerofoil as replaced

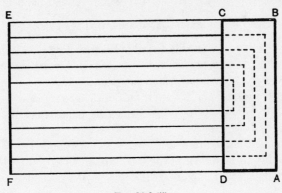

Fig. 10·2 (ii).

by air moving with velocity **V** the vortex lines must be closed. The bound vortices are shown by the dotted lines on the surface S of the aerofoil, where in this schematic diagram the complete vortex lines are depicted as rectangles. As the pattern lengthens with lapse of time the picture becomes that of an aerofoil trailing behind it horseshoe vortices of the type described in 9·52. This lengthening must increase the energy of the wake, so that energy must be supplied to the system. Thus we account for the necessity of a tractive force to maintain the motion, and the resistance thereby implied, which is known as the *induced drag*.

It might be thought that the appearance of circuits with circulation, when no such circuits existed in the initial state (of rest), contradicts Kelvin's theorem (9·21). Such, however, is not the case, for the circuits of that theorem always consist of the same fluid particles, and reflection on fig. 10·2 (ii) will show that no circuit which embraces the aerofoil but penetrates the wake Σ fulfils that condition.

10·21. The impulse of a vortex ring.

The considerations of 10·2 led to the picture of the wake as consisting of the superposition of vortex rings,

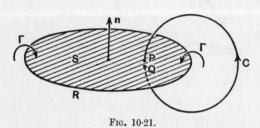

FIG. 10·21.

the rectangles shown in fig. 10·2 (ii), each consisting of a closed vortex filament.

Consider a plane vortex ring R of intensity Γ (see 9·31), the rest of the fluid being in irrotational motion with velocity potential ϕ. Let us suppose the ring to be closed by a plane diaphragm S, shaded in fig. 10·21, and take neighbouring points P and Q, one on each side the diaphragm. Then from 3·32,

$$(1) \qquad \phi_Q - \phi_P = \text{circ } C = \Gamma,$$

where C is any curve joining P to Q and not intersecting the diaphragm.

Now in 3·31 it was proved that any existing irrotational motion with velocity potential ϕ could be generated instantaneously from rest by application of an impulsive pressure $\rho\phi$. To apply this to the present case we imagine the diaphragm S to be a material membrane which immediately disappears after the application of the impulse.

If $\mathbf{n}$ is the unit normal to a small area dS separating P and Q, the impulse on this area due to impulsive pressure is $-\mathbf{n}\rho\phi_P\, dS$ at P and $\mathbf{n}\rho\phi_Q\, dS$ at Q. Thus the resultant linear impulse on the system is the vector

$$(2) \qquad \mathbf{I} = \int_{(S)} \rho(\phi_Q - \phi_P)\mathbf{n}\, dS = \int_{(S)} \rho\Gamma\mathbf{n}\, dS = \rho\Gamma S\mathbf{n},$$

where S is the area of the diaphragm. It should be noted that the second integral of (2) gives the impulse in the general case where neither R nor S is plane. Also it follows from Stokes's theorem (21·7) that the integral, and therefore the impulse, is independent of the particular diaphragm used.

The time rate of change of the impulse gives the force **F** acting,* so that

(3) $$\mathbf{F} = \frac{d\mathbf{I}}{dt} = \rho \Gamma \frac{d}{dt}(\mathbf{n}S),$$

since Γ does not change with time (second theorem of Helmholtz, 9·31).

As a particular application of (2) the magnitude of the impulse of the vortex pair of 4·21 is $4\pi\kappa\rho a$, since here $\Gamma = 2\pi\kappa$. This is, of course, the impulse per unit length of the vortex lines.

10·3. The lift. Consider a closed curve which embraces the aerofoil and which lies in a plane perpendicular to the span. The circulation about such a curve is by Stokes's theorem (21·7) equal to the normal flux of vorticity through it, and is therefore measured by the total intensity of the bound vortex lines which pass through the circuit. Referring to fig. 10·2 (ii), we see that the circulation will be greatest when the curve in question lies in the plane of symmetry of the aerofoil for then the maximum number of bound vortex lines pass through it. The circulation will be less the further the plane of the circuit lies from the plane of symmetry.

Fig. 10·3 shows a circuit of the kind described. If a circulation exists (positive in the sense shown in the diagram), it follows that the average speed above must exceed the average speed below, in other words there will be an excess of pressure below and a defect above, so that the aerofoil will experience a lift. The lift per unit length of span at the median section will be greatest and will fall off towards the tips. It will be remembered that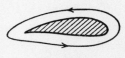

FIG. 10·3.

the vortex system was inferred from the existence of the lift, so that the present remarks are to be regarded as in the nature of a verification and a slight amplification as to the distribution.

10·4. Instability of the vortex sheet. The trailing sheet of free vortices depicted in fig. 10·2 (i) is found to be unstable and cannot persist.

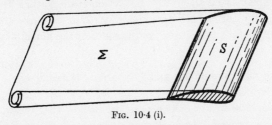

FIG. 10·4 (i).

The sheet tends to roll up somewhat like a sheet of paper and the vortex filaments twist round one another like the strands of a rope. Thus, at a sufficient

---

* Milne-Thomson, *Theoretical Hydrodynamics*, 17·32.

distance behind the aerofoil a section of the wake by a plane perpendicular to the direction of motion would show two cylindrical vortices whose distance apart is less than the span. An approximate calculation (see 11·7) shows that

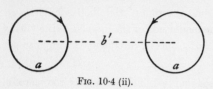

if these cylinders are regarded as having circular sections, the distance apart of their centres is the fraction $\pi/4$ of the span and the diameter of the circles is the fraction 0·171 of the span. This diameter then in the

FIG. 10·4 (ii).

case of a span of 20 ft. would be over 3 ft., so that the vortices are of considerable size. It should be remembered in this connection that the vortex lines which twist up to form these cores become spirals so that the assimilation of the rolled up vortices to cylinders is not exact.

Experiments by Fage and Simmons on an aerofoil of aspect ratio 6 showed that the rolling up was nearly complete at about 13 chords distance behind the trailing edge.

10·41. Tip vortices. The rolling up of the wake to form a helicoidal vortex gives rise to a picture of the wake as consisting solely of these vortices known as wing tip vortices.

The actual state of the wake is of course a compromise between this model * and that which gives the wake as a vortex sheet. The tip vortex model is particu-

FIG. 10·41.

larly useful in discussing the interaction between the planes of a biplane.

The two great contributions of Lanchester (1868–1945) to Aerodynamics and the basis of the modern theory of flight were : (i) the idea of circulation as the cause of lift; (ii) the idea of tip vortices as the cause of induced drag. Lanchester explained his theories to the Birmingham Natural History Society in 1894, but did not publish them until 1907 in his *Aerodynamics*. That Lanchester is in truth the pioneer and founder of the modern science is often overlooked.

10·5. The velocity of the air. Adopting the standpoint of air streaming with general velocity $-\mathbf{V}$ past a fixed aerofoil as depicted in fig. 10·1 (iv), let us suppose that the steady motion is established, so that the wake Σ extends downstream to infinity. As already described in 9·63 we shall suppose the aerofoil to be replaced by air kept at rest by a suitable system of pressures,

---

* See the photographs and methods of J. Valensi, *Application de la méthode des filets de fumée à l'étude des champs aérodynamiques*, Paris (1938).

so that the surface S of the aerofoil can be regarded as a vortex sheet, and the induction law can be applied in the sense described in 9·63 to give the perturbation velocity. The air velocity $\mathbf{q}^P$ at any point P will therefore be expressible in the form

(1) $$\mathbf{q}^P = -\,\mathbf{V} + \mathbf{q}_S{}^P + \mathbf{q}_\Sigma{}^P,$$

where on the right the first term denotes the general velocity of the stream, the second term the velocity induced at P by the vortex sheet S consisting of the bound vortices on the surface of the aerofoil, and the third term the velocity induced at P by the wake Σ. It may be proper to observe at this stage that no particular hypothesis is here made as to the form of the wake. It is irrelevant whether the wake is a sheet or has rolled up in the manner explained in 10·4 The particular form of the wake may be important when performing actual calculations of $\mathbf{q}_\Sigma{}^P$. On the other hand if the point P is on the surface of the aerofoil it is clear that those parts of the wake nearest to the trailing edge will contribute most to the induced velocity and whether the vorticity further downstream is distributed in sheets or cylinders will scarcely matter. But if the point P is situated well astern, for example on the tail plane, the particular form assumed by the wake may be important in calculating its contribution to the velocity.

10·51. The velocity at a point of the aerofoil.

We want an expression for the velocity $\mathbf{q}^P$ with which the air streams past a point P of the aerofoil, that is the velocity just outside the vortex sheet S at P.

Draw the normal at a point P of the replacing vortex sheet S and take points Q and R on this normal, equidistant from P, the point Q being outside the sheet and the point R inside. We then have from 10·5 (1)

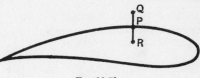

Fig. 10·51.

$$\mathbf{q}^Q = -\,\mathbf{V} + \mathbf{q}_S{}^Q + \mathbf{q}_\Sigma{}^Q,$$
$$\mathbf{q}^R = -\,\mathbf{V} + \mathbf{q}_S{}^R + \mathbf{q}_\Sigma{}^R.$$

But $\mathbf{q}^R$ is zero since R is in air which is by hypothesis at rest. Therefore adding and dividing by 2 we have

$$\tfrac{1}{2}\mathbf{q}^Q = -\,\mathbf{V} + \tfrac{1}{2}(\mathbf{q}_S{}^Q + \mathbf{q}_S{}^R) + \tfrac{1}{2}(\mathbf{q}_\Sigma{}^Q + \mathbf{q}_\Sigma{}^R).$$

If we now let Q, R tend to the point P of the vortex sheet S, and let $\mathbf{v}_S{}^P$ denote the velocity of the sheet at P induced by the rest of the sheet (i.e. excluding the element at P), then from 9·6 (3),

$$\mathbf{v}_S{}^P = \lim \tfrac{1}{2}(\mathbf{q}_S{}^Q + \mathbf{q}_S{}^R).$$

Also there will be no discontinuity in $\mathbf{q}_\Sigma$ when passing through S (only in fact when passing through Σ which does not arise here) so that if we write $\mathbf{v}_\Sigma{}^P = \lim \mathbf{q}_\Sigma{}^Q = \lim \mathbf{q}_\Sigma{}^R$ as Q, R tend to P, we have

$$(1) \qquad \tfrac{1}{2}\mathbf{q}^P = -\mathbf{V} + \mathbf{v}_S{}^P + \mathbf{v}_\Sigma{}^P,$$

which gives the air velocity past the aerofoil at P in terms of the main stream velocity $-\mathbf{V}$, the induced velocity $\mathbf{v}_\Sigma{}^P$ due to the wake and the velocity $\mathbf{v}_S{}^P$ of the vortex sheet at P induced by the rest of the sheet.

10·6. Aerodynamic force.

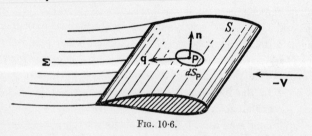

FIG. 10·6.

Let dS_P be an element of surface of the aerofoil at the point P. Let $\mathbf{n}$ be the unit outward normal vector at P and let p be the aerodynamic pressure. Then the aerodynamic force is

$$\mathbf{A} = \int_{(S)} -p\mathbf{n}\, dS_P.$$

By Bernoulli's theorem $p = p_0 - \tfrac{1}{2}\rho q^2$, where p_0 is a constant pressure. Since a constant pressure applied to a closed surface has no resultant effect, we get

$$\mathbf{A} = \tfrac{1}{2}\rho \int_{(S)} \mathbf{n}q^2\, dS_P,$$

where $\mathbf{q}$ is the velocity and q the airspeed at P.

Now (21·2)

$$\mathbf{q} \wedge (\mathbf{n} \wedge \mathbf{q}) = -(\mathbf{n}\mathbf{q})\mathbf{q} + \mathbf{n}(\mathbf{q}\mathbf{q}) = \mathbf{n}q^2,$$

since $\mathbf{n}\mathbf{q} = 0$, for $\mathbf{n}$ and $\mathbf{q}$ are perpendicular. Therefore

$$\mathbf{A} = \tfrac{1}{2}\rho \int_{(S)} \mathbf{q} \wedge (\mathbf{n} \wedge \mathbf{q})\, dS_P.$$

Now from 9·6 we see that $\boldsymbol{\omega}_S{}^P = \mathbf{n} \wedge \mathbf{q}$ is a vector tangential to S, and is the surface vorticity of the vortex sheet S, so that

$$\mathbf{A} = \rho \int_{(S)} \tfrac{1}{2}\mathbf{q}^P \wedge \boldsymbol{\omega}_S{}^P\, dS_P,$$

where we have written $\mathbf{q}^P$ instead of $\mathbf{q}$.

Using the result of 10·51, we have finally

$$\mathbf{A} = \rho \int_{(S)} (-\mathbf{V} + \mathbf{v}_\Sigma{}^P + \mathbf{v}_S{}^P)_\wedge \boldsymbol{\omega}_S{}^P \, dS_P = \mathbf{L} + \mathbf{D}_i + \mathbf{F}, \text{ where}$$

(1) $$\mathbf{L} = -\rho \mathbf{V}_\wedge \int_{(S)} \boldsymbol{\omega}_S{}^P \, dS_P.$$

(2) $$\mathbf{D}_i = \rho \int_{(S)} (\mathbf{v}_\Sigma{}^P {}_\wedge \boldsymbol{\omega}_S{}^P) \, dS_P.$$

(3) $$\mathbf{F} = \rho \int_{(S)} (\mathbf{v}_S{}^P {}_\wedge \boldsymbol{\omega}_S{}^P) \, dS_P.$$

In these results we recall that $\mathbf{v}_\Sigma{}^P$, $\mathbf{v}_S{}^P$, $\boldsymbol{\omega}_S{}^P$ are respectively the velocity induced at P by the wake Σ, the velocity induced at P by the rest of the bound vortex sheet S, and the surface vorticity of this sheet.

We also note that these are exact results, in the sense that no approximations have been made.

The forces $\mathbf{L}$ and $\mathbf{D}_i$ are called the *lift* and the *induced drag* respectively. The appropriateness of this terminology will appear shortly (11·2, 11·21).

Reasons will be adduced later for supposing the force $\mathbf{F}$ to be negligible.

10·61. The force F.

Consider two points P and Q of the surface of the aerofoil and the corresponding surface elements of area dS_P, dS_Q. By the induction law the velocity induced at P by the vortex sheet S is

$$\mathbf{v}_S{}^P = \frac{1}{4\pi} \int_{(S)} \frac{\boldsymbol{\omega}_S{}^Q {}_\wedge \overrightarrow{QP}}{QP^3} \, dS_Q.$$

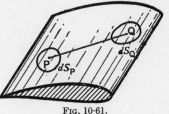

Substitute this in the expression for $\mathbf{F}$ in 10·6 (3). Then

$$\mathbf{F} = \frac{\rho}{4\pi} \int_{(S)} \int_{(S)} \frac{(\boldsymbol{\omega}_S{}^Q {}_\wedge \overrightarrow{QP})_\wedge \boldsymbol{\omega}_S{}^P}{QP^3} \, dS_P \, dS_Q.$$

FIG. 10·61.

Interchanging P and Q gives a similar expression for $\mathbf{F}$ and therefore $\mathbf{F}$ is half the sum of its two expressions, namely

$$\mathbf{F} = \frac{\rho}{8\pi} \int_{(S)} \int_{(S)} \frac{(\boldsymbol{\omega}_S{}^Q {}_\wedge \overrightarrow{QP})_\wedge \boldsymbol{\omega}_S{}^P + (\boldsymbol{\omega}_S{}^P {}_\wedge \overrightarrow{PQ})_\wedge \boldsymbol{\omega}_S{}^Q}{QP^3} \, dS_P \, dS_Q.$$

Now it is easy to prove (21·2) that

$$(\mathbf{a}_\wedge \mathbf{b})_\wedge \mathbf{c} + (\mathbf{b}_\wedge \mathbf{c})_\wedge \mathbf{a} + (\mathbf{c}_\wedge \mathbf{a})_\wedge \mathbf{b} = 0,$$

and applying this to the above expression for $\mathbf{F}$ we get

$$\mathbf{F} = \frac{\rho}{8\pi} \int_{(S)} \int_{(S)} \frac{(\boldsymbol{\omega}_S{}^Q {}_\wedge \boldsymbol{\omega}_S{}^P)_\wedge \overrightarrow{QP}}{QP^3} \, dS_P \, dS_Q.$$

Now the vectors $\boldsymbol{\omega}_S{}^Q$, $\boldsymbol{\omega}_S{}^P$ are tangential to the surface S of the aerofoil and are also tangential to the bound vortex lines. In the case of two-dimensional motion (i.e. of infinite aspect ratio) these vectors are rigorously parallel, so that their vector product is zero, and **F** = 0. For aerofoils of large aspect ratio we should therefore expect **F** to be small. On the hypothesis, which will be made later, that the bound vortex lines are parallel, we shall have **F** = 0.

10·7. Moments. If we take moments with respect to a point O we shall have the vector moment

$$\mathbf{C} = \tfrac{1}{2}\rho \int_{(S)} \overrightarrow{OP}{}_{\wedge}(\mathbf{q}^P{}_{\wedge}\boldsymbol{\omega}_S{}^P)dS_P = \rho \int \overrightarrow{OP}{}_{\wedge}[(-\mathbf{V} + \mathbf{v}_\Sigma{}^P + \mathbf{v}_S{}^P)_{\wedge}\boldsymbol{\omega}_S{}^P]dS_P$$

from 10·51.

The components of this vector will give the rolling, pitching, and yawing moments referred to suitable axes through O.

CHAPTER XI

THE LIFTING LINE THEORY

11·0. In Chapter X we were concerned with a general description of the flow past an aerofoil. In the present chapter special hypotheses will be introduced in order to make numerical calculations possible. These hypotheses, due to Prandtl, will be introduced one by one as required, in order that their exact status in regard to the calculation may be perceived.

Calculations based on the hypotheses just mentioned must be regarded as necessarily approximate. Nevertheless, it has been found that the results are sufficiently accurate for many purposes and have a much wider range of application than would appear, at first sight, to be justifiably expected.

The student should be warned, however, that the investigation on which we are about to embark is one of discussing the deductions to be made from schematization of a very complicated state of affairs and that the "laws of Prandtl" which will be used as a basis are not necessarily laws of nature.

11·01. Geometrical hypotheses. We shall assume that

(*A*) the aerofoil has a median plane of symmetry;

(*B*) the chord of the profile of every section made by a plane parallel to the plane of symmetry is small compared with the span;

(*C*) the trailing edge may be regarded as a straight line.

Assumption (*B*) is equivalent to postulating large aspect ratio, and experimental results indicate that an aspect ratio greater than about 4 may be considered large for this purpose; a rather remarkable conclusion.

Assumption (*C*) means that we can neglect the deviation of the trailing edge from a straight line in the same sense as in 8·01 we neglect the camber of a thin aerofoil and replace it by its chord. We therefore by (*C*) rule out the treatment of aerofoils whose trailing edge has pronounced curvature in plan, and in particular wings with " sweep back (see 17·03) ".

Subject to the above, the profiles and incidence may vary in any manner across the span.

The aerofoil will be treated throughout this chapter as moving with constant velocity in the plane of symmetry, without side slip or rotation.

11·1 Axes of reference. We take rectangular cartesian axes x, y, z.

The y-axis, or *lateral* axis, is normal to the plane of symmetry and along the (straight) trailing edge. Its positive direction is to starboard and the origin is taken at the point where the y-axis meets the plane of symmetry.

The x-axis, or *longitudinal* axis, is in the direction of motion, which has already been stated to be in the plane of symmetry.

The z-axis, or *normal* axis, is perpendicular to the other two axes in such sense that the three axes form a right-handed system. This means, in particular, that in ordinary straight horizontal flight the z-axis will be directed vertically downwards.

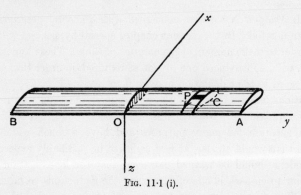

FIG. 11·1 (i).

The axes so chosen are sometimes called *wind-axes*, for the x- and z-axes depend for their exact definition on the incidence. Since the motion of the aerofoil is in the plane of symmetry we have therefore $\mathbf{V} = \mathbf{i}V$, where V is the forward speed, and $\mathbf{i}$ is a unit vector in the positive sense of the x-axis. Another system of axes is sometimes useful, in which the axes are fixed in the aerofoil, the y-axis being defined as before, but the x-axis being drawn towards the leading edge of the chord of the aerofoil. Such axes are known as *chord-axes*.

Whichever axes are used, we denote by $\mathbf{i}$, $\mathbf{j}$, $\mathbf{k}$ unit vectors along the x-, y-, z-axis respectively.

It is often convenient to describe the position of a point P on the trailing

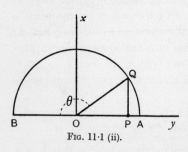

FIG. 11·1 (ii).

edge by an *eccentric angle* similar to that used in connection with thin aerofoils (8·01). Draw a semicircle with the span AB as diameter, and in the (x, y) plane. The eccentric angle θ of the point P is then defined as shown in fig. 11·1 (ii) and the y-coordinate of P is $-\frac{1}{2}b\cos\theta$. Thus P is the point $(0, -\frac{1}{2}b\cos\theta, 0)$, and as we go from the port to the starboard wing tip along the span θ increases from 0 to π.

11·2. Expression for the lift. Referring to fig. 11·1 (i), if P is a point on the surface of the aerofoil, and if we draw the sections at distances y and $y + dy$ from the plane of symmetry, the element of area at P may be written

$$dS_P = dy\, ds,$$

where ds is an element of arc of the profile defined by one of the sections, the positive sense of ds being indicated by the arrow in fig. 11·1 (i).

Thus 10·6 (1) gives for the lift

$$(1) \qquad \mathbf{L} = - \rho \mathbf{V} \wedge \int_{-b/2}^{b/2} dy \int_{(C)} \boldsymbol{\omega}_S{}^P \, ds,$$

where b is the span of the aerofoil and C is the curve bounding the profile at P.

In order to evaluate $\mathbf{L}$ we introduce the first hypothesis.

Hypothesis I. All the bound vorticity vectors $\boldsymbol{\omega}_S$ are parallel to the span. This is strictly true for a cylindrical aerofoil of infinite aspect ratio.

Numerically this hypothesis says that

$$(2) \qquad \boldsymbol{\omega}_S{}^P = \mathbf{j}\omega_S{}^P,$$

where $\omega_S{}^P$ is the magnitude of the bound vorticity vector at P, and $\mathbf{j}$ is a unit vector in the positive sense of the y-axis. Thus

$$(3) \qquad \int_{(C)} \boldsymbol{\omega}_S{}^P \, ds = \mathbf{j} \int_{(C)} \omega_S{}^P \, ds = \mathbf{j} \Gamma(y),$$

where $\Gamma(y)$ measures the total intensity of the vorticity crossing the profile C of the bound vortex sheet in the direction of the span and therefore, by Stokes's theorem, measures the circulation round C.

$$(4) \qquad \Gamma(y) = \text{circ } C.$$

Since (see 21·13) $\mathbf{i} \wedge \mathbf{j} = \mathbf{k}$, where $\mathbf{k}$ is a unit vector in the positive sense of the z-axis, we have from (1) and (3),

$$(5) \qquad \mathbf{L} = - \mathbf{k}L, \qquad L = \rho V \int_{-b/2}^{b/2} \Gamma(y) \, dy,$$

which shows that $\mathbf{L}$ is indeed a lift, in the sense that it is directed perpendicularly to the wind and in the negative direction of the z-axis.

We may justly regard (5) as the extension to aerofoils of finite span of the rule of Kutta and Joukowski for the lift (per unit span) on an aerofoil of infinite span (see 5·5).

We also observe that the section of the aerofoil between the planes y and $y + dy$ contributes to the total lift (see 11·23) the amount $\rho V \Gamma(y) dy$ so that the *rolling moment*, reckoned positive from y to z, that is when the starboard wing tip is urged downwards, is

$$(6) \qquad L_{\text{mom}} = - \rho V \int_{-b/2}^{b/2} y \Gamma(y) \, dy.$$

This moment is of course zero when the circulation $\Gamma(y)$ is symmetrically distributed. If asymmetry is introduced, for example by the use of the ailerons, the rolling moment can be calculated from (6).

Another consequence of hypothesis I is that the force $\mathbf{F}$(10·6 (3)) vanishes.

11·201. Distribution of load. We find in 11·23 that the portion of the aerofoil between the sections y and $y + dy$ exerts the lift $\rho V \Gamma(y)\, dy$; the distribution of lift (and therefore of load) across the span of the aerofoil is proportional to the circulation $\Gamma(y)$. If we draw a curve whose abscissa is y and whose ordinate is $\Gamma(y)$, we get what is called a *load grading curve*.

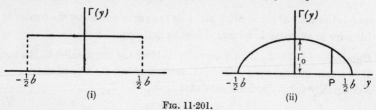

FIG. 11·201.

The diagram shows two particular forms of load grading curve, (i) a straight line parallel to the span corresponding with *uniform loading*, and (ii) a semi-ellipse, corresponding with what is called *elliptic loading*. The area under the load grading curve measures the total load (cf. 11·2 (5)) carried by the aerofoil. As is easily seen from 11·2 (6) the distance of the centroid of the area under the load grading curve from the $\Gamma(y)$ axis is a measure of the rolling moment. When the load grading curve is symmetrical this distance vanishes and the rolling moment is zero.

11·21. Expression for the induced drag. Using the notation of 11·2, we have from 10·6 the induced drag

$$\mathbf{D}_i = \rho \int_{-b/2}^{b/2} dy \int_{(C)} \mathbf{v}_\Sigma{}^P {}_\wedge \boldsymbol{\omega}_S{}^P\, ds.$$

Hypothesis II. The velocity $\mathbf{v}_\Sigma{}^P$ induced at P by the *trailing vortices* may be replaced by the velocity induced at the trailing edge of the profile section through P.

This may be called the *lifting line hypothesis*, for it is tantamount to regarding each profile section as a point to a first approximation (cf. 11·01 (B)). We have for definiteness placed this point at the trailing edge of the profile, and have consequently replaced the aerofoil, or lifting surface, by a lifting line coincident with the trailing edge of the aerofoil. Instead of the trailing edge of the profile its centre of pressure might be used, as is frequently done, but in view of 11·01 (B) the particular point chosen is not really important provided that it is clearly defined, and provided that the locus of such points is a straight line parallel to the span (cf. 11·01 (C)).

Combining this with 11·2 (2), we now get

$$\mathbf{D}_i = \rho \int_{-b/2}^{b/2} \mathbf{v}_\Sigma{}^P {}_\wedge \mathbf{j}\Gamma(y)\, dy.$$

To reduce this further we introduce

Hypothesis III. All the trailing vortices leave the trailing edge in lines which are parallel to the direction of motion of the aerofoil (but in the opposite sense).

Reference to 10·1 and fig. 10·1 (iv) shows that the trailing vortices should leave the trailing edge in the direction of the air velocity there which is, in general (see 11·8 and 12·3), different from that of **V**. Hypothesis III treats the wake Σ as a flat sheet parallel to the xy plane. That the sheet rolls up further down wind (10·4) is unimportant for the present application, since the main contribution to $\mathbf{v}_\Sigma^P$ is from that part of the wake in the immediate neighbourhood of the trailing edge.

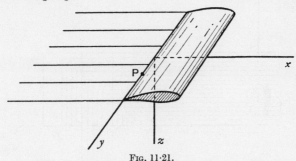

FIG. 11·21.

If we calculate $\mathbf{v}_\Sigma^P$ at a point P of the trailing edge, we see at once from 9·51 that the only component is in the z-direction so that

$$\mathbf{v}_\Sigma^P = w\mathbf{k}. \tag{1}$$

The component of induced velocity which is normal both to the span and the direction of motion is called the *downwash velocity*.*

In the present case the downwash velocity at the trailing edge is w defined by (1).

Since (21·13) $\mathbf{j}_\wedge \mathbf{k} = \mathbf{i}$ we get finally

$$\mathbf{D}_i = -\mathbf{i}D_i, \qquad D_i = \rho \int_{-b/2}^{b/2} w\Gamma(y)\,dy. \tag{2}$$

This expression shows that $\mathbf{D}_i$ opposes the motion and is therefore a drag. Since its existence depends on the normal velocity induced by the trailing vortices it is called the *induced drag*, a term which distinguishes it from the drag due to skin friction.

By considering the induced drag $\rho w\Gamma(y)\,dy$ on the section at distance y from the plane of symmetry (see 11·23) we get for the yawing moment

$$N = \rho \int_{-b/2}^{b/2} wy\Gamma(y)\,dy. \tag{3}$$

---

* If we take into account also the bound vortices, it is clear from 9·52 that just ahead of the aerofoil there will be *upwash*. We are concerned here only with the velocity induced by the free vortices.

The positive sense of the yawing moment is from x to y, that is, when the nose is urged to starboard.

11·22. The downwash velocity at the trailing edge.
Taking the y-axis along the trailing edge, consider two profiles C, C' of the bound vortex sheet at distances η and $\eta + d\eta$ from the plane of symmetry. Let $\Gamma(\eta)$ be the circulation round C. Then the circulation round C' will be $\Gamma(\eta) + d\Gamma(\eta)$.

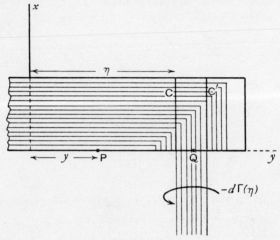

FIG. 11·22.

Fig. 11·22 shows a schematic plan view of the starboard wing with some of the *bound* vortex lines (cf. fig. 10·2 (ii)). The circulation $\Gamma(\eta)$ round C will be equal to the total intensity of the bound vortices which pass through this section. Similarly the circulation $\Gamma(\eta) + d\Gamma(\eta)$ round C' will be equal to the total intensity of the bound vortices which pass through the section C'. In fig. 11·22, as drawn, it is clear that fewer bound vortices pass through C' than through C. Therefore $d\Gamma(\eta)$ is negative and the total circulation of the trailing vortices which escape from the portion of the trailing edge between C and C' will be $-d\Gamma(\eta)$.

Thus the vortex lines which leave the trailing edge between C and C' form a ribbon-like vortex sheet of circulation $-d\Gamma(\eta)$, and this ribbon being arbitrarily narrow may be treated as a line vortex of this circulation when calculating the induced velocity.

Now the velocity induced at P, $(0, y, 0)$, by the semi-infinite line vortex of circulation $-d\Gamma(\eta)$ through the point Q, $(0, \eta, 0)$, is, see 9·51,

$$dw = \frac{-d\Gamma(\eta)}{4\pi . QP}$$

along Oz. Thus the downwash velocity at P is

(1)
$$w = \frac{1}{4\pi} \int_{\eta=-b/2}^{\eta=b/2} \frac{d\Gamma(\eta)}{y - \eta}.$$

As in other integrals of this type the principal value (see 4·7) is to be taken, since the vortex through P induces no velocity on P.

Thus, to evaluate the downwash velocity, we write

$$w = \lim_{\epsilon \to 0} \left\{ \frac{1}{4\pi} \int_{\eta=-b/2}^{\eta=y-\epsilon} \frac{d\Gamma(\eta)}{y - \eta} \right.$$
$$\left. + \frac{1}{4\pi} \int_{\eta=y+\epsilon}^{\eta=b/2} \frac{d\Gamma(\eta)}{y-\eta} \right\}.$$

The further evaluation of this will depend on the precise form of the function $\Gamma(\eta)$.

11·23. The loading law. We have seen that the lift and induced drag on the aerofoil are respectively

(1)
$$L = \int_{-b/2}^{b/2} \rho V\Gamma(y)\, dy, \qquad D_i = \int_{-b/2}^{b/2} \rho w\Gamma(y)\, dy.$$

These were obtained by the quite general method of 10·6, with the addition of Hypotheses I, II, III, by integration of the pressure thrusts over the surface of the aerofoil. Let us call the section of the aerofoil between the adjacent planes y and $y + dy$ the *strip y* (cf. 19·3). Then precisely the same method applied to the strip y would show that it is acted upon by a lift $\rho V\Gamma(y)\, dy$ and an induced drag $\rho w\Gamma(y)\, dy$. This result will be referred to as the *loading law*.

The loading law thus asserts that the integrands of (1) actually represent the elementary loads carried by and localised in the strips.

In this respect the loading law states a physical fact as compared with the superficially similar law of induction (9·4) which merely asserts a quasi-behaviour of the elements of the integral which gives the induced velocity.

11·24. Effect of downwash on incidence. For a given profile of an aerofoil,

Geometrical incidence is the angle between the chord of the profile and the direction of motion of the aerofoil.

Absolute incidence is the angle between the axis of zero lift of the profile and the direction of motion of the aerofoil.

In the sequel incidence will always mean absolute incidence unless the contrary is stated.

When the axes of zero lift of all the profiles of the aerofoil are parallel, each profile meets the wind at the same absolute incidence, the incidence is the same at every point of the span, and the aerofoil is said to be *aerodynamically untwisted*.

An aerofoil is said to have *aerodynamic twist* when the axes of zero lift of its profiles are not all parallel. The incidence is then variable across the span.

We shall have to distinguish between coefficients and incidence of the aerofoil as a whole, and its profiles. For the aerofoil we shall use the notations C_L, C_{D_i}, α; for a profile $C_L{}'$, $C_{D_i}{}'$, α'. The latter are functions of the coordinate y which defines the position of the profile. For a symmetrical aerofoil they are *even* functions of y, that is to say they are the same for y and $-y$.

The coefficients and incidence for a strip are of course the same as those for the profiles which bound the strip.

Let us consider the aerodynamic properties of a strip in the light of the loading law.

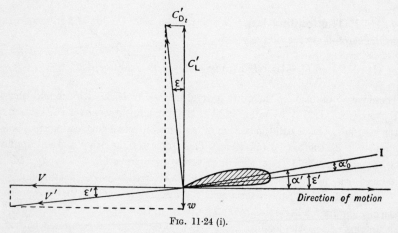

Fig. 11·24 (i).

Fig. 11·24 (i) shows a profile of an aerofoil which is moving in the direction shown with velocity **V** at absolute incidence α' measured from I, the axis of zero lift. From the loading law the forces on the strip whose profile is shown are

$$(1) \qquad \tfrac{1}{2}\rho V^2 c' dy C_L{}' = \rho V \Gamma(y)\, dy, \qquad \tfrac{1}{2}\rho V^2 c' dy C_{D_i}{}' = \rho w \Gamma(y)\, dy,$$

and therefore

$$(2) \qquad \frac{C_{D_i}{}'}{C_L{}'} = \frac{w}{V} = \epsilon'.$$

In a properly designed profile the ratio of induced drag to lift is always small in the working range of incidence, and therefore ϵ', which is called the *angle of downwash*, is a small angle. It follows that if $V'^2 = \surd(V^2 + w^2)$ then $V' = V$, neglecting the second order of small quantities.

It now appears from the diagram that the resultant aerodynamic force on the strip is perpendicular to the direction of V' and not to the direction of V. Since ϵ' is a small angle the coefficient of this force is C_L'. Therefore in respect of lift the strip behaves like a strip of a two-dimensional aerofoil in a relative wind in the direction of V', i.e. at incidence α_0', where

$$(3) \qquad \alpha_0' = \alpha' - \epsilon'.$$

The angle α_0' is called the *effective incidence*.

Thus the effect of downwash is that the downwash velocity combines with the actual relative wind of speed V to produce an effective relative wind in the direction of V'.

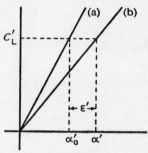

We can now draw two graphs of lift coefficient against incidence (a) and (b) in fig. 11·24 (ii). They are both approximately straight lines (cf. 7·13). Graph (a) shows the lift coefficient as a function of effective incidence and is the graph proper to the profile operating as a two-dimensional aerofoil, slope a_0'; the other graph (b) shows the graph when the profile is operating as part of the actual aerofoil, slope a'.

Fig. 11·24 (ii).

In drawing these graphs the further assumption has been made that the angle of downwash vanishes when the wind is along the axis of zero lift, in other words that this axis is the same in the two- and three-dimensional cases. With this assumption we have

$$(4) \qquad C_L' = a_0'\alpha_0' = a'\alpha'.$$

As to the drag, an actual aerofoil is subject, in addition to induced drag, to *profile* drag due mainly to skin friction, and therefore we can write for the total drag coefficient of the strip, using (2),

$$(5) \qquad C_D' = C_{D_0}' + \epsilon'C_L',$$

where C_{D_0}' is the coefficient of profile drag for the strip.

It may be noted that profile drag is largely independent of incidence in the working range.

Profile drag is really the sum of two effects, drag due to skin friction and drag due to shape.

The part due to skin friction owes its origin to the clinging of the air in the boundary layer to the surface of the body, an effect which must always be present, but which can be lessened by smoothing the surface, and reducing its area.

The part due to shape, otherwise known as *form* drag, is due to lowered pressure in the wake; by shaping the body to reduce the wake to negligible proportions, i.e. by streamlining, form drag can be almost eliminated.

11·3. The integral equation for the circulation.

If c' is the chord of the profile at distance y from the plane of symmetry, the lift coefficient of the profile is given by

$$C_L' = \frac{\rho V \Gamma(y)}{\frac{1}{2}\rho V^2 c'},$$

and therefore *

$$\Gamma(y) = \tfrac{1}{2}c'VC_L' = \tfrac{1}{2}c'Va_0'\alpha_0'.$$

Therefore, from 11·24,

(1)
$$\Gamma(y) = \tfrac{1}{2}c'Va_0'\left(\alpha' - \frac{w}{V}\right).$$

Using the value of w in 11·22 we get

(2)
$$\frac{\Gamma(y)}{\tfrac{1}{2}c'Va_0'} = \alpha' - \frac{1}{4\pi V}\int_{\eta=-b/2}^{\eta=b/2}\frac{d\Gamma(\eta)}{y-\eta}.$$

This is the integral equation from which $\Gamma(y)$ is to be determined. That done, we can find lift, drag, and downwash.

Note that in general α', a_0', c' are all functions of y, since incidence, chord, and profile may vary from section to section. If the profiles are similar curves a_0' is the same at every section, but α' is not the same unless the sections are also similarly situated (untwisted aerofoil).

For a given wing α', a_0', c' are known functions of y, and in particular for thin wings we may take $a_0' = 2\pi$.

There are two fundamental problems connected with aerofoils.

Problem I. Given the distribution of the circulation, i.e. given $\Gamma(y)$, to find the form of the aerofoil, and the induced drag.

Problem II. Given the form of the aerofoil, to find the distribution of the circulation, and the induced drag.

For Problem I we know $\Gamma(\eta)$ the solution of (2) and we have to find a shape of aerofoil to fit it.

For Problem II we have to solve (2), given c', α', a_0'. We proceed to consider the simplest and also the most important case of Problem I, elliptic loading.

11·4. Problem I ; elliptic loading.

Referring to fig. 11·201 (ii) in the case of *elliptic loading*, the $(\Gamma(y), y)$ curve is an ellipse. If P is the point on the span whose eccentric angle is θ, we have $y = -\tfrac{1}{2}b\cos\theta$ and therefore

(1)
$$\Gamma(y) = \Gamma_0\sin\theta,$$

* Cf. 7·13. This is essentially the equation $\kappa = 2aV\sin\beta$, in which, with the present notation, $2\pi\kappa = \Gamma(y)$, $a_0' = 2\pi$, $c' = 4a$, $a_0' = \sin\beta$.

where Γ_0 is the value of $\Gamma(y)$ when $y = 0$. It is easily seen that the elimination of θ gives

$$\left(\frac{\Gamma(y)}{\Gamma_0}\right)^2 + \left(\frac{y}{\frac{1}{2}b}\right)^2 = 1,$$

which is the equation of an ellipse.

Putting $\eta = -\frac{1}{2}b\cos\phi$, we see from 11·22 (1) that the downwash velocity at the trailing edge is

$$(2) \qquad w = \frac{1}{4\pi}\int_0^\pi \frac{\Gamma_0\cos\phi\,d\phi}{\frac{1}{2}b(\cos\phi - \cos\theta)} = \frac{\Gamma_0}{2b},$$

from 4·71. Thus in elliptic loading the downwash velocity is the same at every point of the trailing edge.

If we substitute (1) in the integral equation satisfied by the circulation (11·3), we get

$$(3) \qquad \frac{\Gamma_0}{V}\left\{\frac{2\sin\theta}{c'a_0'} + \frac{1}{2b}\right\} = \alpha',$$

where a_0' and α' refer to the section at distance y from the plane of symmetry.

Here the chord c' and the incidence α' depend, in general, on the particular profile section considered, that is to say on θ. Also Γ_0/V depends on the incidence of the aerofoil. If we increase the incidence of the aerofoil by β, the incidence of each profile section will also increase by β. Thus

$$\left(\frac{\Gamma_0}{V}\right)_1\left\{\frac{2\sin\theta}{c'a_0'} + \frac{1}{2b}\right\} = \alpha' + \beta,$$

where the first term denotes the new value of Γ_0/V. By subtraction we get

$$\left\{\left(\frac{\Gamma_0}{V}\right)_1 - \frac{\Gamma_0}{V}\right\}\left\{\frac{2\sin\theta}{c'a_0'} + \frac{1}{2b}\right\} = \beta.$$

The only term which involves θ is $2\sin\theta/c'a_0'$ and therefore, if we postulate that *the loading shall remain elliptic at all incidences*, we must have

$$(4) \qquad a_0'c' = a_0c_0\sin\theta,$$

where c_0 is the chord of the middle section of the aerofoil, and a_0 is the value of a_0' there.

If, in addition, the profiles of the sections are similar curves, a_0' will be the same at every section and (4) becomes

$$(5) \qquad c' = c_0\sin\theta.$$

This means that the graph of c', the chord, against y is also an ellipse. This situation can be realised by an aerofoil so constructed that its plan form is bounded by two half ellipses whose major axis is equal to the span.

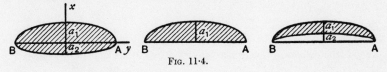

FIG. 11·4.

To prove this, consider the ellipses

$$\frac{x_1^2}{a_1^2} + \frac{4y^2}{b^2} = 1, \quad \frac{x_2^2}{a_2^2} + \frac{4y^2}{b^2} = 1.$$

It follows that

$$\frac{x_1}{a_1} = \frac{x_2}{a_2} = \frac{x_1 \pm x_2}{a_1 \pm a_2},$$

and if $c' = x_1 \pm x_2$, $c_0 = a_1 \pm a_2$, equation (5) is satisfied.

Lastly, it is clear from (3) and (4) that in elliptic loading which remains elliptic for all incidences, the incidence is the same at every profile section, and Γ_0 and therefore, from (2), the downwash is proportional to the incidence.

Another case arises for an aerofoil of rectangular plan. Here the chord c' may be taken as constant and equal to c_0. Retaining the hypothesis that $a_0' = a_0$, which will be true if the sections are similar, or if they are thin, (3) becomes

$$(6) \qquad \frac{\Gamma_0}{V}\left\{\frac{2\sin\theta}{a_0 c_0} + \frac{1}{2b}\right\} = \alpha',$$

which shows that the incidence at each section is different, so that the aerofoil is *twisted*. The incidence at the middle section will be α, got by putting $\theta = \pi/2$ in (6), and therefore

$$(7) \qquad \frac{\alpha'}{\alpha} = \left\{\frac{2\sin\theta}{a_0 c_0} + \frac{1}{2b}\right\} \bigg/ \left\{\frac{2}{a_0 c_0} + \frac{1}{2b}\right\}.$$

From this it is clear that if the loading is elliptic at the incidence α, it ceases to be elliptic at a different incidence.

From these examples of elliptic loading it appears that to make Problem I definite further conditions are necessary.

Unless the contrary is stated, the phrase elliptic loading will be taken to imply that the loading is elliptic at all incidences in the working range.

11·41. Elliptic loading ; lift and induced drag. From 11·2 (5) and 11·4 (1) we have

$$L = \rho V \int_0^\pi \tfrac{1}{2} b \Gamma_0 \sin^2\theta \, d\theta = \tfrac{1}{2}\rho V \Gamma_0 . \tfrac{1}{2}\pi b,$$

whence we get the lift coefficient of the aerofoil

$$C_L = \frac{L}{\tfrac{1}{2}\rho V^2 S} = \frac{\pi b \Gamma_0}{2SV}.$$

Now the aspect ratio is

$$(1) \qquad A = \frac{b^2}{S}, \text{ so that}$$

$$(2) \qquad C_L = \frac{\pi A \Gamma_0}{2bV} = \frac{\pi A w}{V} = \pi A(\alpha - \alpha_0),$$

where w is the constant downwash velocity, 11·4 (2), at the trailing edge.

Here from 11·24 (2), (3) $\alpha' - \alpha_0'$ is the constant angle of downwash and α, α_0 are the averages of α', α_0' across the span.

For the induced drag 11·21 (2) gives

$$D_i = \tfrac{1}{4}\rho \int_0^\pi \Gamma_0{}^2 \sin^2 \theta \, d\theta = \tfrac{1}{8}\pi\rho\Gamma_0{}^2,$$

and the corresponding coefficient is

(3) $$C_{D_i} = \frac{\pi\Gamma_0{}^2}{4SV^2} = \frac{\pi A w^2}{V^2}.$$

From (2) and (3) we get

(4) $$\pi A C_{Di} = C_L{}^2.$$

The graph of lift coefficient against drag coefficient is called the *polar curve* of the aerofoil.   Equation (4) shows that the polar curve of an elliptically loaded aerofoil is a parabola, provided the only source of drag is the induced velocity.

The polar curve can be graduated in incidence as indicated in fig. 11·41. Since α is proportional to C_L, equal increments of incidence graduations on the polar correspond to equal increments of C_L.

In practice, in addition to induced drag, there is profile drag (11·24). The

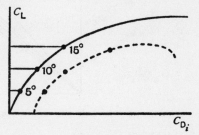

Fig. 11·41.

coefficient of this being denoted by C_{D_0} the complete drag coefficient is

$$C_D = C_{D_0} + C_{D_i}.$$

The corresponding type of polar is the dotted curve in fig. 11·41.

11·42. Slope of the lift graph for elliptic loading.

If, as usual, a_0 is the slope of the lift coefficient graph in two-dimensional motion and a the slope of the lift graph for an elliptic aerofoil of aspect ratio A, differentiation with respect to C_L of the formula, 11·41 (2),

$$\alpha = \alpha_0 + \frac{C_L}{\pi A}$$

gives

$$\frac{1}{a} = \frac{1}{a_0} + \frac{1}{\pi A}.$$

If we take the theoretical value $a_0 = 2\pi$, we get

$$a = \frac{\pi}{\tfrac{1}{2} + \dfrac{1}{A}} \frac{1}{\text{radian}} = \frac{\pi^2}{180\left(\tfrac{1}{2} + \dfrac{1}{A}\right)} \frac{1}{\text{degree}}.$$

A	∞	10	8	6	4
a (per degree)	0·110	0·092	0·088	0·082	0·074

It follows that, for incidence below the stall, the (C_L, α) curves are straight lines whose slopes increase as the aspect ratio increases, fig. 11·42.

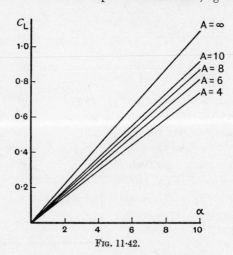

Fig. 11·42.

11·43. Change of aspect ratio in elliptic loading. We have

$$\alpha = \alpha_0 + \frac{C_L}{\pi A}, \qquad C_{D_i} = \frac{C_L^2}{\pi A}.$$

Hence, if the aspect ratio is reduced to A' and if dashes refer to the new aerofoil with the same effective incidence, we have

$$\alpha' - \alpha = \left(\frac{1}{A'} - \frac{1}{A}\right) \frac{C_L}{\pi},$$

$$C_{D_i}' - C_{D_i} = \left(\frac{1}{A'} - \frac{1}{A}\right) \frac{C_L^2}{\pi}.$$

Thus for a given lift coefficient, decrease of aspect ratio increases both the geometrical incidence, and the induced drag coefficient.

These results show that the largest practicable ratio is desirable.

11·5. Problem II. In the foregoing discussion of elliptic loading the circulation $\Gamma(y)$ is given and an appropriate form of the aerofoil is inferred. We now consider the converse problem of finding $\Gamma(y)$ when the form of the aerofoil is given. To do this we must solve the integral equation of 11·3, noting that the symmetry with regard to the median plane of the aerofoil demands that

$\Gamma(y) = \Gamma(-y)$. In terms of the eccentric angle θ we can therefore write the Fourier sine series, since $\Gamma(y)$ vanishes at the tips $\theta = 0$ and $\theta = \pi$.

$$(1) \qquad \Gamma(y) = 2bV\sum_1^\infty A_n \sin n\theta$$

and note that n must be an *odd integer* * to ensure the equality of $\sin n\theta$ and $\sin n(\pi - \theta)$. Thus

$$(2) \qquad \Gamma(y) = 2bV \sum_{n=0}^\infty A_{2n+1} \sin (2n + 1)\theta.$$

This formula for $\Gamma(y)$ is unchanged when $\pi - \theta$ is written for θ, and the value at the centre given by $\theta = \pi/2$ is

$$(3) \qquad \Gamma_0 = 2bV \sum_{n=0}^\infty A_{2n+1}(-1)^n.$$

Substitution of (2) in 11·3 (2) gives

$$(4) \quad \frac{4b}{c'a_0}\sum_0^\infty A_{2n+1} \sin (2n + 1)\theta = \alpha' - \frac{b}{2\pi}\int_0^\pi \frac{\sum_0^\infty (2n+1) A_{2n+1} \cos (2n+1)\phi\, d\phi}{\tfrac12 b(\cos \phi - \cos \theta)}.$$

The values of the integrals on the right are given in 4·71. Let us write α_θ for α' to emphasise that α' is a function of θ and put

$$(5) \qquad \mu_\theta = \frac{c'a_0'}{4b}.$$

Then we get from (4)

$$(6) \qquad \sum_0^\infty A_{2n+1}[(2n + 1)\mu_\theta + \sin \theta] \sin (2n + 1)\theta = \mu_\theta\alpha_\theta \sin \theta.$$

To find the coefficients A_{2n+1} from (6) would necessitate the expansion of each side, and of each term of the left side, in a Fourier series, thus leading to infinitely many equations in infinitely many unknowns. The solution of these being an obvious difficulty we have recourse to a practical method of solution, due to Glauert. Replace the infinite series of (1) by a finite series of, say, $m + 1$ terms, where m is a given integer, thus giving

$$(7) \qquad \sum_0^m A_{2n+1}[(2n+1)\mu_\theta + \sin \theta] \sin (2n+1)\theta = \mu_\theta\alpha_\theta \sin \theta.$$

This equation cannot be satisfied identically. If, however, we take a particular value of θ we get a linear equation in the coefficients $A_1, A_3, \ldots, A_{2m+1}$. If $m + 1$ particular values are ascribed to θ, we get $m + 1$ linear equations from which the coefficients A_{2n+1} can be calculated, and the values so obtained will satisfy (7), not identically, but only at the selected points. The solution will

---

* The same method could be used to investigate the effect of moving the ailerons, but even values of n would then also have to be included.

be satisfactory if the coefficients so determined satisfy (7) at other points within the standard of accuracy demanded in any particular case.

Since (7) is satisfied in any case when $\theta = 0$ or π we have $(m + 1)$ points other than these available. The chosen points are usually taken as equally spaced in θ over the half-span. Thus if $m = 3$ we should take

$$\theta = \frac{\pi}{2}, \ \frac{3\pi}{8}, \ \frac{\pi}{4}, \ \frac{\pi}{8},$$

and with these values we could determine four coefficients, namely

(8) $\qquad\qquad\qquad\qquad A_1, \ A_3, \ A_5, \ A_7.$

These usually suffice.

A rough approximation is obtained by taking $m = 1$, and $\theta = \pi/4, \ \pi/2$. This will determine two coefficients $A_1', \ A_3'$ but it must not be inferred that, comparing with (8), $A_1' = A_1, \ A_3' = A_3$. Indeed, the process is such that the coefficients are functions of m and the validity of the process depends on these values tending rapidly to limits suited to the form (1) as m increases. Glauert has tested this numerically in the case of a rectangular aerofoil and has concluded that in this case four coefficients should suffice.

If the incidence α_θ has the same value α at each point of the span, (6) shows that A_{2n+1} is proportional to α, and if we write $A_{2n+1} = \alpha B_{2n+1}$, the coefficients B_{2n+1} are independent of incidence and may therefore be determined once for all.

An ingenious electrical analogy has led Malavard * to devise a method of solving the integral equation of 11·3 by measurements of electrical potentials in a suitable apparatus.

When the incidence is variable across the span (aerodynamic twist), we may write

$$\alpha_\theta = \alpha_m + f(\theta),$$

where α_m is the incidence at the middle section. If then in the equation (6) we write

$$A_{2n+1} = A'_{2n+1} \alpha_m + B_{2n+1},$$

the equation is equivalent to the two equations

$$\Sigma A'_{2n+1} \sin{(2n + 1)}\theta[(2n+1)\mu + \sin\theta] = \mu \sin\theta,$$

$$\Sigma B_{2n+1} \sin{(2n + 1)}\theta[(2n+1)\mu + \sin\theta] = \mu \sin\theta f(\theta),$$

and all the numbers so determined are independent of the incidence. Thus

$$C_L = \pi A A_1 = \pi A [A_1' \alpha_m + B_1] = a\alpha_m + b$$

where a and b are constants of the aerofoil.

---

* L. Malavard, *Applications des analogies électriques à la solution de quelques problèmes de l'Hydrodynamique*, Paris, 1936.

11·51. The lift. We have from 11·5 (2)

(1) $$\Gamma(y)\,dy = b^2 V \overset{\infty}{\underset{0}{\Sigma}} A_{2n+1} \sin(2n+1)\theta \sin\theta\,d\theta.$$

Now $\int_0^\pi \sin(2n+1)\theta \sin\theta\,d\theta = 0$, unless $n = 0$ when the integral is $\frac{1}{2}\pi$.
Therefore using 11·2 (5)

$$L = \tfrac{1}{2}\rho V^2 b^2 \pi A_1.$$

Thus the lift coefficient for the whole aerofoil is

(2) $$C_L = \frac{\pi b^2 A_1}{S} = \pi A A_1,$$

where A is the aspect ratio.

Thus $A_1 = C_L/\pi A$, and this gives a check on the theoretical value of A_1,
for C_L can be determined by wind tunnel measurements.

In the case of elliptic loading all the A_{2n+1} are zero except A_1, and 11·5 (3)
gives

(3) $$\Gamma_0 = 2bVA_1,$$

and elimination of A_1 between this and (2) gives the result already obtained
in 11·41 (2).

If the incidence α is the same at every point of the span, we have seen (11·5)
that A_1 is proportional to α.

Since $C_L = \dfrac{\pi A A_1}{\alpha}\,.\,\alpha$ we have, in this case, $a = (\pi A)\dfrac{A_1}{\alpha}$, where a is the
slope of the (C_L, α) graph. If a_0, α_0 are the corresponding slope and incidence
in two dimensional motion we have

$$C_L = a\alpha = a_0\alpha_0,$$

and therefore

(4) $$\alpha - \alpha_0 = C_L\left(\frac{1}{a} - \frac{1}{a_0}\right) = \frac{(1+\tau)}{\pi A}\,C_L,$$

where

$$1 + \tau = \pi A\left(\frac{1}{a} - \frac{1}{a_0}\right) = \pi A\left(\frac{\alpha}{\pi A A_1} - \frac{1}{a_0}\right) = \frac{\alpha}{A_1} - \frac{\pi A}{a_0}.$$

For elliptic loading $\tau = 0$, see 11·42, so that τ may be regarded as a measure
of the departure of the actual loading from elliptic loading taken as standard.

11·52. The downwash velocity. We have at the point of the trailing
edge, whose eccentric angle is θ, from 11·22 (1),

$$w_\theta = \frac{1}{4\pi}\int_0^\pi \frac{2bV\Sigma(2n+1)A_{2n+1}\cos(2n+1)\phi\,d\phi}{\frac{1}{2}b\cos\phi - \frac{1}{2}b\cos\theta}.$$

Using the values of the integrals given in 4·71 we have

(1) $$w_\theta = V\overset{\infty}{\underset{0}{\Sigma}}(2n+1)A_{2n+1}\frac{\sin(2n+1)\theta}{\sin\theta}.$$

In the case of elliptic loading we get from 11·51 (3)

$$w_\theta = VA_1 = \Gamma_0/2b,$$

which is constant across the span.

11·53. The induced drag. We have, from 11·21,

$$D_i = \int_{-b/2}^{b/2} \rho w \Gamma(y)\,dy,$$

and from our previous calculations, 11·51 (1), 11·52 (1),

$$w\Gamma(y)dy = b^2 V^2 (\overset{\infty}{\underset{0}{\Sigma}} A_{2n+1} \sin (2n+1)\theta)(\overset{\infty}{\underset{0}{\Sigma}} (2n+1) A_{2n+1} \sin (2n+1)\theta)\,d\theta.$$

Now

$$\int_0^\pi \sin (2n+1)\theta \sin (2r+1)\theta\,d\theta = 0,$$

where r is any integer other than n, but if $r = n$ the value of the integral is $\pi/2$, so that

$$D_i = \tfrac{1}{2}\pi\rho b^2 V^2 \overset{\infty}{\underset{0}{\Sigma}} (2n+1) A_{2n+1}^2.$$

Therefore

(1) $$C_{D_i} = \pi A \overset{\infty}{\underset{0}{\Sigma}} (2n+1) A_{2n+1}^2 = \pi A A_1^2 (1+\delta),$$

where

(2) $$\delta = \frac{3A_3^2 + 5A_5^2 + \ldots + (2m+1) A_{2m+1}^2 + \ldots}{A_1^2}.$$

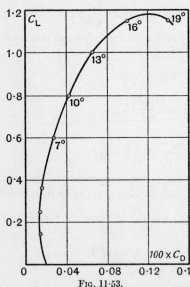

FIG. 11·53.

Note that δ is never negative, and is zero only in the case of elliptic loading.

The total drag coefficient is

(3) $$C_D = C_{D_0} + C_{D_i}$$
$$= C_{D_0} + \frac{1}{\pi A} C_L^2 (1 + \delta),$$

where C_{D_0} is the profile drag coefficient of the aerofoil and (3) is now the equation of the polar curve of the aerofoil.

A typical polar curve is shown in fig. 11·53.

If the profile drag coefficient for each profile section is a function of the position of the section, the profile drag coefficient of the aerofoil is

$$C_{D_0} = \frac{1}{S} \int_{-b/2}^{b/2} C_{D_0}' c'\,dy,$$

where the dashes refer to the section at distance y.

11·6. Minimum induced drag. Given the span and the lift, in what conditions is the induced drag least? We have

$$L = \tfrac{1}{2}\rho V^2 \pi b^2 A_1, \qquad D_i = \tfrac{1}{2}\rho V^2 \pi b^2 A_1^2 (1 + \delta).$$

Thus when L is given D_i is least when $\delta = 0$, i.e. when

$$3A_3^2 + 5A_5^2 + \ldots = 0,$$

which implies that $A_3 = A_5 = A_7 = \ldots = 0$, and therefore the loading is elliptic. Thus, of all wings of given span and lift the elliptically loaded wing gives the least induced drag.

This result shows the practical importance of elliptic loading. It also shows that wings which are found efficient in practice are probably so found because their loading approaches the theoretically best distribution. It is therefore reasonable and often convenient to assume elliptic loading when deducing aerofoil properties.

11·7. The wake far down wind. In 10·4 it was stated that the vortex sheet wake is unstable and rolls up into concentrated vortices.

Let Γ_0 be the circulation of each vortex of the resulting pair treated as circular cylinders of radius a whose centres are b' apart, see fig. 10·4 (ii). The circulation Γ_0 is of course the same as the circulation round the central section of the aerofoil. From the method of 10·21 the impulse of this pair per unit length of cylinder will be $\rho \Gamma_0 b'$. When the aerofoil travels unit distance the ribbon vortices of 11·22 will increase their impulse (see 11·8) by $\rho(-d\Gamma)2y$, and therefore the total increase will be

$$\rho \Gamma_0 b' = \int_{y=0}^{y=b/2} \rho(-d\Gamma)2y.$$

If we assume elliptic loading, $\Gamma = \Gamma_0 \sin \theta$, we get

$$\rho \Gamma_0 b' = \int_0^\pi \tfrac{1}{2}\rho b \Gamma_0 \cos^2 \theta \, d\theta = \tfrac{1}{4}\pi b \rho \Gamma_0,$$

so that $b' = \pi b/4 = 0.785b$.

Again, if E is the energy per unit length of the wake, the increase of energy EV of the wake per unit time due to the lengthening must be equal to the work done against the induced drag, i.e. $D_i \times V$. Thus, using 11·41 and the value of E from 4·6, we get

$$\tfrac{1}{8}\pi \rho \Gamma_0^2 = \frac{1}{2\pi} \rho \Gamma_0^2 \left\{ \log \frac{b'}{a} + \tfrac{1}{4} \right\},$$

whence $2a/b = 0.171$.

The velocity distribution due to such a pair of vortices is shown in fig. 4·11 (b). It will be seen that there is downwash in the central part and upwash outside. When aircraft fly in V-formation the leading aircraft creates this upwash in which the two astern on either side fly, and so on. This formation is also adopted by wild geese who apparently have discovered its advantage naturally.

11·8. Lift and drag deduced from the impulse.
Consider an aerofoil, regarded as a lifting line AB, started from rest, and moved in a straight

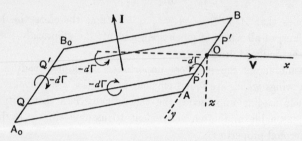

FIG. 11·8.

line. Let the velocity be $\mathbf{V}$ at time t. Then at time t the starting vortex will be at, say, A_0B_0.

We shall assume that the wake ABA_0B_0 remains a rectangular vortex sheet.

If P is the point $(0, y, 0)$ and Q is the point $(-l, y, z)$, there springs (11·22) from P a trailing vortex of circulation $-d\Gamma(y)$, and so the whole wake may be regarded as resulting from the superposition of rectangular vortex rings, a typical one being $PP'Q'Q$ of circulation $-d\Gamma(y)$.

Let $h = PQ$. The area of the ring is $2yh$ and its impulse from 10·21 is therefore, if $\mathbf{n}$ is the unit normal to the plane of the rectangle,

$$d\mathbf{I} = \rho(-d\Gamma(y)) \, 2yh\mathbf{n} = \rho d\Gamma(y) \, . \, 2y \, (\mathbf{k}l + \mathbf{i}z),$$

since $-h\mathbf{n} = \mathbf{k}l + \mathbf{i}z$, by geometry. The impulse $\mathbf{I}$ of the whole wake is got by integration. Now

$$\int_{(OA)} 2y \, d\Gamma(y) = \int_{(BA)} y \, d\Gamma(y) = \left[y\Gamma(y) \right]_{-b/2}^{b/2} - \int_{-b/2}^{b/2} \Gamma(y) \, dy,$$

which reduces to the last integral since $\Gamma(y)$ vanishes at the tips. Thus

$$\mathbf{I} = -\int_{-b/2}^{b/2} \rho(\mathbf{k}l + \mathbf{i}z) \, \Gamma(y) \, dy.$$

The aerodynamic force (see 10·21) is $d\mathbf{I}/dt$ and $dl/dt = V$, while $dz/dt = w$, the normal velocity at Q, which by the symmetry of the ring is equal to the downwash velocity at P. Thus the aerodynamic force is

$$-\mathbf{k}\int_{-b/2}^{b/2} \rho V \Gamma(y) \, dy - \mathbf{i}\int_{-b/2}^{b/2} \rho w \Gamma(y) \, dy,$$

which consists of the same lift and induced drag as were calculated by Prandtl's hypotheses.

It is interesting to note that the above investigation denies Prandtl's Hypothesis III and replaces it by the assumption of a rectangular wake inclined to the x, y plane at the angle

$$\epsilon = \frac{z}{l} = \frac{\int_0^t w \, dt}{\int_0^t V \, dt}.$$

Since ϵ is of the order w/V, it is anyway small. This new hypothesis is tantamount to assuming that the downwash velocity is constant across the span, a characteristic property of elliptic loading (11·41). Since this form of loading is desirable, the hypothesis is reasonable for well-behaved wing shapes.

The method of the impulse is of course applicable whatever the form of the wake.

It is also noteworthy that the above method does not depend on any assumption that steady motion is already established.

11·9. The rectangular aerofoil.
By this term is intended an aerofoil whose plan form is a rectangle. An aerofoil whose shape is that of a cylinder erected on an aerofoil profile satisfies this requirement. But so do more general shapes, for example an aerofoil whose two wings are inclined to the plane of symmetry, or an aerofoil whose wings are appropriately twisted.

11·91. Cylindrical rectangular aerofoil.
This is the simplest type, of span b and chord c, which is also the chord of all the sections. All the sections are similar and similarly situated and therefore with the notation of 11·5, taking $m = 3$, we have

$$\sum_0^3 A_{2n+1}[(2n+1)\mu + \sin\theta]\sin(2n+1)\theta = \mu\alpha\sin\theta.$$

Put $A_{2n+1} = B_{2n+1}\alpha$. Then

(1) $B_1(\mu + \sin\theta)\sin\theta + B_3(3\mu + \sin\theta)\sin 3\theta + B_5(5\mu + \sin\theta)\sin 5\theta$
$\qquad\qquad + B_7(7\mu + \sin\theta)\sin 7\theta = \mu\sin\theta,$

(2) $\mu = a_0 c/(4b) = a_0/(4A)$ where $A = b/c$ is the aspect ratio.

Glauert * has calculated the values of the A_n, taking $\theta = 22\frac{1}{2}°, 45°, 67\frac{1}{2}°, 90°$.

11·92. Method of Betz.
This consists in assuming for the circulation Γ at distance y from the plane of symmetry the formula

(1) $\qquad\qquad \Gamma = \sqrt{(1-\eta'^2)}[\Gamma_0 + \Gamma_2\eta'^2 + \Gamma_4\eta'^4 + \dots]$

where in our previous notation

(2) $\qquad\qquad \eta' = y/\tfrac{1}{2}b = -\cos\theta,$

---

* H. Glauert, *Elements of Aerofoil and Airscrew Theory*, Cambridge (1930).

and the coefficients are then determined by substitution in 11·3 (2). The actua calculations are laborious and we shall content ourselves with a summary of the results. For a rectangular wing the aspect ratio is

(3) $$A = b/c.$$

There are two limiting cases. When the chord is large compared with the span, A approaches zero. This is the case for example in a long fin of small height, or for an aeroplane fuselage regarded as of rectangular plan. In the case $A = 0$ Betz finds an elliptical distribution of loading. As A increases to infinity the loading approaches the rectangular distribution (see fig. 11·201). For intermediate values of A the load grading curve has the form shown in

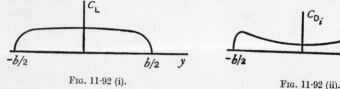

FIG. 11·92 (i).

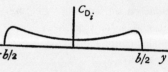

FIG. 11·92 (ii).

fig. 11·92 (i) while the form of the drag grading curve is shown in fig. 11·92 (ii).

The graphs of τ and δ are shown for values of $4A/a_0$ between 0 and 10 in fig. 11·92 (iii).

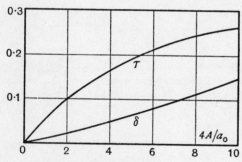

FIG. 11·92 (iii).

The method of Betz leads to integrals of the type

$$J_n = \int_{-1}^{1} \frac{\eta'^n \, d\eta'}{(\eta'_1 - \eta')\sqrt{(1 - \eta'^2)}}$$
$$= (-1)^n \int_0^{\pi} \frac{\cos^n \theta \, d\theta}{\cos \theta - \cos \phi},$$

where

$$\eta' = -\cos \theta, \quad \eta_1' = -\cos \phi.$$

The evaluation of this can be made to depend on the integrals I_n of 4·71, by observing that

$$(2 \cos \theta)^n = (e^{i\theta} + e^{-i\theta})^n$$
$$= 2 \cos n\theta + 2 \binom{n}{1} \cos (n-2)\theta + 2 \binom{n}{2} \cos (n-4)\theta + \dots,$$

and so

$$2^{n-1}(-1)^n J_n = \pi \frac{\sin n\phi}{\sin \phi} + 2\pi \binom{n}{1} \frac{\sin (n-2)\phi}{\sin \phi} + \dots.$$

Again, $\sin n\phi/\sin \phi$ can be expanded in descending powers of $\cos \phi$, the leading term being $2^{n-1} \cos^{n-1} \phi$. Thus

$$J_n = -\pi [\eta_1'^{n-1} + k_3 \eta_1'^{n-3} + \dots],$$

where the coefficients $k_3, k_5, \dots$ can easily be calculated.

11·93. Method of Fuchs. If in 11·92 (1) we restrict the number of terms, we may write

(1) $\Gamma = \surd(1 - \eta'^2)[\Gamma_0 + \Gamma_2\eta'^2 + \Gamma_4\eta'^4 + \ldots + \Gamma_{2n}\eta'^{2n}].$

From this we can calculate the downwash angle w/V and so write for the apparent incidence

$$\alpha = \frac{w}{V} + \frac{2\Gamma}{ca_0V},$$

which gives α as a function of η'.

It is then postulated that the graph of α against η' shall have contact of the $(2n + 1)$th order with its tangent at $\eta' = 0$, that is to say the first $2n + 1$ derivatives $d\alpha/d\eta'$, $d^2\alpha/d\eta'^2, \ldots,$ shall vanish at $\eta' = 0$. This procedure gives n equations to determine the ratios of $\Gamma_0, \Gamma_2, \ldots, \Gamma_{2n}$. The method leads to the following formula for δ and τ applicable to *corrected aspect ratios* A' from 6 to 10.

$$\delta = - 0\cdot04 + 0\cdot014A',$$
$$\tau = 0\cdot05 + 0\cdot02A'.$$

Here the corrected aspect ratio is defined by

$$A' = 2\pi A/a_0,$$

so that $A' = A$ when $a_0 = 2\pi$.

EXAMPLES XI

1. A flat aerofoil of span b is in the form of that part of the parabola $y^2 = \frac{1}{4}b^2\left(1 - \frac{x}{h}\right)$ for which x is positive. Calculate the mean chord, the plan area, and the aspect ratio. Find h if the aspect ratio is 6.

2. Assuming uniform loading, show that the wake consists of two trailing vortices only, one from each wing tip, and that the vortex system then consists of a single horseshoe vortex.

Show that uniform loading is physically unacceptable on account of infinite velocity at the tips, but that the assumption of trailing vortices of diameter different from zero will remove the difficulty to a first approximation.

3. Draw a graph of the downwash velocity at points of the span both inside and outside the tips on the assumption of uniform loading.

4. In Ex. I, 21, C_L, C_D are given for the aerofoil Clark YH. Use the following additional values of C_{D_0}, the coefficient of profile drag, to calculate C_{D_i} for the incidences given below and draw a graph of both coefficients.

$\alpha°$	$-2\cdot9$	$-1\cdot7$	$+0\cdot6$	$2\cdot8$	$5\cdot1$	$7\cdot4$	$9\cdot6$
C_{D_0}	0·0088	·0089	·0081	·0083	·0072	·0094	·0110

On the assumption that, in this range, the drag coefficient due to skin friction is constant and equal to 0·0065, estimate and exhibit graphically, the form drag coefficient.

5. Given the circulation $\Gamma(y)$, show that the aerofoil has ininfitely many possible forms subject to the condition $C_L'c'V = 2\Gamma(y)$.

Discuss the case where the loading is elliptic and the chord c' is the same length at every section.

6. Prove that in elliptic loading the local lift coefficient C_L' is equal to the lift coefficient of the aerofoil.

7. Show that in elliptic loading Γ_0/V is a linear function of incidence.

8. Use Ex. IX, 13, to show that, for uniform loading, at a point on the z-axis the downwash velocity is

$$\frac{\Gamma}{\pi}\frac{b}{4z^2 + b^2}.$$

9. Use the preceding example to show that in the case of elliptic loading the downwash velocity at a point of the z-axis is

$$w = \frac{1}{2\pi}\int_0^{b/2}\frac{4w_0\,y}{\sqrt{(\tfrac14 b^2 - y^2)}}\frac{y}{z^2 + y^2}\,dy,$$

and by means of the substitution $y = \tfrac12 b \sin\chi$, prove that

$$w = w_0\left(1 - \frac{2z}{\sqrt{(b^2 + 4z^2)}}\right),$$

where w_0 is the downwash velocity at the trailing edge.

10. Use the fact that in elliptic loading the downwash velocity has a constant value w_0 across the span to show that the flow in the lateral plane ($x = 0$) is the same as the two-dimensional flow caused by a line of length b moving normally to itself with velocity $w_0 = \Gamma_0/2b$.

11. Prove that the complex potential of the two-dimensional motion described in the preceding example is

$$-iw_0(\sqrt{\{(y + iz)^2 - b^2/4\}} - (y + iz))$$

and therefore that the downwash velocity is given by

$$w = \text{real part of } w_0\left\{1 - \frac{y + iz}{[(y + iz)^2 - b^2/4]^{\frac12}}\right\}.$$

12. By means of the preceding example show that at any point of the lateral plane

$$\frac{w}{w_0} = 1 - \frac{\sinh\mu\,\cosh\mu}{\cosh^2\mu - \sin^2\lambda},$$

where $y = \tfrac12 b \sin\lambda\cosh\mu, \quad z = \tfrac12 b \cos\lambda\sinh\mu,$

and exhibit w/w_0 graphically as a function of $2y/b$, for different given values of $2z/b$.

13. Use Ex. 11 to prove that far from the aerofoil the value of w in the lateral plane is

$$-\frac{b^2}{8}\frac{y^2 - z^2}{(y^2 + z^2)^2}\,w_0,$$

and that this would have had the same form, had the loading been assumed uniform.

14. Draw the graph of $dC_L/d\alpha$ as a function of aspect ratio in the case of elliptic loading.

15. In the case of elliptic loading show that

$$\frac{d\alpha}{dA} = -\frac{C_L}{\pi A^2}, \quad \frac{dC_{D_i}}{dA} = -\frac{C_L^2}{\pi A^2}.$$

Explain the significance of the negative signs.

16. Verify the following results due to Glauert which give the coefficients A_1, A_3, etc. when 1, 2, etc. terms are taken in the method of 11·5, i.e. when $m = 0$, 1, etc., for a rectangular aerofoil of aspect ratio $A = a_0$:

m	$A_1/\mu\alpha$	$A_3/\mu\alpha$	$A_5/\mu\alpha$	$A_7/\mu\alpha$	τ	δ
0	0·800	—	—	—	·86	0
1	·917	·084	—	—	·22	·025
2	·926	·110	·016	—	·18	·044
3	·928	·115	·023	·004	·17	·049

Extend the table to $m = 4$.

17. Prove that in the case of a two-dimensional thin aerofoil the pitching moment coefficient about the leading edge is related to the pitching moment coefficient about the quarter point by the formula

$$C_m' = -\tfrac{1}{4}C_L' + C_{mQ}'.$$

Prove that in an untwisted rectangular aerofoil the same relation holds in the form

$$C_m = -\tfrac{1}{4}C_L + C_{mQ}.$$

18. Draw the polar curve of the aerofoil Clark YH (see Ex. I, 21) and graduate it for absolute incidence.

19. Plot τ and δ in terms of the corrected aspect ratio from the formulae in 11·93.

20. Deduce 11·21 (2) by applying the Kutta-Joukowski rule to the circulation and downwash velocity.

CHAPTER XII

LIFTING SURFACE THEORY

12·0. In this chapter we discuss some aspects of the aerofoil regarded as composed of lifting linear elements, and then give a brief account of theories which treat the aerofoil as a vortex sheet over which vorticity is spread at a given rate. Finally, some applications of the acceleration potential are considered.

12·1. Velocity induced by a lifting line element.

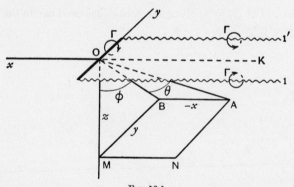

Fig. 12·1.

Consider a horseshoe vortex of infinitesimal span ds and circulation Γ. Taking the origin at the mid-point of the span and the x-axis parallel, but opposed in sense, to the arms 1, 1′ of the horseshoe we calculate the induced velocity at the point A (x, y, z). To this end consider first a single semi-infinite vortex OK of circulation Γ, in the same sense as the circulation about 1. Let this vortex induce at A the velocity (u_1, v_1, w_1). If this vortex OK were shifted to coincide with 1 the induced velocity * would be, by Taylor's theorem,

$$u_1 + \frac{\partial u_1}{\partial y} \cdot \tfrac{1}{2}ds, \quad v_1 + \frac{\partial v_1}{\partial y} \cdot \tfrac{1}{2}ds, \quad w_1 + \frac{\partial w_1}{\partial y} \cdot \tfrac{1}{2} ds,$$

---

* Observe that the effect on the velocity is the same as if no shift were made, and the y-co-ordinate of A were increased by $\tfrac{1}{2}ds$.

while 1′, being of opposite vorticity, would induce at A the velocity

$$- \left(u_1 - \frac{\partial u_1}{\partial y} \cdot \tfrac{1}{2} ds \right), \quad - \left(v_1 - \frac{\partial v_1}{\partial y} \cdot \tfrac{1}{2} ds \right), \quad - \left(w_1 - \frac{\partial w_1}{\partial y} \cdot \tfrac{1}{2} ds \right),$$

so that the total velocity induced at A by the pair 1, 1′ is

$$\frac{\partial u_1}{\partial y} ds, \quad \frac{\partial v_1}{\partial y} ds, \quad \frac{\partial w_1}{\partial y} ds.$$

Projecting A on the plane $x = 0$, the plane $y = 0$ and the z-axis we get the points B, N, M of the figure. Let $OB = n, OA = r$.

The vortex OK induces a velocity q_1 perpendicular to the plane OAB. Thus

$$u_1 = 0, \quad v_1 = - q_1 \cos \phi, \quad w_1 = q_1 \sin \phi,$$

where, from 9·51,

$$q_1 = \frac{\Gamma}{4\pi n} (1 + \cos KOA) = \frac{\Gamma}{4\pi n} (1 + \sin \theta).$$

Hence

$$v_1 = - \frac{\Gamma}{4\pi n^2} z(1 + \sin \theta) = - \frac{\Gamma z}{4\pi(y^2 + z^2)} \left(1 - \frac{x}{r} \right),$$

$$w_1 = \frac{\Gamma y}{4\pi n^2} (1 + \sin \theta) = \frac{\Gamma y}{4\pi(y^2 + z^2)} \left(1 - \frac{x}{r} \right).$$

Thus the induced velocity at A due to 1 and 1′ has components $du_1 = 0$, and

$$dv_1 = - \frac{\Gamma z \, ds}{4\pi} \left\{ - \frac{2y}{(y^2 + z^2)^2} \left(1 - \frac{x}{r} \right) + \frac{xy}{r^3} \frac{1}{y^2 + z^2} \right\},$$

$$dw_1 = \frac{\Gamma \, ds}{4\pi} \left\{ \frac{1}{y^2 + z^2} \left(1 - \frac{x}{r} \right) - \frac{2y^2}{(y^2 + z^2)^2} \left(1 - \frac{x}{r} \right) + \frac{xy^2}{r^3} \frac{1}{y^2 + z^2} \right\}.$$

In addition we have to take account of the velocity induced at A by the vortex ds. This velocity (du_2, dv_2, dw_2) is of magnitude $dq_2 = \Gamma \, ds \sin \alpha/(4\pi r^2)$, where $\alpha = \angle OAN$, and is perpendicular to the plane OAN and therefore parallel to the plane OMN. Thus $du_2 = dq_2 \cos \gamma, \ dv_2 = 0, \ dw_2 = dq_2 \sin \gamma$ where $\gamma = \angle MON$. Now

$$\sin \alpha \cos \gamma = \frac{ON}{r} \cdot \frac{z}{ON} = \cos \theta \cos \phi, \quad \sin \alpha \sin \gamma = \frac{ON}{r} \cdot \frac{-x}{ON} = \sin \theta.$$

Thus, finally the components of velocity induced at A, after an easy reduction, are found to be

$$(1) \quad du = \frac{\Gamma \, ds}{4\pi r^2} \cos \theta \cos \phi, \quad dv = \frac{\Gamma \, ds \sin 2\phi}{4\pi} \left\{ \frac{1 + \sin \theta}{n^2} + \frac{\sin \theta}{2r^2} \right\},$$

$$dw = \frac{\Gamma \, ds}{4\pi} \left\{ \frac{(1 + \sin \theta) \cos 2\phi}{n^2} + \frac{\sin \theta(1 + \cos 2\phi)}{2r^2} \right\}.$$

12·11. Munk's theorem of stagger.

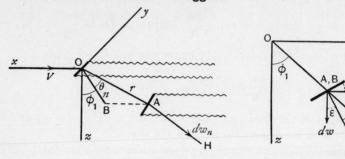

Fig. 12·11.

Consider the lifting element ds_1 placed as before at O, and a lifting element ds_2 placed at A in a *plane parallel to* $x = 0$ making an angle ϵ with ds_1. Let the circulations be Γ_1, Γ_2, and let ϕ_1 denote the angle formerly called ϕ. The normal AH to ds_2 drawn in the plane parallel to $x = 0$ makes an angle ϕ_2 with OB and

$$\epsilon = \phi_1 - \phi_2.$$

If dw_n denotes the induced velocity at A along AH,

$$dw_n = dv \sin \epsilon + dw \cos \epsilon$$

$$= \frac{\Gamma_1 \, ds_1}{4\pi} \left\{ \frac{(1+\sin \theta)\cos (2\phi_1 - \epsilon)}{n^2} + \frac{\sin \theta \cos (2\phi_1 - \epsilon)}{2r^2} + \frac{\sin \theta \cos \epsilon}{2r^2} \right\}$$

$$= \frac{\Gamma_1 \, ds_1}{4\pi} \left\{ \frac{(1+\sin \theta)\cos (\phi_1 + \phi_2)}{n^2} + \frac{\sin \theta \cos (\phi_1 + \phi_2)}{2r^2} + \frac{\sin \theta \cos (\phi_1 - \phi_2)}{2r^2} \right\},$$

using 12·1 (1).

If there is a wind V along xO, the drag induced by ds_1 on ds_2 is $d^2D_{12} = \rho \Gamma_2 \, ds_2 \, dw_n$, so that

$$d^2D_{12} = \rho \, \frac{\Gamma_1 \Gamma_2 \, ds_1 \, ds_2}{4\pi}$$

$$\times \left\{ \frac{(1+\sin \theta)\cos (\phi_1 + \phi_2)}{n^2} + \frac{\sin \theta \cos (\phi_1 + \phi_2)}{2r^2} + \frac{\sin \theta \cos (\phi_1 - \phi_2)}{2r^2} \right\}.$$

To get the drag induced on ds_1 by ds_2 we write $-\theta$ for θ, the angle of stagger. Then

$$d^2D_{21} = \frac{\rho \Gamma_1 \Gamma_2 \, ds_1 \, ds_2}{4\pi}$$

$$\times \left\{ \frac{(1-\sin \theta)\cos (\phi_1 + \phi_2)}{n^2} - \frac{\sin \theta \cos (\phi_1 + \phi_2)}{2r^2} - \frac{\sin \theta \cos (\phi_1 - \phi_2)}{2r^2} \right\}.$$

By addition, the total drag mutually induced on the pair is

$$d^2D_{12} + d^2D_{21} = \frac{\rho \Gamma_1 \Gamma_2 \, ds_1 \, ds_2}{2\pi} \, \frac{\cos (\phi_1 + \phi_2)}{n^2},$$

which is independent of the angle of stagger. This yields

Munk's theorem of stagger.

The *total* induced drag of a multiplane system does not change when the elements are translated parallel to the direction of the wind, provided that the circulations are left unchanged. Thus the total induced drag depends only on the frontal aspect.

Again, when the system is unstaggered ($\theta = 0$),

$$d^2D_{12} = d^2D_{21},$$

and thus if the lifting systems are in the same plane normal to the wind, the drag induced in the first by the second is equal to the drag induced in the second by the first. This result constitutes *Munk's reciprocal theorem.*

The total mutual induced drag is

$$\iint \frac{\rho}{2\pi} \frac{\Gamma_1\Gamma_2 \cos(\phi_1 + \phi_2)}{n^2} \, ds_1 \, ds_2,$$

where ϕ_1 is the angle between the plane containing the normals to the element ds_1 and the projection of the line joining the elements on a plane normal to the wind. (Similarly for ϕ_2 with ds_2.)

12·12. The induced lift.

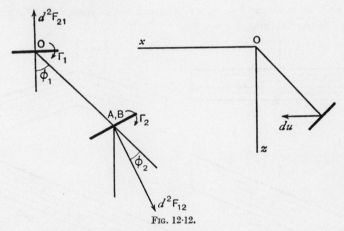

FIG. 12·12.

The velocity induced at ds_2 by ds_1 along Ox is, see 12·1 (1),

$$du = \frac{\Gamma_1 ds_1}{4\pi r^2} \cos\theta \cos\phi_1$$

against the wind. This induces in ds_2 a lift

$$d^2F_{12} = \rho\Gamma_2 ds_2 \, du = \frac{\rho\Gamma_1\Gamma_2}{4\pi r^2} ds_1 \, ds_2 \cos\theta \cos\phi_1.$$

The velocity induced by ds_2 at ds_1 is with the wind; and the induced lift is

$$d^2F_{21} = \frac{\rho \Gamma_1 \Gamma_2}{4\pi r^2} ds_1 \, ds_2 \cos \theta \cos \phi_2.$$

Resolving along n the projection of the line joining the elements on a plane normal to the wind, we get

$$d^2F_{12} \cos \phi_2 - d^2F_{21} \cos \phi_1 = 0.$$

Resolving perpendicularly to n, we get

$$d^2F_{21} \sin \phi_1 - d^2F_{12} \sin \phi_2 = \frac{\rho \Gamma_1 \Gamma_2 \, ds_1 \, ds_2}{4\pi r^2} \cos \theta \sin (\phi_1 - \phi_2).$$

This vanishes if $\phi_1 = \phi_2$ and is small in general. This investigation provides a theoretical method of obtaining the complete interaction between two wings by dividing them into pairs of elements.

12·2. Blenk's method. All the methods hitherto considered for wings of finite aspect ratio have been based on the lifting line theory of Prandtl, and that implied a limitation to aerofoils moving in the plane of symmetry and with a trailing edge which could be regarded as approximately straight. Blenk considers the wing as a lifting surface, that is to say the wing is replaced by a system of bound vortices distributed over its surface rather than along a straight line coinciding with the span. His theory, however, has the limitation that the wing is assumed to be *thin* and practically *plane*. The shapes considered are the following :

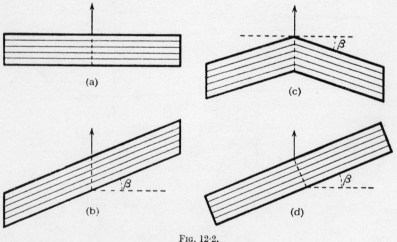

Fig. 12·2.

In all cases the arrow shows the direction of motion.

(a) is a rectangular wing moving in the plane of symmetry ;

(b) is a skew wing in the form of a parallelogram moving parallel to its shorter sides ;

(c) is a symmetrical arrow-shaped wing, i.e. the case of sweep-back;

(d) is the same form as (a), side-slipping, i.e. not moving in the plane of symmetry.

The angle β, which will be regarded as small, is the angle between a leading edge and the normal, in the plane of the wing, to the direction of motion.

In all four cases the hypothesis is made that the bound vortices are parallel to the leading edge, so that in particular for (c) the bound vortex lines are also arrow-shaped.

There are here, as before, two main problems.

I. Given the load distribution and the plan, to find the profiles of the sections.

II. Given the plan and the profiles, to find the load distribution.

We remind the reader that load distribution is proportional to vorticity distribution.

12·21. Rectangular aerofoil.

We consider problem I for case (a) of 12·2, namely the rectangular aerofoil moving in the plane of symmetry.

Taking chord axes with the origin at the centre of the rectangle, we suppose the profiles to be thin (see 8·01), so that the whole aerofoil may be considered to lie in the xy plane. Let $\gamma_1(x, y)$ be the circulation *per unit length of chord* at the point $(x, y, 0)$ so that the circulation round the profile at distance y from the plane of symmetry is

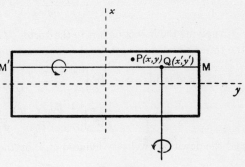

Fig. 12·21.

$$(1) \qquad \Gamma_1(y) = \int_{-c/2}^{c/2} \gamma_1(x, y)\, dx.$$

Introducing " dimensionless " coordinates

$$\xi = 2x/c, \qquad \eta = 2y/b,$$

we can write the above circulation in the form

$$(2) \qquad \Gamma(\eta) = \int_{-1}^{1} \gamma(\xi, \eta)\, d\xi,$$

where $\gamma(\xi, \eta) = c\gamma_1(x, y)/2$ has the dimensions of a circulation.

For $\gamma(\xi, \eta)$ we choose an elliptic distribution over the *span*

$$\gamma(\xi, \eta) = \gamma_0(\xi)\, \sqrt{(1 - \eta^2)},$$

and for $\gamma_0(\xi)$ we shall consider three different functions,

(3)   (i) $\gamma_0(\xi) = a_0 \sqrt{\dfrac{1+\xi}{1-\xi}}$;   (ii) $\gamma_0(\xi) = b_0\sqrt{(1-\xi^2)}$;

(iii) $\gamma_0(\xi) = c_0\xi\sqrt{(1-\xi^2)}$,

where a_0, b_0, c_0 are arbitrary constants. Observe that (i) is the distribution for a thin flat aerofoil in two-dimensional motion (see 8·1). The most general distribution here considered will then be of the form

(4) $\gamma(\xi, \eta) = \left[a_0 \sqrt{\dfrac{1+\xi}{1-\xi}} + b_0\sqrt{(1-\xi^2)} + c_0\xi\sqrt{(1-\xi^2)} \right]\sqrt{(1-\eta^2)}.$

12·22. Calculation of the downwash velocity.

Consider first the velocity induced at $P(x, y, 0)$ by a single bound vortex $M'M$ parallel to the span (lifting line) and the trailing vortices which spring from it, fig. 12·21. Let $Q(x', y', 0)$ be a point on MM'. The circulation at Q is then $\gamma_1(x', y')\, dx'$ and from Q there trails a vortex of circulation (see 11·22)

$$-\frac{\partial \gamma_1(x', y')}{\partial y'}\, dy'\, dx'.$$

Applying the induction law to the vortex MM', and the formulae of 9·51 to the trailing vortex, we get for the downwash velocity at P

(1) $4\pi\, w_1(x')dx' = -\, dx' \displaystyle\int_{-b/2}^{b/2} \frac{\gamma_1(x', y')(x-x')dy'}{QP^3}$

$$-\, dx' \int_{-b/2}^{b/2} \frac{\partial \gamma_1(x', y')}{\partial y'} \frac{dy'}{y'-y}\left(1 - \frac{x-x'}{QP}\right),$$

and the downwash velocity induced at P by the whole aerofoil is

(2) $w(x, y) = \displaystyle\int_{-c/2}^{c/2} w_1(x')dx',$

and therefore, in terms of ξ and η,

(3) $w(\xi, \eta) = \displaystyle\int_{-1}^{1} w(\xi')d\xi',   \quad w(\xi') = \frac{c}{2}\, w_1(x').$

Thus from (1) we get

(4) $w(\xi') = \dfrac{1}{2\pi b}\displaystyle\int_{-1}^{1} \frac{\partial \gamma(\xi', \eta')}{\partial \eta'} \frac{d\eta'}{\eta-\eta'} - \frac{c(\xi-\xi')}{2\pi b^2}\int_{-1}^{1} \frac{\partial \gamma(\xi', \eta')}{\partial \eta'} \frac{d\eta'}{\lambda(\eta-\eta')}$

$$-\, \frac{c(\xi-\xi')}{2\pi b^2}\int_{-1}^{1} \frac{\gamma(\xi', \eta')d\eta'}{\lambda^3},$$

where, A being the aspect ratio,

(5) $\lambda^2 = \dfrac{(\xi-\xi')^2}{A^2} + (\eta-\eta')^2.$

It may be noted that if $\xi' = \xi$ the expression (4) reduces to its first term and agrees with the result given by the lifting line theory (see 11·22).

If now in (4) we put $\gamma(\xi', \eta') = \gamma_0(\xi')\sqrt{(1 - \eta'^2)}$, the elliptic distribution across the span, we get

$$(6) \qquad w(\xi') = \frac{\gamma_0(\xi')}{2b} + \frac{c(\xi - \xi')\gamma_0(\xi')}{2\pi b^2} \int_{-1}^{1} \frac{\eta'\, d\eta'}{\lambda(\eta - \eta')\sqrt{(1 - \eta'^2)}}$$

$$- \frac{c(\xi - \xi')\gamma_0(\xi')}{2\pi b^2} \int_{-1}^{1} \frac{d\eta'}{\lambda^3}\sqrt{(1 - \eta'^2)}.$$

Substitution of (6) in (3) leads to the downwash velocity. The induced drag is then

$$D_i = \tfrac{1}{2}b\rho \int_{-1}^{1}\int_{-1}^{1} \gamma(\xi, \eta)\, w(\xi, \eta)\, d\xi\, d\eta,$$

which includes the suction force (7·5) at the leading edge should it not be rounded.

The integrals in (6) are of elliptic type and cannot be evaluated in terms of elementary functions. Blenk therefore adopts an ingenious but lengthy method of approximation, which unfortunately applies only to the middle part of the wing so that the tip effects are uncertain. The approximation is better for larger aspect ratios. The method leads to replacing (6) by

$$(7)\ \ w(\xi') = \gamma_0(\xi')$$

$$\times \left[A_1 + \frac{B_1}{\xi - \xi'} + C_1(\xi - \xi') + D_1 \log(\xi - \xi') + E_1(\xi - \xi')\log(\xi - \xi') \right],$$

where the coefficients A_1, etc. are functions of η which depend on the particular case of the four considered. For the rectangular wing moving in the plane of symmetry

$$A_1 = \frac{1}{2b}, \qquad B_1 = -\frac{1}{\pi c}\sqrt{(1 - \eta^2)},$$

$$C_1 = \frac{-c}{2\pi b^2 \sqrt{(1 - \eta^2)}}\left(\tfrac{1}{2} - \eta^2 - \log 4A(1 - \eta^2)\right),\ D_1 = 0,\ E_1 = \frac{c}{4\pi b^2\sqrt{(1 - \eta^2)}}.$$

The downwash velocity may now be calculated from (3) with the aid of (7).

To determine the profile of the section at distance y from the centre, we suppose the relative wind to blow along the x-axis. Since our hypothesis is such that the disturbance of the general stream is reasonably small, and since the air must stream along the profile, we have

$$\frac{\partial z}{\partial x} = \frac{w}{V}, \text{ and therefore } z = \frac{1}{V}\int^{x} w(x, y)\, dx.$$

Comparison of this result with the ordinary theory of the lifting line gives the following mean additions to the incidence and curvature for the rectangular wing

$$\Delta\alpha = 0\!\cdot\!059\,\frac{C_L}{A}, \qquad \Delta\frac{1}{R} = 0\!\cdot\!056\,\frac{C_L}{b}.$$

In the case of the swept-back wing the mean increase of incidence, according to Blenk, should be $1·6\beta(\frac{1}{4} - |\,\eta\,|)$ per cent. of the absolute incidence without sweep-back.

12·23. Side-slipping rectangular aerofoil.

Blenk has also applied his method to problem II (see 12·2) for the case (d) with aspect ratio 6. Here we are given a flat rectangular wing so that the downwash velocity must be constant. If we put

$$\gamma(\xi, \eta) = \sqrt{\frac{1 + \xi}{1 - \xi}} \left(a_0\sqrt{(1 - \eta^2)} + a_1\eta + a_2\eta^2 + a_3\eta^3 \right)$$
$$+ \sqrt{(1 - \xi^2)}\left(b_0\sqrt{(1 - \eta^2)} + b_1\eta + b_2\eta^2 + b_3\eta^3 \right)$$
$$+ \xi\sqrt{(1 - \xi^2)}\left(c_0\sqrt{(1 - \eta^2)} + c_1\eta + c_2\eta^2 + c_3\eta^3 \right),$$

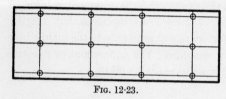

FIG. 12·23.

which contains 12 arbitrary constants, we can calculate $w(\xi, \eta)$ by the method already described, and by equating this to a constant, which is a known multiple of the incidence α, we can choose the 12 constants so that this equation is certainly satisfied at 12 points. The points chosen are the intersections of the lines $\xi = 0$, $\xi = \pm\frac{1}{2}\sqrt{3}$ with the lines $\eta = \pm\frac{1}{4}$, $\eta = \pm\frac{3}{4}$, indicated in fig. 12·23. The lift, drag, and rolling moment coefficient about the x-axis can be calculated when the constants are known.

The results of the calculation are shown in the following table in which $\lambda = 10^3/(2bV\tan\alpha)$, and β is in radians.

β	$a_0\lambda$	$b_0\lambda$	$c_0\lambda$	$a_1\lambda$	$b_1\lambda$	$c_1\lambda$	$a_2\lambda$	$b_2\lambda$	$c_2\lambda$	$a_3\lambda$	$b_3\lambda$	$c_3\lambda$
0	66·8	−1·46	0·23	0	0	0	29·5	−20·5	−7·35	0	0	0
0·1	66·8	−1·46	0·17	−1·58	0·09	−1·07	29·4	−20·1	−6·68	−1·77	3·72	1·95
0·3	66·7	−1·14	0·25	−4·58	−1·01	−1·43	30·5	−20·1	−7·30	−5·40	11·4	2·90
0·5	66·5	−0·48	0·29	−7·54	−1·76	−1·68	31·5	−20·0	−7·37	−9·02	19·0	3·48

Blenk has also investigated the distribution of lift over an aerofoil of finite thickness. The procedure then adopted is the comparison of the given profiles with a series of crescent-shaped Kármán-Trefftz profiles (see 7·6) having equal chords and various thickness but all giving the same lift. By this means an infinitely thin profile is deduced which can be regarded as replacing the given profile and to the resulting thin aerofoil the foregoing method may be applied. The calculated values of the lift coefficient agree very satisfactorily with the measured values.

The reader interested in the above theory should amplify this **short** outline by consulting the original paper,* noting, however, that the notation here adopted differs from Blenk's.

12·3. Aerofoils of small aspect ratio.

For aspect ratios less than about unity the agreement between theoretical and experimental lift distribution breaks down. The matter has been investigated by W. Bollay † who attributes the discrepancy largely to the consequences of Prandtl's Hypothesis III (11·21) that the free vortex lines leave the trailing edge in the same line as the main stream. This assumption leads to a linear integral equation for the circulation (cf. 11·3) whereas Bollay's theory leads to a non-linear equation.

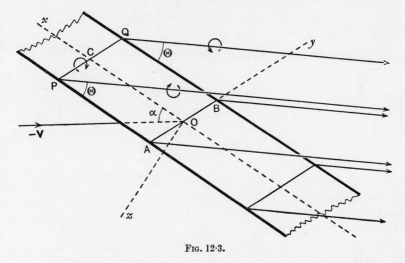

FIG. 12·3.

Fig. 12·3 shows a portion of a flat rectangular aerofoil whose chord c is large compared with the span b. We take the usual chord axes with the origin at the centre of the rectangle.

The bound vorticity $\gamma(x)$ is assumed to be independent of y, that is to say is constant across a span such as PQ but is variable along the chord.

The downwash is also assumed to be independent of y and may therefore be calculated at the centre of each span.

The main point of Bollay's theory, however, is that the trailing vortices which leave the tips of each span such as PQ, make an angle Θ with the chord which is different from α, the incidence. Since the trailing vortices follow the fluid particles which leave the edges of the aerofoil the angle Θ will presumably be a function of x. To a first approximation it is assumed to be constant.

---

* H. Blenk, Der Eindecker als tragender Wirbelfläche. *Z.a.M.M.* 5 (1925).
† *Z. a. M. M.*, Vol. 19 (1939).

12·31. The integral equation. We begin by calculating the velocity induced at the centre C $(x, 0, 0)$ of the span PQ. Consider then the span RS,

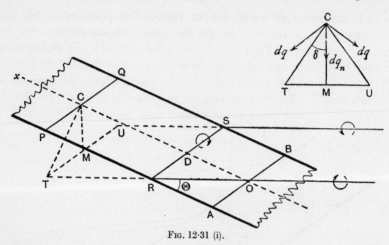

Fig. 12·31 (i).

centre D $(\xi, 0, 0)$. The bound vortex associated with RS is of circulation $\gamma(\xi) d\xi$ and induces at C a velocity, in the z-direction,

$$dw_1 = - \frac{\gamma(\xi)d\xi}{4\pi \cdot CD} (\cos C\hat{R}S + \cos C\hat{S}R),$$

and so the downwash (i.e. the induced velocity in the z-direction) due to the whole set of bound vortices is

$$(1) \qquad w_1 = \int_{-c/2}^{c/2} \frac{- \gamma(\xi)d\xi}{4\pi(x - \xi)} \frac{b}{\sqrt{(\frac{1}{4}b^2 + (x - \xi)^2)}}.$$

Also if u_1 is the induced velocity at C in the x-direction we see from 9·61 that

$$(2) \qquad u_1 = - \tfrac{1}{2}\gamma(x) \quad \text{or} \quad + \tfrac{1}{2}\gamma(x)$$

according as we ascribe to z an infinitesimal negative or positive value.

To get the velocity induced at C by the vortices trailing from R and S, let T, M, U be the projections of P, C, Q on the plane of these vortices. Then the vortex trailing from R induces at C a velocity of magnitude dq perpendicular to the plane RCT, and the vortex trailing from S induces at C a velocity of the same magnitude perpendicular to the plane SUC, see fig. (i). Let dq_n be the resultant induced velocity; direction CM. Then, if the angle $TCM = \delta$, we have

$$dq_n = 2dq \sin \delta = \frac{2\gamma(\xi)d\xi \times \frac{1}{2}b}{4\pi CT^2} (1 - \cos C\hat{R}T),$$

and therefore, for all the trailing vortices, the resultant is

$$(3) \qquad q_n = \frac{b}{4\pi} \int_{-c/2}^{c/2} \frac{\gamma(\xi)d\xi}{\frac{1}{4}b^2 + (x - \xi)^2 \sin^2 \Theta} \left(1 - \frac{(x - \xi)\cos \Theta}{\sqrt{(\frac{1}{4}b^2 + (x - \xi)^2)}}\right).$$

If u_2, w_2 are the components of q_n in the x- and z-directions,

(4) $w_2 = q_n \cos \Theta, \quad u_2 = - q_n \sin \Theta = - w_2 \tan \Theta.$

Now the boundary condition is that there shall be no flow through the aerofoil, i.e. that the normal induced velocity just cancels the normal velocity due to the stream. Therefore

(5) $w_1 + w_2 = V \sin \alpha.$

The required integral equation then follows by substituting the values from (1) and (4). Before doing this it is useful to employ "dimensionless" coordinates as in Blenk's method, i.e. we replace $2x/c$ by x, $2\xi/c$ by ξ, and introduce the aspect ratio $A = b/c$. With these adjustments we get finally

(6) $$\frac{2\pi V \sin \alpha}{A} = - \int_{-1}^{1} \frac{\gamma(\xi)}{(x - \xi)} \frac{d\xi}{\sqrt{(A^2 + (x - \xi)^2)}}$$

$$+ \int_{-1}^{+1} \frac{\gamma(\xi) \cos \Theta \, d\xi}{A^2 + (x - \xi)^2 \sin^2 \Theta} - \int_{-1}^{1} \frac{\gamma(\xi) \cos^2 \Theta (x - \xi) d\xi}{(A^2 + (x - \xi)^2 \sin^2 \Theta) \sqrt{(A^2 + (x - \xi)^2)}},$$

and this is a non-linear equation since Θ is itself a function of $\gamma(\xi)$.

The tentative method of solution which Bollay adopts is to put

(7) $$\gamma(\xi) = \gamma_0 \sqrt{\frac{1 + \xi}{1 - \xi}},$$

which holds for large aspect ratios (see 8·1) and then to use (5) to determine γ in terms of Θ and subsequently to approximate to a suitable mean value for Θ_0.

Into these laborious calculations we shall not enter, but will content ourselves with reproducing a graph of (C_N, α) where C_N is the coefficient of normal force on the aerofoil, i.e. the normal force divided by $\frac{1}{2}bc\rho V^2$.

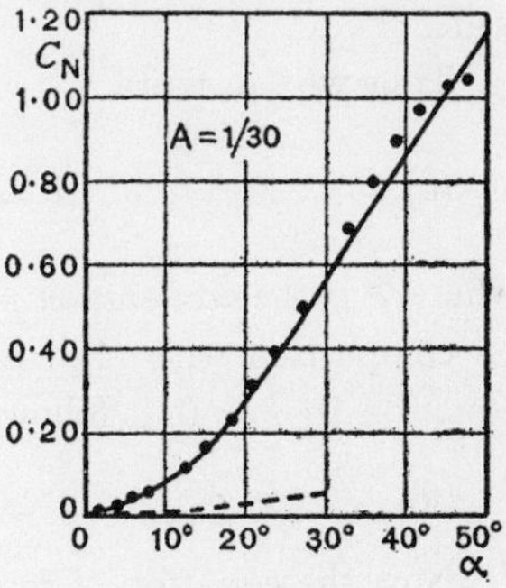

The full line represents C_N as calculated for $A = 1/30$. The line — — — represents the same calculations on the lifting line theory, and the dots represent certain experimental values obtained by H. Winter in 1935.

It is worth observing that aerofoils of very small aspect ratio may have a stalling incidence as high as 45° so that α is not necessarily a small angle.

Fig. 12·31 (ii).

12·32. Zero aspect ratio. The limiting case of zero aspect ratio ($c \to \infty$) is particularly interesting, for Bollay finds that here

$$C_N = 2 \sin^2 \alpha,$$

and this is precisely the behaviour predicted by Isaac Newton for a flat plate which experiences a normal force proportional to the time rate of change of momentum of inelastic fluid particles impinging on it. In fact here we should have

$$N = \rho V \sin \alpha \cdot V \sin \alpha \cdot S,$$

whence the above value of C_N.

12·4. The acceleration potential.

Consider an aerofoil placed in a uniform wind $-\mathbf{V}$ in the negative direction of the x-axis. We can as usual consider the aerofoil replaced by bound vortices at its surface enclosing air at rest and accompanied by a wake of free trailing vortices. Outside the region consisting of the bound vortices and the wake the motion is irrotational and there is a velocity potential ϕ such that the air velocity is

(1) $$\mathbf{q} = -\nabla\phi.$$

If $\mathbf{v}$ is the velocity induced by the vortex system, we have

(2) $$\mathbf{q} = -\mathbf{V} + \mathbf{v},$$

and since the motion is steady the acceleration (21·31) is

(3) $$\mathbf{a} = \frac{d\mathbf{q}}{dt} = (\mathbf{q}\,\nabla)\,\mathbf{q}.$$

If we make the hypothesis that the magnitude of $\mathbf{v}$ is small * compared with that of $\mathbf{V}$, (3) can be written

(4) $$\mathbf{a} = -(\mathbf{V}\,\nabla)\,\mathbf{q} = -V\frac{\partial\mathbf{q}}{\partial x} = V\frac{\partial}{\partial x}(\nabla\phi),$$

using (1).

Thus we can write

(5) $$\mathbf{a} = -\nabla\Phi, \quad \Phi = -V\frac{\partial\phi}{\partial x},$$

where Φ is the *acceleration potential*, the appropriateness of the term being seen on comparison with (1). Also since the velocity potential satisfies Laplace's equation $\nabla^2\phi = 0$, it follows from (5) that

(6) $$\nabla^2\Phi = 0.$$

Now the equation of motion of the air is (9·2), assuming incompressibility and neglecting external forces,

(7) $$\mathbf{a} = \frac{d\mathbf{q}}{dt} = -\nabla\left(\frac{p}{\rho}\right),$$

and this shows that an acceleration potential always exists. Only, however, with our special hypothesis does it satisfy Laplace's equation. Comparing

---

* This hypothesis will fail at a stagnation point, for then $\mathbf{v} = \mathbf{V}$. This will, however, give rise to no subsequent difficulty.

(5) and (7) we see that Φ and p/ρ can differ only by a constant, and we can take

$$(8) \qquad \Phi = \frac{p - \Pi}{\rho},$$

where Π is the pressure at infinity. There is no importance beyond physical definiteness in this particular choice of the constant.

12·41. Lifting surface. One of the most successful applications of the acceleration potential has been found in the theory of thin aerofoils which can be sufficiently approximated by replacing them by their plan areas in the xy plane.

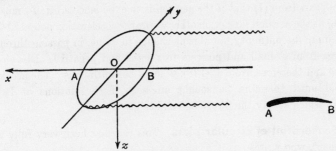

Fig. 12·41.

Considering such an aerofoil, we replace it by its plan area but remark that the sections may be in fact quite reasonable aerofoil profiles as shown by the section AB on the right of fig. 12·41. If we denote by p_U the pressure at a point on the upper or suction face and by p_L the pressure at the corresponding point on the lower or pressure face, we have from 12·4 (8)

$$(1) \qquad p_L - p_U = \rho(\Phi_L - \Phi_U),$$

where Φ_L, Φ_U are the corresponding values of the acceleration potential. Thus we have for the lift and pitching moment

$$(2) \qquad L = \rho \int_{(S)} (\Phi_L - \Phi_U) dS, \qquad M = \rho \int_{(S)} x(\Phi_L - \Phi_U) dS,$$

where the integrals are taken over the surface of the plan.

From (2) it follows that the centre of pressure is at distance $x_p = M/L$ from the origin.

The downwash velocity w is obtained by equating the values of the z-component of the acceleration as given by 12·4 (4) and (5). Thus

$$V \frac{\partial w}{\partial x} = \frac{\partial \Phi}{\partial z},$$

and therefore since w vanishes when $x = \infty$, i.e. far upstream,

$$(3) \qquad w = \frac{1}{V} \int_{\infty}^{x} \frac{\partial \Phi}{\partial z} dx.$$

The induced drag is then

$$(4) \qquad D_i = \int_{(S)} (p_L - p_U) \frac{w}{V} \, dS = \frac{\rho}{V} \int_{(S)} (\Phi_L - \Phi_U) w \, dS.$$

Should the leading edge be sharp instead of rounded this expression for D_i will include a suction force (cf. 7·5), for which due allowance must be made.

As in 12·22 the profile $z = z(x, y)$, when y is given, is determined by

$$(5) \qquad z = \frac{1}{V} \int^x w(x, y) \, dx.$$

It appears from (1) that if the aerofoil is to experience a lift, Φ_L must differ from Φ_U at the aerofoil surface, in other words the acceleration potential Φ must have a jump discontinuity, an abrupt change in value, in passing through the plan area from z small and positive to z small and negative. Elsewhere the pressure and therefore, from 12·4 (8), Φ must be continuous. The acceleration potential must therefore be sought amongst those solutions of Laplace's equation which satisfy the foregoing conditions.

12·5. Aerofoil of circular plan.

This case has been very fully treated by Kinner[*], who writes

$$(1) \qquad x = a\sqrt{(1 - \mu^2)} \cdot \sqrt{(1 + \eta^2)} \cos \phi,$$
$$y = a\sqrt{(1 - \mu^2)} \cdot \sqrt{(1 + \eta^2)} \sin \phi, \qquad z = a\mu\eta,$$

so that the surfaces $\mu = $ constant are hyperboloids of revolution of one sheet, the surfaces $\eta = $ constant are ellipsoids of revolution, and the surfaces $\phi = $ constant are planes through the z-axis. Every point of space is included exactly once if

$$-1 \leqslant \mu \leqslant 1, \qquad 0 \leqslant \eta < \infty, \qquad 0 \leqslant \phi < 2\pi,$$

and $\eta = 0$, $\mu = 0$ is the circle $z = 0$, $x^2 + y^2 = a^2$, which is taken as the plan form.

With these coordinates Laplace's equation becomes

$$\frac{\partial}{\partial \mu}\left[(1 - \mu^2)\frac{\partial \Phi}{\partial \mu}\right] + \frac{\partial}{\partial \eta}\left[(1 + \eta^2)\frac{\partial \Phi}{\partial \eta}\right] + \left(\frac{1}{1 - \mu^2} - \frac{1}{1 + \eta^2}\right)\frac{\partial^2 \Phi}{\partial \phi^2} = 0.$$

Solutions of this equation are constructed which have the necessary mathematical properties described in 12·41.

Space will not allow a detailed discussion of Kinner's results but some of his conclusions for the circular plate aerofoil are interesting. For this he gets (see 12·41)

$$C_L = 1 \cdot 82 \tan \alpha, \qquad C_m = 0 \cdot 44 \tan \alpha, \qquad x_p = 0 \cdot 515 a.$$

* W. Kinner. Die kreisförmige Tragfläche auf potentialtheoretischer Grundlage. *Z.a.M.M.* 18 (1937).

$C_{D_i} = 1.82 \tan^2 \alpha$ or $0.82 \tan^2 \alpha$ according as suction is or is not included. Fig. 12·5 shows the polar curves, the full line corresponding with the case of no suction.

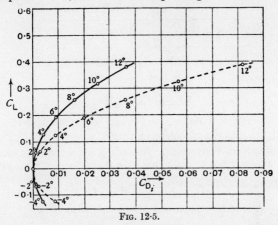

Fig. 12·5.

It is interesting to note that x_p being independent of incidence there is a fixed centre of pressure at a distance 0·243 of the chord ($=2a$) from the leading edge, i.e. practically at the quarter point.

12·6. Aerofoil of elliptic plan.

This case provides an obvious but not too easy extension of the method used for the circular plan form and has been discussed by Krienes[*], who transforms Laplace's equation into other coordinates by the use of Weierstrassian elliptic functions. Appropriate solutions are then constructed. The following results are obtained in the case of an elliptic plate semi-axes a, b, moving in the direction of the axis b. A is the aspect ratio and C_p is the centre of pressure coefficient.

b/a	A	$C_L/\tan \alpha$	$C_m/\tan \alpha$	C_p
0	∞	2π	—	0·288
0·2	6·37	4·55	1·98	0·283
0·5	2·55	2·99	1·397	0·267
1	1·272	1·82	0·44	0·243
2	0·637	0·99	—	0·208

When applied to the elliptic plate the method of Krienes also allows the investigation of the effect of side slip. The following results are quoted for the case $b/a = 0.2$, $A = 6.37$, β being the angle of side slip (18·33).

---

[*] K. Krienes. Die elliptische Tragfläche auf potentialtheoretischer Grundlage. *Z.a.M.M.* 20 (1940).

β	$C_L/\tan\alpha$	$C_m/\tan\alpha$	$C_l/\tan\alpha$	x_p/b	y_p/a
15°	4·16	1·83	0·0384	0·440	0·00925
30°	3·26	1·43	0·074	0·439	0·0277

There is now a rolling moment (see 19·2) and (x_p, y_p) are the coordinates of the centre of pressure.

EXAMPLES XII

1. Supply the intermediate steps which lead to the formulae 12·1 (1).

2. A biplane consists of two parallel lifting lines A_1B_1, A_2B_2 each accompanied by a single trailing vortex at each tip. Neglecting interference between the wings, show that the drag induced on the upper wing (1) by the lower wing (2) is

$$D_{21} = \rho \frac{\Gamma_1\Gamma_2}{2\pi}\left[\log\frac{A_2B_1 + f}{A_2A_1 + f} - \frac{f}{f^2 + h^2}(A_2B_1 - A_2A_1)\right],$$

where f is the aerodynamic stagger and h is the aerodynamic gap.

3. Show that in a parallel biplane with aerofoils of equal span the induced drag is least when each aerofoil exerts the same lift.

4. Show that Munk's theorem of stagger can be inferred from the principle that the work done by the induced drag reappears as energy in the wake.

5. Show that for zero aspect ratio the distribution $\gamma(\xi)$ of 12·31 (7) is constant along the chord.

6. Show, in the case of zero aspect ratio, that the velocity q_n of 12·31 (3) tends at the centre of the aerofoil to the value $\gamma_0/2 \sin\Theta$ where γ_0 is the constant value of the vorticity distribution.

7. Show that for an aerofoil of zero aspect ratio $\gamma_0 = 2V \sin\alpha \tan\Theta$.

8. Derive the form of Laplace's equation given in 12·5.

9. Show that in Blenk's method, 12·22, the drag without suction force can be obtained by replacing, in the expression for the circulation, $(1 + \xi)/(1 - \xi)$ by $(1 + \xi)/(1 - \xi - \theta)$, and then letting θ tend to zero.

CHAPTER XIII

PROPELLERS

13·0. In this chapter we consider the elementary theory of the screw propeller, and endeavour to show what assumptions are usually made in arriving at methods for numerical calculation.

13·1. Propellers. A propeller consists of a certain number of *blades* rotating about an axis.

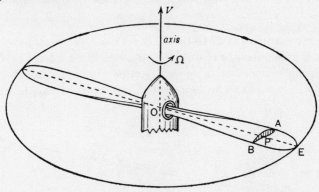

Fig. 13·1.

Propellers are designed to exert thrust to drive the aircraft forward. If T is the thrust in the direction of the axis of rotation, Ω the angular speed of the propeller shaft, Q the torque exerted by the engine and V the forward speed in the direction of the axis of rotation, the work done per unit time by the engine is $Q\Omega$ and by the propeller is TV. Thus a propeller converts torque power into thrust power and the *efficiency* is

$$\eta = \frac{TV}{Q\Omega}.$$

Thrust is obtained by proper shaping of the blades, which are in fact twisted aerofoils.

Every point of a blade lies on a circular cylinder whose axis is the axis of rotation, and therefore as the aircraft advances each point describes a helix or spiral curve on the cylinder on which that point always lies. Of these cylinders there is one of maximum radius. The point of the propeller blade which lies on this cylinder is the *tip*, E in fig. 13·1, of the blade. From the tip E we can draw a perpendicular OE to the axis of rotation. This line may be called the *axis of the blade*. The section of the cylinder by the plane through OE perpendicular to the axis of rotation is called the *propeller disc*, or simply the disc.

If we take a point P on OE such that $OP = r$ and describe a cylinder whose axis is the axis of rotation and whose radius is OP, the points of the surface of the blade which lie on this cylinder will constitute a curve resembling an aerofoil profile ; the totality of such curves defines the shape of the blade. It is, however, customary to define the shape of the blade by plane sections. Thus at P the section of the blade will be that made by the plane through P perpendicular to the blade axis, giving, for example, the profile marked AB in fig. 13·1. Such a section is called a *blade profile*.

The portion of a blade between the blade profiles at distances r and $r + dr$ from the axis of rotation is called a *blade element*.

13·2. How thrust is developed.

Each blade element behaves like an aerofoil and undergoes lift and drag, and of course leaves a wake behind as it moves. In the present section we shall omit all consideration of the velocity induced by the wake as we are here concerned only with a main principle.

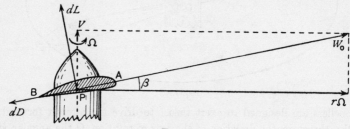

FIG. 13·2.

Fig. 13·2 shows the blade element one of whose bounding profiles is the blade profile of fig. 13·1.

We now introduce two assumptions which will be retained throughout this chapter.

Assumption I. The aircraft is moving in the direction of the axis of rotation of the propeller.

This is only rigorously true for a certain particular incidence of the main lifting system.

Assumption II. Every point of the blade element between the planes r and $r + dr$ has, due to the rotation, a velocity $r\Omega$.

This is clearly the more nearly true, the greater the value of r.

The resultant velocity is then $\mathbf{W}_0$ where $W_0{}^2 = V^2 + r^2\Omega^2$.

If then the blade profile is disposed as in the figure it will undergo a lift dL perpendicular to $\mathbf{W}_0$ and a drag dD opposed to $\mathbf{W}_0$.

If β is the angle which $\mathbf{W}_0$ makes with the direction of $r\Omega$ the blade element has a thrust in the direction of $\mathbf{V}$ of amount

$$(1) \qquad\qquad dT = dL \cos \beta - dD \sin \beta,$$

and the whole blade undergoes a forward thrust equal to the sum of the dT arising from its various elements.

At the same time there is a torque

$$(2) \qquad\qquad dQ = r(dL \sin \beta + dD \cos \beta)$$

opposing the rotation of the blade, and the sum of the dQ is the total torque which the engine must exert to turn the blade.

It appears from this elementary exposition that, if R is the radius of the propeller disc, the maximum speed of a blade profile is $\sqrt{(V^2 + R^2\Omega^2)}$ and that, if compression waves are not to develop, this maximum must be kept below the speed of sound (see 16·1). This places a limitation on the radius of the propeller disc when maximum values of V and Ω are assigned. Again, in order that the speed of sound may be more nearly approached without adverse effects the tip profiles must be made thin (see 16·6). This is also dynamically desirable to avoid too great a thickness at the root of the blade which, for reasons of strength, would be necessary if the blade were unduly massive towards the tip.

We also note that if $\Omega = 0$, then $\mathbf{W}_0 = \mathbf{V}$, $\beta = 90°$ and $dQ = r\,dL$. Thus, if the blade profile is set so that its axis of zero lift is in the direction of $\mathbf{V}$, we shall have $dQ = 0$ and there will be no torque on the propeller shaft from this element. If all the blade profiles are set in this way the propeller is said to be *feathered*. The feathered attitude is usually a possible setting with propellers of variable pitch (see 13·42).

In postulating the existence of lift and drag on the blade elements we are tacitly assuming that there is circulation round these elements, and therefore that the surface of the blade is equivalent to a sheet of bound vortices. These will give rise to a wake and therefore to induced velocity additional to $\mathbf{W}_0$. We proceed to discuss the nature of the wake.

13·3. The slipstream.
This is constituted by the air which has passed through the propeller disc.

Consider the portion of the blade between P on the blade axis and the tip E. This portion behaves as an aerofoil, and, if we adopt the lifting line theory, from the trailing edge PE there escapes a vortex sheet. As the blade rotates this sheet assumes a helicoidal or spiral form, as indicated in the figure, and the slipstream consists of an assemblage of such surfaces, one for each blade. In the wake of the trailing edge there is therefore a downwash, and just ahead there is an upwash from the bound vortices. As the air passes through the propeller disc its axial velocity must be continuous. The downwash velocity at a point of the wake has also a component tangential to the cylinder on which that point lies and the air in the slipstream is therefore rotating about

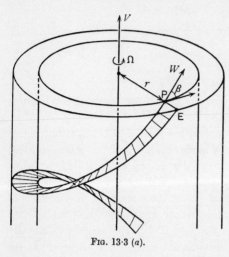

FIG. 13·3 (a).

the line of advance of the aircraft in the same sense as the angular velocity of the propeller.

Ahead of the propeller disc the bound vortices induce a tangential component in the opposite sense, but the undisturbed air into which the propeller disc is advancing can have no axial rotation, so this must be cancelled by the velocity induced by the wake.

The vortex sheet which springs from the trailing edge of a blade is unstable and rolls up into a spiral of concentrated vorticity. In the particular case in which the lift is uniformly distributed along the blade axis, the vortex system due to the blade will consist of (i) a bound vortex along the blade axis, (ii) a spiral vortex springing from the tip of the blade, and (iii) a rectilinear vortex along the part of the axis of rotation which is on the downstream side of the disc. (See the frontispiece.)

Fig. 13·3 (b) illustrates the scheme, and in the case mentioned the circulation round each part has the same value Γ, say. When the lift is not uniformly distributed along the blade axis, the sheet will still roll up almost immediately into concentrated vortices * whose cores will be represented by lines of the spiral type (ii) and the axial type (iii). It is this arrangement which replaces the horseshoe vortices of the lifting line theory of aerofoils. The vortex (ii) has been described as spiral in form. This must not

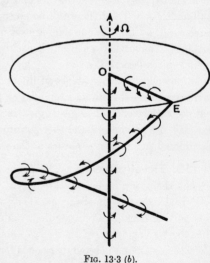

FIG. 13·3 (b).

---

* J. Valensi, Étude de l'air autour d'une hélice, *Thèse*, Paris (1935), has obtained some beautiful photographs of the vortices and developed a method of obtaining quantitative results from their measurement.

be taken to mean that it is a regular helix drawn on a circular cylinder. For some calculations it is convenient to make that an assumption, but in general the diameter of the slipstream contracts (see fig. 13·31) as we proceed downstream and the slipstream only ultimately assumes the cylindrical form in an ideal incompressible inviscid fluid.

Observe that if there are several blades,* each will contribute a vortex of the type (iii) so that in the case of B blades, each with circulation Γ, the axial vortex will be of circulation $B\Gamma$.

13·31. Velocity and pressure in the slipstream.

When the propeller advances with constant velocity $\mathbf{V}$ and rotates with constant angular velocity Ω, the motion of the air at a point fixed in space is not steady, the pressure p and the velocity $\mathbf{q}$ depend on the time. If, however, we take a system of axes of reference fixed in the propeller, and therefore rotating and advancing with it, the motion is steady with regard to these axes. If $\mathbf{q}'$ is the air velocity measured with respect to these moving axes, Bernoulli's equation (see 2·11) becomes

$$(1) \qquad \frac{p}{\rho} + \tfrac{1}{2}q'^2 - \tfrac{1}{2}\Omega^2 r^2 = \text{constant},$$

the last term on the left representing the potential energy of the fictitious field of force introduced by the rotation. If we denote by (q_a, q_r, q_t) the axial, radial, and tangential components of the absolute air velocity $\mathbf{q}$, the components of the relative velocity will be $(q_a, q_r, q_t - r\Omega)$ and therefore

$$q'^2 = q_a{}^2 + q_r{}^2 + (q_t - r\Omega)^2 = q^2 + r^2\Omega^2 - 2q_t r\Omega$$

and so (1) becomes

$$(2) \qquad \frac{p}{\rho} + \tfrac{1}{2}q^2 - q_t r\Omega = \text{constant},$$

where q is now the absolute air speed.

Fig. 13·31 shows schematically a section through the centre O of the propeller disc DD', the hatched part representing the slipstream, the point of view being that of an observer moving with the axes of reference. Outside the slipstream the motion is everywhere irrotational. Far ahead of the propeller the air appears to form a uniform stream $-\mathbf{V}$, as it also does far astern *except in the slipstream.*

The slipstream itself contracts from its greatest diameter DD' at the disc, and asymptotically approaches the form of a circular cylinder typified by the diameter GG' in fig. 13·31.

It is useful to regard the air ahead of the disc and bounded by the surface whose sections are indicated by ADG, $A'D'G'$ in fig. 13·31 as an " extension "

---

* See the frontispiece. The photograph was taken in water in a cavitation tank with the propeller axis horizontal. The vortices are made visible by the cavitation bubbles (which contain water vapour) escaping from the blades.

of the slipstream, but it should be carefully noted that in crossing say DG from inside to outside the slipstream there is an abrupt change of velocity so that the boundary of the actual slipstream is a vortex sheet, whereas in crossing AD

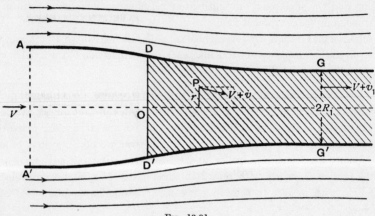

Fig. 13·31.

there is no abrupt change and therefore the boundary of the above " extension " is not a vortex sheet.

Now consider a point P of the slipstream. Let us put $q_t = r\omega$, so that ω is the angular speed of a plane containing the air particle which is at P, and the axis of propeller rotation. Then outside the slipstream $\omega = 0$, for the motion is irrotational and there can therefore be no circulation. Thus (2) can be written

$$(3) \qquad \frac{p}{\rho} + \tfrac{1}{2}q^2 - r^2\omega\Omega = \frac{\Pi}{\rho} + \tfrac{1}{2}V^2,$$

where Π is the pressure at infinity ahead of the disc ; and so (3) is valid throughout the field of flow.

Finally, if the projection of $\mathbf{q}$ on the plane of the section of fig. 13·31 is denoted by $(V + v)$ we have $q^2 = (V + v)^2 + r^2\omega^2$ and therefore

$$(4) \qquad \frac{p - \Pi}{\rho} = r^2\omega \left(\Omega - \frac{\omega}{2} \right) - v \left(V + \frac{v}{2} \right).$$

This is an exact equation which applies throughout the fluid.

If we describe the position of P by cylindrical coordinates (r, θ, z) where z is the distance downstream from O, the pressure p and the velocities v, ω are functions of all three coordinates.

13·4. Interference velocity.
The trailing vortex system described in 13·3 gives rise to induced velocity, known as *interference velocity*.

This velocity will have three components, axial, radial, and tangential, the latter term referring to the tangent to that circular section of the slipstream which passes through the point which we are considering.

Assumption III. The radial component may be neglected.

In the plane of the disc, on the downstream face, the interference velocity will, see fig. 13·3 (*a*), have its axial component opposite to the direction of advance of the propeller and its tangential component in the same sense as the rotation.* Thus relatively to the propeller the total axial component is increased to, say, $V(1 + a)$, and the total tangential component is decreased to, say, $r\Omega(1 - a')$. The numbers a and a' are called *interference factors*. For a propeller they are both positive.

Now consider points P_2, P_1 at radial distance r just ahead and just astern of the disc. By symmetry the bound vortices induce no axial velocity at these points, so the total induced axial velocity is the same at P_1 and P_2 and may be written

$$v = aV = \tfrac{1}{2}v_1.$$

If the bound vortices induce angular velocity ω' at P_1, by symmetry, they must induce angular velocity $-\omega'$ at P_2. Thus the total induced angular velocity can be written as $\tfrac{1}{2}\omega_1 + \omega'$ at P_1 and $\tfrac{1}{2}\omega_1 - \omega'$ at P_2. But P_2 is in the irrotational flow and therefore $\tfrac{1}{2}\omega_1 - \omega' = 0$. Therefore the total angular velocity induced at P_1 by both bound and free vortices is $\omega_1 = 2\omega' = 2a'\Omega$.

13·41. The force on a blade element. To calculate the force we introduce

Assumption IV. Each blade element may be treated as a two-dimensional aerofoil moving with the relative velocity calculated at the downstream face of the disc.

The relative velocity here referred to is the velocity whose axial and radial components at distance r are $V(1 + a)$, $r\Omega(1 - a')$.

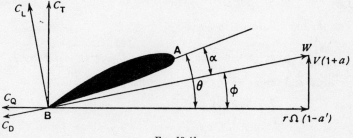

Fig. 13·41.

Let AB be the chord of the blade element.

The angle θ which AB makes with the direction of the tangential velocity is called the *blade angle*. If ϕ is the angle which the relative air velocity W makes with the tangential velocity the incidence is $\alpha = \theta - \phi$, and the cor-

---

* On the upstream face the tangential component vanishes, see 13·3. The axial component is the same on both faces.

responding lift and drag coefficients may be found from the graphs appropriate to the blade profile. If c is the chord, the lift and drag in the blade element are

$$dL = C_L \cdot \tfrac{1}{2}\rho W^2 c \, dr, \quad dD = C_D \cdot \tfrac{1}{2}\rho W^2 c \, dr,$$

where C_L and C_D are the lift and drag coefficients of the blade element.

If we now write

(1) $\qquad C_T = C_L \cos\phi - C_D \sin\phi, \quad C_Q = C_L \sin\phi + C_D \cos\phi,$

we get for the thrust and torque due to the blade element

$$dT^{(1)} = C_T \cdot \tfrac{1}{2}\rho W^2 c \, dr, \quad dQ^{(1)} = C_Q \cdot \tfrac{1}{2}\rho W^2 c r \, dr.$$

If there are B blades the contributions of all the blade elements at distance r will be

$$dT_r = B \, dT^{(1)}, \quad dQ_r = B \, dQ^{(1)}.$$

The projected area of all the blade elements on their chords is $Bc \, dr$ and the area of the annulus between radii r and $r + dr$ is $2\pi r \, dr$. The ratio of these areas is termed the *solidity* of the blade element and is denoted by $\sigma = Bc/(2\pi r)$. Also, from fig. 13·41,

(2) $$\tan\phi = \frac{V}{r\Omega} \cdot \frac{1 + a}{1 - a'},$$

and therefore

(3) $\dfrac{dT_r}{dr} = C_T \cdot \pi\rho\sigma r V^2 (1 + a)^2 \operatorname{cosec}^2 \phi = C_T \cdot \pi\rho\sigma r^3 \Omega^2 (1 - a')^2 \sec^2 \phi,$

(4) $\dfrac{dQ_r}{dr} = C_Q \cdot \pi\rho\sigma r^2 V^2 (1 + a)^2 \operatorname{cosec}^2 \phi = C_Q \cdot \pi\rho\sigma r^4 \Omega^2 (1 - a')^2 \sec^2 \phi.$

13·42. Characteristic coefficients. If we write

(1) $$T_r = \tau_\xi \rho R^4 \Omega^2, \quad Q_r = \kappa_\xi \rho R^5 \Omega^2, \quad \xi = \frac{r}{R},$$

equations (3) and (4) of 13·41 become

(2) $\qquad \dfrac{d\tau_\xi}{d\xi} = C_T \pi\sigma\xi^3 (1 - a')^2 \sec^2 \phi, \quad \dfrac{d\kappa_\xi}{d\xi} = C_Q \pi\sigma\xi^4 (1 - a')^2 \sec^2 \phi.$

The thrust and torque on the whole propeller are then

(3) $\qquad T = \tau\rho R^4 \Omega^2, \quad Q = \kappa\rho R^5 \Omega^2, \quad$ where

(4) $$\tau = \int_0^1 \frac{d\tau_\xi}{d\xi} \, d\xi, \quad \kappa = \int_0^1 \frac{d\kappa_\xi}{d\xi} \, d\xi.$$

In terms of n, the number of revolutions per unit time, and D, the diameter of the disc, we define the *rate of advance coefficient*

(5) $$J = \frac{V}{nD} = \frac{\pi V}{R\Omega} = \pi\xi \frac{1 - a'}{1 + a} \tan\phi,$$

the last term being obvious from an inspection of fig. 13·41.

The efficiency of the propeller is then

$$(6) \qquad \eta = \frac{TV}{Q\Omega} = \frac{J\tau}{\pi\kappa}.$$

For a given propeller the geometrical quantities σ, θ are known for each value of ξ, and also the aerodynamic quantities C_L, C_D for each value of α. Thus, if we know the interference factors a, a' (see 13·7), we can calculate ϕ, J, C_T, C_Q and so obtain the differential coefficients $d\tau_\xi/d\xi$, $d\kappa_\xi/d\xi$ from (2).

Graphical integration will now give *the characteristic coefficients* τ, κ ana therefore η for different values of J. Typical graphs are shown in fig. 13·42.

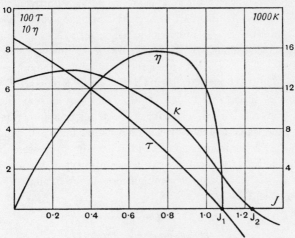

FIG. 13·42.

We observe from (6) that if τ vanishes for a value J_1 of J, η will also vanish for this value, and the graph shows that κ vanishes for a value J_2 of J.

If $J = J_1$, there is no thrust and the propeller is feathered.

If $J_1 < J < J_2$, τ and therefore the thrust is negative but the torque remains positive. Thus the propeller is acting as a *brake*.

If $J > J_2$, both thrust and torque are negative and the propeller acts as a *windmill*, i.e. supplies power instead of consuming it. The efficiency is then $1/\eta$ taken positively.

If $J = J_2$ the propeller is capable of *autorotation*, i.e. of rotating without demanding power from the engine, as in the autogyro.

With regard to J_1, if the propeller makes one revolution in a unit of time it advances the distance $V = J_1 D$. The length $J_1 D$ is the *experimental mean pitch*, and is the distance the propeller advances per complete turn of the blades when no thrust is exerted.

In variable pitch propellers it is possible to rotate the blades, each about the blade axis, and thus obtain a different experimental mean pitch for each setting.

This turning of the blades will of course alter the incidence of every blade element.

13·5. Infinitely many blades.

If we suppose the propeller to have an infinite number of equal blades each carrying an infinitesimal proportion of the total thrust, the situation undergoes a notable simplification in that the principle of momentum is easily applied. Referring to fig. 13·31 we shall consider the air which occupies the slipstream and its upstream " extension " and bounded by the section AA', GG', the former being so far upstream that the velocity is V parallel to the axis of rotation and the pressure is Π, the latter being a long way downstream at a point where the slipstream has become sensibly cylindrical. To this part of the slipstream we apply the suffix unity so that the quantities p, ω, v are denoted by p_1, ω_1, v_1, and are functions of r only and not of the azimuth θ. We denote by $2R_1$ the diameter of this part of the slipstream. The forces acting on the body of fluid here considered are (i) the thrust T, (ii) the pressure thrust due to uniform pressure Π over AA', (iii) the pressure thrust due to p_1 over GG', (iv) the pressure thrust due to pressure p over the curved boundary AG, $A'G'$. A uniform pressure $-\Pi$ applied over the whole boundary yields no resultant force and, supposing this to be applied, (ii) is eliminated and (iii) and (iv) are due to pressures $p_1 - \Pi$ and $p - \Pi$ respectively. If we denote by X the component of the new (iv) in the direction of T, we can equate the resultant force to the net flux of momentum out of the volume $AA'\,GG'$.

Thus we get

$$(1) \quad T + X - \int_0^{R_1} (p_1 - \Pi)\,2\pi r\,dr = \rho \int_0^{R_1} 2\pi r\,dr\,(V + v_1)^2$$
$$- \rho V \int_0^{R_1} 2\pi r\,dr\,(V + v_1),$$

the second integral on the right giving the volume flux out of GG' and therefore by the equation of continuity the corresponding flux at AA'. Thus, taking the value of $p_1 - \Pi$ from 13·31 (4) we get

$$(2) \qquad T + X = 2\pi\rho \int_0^{R_1} [r^2 \omega_1(\Omega - \tfrac{1}{2}\omega_1) + \tfrac{1}{2}v_1^2]\,dr.$$

In 13·51 we shall prove that $X = 0$.

To evaluate p_1 observe that in the cylindrical part of the slipstream, resolving radially,

$$(3) \qquad \frac{dp_1}{dr} = r\omega_1^2\,\rho,$$

which means that p_1 decreases as we move towards the axis of rotation, and since $p_1 = \Pi$ when $r = R_1$ we have

$$(4) \qquad p_1 - \Pi = -\rho \int_r^{R_1} r\omega_1^2\,dr.$$

13·51. The encased propeller. If we consider the propeller with infinitely many blades to be operating in an infinite coaxial cylindrical tube of diameter $2h$, equation 13·5 (1) will still hold. Outside the cylindrical part of the slipstream the velocity will be constant and equal to, say, $V - v_2'$, the fact that it is less than V following from the equation of continuity. If we consider the air outside the slipstream and its " extension ", we shall have by the same argument as that which yields 13·5 (1)

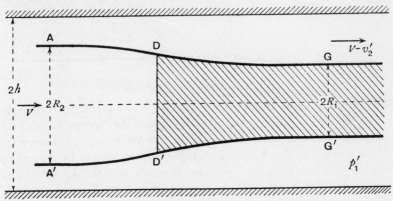

FIG. 13·51.

$$(1) \qquad X + \int_{R_1}^{h} (p_1' - \Pi)\, 2\pi r\, dr = 2\pi\rho \int_{R_1}^{h} (V - v_2')v_2' r\, dr,$$

where p_1' is the pressure over the section outside the slipstream. Now by Bernoulli's theorem (or by 13·31 (4)), noting that $\omega = 0$, $v = -v_2'$)

$$p_1 - \Pi = \rho v_2'(V - \tfrac{1}{2}v_2').$$

Therefore from (1), since v_2' is constant,

$$(2) \qquad X = 2\pi\rho \int_{R_1}^{h} - \tfrac{1}{2}v_2'^2\, r\, dr = -\tfrac{1}{2}\pi\rho(h^2 - R_1^2)v_2'^2,$$

and thus it appears that X is negative, i.e. opposes the thrust.

Now by the equation of continuity, if $2R_2$ is the diameter AA' of the extension of the slipstream in fig. 13·51, we have

$$\pi(h^2 - R_2^2)\, V = \pi(h^2 - R_1^2)(V - v_2')$$

and therefore

$$v_2' = (R_2^2 - R_1^2)\, V/(h^2 - R_1^2),$$

so that from (2)

$$X = -\tfrac{1}{2}\pi\rho V^2 \frac{(R_2^2 - R_1^2)^2}{h^2 - R_1^2},$$

and when $h \to \infty$, i.e. when the casing is absent, $X \to 0$, which proves the assertion made in 13·5.

The problem envisaged in this section may be regarded as approximating to the case of a propeller in a wind tunnel of circular section.

13·6. Froude's law. Consider a propeller with infinitely many blades and introduce

Assumption V. The contraction of the slipstream may be neglected.

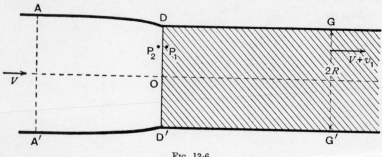

FIG. 13·6.

The diagram now shows a slipstream of radius R, which is cylindrical throughout. As before, AA', GG' represent cross-sections of the "extension" and the slipstream at *infinite* distance from the disc. On GG' at radial distance r the axial velocity will be $V + v_1$ so that v_1 is the velocity induced by the *trailing* vortex system constituting the slipstream. Similarly, ω_1 is the angular velocity induced at the same point by the trailing vortex system. Thus, at a point of GG' each half of the infinite trailing vortex system induces the velocities $\frac{1}{2}v_1$, $\frac{1}{2}\omega_1$. Thus the corresponding velocities induced at a point of the disc at radius r are $\frac{1}{2}v_1$ and $\frac{1}{2}\omega_1$.

This is Froude's law.

13·7. Interference factors. Considering still the propeller with infinitely many blades we introduce

Assumption VI. The induced angular velocity is insufficient to produce appreciable variation of pressure across a section of the slipstream.

With this assumption 13·5 (4) shows that $p_1 = \Pi$.

Now the flux of mass through the annulus of the disc comprised between radii r and $r + dr$ is $2\pi r \, dr \, \rho(V + aV)$. Therefore if dT_r and dQ_r are the thrust and torque on this annulus,

$$dT_r = 2\pi r \, dr \rho(V + aV)v_1, \quad dQ_r = r \cdot 2\pi r \, dr \rho(V + aV)\omega_1 r^2,$$

by the principles of linear and angular momentum. Therefore

(1) $$\frac{dT_r}{dr} = 4\pi r \rho V^2 a(1 + a), \quad \frac{dQ_r}{dr} = 4\pi r^3 \rho V \Omega(1 + a)a',$$

since from 13·4, $v_1 = 2aV$, $\omega_1 = 2a'\Omega$.

Assumption VII. The formulae (1) can be applied to a propeller with a finite number of blades.

If we equate the values (1) to the corresponding values of 13·41 (3), (4), we get

$$\frac{a}{1 + a} = \frac{C_T \sigma}{2(1 - \cos 2\phi)}, \qquad \frac{a'}{1 - a'} = \frac{C_Q \sigma}{2 \sin 2\phi},$$

and from 13·42 (5)

$$\tan \phi = \frac{1 + a}{1 - a'} \frac{J}{\pi \xi}.$$

These three equations then determine a, a', and ϕ. Graphical methods can be applied to finding the solution.

EXAMPLES XIII

1. If $k_T = T/\rho n^2 D^4$, $k_Q = Q/\rho n^2 D^5$, show that, with the definitions of 13·42,

$$k_T = \frac{\pi^2}{4} \tau, \quad k_Q = \frac{\pi^2}{8} \kappa.$$

Show also that the efficiency is $J k_T / 2\pi k_Q$.

2. If T, Q, P are the thrust, torque and power of a propeller, show that

$$\frac{1}{\rho} T = \frac{\pi^2 \tau}{J^2} R^2 V^2 = \frac{\pi^4 \tau}{J^4} \frac{V^4}{\Omega^2},$$

$$\frac{1}{\rho} Q = \frac{\pi^2 \kappa}{J^2} R^3 V^2 = \frac{\pi^5 \kappa}{J^5} \frac{V^5}{\Omega^3},$$

$$\frac{1}{\rho} P = \frac{\pi^3 \kappa}{J^3} R^2 V^3 = \frac{\pi^5 \kappa}{J^5} \frac{V^5}{\Omega^2}.$$

3. Prove that the free vortex lines of the absolute motion coincide with the streamlines of the relative motion in the slipstream.

4. Prove that the total circulation round the blade elements at radius r is $\pi \sigma C_L r W$.

5. Show that the loss of energy for the blade elements at radius r is, in unit time,

$$dE = (1 - a') \Omega \, dQ_r - (1 + a) V \, dT_r.$$

Hence prove that

$$dE = \tfrac{1}{2} C_D \rho W^3 Bc \, dr,$$

which is the work done against the drag of the blade elements in unit time.

6. If ϵ is the angle between the apparent and effective relative velocities at the blade elements at radius r, prove that

$$\epsilon = \frac{Bc}{8\pi r} \left(\frac{C_L}{\sin \phi} - \frac{C_D}{\cos \phi} \right).$$

7. If η_r is the efficiency of the blade elements at radius r, defined by

$$\eta_r = \frac{V}{\Omega} \frac{dT_r}{dQ_r},$$

prove that $\qquad \eta_r = \dfrac{1 - a'}{1 + a} \dfrac{1 - \epsilon \tan \phi}{1 + \epsilon \cot \phi}$, where $\epsilon = \dfrac{C_D}{C_L}$.

8. If a free vortex of circulation $\Gamma(r)$ issues from a point P, at radial distance r, of the disc and proceeds downstream as a regular helix, prove that, at the point P' of the disc at radial distance r', and at angular distance β from P, the components of velocity induced by the helical vortex are

$$q_a = \frac{\Gamma(r)}{4\pi} r \int_0^\infty l^{-3}[r - r' \cos(\theta + \beta)]\, d\theta,$$

$$q_t = \frac{\Gamma(r)}{4\pi} \frac{V}{\Omega} \int_0^\infty l^{-3}[r' - r \cos(\theta + \beta) - r\theta \sin(\theta + \beta)]\, d\theta,$$

$$q_r = \frac{\Gamma(r)}{4\pi} \frac{V}{\Omega} \int_0^\infty l^{-3}[r \sin(\theta + \beta) - r\theta \cos(\theta + \beta)]\, d\theta,$$

where

$$l^2 = r^2 + r'^2 - 2rr' \cos(\theta + \beta) + V^2\theta^2/\Omega^2.$$

9. Draw graphs to show, for the propeller with infinitely many blades, (i) the axial incremental speed, and (ii) the incremental angular speed ω at radius r, in proceeding from far ahead to far astern of the propeller.

Add to (ii) a graph to show the part due to the bound vortices.

10. In a propeller with infinitely many blades, prove that the pressure on the downstream face of the disc exceeds the pressure on the upstream face by $\rho v_1(V + \frac{1}{2}v_1)$, where v_1 is the axial incremental velocity far down in the slipstream and the other incremental velocities are neglected.

Prove that the efficiency is

$$\left(1 + \frac{v_1}{2V}\right)^{-1}.$$

11. In a propeller with infinitely many blades, prove that the pressure jump in passing through the disc is at radius r

$$\rho r^2\omega(\Omega - \tfrac{1}{2}\omega),$$

where $r\omega$ is the tangential velocity at radius r.

CHAPTER XIV

WIND TUNNEL CORRECTIONS

14·0. Aircraft are designed to perform the major part of their flying in natural air, which is but little turbulent, and at a height from the earth which is usually great enough for the boundary effect to be negligible. On the other hand much of the data, on which the designer relies, is obtained from measurements made on a model in a wind tunnel. In such conditions there is a pressure gradient, turbulence, scale effect due to the use of a different Reynolds' number, and a modification of the flow due to the presence of boundaries. This latter modification is accessible to mathematical analysis and it is here proposed to consider the first order approximation to the necessary corrections.

The problem will, as usual, be idealised by making certain simplifying assumptions. In the first place it will be assumed that the windstream in the tunnel, before the model is inserted, is uniform parallel flow, of unlimited length up and down stream ; the wind stream is in fact cylindrical or prismatic, or perhaps confined between parallel planes.

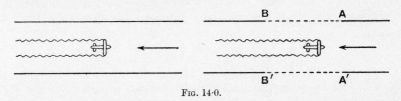

Fig. 14·0.

It will also be assumed that the insertion of the model does not alter the boundaries of the wind stream.

Fig. 14·0 is intended to illustrate the two main cases which arise, the *closed* working section, and the *open* working section. In the latter case our assumption states that the wind stream or jet will be presumed to leap the gap between AA', BB', streaming along the dotted lines AB, $A'B'$ as if the gap in the walls did not exist. We shall, however, find it necessary to go further than this on the road to simplification. In the case of the open working section we shall assume that the gap itself is of infinite length, in short that A, A' and B, B' are points at infinity up and down stream, and that the model is thus exposed to an open cylindrical or prismatic jet unconfined by solid walls.

14·1. Nature of the corrections.

Without entering into any question of transition from model to full-scale, we primarily seek from wind tunnel measurements the values which the measured quantities would have, had the measurements been made in the absence of the tunnel boundaries.* We shall see presently that the quantities mainly affected are the apparent incidence and the induced drag.

Let us denote by C_L', C_D', α' numbers measured in the tunnel, and by C_L, C_D, α the corresponding numbers measured in free flight. Then it will appear (see e.g. 14·4) that there exists an *interference angle* ϵ_I such that

$$C_L = C_L', \quad \alpha = \alpha' + \epsilon_I, \quad C_D = C_D' + \epsilon_I C_L.$$

Our first problem is to evaluate the interference angle. We can then calculate C_D, α and draw the appropriate graphs and polars for the Reynolds' number at which the measurements were made.

Although we have described the situation in terms of a model, exactly the same considerations apply to full-scale tests in a tunnel.

14·2. Boundary conditions.

In the case of a *closed* working section the boundary condition is simply that the velocity component normal to the boundary shall be zero ; $q_n = 0$, where q_n is the normal component.

In the case of an *open* working section the condition is that the pressure shall be continuous as we pass from inside the air stream to the air outside. We shall take the outside air to be at rest, but it may be observed that the case of moving outside air has some aerodynamic interest as in the case of a fuselage or tail plane moving through the air, and at the same time experiencing the influence of the slipstream, where the velocity is different. We have by Bernoulli's theorem

$$\frac{p}{\rho} + \tfrac{1}{2}(V + q_t)^2 + \tfrac{1}{2}q_s^2 = \text{constant},$$

where q_t is the component of the perturbation velocity parallel to V and q_s the component in the plane of a cross-section. At the boundary of the open jet p is constant and therefore $V q_t + \tfrac{1}{2}(q_s^2 + q_t^2)$ is constant. To a first approximation we neglect q_s^2 and q_t^2, and therefore q_t is constant along the boundary. But clearly at infinity $q_t = 0$. Therefore the condition for the open jet is $q_t = 0$ at the boundary.

Before the model is inserted the velocity potential is, say, ϕ_0, and if the direction of the z-axis is taken downstream (see 14·31) we shall have

$$- \partial\phi_0/\partial z = V.$$

If the model were in an unlimited stream V, the velocity potential would

---

* Were the tunnel walls so shaped as to follow streamlines of the motion which would obtain in their absence, no correction would be required for their presence.

be, say, $\phi_0 + \phi_F$, ϕ_F thus denoting the disturbance of the free or unlimited stream due to the presence of the model.

If the tunnel jet is present there will have to be added an interference potential ϕ_I so that the complete velocity potential will be

$$\phi = \phi_0 + \phi_F + \phi_I.$$

In this expression ϕ_0 and ϕ_F are known, and ϕ_I must be determined to satisfy the boundary conditions.

In the case of the open jet we have seen that $q_t = 0$, so that $-\partial\phi/\partial z = V$ at the boundary and therefore $\phi_F + \phi_I$ must be independent of z and so constant, since there is no perturbation at infinity upstream. The boundary condition can therefore be satisfied if we choose $\phi_I = -\phi_F$ at the boundary.

14·3. Reduction to two dimensions.

Consider a model of a monoplane aerofoil placed in the idealised wind tunnel described in 14·0. From the trailing edge free vortices spring and go down-stream as a wake. We shall assume that these vortices are parallel to the generating lines of the tunnel walls, i.e. Hypothesis III of 11·21. We shall further assume, as in the lifting line theory, that the bound vortices lie on a line along the span of the aerofoil. Consider the cross-section C of the tunnel which contains this line. If we *imagine* the trailing vortices to be con-tinued indefinitely *upstream* the section C would be at right angles to a set of ordinary rectilinear vortices and the calculation of

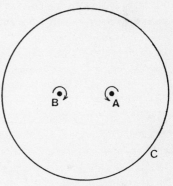

Fig. 14·3.

the velocity components in the plane C would be a two-dimensional problem. Moreover, any velocity component in this plane would be induced in equal amounts by the upstream and downstream parts of the rectilinear vortices in question. It therefore follows that the induced velocity components in the plane of C can be calculated as a two-dimensional problem in rectilinear vortices whose strength is *exactly half the strength of the free vortices* trailing behind the aerofoil (cf. 9·52). Thus, to take the simplest possible case, if we suppose the lift to be uniformly distributed across the span, the aerofoil is equivalent to a horseshoe vortex of constant circulation Γ, the two " arms " of the horseshoe springing from the wing tips. In the plane of C we therefore have a vortex pair A, B, the circulation of each vortex being $\tfrac{1}{2}\Gamma$, and the strength of each therefore being $\Gamma/4\pi$. If the boundary C were absent, the complex potential due to this vortex pair would be, say, w_F. Our problem is therefore to find an " interference " com-plex potential w_I such that the complete potential $w = w_F + w_I$ satisfies the appropriate condition on the boundary C.

In the case of a *closed* jet the boundary C is a streamline $\psi = $ constant, and without loss of generality we may take it to be $\psi = 0$.

In the case of an open jet the discussion in 14·2 shows that on C we must have $\phi = 0$ (or a constant).

It may be remarked that, as far as calculations of induced drag are concerned, we may suppose all the lifting elements to be moved parallel to **V** into the plane of C, for by Munk's stagger theorem (12·11) the total induced drag is not thereby altered, provided the distribution of circulation remains unchanged.

We have therefore reduced the problem of tunnel interference to one of two-dimensional motion.

14·31. Axes of Reference.

As in this chapter we shall be using the complex variable, it is expedient, in order to avoid confusion with the customary notation, $x + iy$, to take as axes of reference the following system ; x-axis to starboard, z-axis downstream and the y-axis so as to complete a right-handed set. If the model is mounted " right-way up " in a horizontal wind tunnel, the y-axis will then be vertically upwards when the span is horizontal. (In experimental work it is often convenient to suspend the model upside-down, and in that case the y-axis would be vertically downwards.) The results which we shall obtain will not involve mention of the axes so that this temporary departure from our standard convention need entail no difficulty, and we can use the customary notation $z = x + iy$ for the complex variable, since the z-coordinate will not enter into our calculations. The velocity components will be denoted by u in the x-direction and v in the y-direction, so that downwash velocity will be denoted by v, which will be negative for downwash and positive for upwash. We shall reserve w for the complex potential, for the component parallel to the z-axis is V to the order of approximation here adopted.

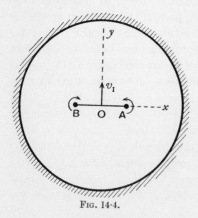

14·4. Circular closed section.

Let the section of the tunnel be a circle C of radius R. To take a specific case, consider a closed tunnel in which an aerofoil AB is placed along the x-axis, the centre of the span coinciding with the centre of the tunnel. Let us replace the aerofoil by a pair of vortices whose strengths are $\kappa = \Gamma/4\pi$. Then if the tunnel were absent, we should have

(1)

$$w_F = i\kappa \log (z - \tfrac{1}{2}b) - i\kappa \log (z + \tfrac{1}{2}b),$$

and therefore, using the method of the circle theorem (5·2), the interference potential is (see 5·2, (vi))

$$(2) \qquad w_I = - i\kappa \log\left(\frac{R^2}{z} - \tfrac{1}{2}b\right) + i\kappa \log\left(\frac{R^2}{z} + \tfrac{1}{2}b\right),$$

and the complete complex potential is

$$(3) \qquad w = w_F + w_I.$$

It is easy to prove that (2) is the complex potential of vortices at the inverse points of A and B with respect to the circle, and of opposite circulations.

The interference velocity is given by

$$(4) \qquad u_I - iv_I = - \frac{dw_I}{dz} = \frac{- bi\kappa R^2}{R^4 - \tfrac{1}{4}b^2z^2} = \frac{- ib\Gamma}{4\pi R^2}\left(1 + \frac{b^2z^2}{4R^4} + \dots\right),$$

and therefore at the centre of the span $u_I = 0$, and $v_I = b\Gamma/(4\pi R^2)$ which is positive. Now if C_L is the lift coefficient in free flight, we have

$$C_L \times \tfrac{1}{2}\rho V^2 S = b\Gamma\rho V$$

and therefore

$$(5) \qquad \Gamma = \frac{SC_LV}{2b}, \qquad b\kappa = \frac{SC_LV}{8\pi},$$

so that

$$(6) \qquad v_I = \frac{C_LSV}{8\pi R^2} = \tfrac{1}{8}\frac{C_LSV}{C},$$

where C is the area of the cross-section of the tunnel. The general effect of the interference is therefore to *decrease* the downwash velocity by the amount v_I, and therefore to *decrease* the downwash angle by the *interference angle*

$$(7) \qquad \epsilon_I = \frac{v_I}{V} = \tfrac{1}{8} \cdot \frac{S}{C} \cdot C_L.$$

Now let α' be the absolute incidence as measured in the wind tunnel and let α be the absolute incidence in free flight which would give the *same* lift coefficient C_L as that found from the tunnel measurements. If ϵ is the angle of downwash in the free flight, we have from 11·24

$$C_L = a_0(\alpha - \epsilon), \quad C_L = a_0(\alpha' - \epsilon + \epsilon_I), \quad C_{D_i} = \epsilon C_L, \quad C_{D_i}' = (\epsilon - \epsilon_I)\,C_L,$$

where C_{D_i} refers to free flight and C_{D_i}' to the tunnel. Comparing these we have

$$(8) \qquad \alpha = \alpha' + \epsilon_I, \qquad C_{D_i} = C_{D_i}' + \epsilon_I C_L.$$

From these results it appears that the wind-tunnel measurements give values of incidence and induced drag which are too small. Thus the corrections are, for the incidence *add* $C_LS/8C$; for the induced drag coefficient *add* $C_L{}^2S/8C$; to the observed values.

Actually tunnel measurements will yield the total drag coefficient C_D but since profile drag is largely independent of incidence the above correction is the only one required.

The above calculation has proceeded on the tacit assumptions that (i) we may use only the value of ϵ_I at the centre of the span, and (ii) that the loading is uniform.

As to (i), (4) shows that the maximum variation of ϵ_I is about $b^4/(16R^4)$ of its value at the centre of the span ; as to (ii) see 14·41.

14·41. Elliptic loading.

In 14·4 we made the crudest hypothesis ; that the loading is uniform. When the loading is *elliptic* we have

$$\Gamma = \Gamma_0 \sqrt{(1 - 4x^2/b^2)},$$

where b is the span. We can now consider the span to be made up of vortex pairs of circulations $- d\Gamma/2$, $d\Gamma/2$ (see 11·22) and therefore of strengths $\mp d\Gamma/4\pi$. If therefore in 14·4 (4) we put $\kappa = - d\Gamma/4\pi$, $z = X$, $b = 2x$ we get, for this vortex pair, the induced velocity dv_I at the point $(X, 0)$,

$$dv_I = \frac{- 2xR^2 d\Gamma}{4\pi(R^4 - x^2X^2)}.$$

In this put $x = \frac{1}{2}b \sin\theta$, and integrate across the span. Then we easily find that

(1) $$v_I = \frac{b\Gamma_0}{4\pi R^2} \int_0^{\pi/2} \frac{\sin^2\theta \, d\theta}{1 - \eta^2 \sin^2\theta}, \qquad \eta = \frac{bX}{2R^2}.$$

Now $$\frac{\sin^2\theta}{1 - \eta^2\sin^2\theta} = \frac{1}{\eta^2}\left(\frac{1}{1 - \eta^2\sin^2\theta} - 1\right),$$

and therefore the integration gives *

(2) $$\epsilon_I = \frac{v_I}{V} = \frac{SC_L}{4\pi R^2}\frac{1}{\eta^2}\left(\frac{1}{\sqrt{(1 - \eta^2)}} - 1\right) = \frac{SC_L}{8C}(1 + \tfrac{3}{4}\eta^2 + \ldots),$$

where we have used the result $\pi b\Gamma_0 = 2SVC_L$ from 11·41.

This is the correction at $(X, 0)$ and agrees with the result of 14·4 if we neglect the term in η^2. This neglected term will be less than 10 per cent. of the leading term if

$$\tfrac{3}{4}\eta^2_{\max} = \tfrac{3}{4}\left(\frac{b}{2R}\right)^4 < \tfrac{1}{10}, \text{ i.e. if } \frac{b}{2R} < 0·6.$$

The correction to induced drag at $(X, 0)$ will be, from 11·23, $\epsilon_I\rho VГ dX$, and therefore the total correction to the drag coefficient will be

$$\Delta C_D = \frac{1}{\frac{1}{2}\rho V^2 S}\int_{-b/2}^{b/2} \epsilon_I\rho\Gamma_0 V\sqrt{(1 - 4X^2/b^2)} \, dX$$

$$= \frac{SC_L{}^2}{8C} \cdot \frac{4}{\pi b}\int_{-b/2}^{b/2} (1 + \tfrac{3}{4}\eta^2 + \ldots)\sqrt{(1 - 4X^2/b^2)} \, dX.$$

---

* Milne-Thomson & Comrie, *Standard Four-Figure Tables* (1945), p. 223, No. 137.

Putting $2X = b \sin \theta$, the integral becomes

$$b \int_0^{\pi/2} (\cos^2 \theta + \tfrac{3}{16} \left(\frac{b}{2R}\right)^4 \sin^2 2\theta + \ldots) d\theta = \frac{\pi b}{4} (1 + \tfrac{3}{16} \left(\frac{b}{2R}\right)^4 + \ldots),$$

so that

$$\Delta C_D = \frac{S C_L^2}{8C} (1 + \tfrac{3}{16} \left(\frac{b}{2R}\right)^4 + \ldots),$$

which again agrees with 14·4 when the second term is neglected, and neglect of this term when $b/2R = 0.75$ entails an error of less than 6 per cent. We can infer from these results that the exact load distribution is not very important in the circular tunnel.

A similar investigation should in strictness be made in examining the tunnel corrections for forms of cross-section other than circular ; but having made this remark we shall be content with the evidence afforded by the present case, and proceed in other cases to assume uniform loading and to use the value of ϵ_J at the centre of the span. From such calculations the correction for other distributions can be inferred by integration.

14·42. Open jet circular section.

Theorem. The corrections for an open jet tunnel of circular section are the same, with reversed signs, as the corrections for a closed jet of the same section.

Proof. Let R be the radius of the section and take the origin at the centre. Let $w_F(z)$ be the complex potential when the tunnel walls are absent.

Now consider the two complex potentials

$$w_C = w_F(z) + \bar{w}_F \left(\frac{R^2}{z}\right), \qquad w_O = w_F(z) - \bar{w}_F \left(\frac{R^2}{z}\right).$$

On the circle $|z| = R$, w_C is the sum of two conjugate complex quantities, and w_O is their difference. Thus on the circle $|z| = R$ we have $\psi = 0$ in the case of w_C, and $\phi = 0$ in the case of w_O. These are therefore the complex potentials for the closed and open sections. The interference potentials are thus $\pm \bar{w}_F(R^2/z)$ and the corresponding corrections differ only in sign. Q.E.D.

Thus, for example, the corrections of 14·4, 14·41 with the signs changed apply to the open jet case.

The above theorem applies without restriction to every vorticity distribution.

14·43. Glauert's theorem.

The theorem proved in 14·42 is capable of the following generalisation due to Glauert.

Theorem. The corrections for an aerofoil whose span is small compared with the dimensions of a tunnel which consists partly of solid walls and partly of open jets are the same, with the signs changed, for the same aerofoil turned through a right angle in a tunnel in which the parts previously open are closed and the parts previously closed are open.

Proof. *ABCD* in fig. 14·43 indicates parts of the tunnel walls, the dotted portions being open. Let the aerofoil, of span b, have its centre at the origin, and in (i) lie along the x-axis, and in (ii) along the y-axis.

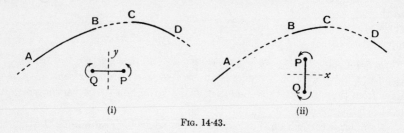

FIG. 14·43.

Then in the two cases, if w_F, w_F' are the free complex potentials

$$w_F = i\kappa \log\left(z - \tfrac{1}{2}b\right) - i\kappa \log\left(z + \tfrac{1}{2}b\right),$$

$$w_F' = i\kappa \log\left(z - \tfrac{1}{2}ib\right) - i\kappa \log\left(z + \tfrac{1}{2}ib\right),$$

the aerofoil being supposed replaced by a vortex pair. Expanding the logarithms

$$w_F = -\frac{i\kappa b}{z} - \frac{i\kappa b^3}{12z^3} + \dots, \qquad w_F' = \frac{\kappa b}{z} - \frac{\kappa b^3}{12z^3} + \dots.$$

Thus if the terms of order b^3/z^3 are negligible in the potentials (and therefore the terms of order b^3/z^4 are negligible in the velocities)

$$w_F = -iw_F' = -\frac{i\kappa b}{z}.$$

Now if w, w' are the complete complex potentials, we have

$$w = w_F + w_I, \qquad w' = w_F' + w_I',$$

so that w_I, w_I' have to be determined from

$$0 = w \pm \bar{w} = -i\kappa b\left(\frac{1}{z} \mp \frac{1}{\bar{z}}\right) + w_I \pm \bar{w}_I,$$

$$0 = w' \mp \bar{w}' = \kappa b\left(\frac{1}{z} \mp \frac{1}{\bar{z}}\right) + w_I' \mp \bar{w}_I',$$

along the boundaries. The upper signs correspond to an open wall in fig. (i) on which $\phi = 0$ and therefore to a closed wall in fig. (ii) on which $\psi = 0$. It is clear, on multiplying the second set of equations by i and comparing with the first set, that

$$w = -iw', \quad \text{and} \quad w_I = -iw_I'.$$

The first relation gives

$$\phi' = -\psi. \quad \psi' = \phi,$$

and therefore when $\psi = 0$, $\phi' = 0$, and when $\psi' = 0$, $\phi = 0$, which proves that open and closed parts correspond in the two cases.

The second relation gives $u_I - iv_I = -i(u_I' - iv_I')$,

and so $$v_I = u_I',$$

which proves that the interference velocities have the same magnitudes but are oppositely directed when the aerofoil is turned through a right angle.   Q.E.D.

14·5. Parallel plane boundaries.

When the aerofoil is between parallel planes we may use the conformal transformation of 4·3.

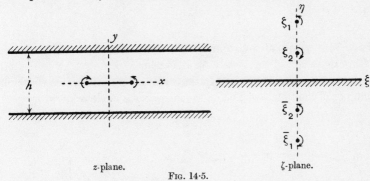

z-plane.   ζ-plane.

FIG. 14·5.

Suppose the aerofoil is placed midway between and parallel to the planes $y = \pm \frac{1}{2}h$. Taking the origin at the centre of the span, the region between the planes is mapped on the upper half of the ζ-plane by

(1) $$\zeta = ie^{\pi z/h}.$$

Replacing the aerofoil by a vortex κ at $(\frac{1}{2}b, 0)$ and $-\kappa$ at $(-\frac{1}{2}b, 0)$ the vortex κ at $z = \frac{1}{2}b$ gives rise to a corresponding vortex κ at $\zeta_1 = ie^{\pi b/2h}$, while the vortex $-\kappa$ at $z = -\frac{1}{2}b$ corresponds with a vortex $-\kappa$ at $\zeta_2 = ie^{-\pi b/2h}$. The image system which makes the real axis in the ζ-plane a boundary consists of a vortex $-\kappa$ at $\overline{\zeta}_1$ and κ at $\overline{\zeta}_2$. Therefore

$$w = w_F + w_I = i\kappa \log \frac{\zeta - \zeta_1}{\zeta - \overline{\zeta}_1} - i\kappa \log \frac{\zeta - \zeta_2}{\zeta - \overline{\zeta}_2},$$

and $w_F = i\kappa \log (z - \frac{1}{2}b) - i\kappa \log (z + \frac{1}{2}b)$, so that since $w_I = w - w_F$,

$$-u_I + iv_I = \frac{dw}{d\zeta} \cdot \frac{d\zeta}{dz} - \frac{dw_F}{dz}$$

(2) $$= \frac{\pi i\kappa \zeta}{h} \left\{ \frac{1}{\zeta - \zeta_1} - \frac{1}{\zeta - \overline{\zeta}_1} - \frac{1}{\zeta - \zeta_2} + \frac{1}{\zeta - \overline{\zeta}_2} \right\} - \frac{i\kappa}{z - \frac{1}{2}b} + \frac{i\kappa}{z + \frac{1}{2}b}.$$

At the centre of the span $z = 0$ and therefore, from (1), $\zeta = i$. With these values we get $u_I = 0$ and

$$v_I = \frac{4\pi\kappa}{h} \frac{e^{\pi b/2h}}{1 - e^{\pi b/h}} + \frac{4\kappa}{b}.$$

If b/h is small we get approximately (see 1·9 (i)) $v_I = \kappa \pi^2 b/6h^2$. Putting $\kappa b = V S C_L / 8\pi$ (see 14·4 (5)) the interference angle is

$$(3) \qquad \epsilon_I = \frac{SC_L}{h^2} \cdot \frac{\pi}{48}.$$

When the aerofoil is placed perpendicularly to the plane boundaries with its centre midway between them we have $\zeta_1 = ie^{i\pi b/2h}$, and $\zeta_2 = ie^{-i\pi b/2h}$. This leads by similar steps to the approximation

$$(4) \qquad \epsilon_I = \frac{SC_L}{h^2} \cdot \frac{\pi}{24}.$$

14·6. The general problem. We have seen that the interference problem can be solved completely for the circular section. The case of other sections can be reduced to this when the conformal transformation which maps the interior of the tunnel section on the interior of a circle is known.

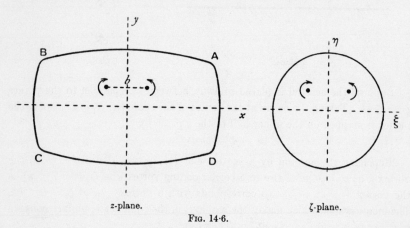

z-plane.   ζ-plane.

Fig. 14·6.

Suppose that the transformation

$$(1) \qquad z = f(\zeta)$$

maps the interior of the tunnel section in the z-plane on the interior of the unit circle $|\zeta| = 1$ in the ζ-plane, everywhere conformally, so that $f'(\zeta)$ does not vanish or become infinite within the unit circle. We shall continue to consider the case of uniform loading, so that the aerofoil is replaced by a vortex pair $(\kappa, -\kappa)$ at distance b apart. The correction for other types of loading can be deduced from this by integration, as already exemplified in 14·41 for the case of elliptic loading.

If the vortices are situated at the points z_1, z_2 in the z-plane, there will be vortices of strengths κ, $-\kappa$ at the corresponding map points ζ_1, ζ_2 in the ζ-plane.

Take first a *closed* working section. Then using the circle theorem (5·2) we get, as in 14·4, for the complex potential

$$(2) \qquad w = i\kappa \log \frac{\zeta - \zeta_1}{\zeta - \zeta_2} - i\kappa \log \frac{\zeta^{-1} - \overline{\zeta_1}}{\zeta^{-1} - \overline{\zeta_2}}.$$

Since $w = w_I + w_F$, and $w_F = i\kappa \log (z - z_1) - i\kappa \log (z - z_2)$, we get

$$(3) \qquad w_I = i\kappa \log \frac{\zeta - \zeta_1}{\zeta - \zeta_2} - i\kappa \log \frac{\zeta - 1/\overline{\zeta_1}}{\zeta - 1/\overline{\zeta_2}} - i\kappa \log \frac{z - z_1}{z - z_2},$$

where the middle term on the right differs from the last term of (2) by the irrelevant constant $i\kappa \log \overline{\zeta_2}/\overline{\zeta_1}$. We now have

$$- u_I + iv_I = \frac{dw}{d\zeta} \frac{d\zeta}{dz} - \frac{dw_F}{dz}$$

$$(4) \qquad = \frac{i\kappa}{f'(\zeta)} \left\{ \frac{1}{\zeta - \zeta_1} - \frac{1}{\zeta - \zeta_2} - \frac{1}{\zeta - 1/\overline{\zeta_1}} + \frac{1}{\zeta - 1/\overline{\zeta_2}} \right\}$$

$$- i\kappa \left\{ \frac{1}{z - z_1} - \frac{1}{z - z_2} \right\},$$

from which the interference angle at the centre or at any other point of the span can be at once deduced.*

When the working section is *open*, we have to arrange that the velocity potential given by (2) is constant on the boundary, to which end we change the sign between the two terms on the right of (2), thus giving, on the model of (4),

$$(5) \qquad - u_I + iv_I = \frac{i\kappa}{f'(\zeta)} \left\{ \frac{1}{\zeta - \zeta_1} - \frac{1}{\zeta - \zeta_2} + \frac{1}{\zeta - 1/\overline{\zeta_1}} - \frac{1}{\zeta - 1/\overline{\zeta_2}} \right\}$$

$$- i\kappa \left\{ \frac{1}{z - z_1} - \frac{1}{z - z_2} \right\}.$$

When the real axes and also the origins correspond in the two planes and the span is on the real axis with its centre at the origin, we have $z = 0$, $\zeta = 0$ at the centre of the span and $z_1 = -z_2 = \frac{1}{2}b$, $\zeta_1 = -\zeta_2 = \frac{1}{2}l$, where

$$(6) \qquad \tfrac{1}{2}b = f(\tfrac{1}{2}l).$$

In the case of a *closed* section, (4) then gives

$$v_I = \frac{4\kappa}{b} - \frac{2\kappa}{f'(0)} \left(\frac{2}{l} - \frac{l}{2} \right).$$

Putting $b\kappa = SC_L V/8\pi$ (see 14·4 (5)) we get

$$(7) \qquad \epsilon_l = \frac{SC_L}{2\pi b^2} \left\{ 1 - \frac{b}{2f'(0)} \left(\frac{2}{l} - \frac{l}{2} \right) \right\}.$$

---

* Had the restriction $f'(\zeta) \neq 0$ not been made, we should get a non-physical infinity at a zero of $f'(\zeta)$.

For an *open* section, (5) leads similarly to

$$(8) \qquad \epsilon_I = \frac{SC_L}{2\pi b^2}\left\{1 - \frac{b}{2f'(0)}\left(\frac{2}{l} + \frac{l}{2}\right)\right\}.$$

We shall apply these general results to two particular cases, (i) the rectangular section, (ii) the elliptic section. This will be found to lead to mapping functions which are doubly periodic. In 14·7 we shall outline the properties of such functions in so far as they concern this problem.

14·7. Jacobian elliptic functions.

The *circular* function $\sin z$ has the period 2π, that is to say, $\sin(z + 2\pi) = \sin z$. It follows that $\sin z$ takes the same value at all the points $z + p \cdot 2\pi$ where p is zero or any integer positive or negative. The number $\frac{1}{2}\pi$ is a *quarter period* for $\sin z$. The circular function $\tan z$ has the smaller period π, and therefore also the period 2π, so that $\frac{1}{2}\pi$ is not only a quarter period of $\tan z$ but also a half period.

The function $\tan z$ has a zero at $z = 0$ and therefore at all the points $p \cdot \pi$. Near $z = 0$ we have an expansion $\tan z = z + \frac{1}{3}z^3 + \ldots$, so that the *leading coefficient* at $z = 0$ is unity. Moreover $\tan z$ has a simple pole at $z = \frac{1}{2}\pi$, for

$$\tan z = \frac{\cos(\frac{1}{2}\pi - z)}{\sin(\frac{1}{2}\pi - z)} = \frac{-1}{z - \frac{1}{2}\pi} \text{ nearly},$$

when z is near the value $\frac{1}{2}\pi$. Thus $\tan z$ has simple poles at all the points $\frac{1}{2}\pi + p \cdot \pi$.

Consider the following linear arrangement of points labelled alternately s and c at intervals or *steps* of $\frac{1}{2}\pi$.

$$(1) \qquad \cdot \text{s} \quad \cdot \text{c} \quad \cdot \text{s} \quad \cdot \text{c} \quad \cdot \text{s} \quad \cdot \text{c} \quad \cdot \text{s} \quad \cdot \text{c} \quad \cdot \text{s},$$

the arrangement being supposed continued indefinitely in both directions. If we regard these points as being marked in an Argand diagram, and denote an arbitrarily chosen point labelled s by K_s, we thereby define the adjacent point c east of s which will be denoted by K_c. If the points are considered to lie on the x-axis and if $K_s = 0$, it would be possible to *define* $\tan z$ by the following requirements:

    (i) $\tan z$ is a periodic function with a simple zero at K_s and a simple pole at K_c.

    (ii) The step $K_c - K_s$ ($= \frac{1}{2}\pi$) from a zero to a pole is a half period.

    (iii) The coefficient of the leading term in the expansion near $z = 0$ is unity (i.e. near $z = 0$, $\tan z = z$ nearly).

Similarly we can define $\cot z$ as the periodic function with a simple zero at K_c, a simple pole at K_s, with the step from the zero to the pole as a half period and having unity for the leading coefficient near $z = 0$ [$\cot z = 1/z$ nearly at the origin]. That the function $\tan z \cot z$ has no poles follows from these definitions and it is therefore a constant $(c,$ say). * Near the origin the product is $z \times 1/z = 1$, so that $c = 1$, and therefore $\tan z$ and $\cot z$ are reciprocals. This well-known property therefore follows from the new definitions.

Again, from the diagram (1), we see that $\tan (z + K_c) = \tan (z + \frac{1}{2}\pi)$ has a pole at $z = K_s$ and a zero at $z = K_c$. Thus $\tan (z + \frac{1}{2}\pi) = $ constant $\times \cot z$, and if we let $z \to 0$ we see that the constant is $- 1$.

We could define the *hyperbolic* function $\tanh z$ in like manner, by considering an arrangement like (1) but along the y-axis, as the periodic function with a zero at K_s and a pole at K_c, with the half period equal to the step $\frac{1}{2}i\pi$ from the zero to the pole, and having unity for leading coefficient near the origin $(\tanh z = z)$.

Now consider the integrals

$$(2) \qquad K = K(m) = \int_0^{\frac{1}{2}\pi} \frac{d\theta}{\sqrt{(1 - m \sin^2 \theta)}} ,$$

$$iK' = iK'(m) = i \int_0^{\frac{1}{2}\pi} \frac{d\theta}{\sqrt{(1 - m_1 \sin^2 \theta)}} ,$$

where $\qquad\qquad 0 \leqslant m \leqslant 1, \quad m_1 = 1 - m.$

We call m the *squared modulus* and observe that $K(m)$ is the same function of m as $K'(m)$ is of m_1, or $K(m) = K'(1 - m)$. When m is given, both K and K' are uniquely determined.

Now when $m = 0$ we have $K(0) = \frac{1}{2}\pi$, $iK'(0) = \infty$, so that $K(0)$ is a quarter period of all the *circular* functions. Similarly when $m = 1$, we have $K(1) = \infty$, $iK'(1) = \frac{1}{2}i\pi$, so that $iK'(1)$ is a quarter period of all the *hyperbolic* functions. A natural extension of these ideas is afforded by the Jacobian *elliptic* functions, which have both $K(m)$ and $iK'(m)$ for quarter periods.

The Jacobian elliptic functions are 12 in number and may be readily defined † with respect to the following doubly infinite rectangular array of *lattice points*, analogous to the arrangement (1).

·s	·c	·s	·c	·s	·c	·s	·c	·s
·n	·d	·n	·d	·n	·d	·n	·d	·n
·s	·c	·s	·c	·s	·c	·s	·c	·s
·n	·d	·n	·d	·n	·d	·n	·d	·n

The pattern is repeated indefinitely on all sides. If we denote by the (complex) number K_s an arbitrary point labelled s, we thereby define adjacent points labelled c, n, d situated respectively east, north, and southwest from s. We

---

* A function which is holomorphic at all points of the plane including infinity must be a constant (Liouville's theorem).

† This account is based on the original and attractive treatment of E. H. Neville, *Jacobian elliptic functions*, Oxford (1951).

denote these points by the complex numbers K_c, K_n, K_d. If we take K_s as origin it is plain that $K_s = 0$ and that

$$K_s + K_c + K_d + K_n = 0.$$

We shall suppose that the scale of the above lattice is such that, with properly chosen axes of reference,

(3) $$K_c = K, \quad K_n = iK', \quad K_d = -K - iK'.$$

Now let the letters q, r, t, v be any permutation of the letters s, c, d, n. Then the elliptic function qr z is defined by the following statements :

(i) qr z is a doubly periodic function with a simple zero at K_q and a simple pole at K_r.

(ii) The step $K_r - K_q$ from the zero to the pole is a half period ; those of the numbers K_c, K_n, K_d, which are different from $K_r - K_q$, are only quarter periods.

(iii) The coefficient of the leading term in the expansion of qr z near $z = 0$ is unity.

This definition should be compared with that given above for tan z.

With regard to (iii), the leading term at the origin is $z, 1/z, 1$, according as the origin is a zero, a pole, or an ordinary point.

The following table shows the poles and periods of all 12 functions :

Periods	Pole K_s	Pole K_c	Pole K_n	Pole K_d
$2K, 4iK'$	cs z	sc z	dn z	nd z
$4K, 2iK'$	ns z	dc z	sn z	cd z
$4K, 2K + 2iK'$	ds z	nc z	cn z	sd z

It is a known theorem that a doubly periodic function devoid of poles is simply a constant.* Now the product (pq z)(qp z) is such a function, since each factor has a simple zero at the (simple) poles of the other. Thus the product (pq z)(qp z) is a constant, and, in view of (iii) above, this constant is unity. Therefore the functions pq z, qp z are reciprocals. If we agree that pp z is to be replaced always by unity, then a similar argument shows that

(4) $$\text{pq } z = \text{pr } z/\text{qr } z$$

however p, q, r may be chosen from s, c, d, n.

A more ambitious application of the same principle is to show that

(5) $$\left(\frac{\text{sn } z \, \text{dn } z}{\text{cn } z}\right)^2 = \frac{1 - \text{cn } 2z}{1 + \text{cn } 2z},$$

which can be done by showing that the quotient of the two sides is a function devoid of poles.

---

* Liouville's theorem. See the footnote on p. 256.

Of the above 12 functions, the 6 which have a pole or a zero at the origin are odd functions of z, the remaining 6 are even functions. Thus, for example,

$$(6) \qquad \operatorname{cs} z = - \operatorname{cs}(-z), \qquad \operatorname{cn} z = \operatorname{cn}(-z).$$

To see this, observe that $\operatorname{cs} z$ and $\operatorname{cs}(-z)$ have the same poles and zeros, and therefore $\operatorname{cs} z \div \operatorname{cs}(-z)$ is a constant which must be -1 in virtue of property (iii) in the definition.

Just as a circular function such as $\sin z$ repeats its values in infinite strips of breadth 2π, so an elliptic function such as $\operatorname{sn} z$ repeats its values in a chequer pattern of rectangles whose sides are of lengths $4K$ and $2K'$. (A similar remark applies to all 12 functions). Thus $\operatorname{sn} z$ assumes the same value at the *congruent* points $z + p \cdot 4K + n \cdot 2iK'$. Within such a rectangle $\operatorname{sn} z$ has a simple pole at the point which is congruent with iK', residue $1/\sqrt{m}$, and a simple pole at the point congruent with $2K + iK'$, residue $-1/\sqrt{m}$.

The following list of critical values will be found useful (proofs omitted).

$$(7) \quad \operatorname{cs}(iK') = -i, \qquad \operatorname{ns} K = 1, \qquad \operatorname{ns}(K + iK') = m^{\frac{1}{2}},$$
$$\operatorname{ds}(iK') = -im^{\frac{1}{2}}, \qquad \operatorname{ds} K = m_1^{\frac{1}{2}}, \qquad \operatorname{cs}(K + iK') = -im_1^{\frac{1}{2}}.$$

The effect of a quarter period step in the lattice is clear from the diagram. Thus, for example,

$$\operatorname{cn}(K + z) = \text{constant} \times \operatorname{sd} z,$$

for the step K (or K_c) to the right changes c to s and n to d.

To discover the constant put $z = -K$, then

$$1 = \text{constant} \times \operatorname{sd}(-K) = - \text{constant} \times \operatorname{sd}(K),$$

so that from the above table of critical values we get

$$(8) \qquad \operatorname{cn}(K + z) = - m_1^{\frac{1}{2}} \operatorname{sd} z.$$

Similarly, we can show that

$$(9) \qquad \operatorname{cn}(iK' + z) = - im^{-\frac{1}{2}} \operatorname{ds} z.$$
$$(10) \qquad \operatorname{sn}(iK' + z) = m^{-\frac{1}{2}} \operatorname{ns} z.$$
$$(11) \qquad \operatorname{sn}(K + z) = \operatorname{cd} z.$$

Lastly, we state, without proof, some further results which will be needed.

It appears from (2) that the quarter periods K, iK' and therefore all the 12 functions depend on the single parameter m. When it is necessary to show this dependence on m we write * for example, $\operatorname{sn}(z \mid m)$ instead of $\operatorname{sn} z$. We then have Jacobi's imaginary transformation

$$(12) \quad \operatorname{sn}(iz \mid m) = i \operatorname{sc}(z \mid m_1), \quad \operatorname{cn}(iz \mid m) = \operatorname{nc}(z \mid m_1), \quad \operatorname{dn}(iz \mid m) = \operatorname{dc}(z \mid m_1).$$

In the neighbourhood of $z = 0$ we have the expansions

$$\operatorname{sn} z = z - (1 + m) \frac{z^3}{3!} + (1 + 14m + m^2) \frac{z^5}{5!} - \dots .$$

---

* The Jacobian functions are usually regarded as dependent on k rather than on m, where $m = k^2$, $m_1 = 1 - k^2 = k'^2$. In this notation $\operatorname{sn} z = \operatorname{sn}(z \mid m) = \operatorname{sn}(z, k)$, and for example $\operatorname{sn}(iz \mid m) = \operatorname{sn}(iz, k) = i \operatorname{sc}(z, k') = i \operatorname{sc}(z \mid m_1)$.

(13) $\operatorname{cn} z = 1 - \dfrac{z^2}{2!} + (1 + 4m)\dfrac{z^4}{4!} - \ldots$

$\operatorname{dn} z = 1 - m\,\dfrac{z^2}{2!} + m(m+4)\dfrac{z^4}{4!} - \ldots$

Note also :

$\operatorname{sn}(z\,|\,0) = \sin z, \qquad \operatorname{cn}(z\,|\,0) = \cos z, \qquad \operatorname{dn}(z\,|\,0) = 1 \ ;$

$\operatorname{sn}(z\,|\,1) = \tanh z, \qquad \operatorname{cn}(z\,|\,1) = \operatorname{dn}(z\,|\,1) = \operatorname{sech} z.$

14·8. Rectangular section.

Let the cross-section of the tunnel be a rectangle of breadth a and height h. Take axes of reference as shown in fig. 14·8.

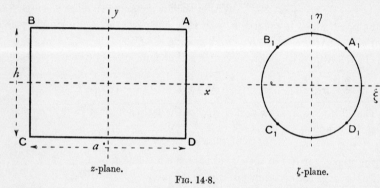

z-plane.   ζ-plane.

Fig. 14·8.

We take Jacobian elliptic functions whose quarter periods K and iK' satisfy the relation

(1) $$\frac{K}{a} = \frac{K'}{h} = \frac{\lambda}{2}.$$

Thus when a and h are given, K, K' and the squared modulus m are uniquely determined.

We begin by showing that the interior of the rectangle in the z-plane is mapped on the interior of the circle $|\zeta| = 1$ by

(2) $$\zeta = \frac{\operatorname{sn}\tfrac12\lambda z \,\operatorname{dn}\tfrac12\lambda z}{\operatorname{cn}\tfrac12\lambda z}.$$

Proof. From 14·7 (5) we see that

(3) $$\zeta^2 = \frac{1 - \operatorname{cn}\lambda z}{1 + \operatorname{cn}\lambda z}.$$

The form (3) is more convenient for our proof, but observe that (3) gives two values of ζ to each value of z, whereas (2) gives only one.

On AD we have $z = \tfrac12 a + iy$ and therefore $\lambda z = K + i\lambda y$. Now from 14·7 (8),

(4) $$\operatorname{cn}(K + i\lambda y) = -\,m_1^{\frac12}\operatorname{sd}(i\lambda y) = i\mu, \text{ say,}$$

where μ is real (but not constant), since sd z is an odd function of z. Therefore when z is on AD,

$$(5) \qquad \zeta^2 = \frac{1 - i\mu}{1 + i\mu}, \qquad (\zeta\bar{\zeta})^2 = \frac{1 + \mu^2}{1 + \mu^2} = 1,$$

so that $|\zeta| = 1$, and as z moves along AD, ζ describes an arc of the circle $|\zeta| = 1$.

Similarly, on AB, $z = x + \frac{1}{2}ih$ and therefore $\lambda z = \lambda x + iK'$. Now from 14·7 (9),

$$\mathrm{cn}(iK' + \lambda x) = -\frac{i}{\sqrt{m}} \, \mathrm{ds} \, \lambda x,$$

which again leads to a relation of the form (5), so that as z moves along AB, ζ describes an arc of the circle $|\zeta| = 1$. The discussion for BC, CD leads to the same conclusion. Also when $z = 0$, we see from (2) that $\zeta = 0$, so that the interiors correspond. It follows that (2) has the required mapping property.   Q.E.D.

Now near $z = 0$, $\mathrm{sn} \frac{1}{2}\lambda z = \frac{1}{2}\lambda z$, $\mathrm{dn} \frac{1}{2}\lambda z = 1$, $\mathrm{cn} \frac{1}{2}\lambda z = 1$, and therefore from (2) $\zeta = \frac{1}{2}\lambda z$ approximately so that

$$(6) \qquad \left(\frac{d\zeta}{dz}\right)_{z=0} = \tfrac{1}{2}\lambda.$$

Thus from 14·6 (7) and (8),

$$(7) \qquad \epsilon_I = \frac{SC_L}{2\pi b^2}\left\{1 - \frac{\lambda b}{4}\left(\frac{2}{l} \mp \frac{l}{2}\right)\right\}.$$

Writing $x = \frac{1}{2}\lambda b$, we get $\frac{1}{2}l$ from (3), and noting that $\mathrm{sn}^2 x + \mathrm{cn}^2 x = 1$,

$$(8) \qquad \frac{2}{l} - \frac{l}{2} = 2\,\mathrm{cs}\,x, \qquad \frac{2}{l} + \frac{l}{2} = 2\,\mathrm{ns}\,x.$$

To compute (7) we need tables of elliptic functions.*
The procedure is as follows. From (1),

$$(9) \qquad \frac{K}{K'} = \frac{a}{h} = \frac{\text{breadth of tunnel}}{\text{height of tunnel}}.$$

Thus, given tables of K, K' as functions of the squared modulus m, we can obtain K/K' as a function of m, and, by comparison with the known value of a/h, deduce the values of m, K, K' separately from the same tables; (1) then gives λ and the tables give the values of (8) and so of (7).

When b is small, the series expansions of 14·7 give the approximations

$$(10) \qquad 1 - x\,\mathrm{cs}\,x = (2-m)x^2/6, \qquad 1 - x\,\mathrm{ns}\,x = -(1+m)x^2/6,$$

---

* Milne-Thomson, *Die elliptischen Funktionen von Jacobi*, Berlin (1931); *Jacobian Elliptic Function Tables*, Dover Publications, New York (1960).

whence, taking the negative sign in (7), we get for a *closed* working section

$$(11) \qquad \epsilon_I = \frac{SC_L}{2\pi b^2} \frac{1 + m_1}{24} \lambda^2 b^2 = \frac{SC_L}{C} \frac{(1 + m_1)KK'}{12\pi},$$

using (1), where $C = ah$ is the area of the section, and $m_1 = 1 - m$ is the complementary squared modulus.

The correction for the *open* working section (positive sign in (7))

$$(12) \qquad \epsilon_I = - \frac{SC_L}{C} \frac{(1 + m) K'K}{12\pi}.$$

Notes. (i) We remark that K is the same function of m as K' is of m_1, so that (11) and (12) have the same mathematical structure. Now it can be proved from the transformation theory of elliptic functions * that $KK' (1 + m_1)$ has the same value whether K/K' is determined from (9) or from $K/K' = 2h/a$. This result means that the correction (11) is the same for a tunnel of height h and breadth a as for one of height $a/\sqrt{2}$ and breadth $h\sqrt{2}$. Correspondingly for the open section we can replace the height by $a \sqrt{2}$ and the breadth by $h/\sqrt{2}$ without altering the correction (12).

(ii) We may regard (11) and (12) as verifications of Glauert's theorem in this case, for if the aerofoil is turned through a right angle in the closed tunnel the correction is (12) with the sign changed.

(iii) The results of 14·5 may be regarded as limiting cases for (11) can be written in either of the forms

$$\frac{SC_L}{h^2} \cdot \frac{K'^2}{12\pi}(1 + m_1), \quad \frac{SC_L}{a^2} \frac{K^2}{12\pi} (1 + m_1).$$

If we let $a \to \infty$, we must have from (1) $K \to \infty$, and therefore $K' \to \pi/2$, $m \to 1$, $m_1 \to 0$, which gives 14·5 (3). Similarly, if we let $h \to \infty$, we get $K' \to \infty$, $K \to \pi/2$, $m \to 0$, $m_1 \to 1$, whence we get the equivalent of 14·5 (4).

(iv) The above results can also be got by the method of images. For example, in the case of the closed section, in order that the normal velocity shall vanish at the walls, suitable image vortices must be introduced, leading to a doubly infinite system of images, whence $u - iv$ can be expressed in terms of elliptic functions with poles at the images (see Ex. XIV, 11).

(v) Rosenhead † has discussed, with the aid of theta functions, the case of elliptic loading, and also the tunnel open on two sides and closed on two sides.

14·9. Mapping an ellipse on a circle.

The interior of the ellipse $(x/\alpha)^2 + (y/\beta)^2 = 1$ is mapped on the interior of the circle $|\zeta| = 1$ by the transformation

$$(1) \qquad z = c \sin \lambda\sigma, \quad \zeta = m^{\frac{1}{4}} \operatorname{sn} \sigma, \quad \lambda = \pi/2K,$$

---

* By the use of Jacobi's imaginary transformation and Landen's transformation, *op. cit.*, p. 259.

† *Proc. Roy. Soc. (A)*, 142 (1933).

where m is the squared modulus of elliptic functions whose quarter periods are determined by

$$(2) \qquad \alpha = c \cosh \frac{\pi K'}{4K}, \quad \beta = c \sinh \frac{\pi K'}{4K}, \quad c^2 = \alpha^2 - \beta^2.$$

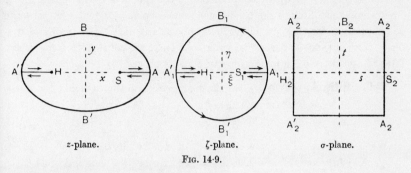

z-plane.   ζ-plane.   σ-plane.

FIG. 14·9.

Proof. Consider the three planes $z = x + iy$, $\zeta = \xi + i\eta$, $\sigma = s + it$. In the σ-plane, draw the rectangle whose corners are the points $\pm K \pm \frac{1}{2}iK'$. The transformation $z = c \sin \lambda\sigma$ gives

$$(3) \qquad x = c \sin \lambda s \cosh \lambda t, \quad y = c \cos \lambda s \sinh \lambda t.$$

Then if $t = \pm \frac{1}{2}K'$, the point (x, y) lies on the ellipse $(x/\alpha)^2 + (y/\beta)^2 = 1$.

Thus on the side $A_2 B_2 A_2'$ of the rectangle we have

$$(4) \qquad x = \alpha \sin \lambda s, \quad y = \beta \cos \lambda s,$$

so that as σ goes from A_2 through B_2 to A_2', z goes along the arc ABA' of the ellipse. Similarly, as σ describes the lower side $A_2'A_2$, z describes the arc $A'B'A$ of the ellipse.

On the side $A_2 S_2 A_2$ of the rectangle $\lambda s = \frac{1}{2}\pi$, and therefore, from (3), $x = c \cosh \lambda t$, $y = 0$, so that as t increases from $-\frac{1}{2}K'$ to $\frac{1}{2}K'$, z moves from A to S and thence back from S to A.

Thus as σ moves round the rectangle, z moves round the ellipse and to and fro along the two slits AS, $A'H$, as shown in fig. 14·9.

Now consider $\zeta = m^{\frac{1}{4}} \operatorname{sn} \sigma$. From 14·7 (10) we have $\operatorname{sn} \sigma \operatorname{sn} (\sigma + iK') = m^{-\frac{1}{2}}$, and writing $s - \frac{1}{2}iK'$ for σ, we get

$$(5) \qquad \operatorname{sn} (s - \tfrac{1}{2}iK') \operatorname{sn} (s + \tfrac{1}{2}iK') = m^{-\frac{1}{2}}.$$

When s is *real* the two factors of the left side of (5) are conjugate complex quantities, and therefore

$$|\zeta| = |m^{\frac{1}{4}} \operatorname{sn} (s + \tfrac{1}{2}iK')| = 1.$$

Therefore as σ moves along $A_2 B_2 A_2'$, ζ moves along the semicircular arc $A_1 B_1 A_1'$ of the circle $|\zeta| = 1$. On the side $A_2 S_2 A_2$ of the rectangle $\sigma = K + it$ and therefore (14·7 (11), (12)) $\zeta = \operatorname{nd}(t \mid m_1)$, so that $\xi = \operatorname{nd}(t \mid m_1)$, $\eta = 0$.

Thus as σ goes along $A_2 S_2 A_2$, z goes from A_1 to S_1 and thence back to A_1. A similar discussion applies to the side $A_2' H_2 A_2'$, and therefore when σ describes the rectangle counterclockwise, ζ moves round the circle $|\zeta| = 1$ and to and fro along the slits $A_1 S_1$, $A_1' H_1$. Therefore the interiors of the circle in the ζ-plane and the ellipse in the z-plane correspond point by point. The slits can be suppressed, for they are occasioned only by discontinuities in σ, and it is easily verified that the functional values of z and ζ are continuous across them. Thus the postulated mapping property of (1) is proved.*

14·91. Elliptic section.

Let the cross-section of the tunnel be the ellipse

$$\frac{x^2}{\alpha^2} + \frac{y^2}{\beta^2} = 1.$$

When the aerofoil is replaced by a vortex κ at $(\tfrac{1}{2}b, 0)$, $-\kappa$ at $(-\tfrac{1}{2}b, 0)$, the general formulae of 14·6 may be applied, using the mapping of 14·9 (1), which gives, near the origin,

$$(1) \qquad z = c\lambda\sigma, \qquad \zeta = m^{\frac{1}{4}}\sigma, \qquad \lambda = \pi/2K, \qquad c^2 = \alpha^2 - \beta^2,$$

so that

$$(dz/d\zeta)_0 = c\lambda m^{-\frac{1}{4}}.$$

Also l is determined from

$$(2) \qquad \tfrac{1}{2}b = c \sin \lambda\sigma_0, \qquad \tfrac{1}{2}l = m^{\frac{1}{4}} \operatorname{sn} \sigma_0,$$

where σ_0 is the corresponding (real) value of σ.

With these determinations we have from 14·6 (7), (8),

$$(3) \qquad \epsilon_I = \frac{SC_L}{2\pi b^2}\left\{1 - \frac{bKm^{\frac{1}{4}}}{c\pi}\left(\frac{2}{l} \mp \frac{l}{2}\right)\right\},$$

the *upper sign* for a *closed* working section, and the lower sign for an open section.

The quarter periods of the elliptic functions are determined from 14·9 (2) by

$$(4) \qquad \tanh \frac{\pi K'}{4K} = \frac{\beta}{\alpha}.$$

When the aerofoil stretches from one focus to the other, we have $b = 2c$, and therefore from (2)

$$\lambda\sigma_0 = \tfrac{1}{2}\pi, \qquad \sigma_0 = K, \qquad \tfrac{1}{2}l = m^{\frac{1}{4}},$$

so that

$$(5) \qquad \epsilon_I = \frac{SC_L}{8\pi c^2}\left\{1 - \frac{2K}{\pi}(1 \mp \sqrt{m})\right\},$$

but with an aerofoil of span comparable with the focal interval it is doubtful whether replacement by a vortex pair is sufficiently accurate.†

---

* This elegant argument is due to Neville, *op. cit.*, p. 259.

† Glauert, *R. and M.*, 1470 (1932), gives the corrections in this case assuming *elliptic*, not uniform, loading in the form $\delta SC_L/C$, where $4\delta = \beta/(\alpha + \beta)$ for the closed and $-\alpha/(\alpha + \beta)$ for the open working section. See also Milne-Thomson, " Application of elliptic functions to wind tunnel interference ", *Proc. Roy. Soc. Edin.*, (*A*), LXIII, 1947.

When b/c is small, equations (2) can be written in the approximate forms,

(6) $\qquad \frac{1}{2}b = c(\lambda\sigma_0 - \frac{1}{2}\lambda^3\sigma^3), \qquad \frac{1}{2}l = m^{\frac{1}{4}}(\sigma_0 - \frac{1}{6}(1+m)\sigma_0{}^3).$

Thus a first approximation is $\lambda\sigma_0 = b/2c$, and so, from 1·9 (iii), the second approximation is

$$\lambda\sigma_0 = \frac{b}{2c}\left(1 + \frac{b^2}{24c^2}\right).$$

With this value (6) gives

$$\frac{1}{2}l = \frac{bm^{\frac{1}{4}}}{2c\lambda}\left\{1 + \frac{b^2}{24c^2}\left(1 - \frac{1+m}{\lambda^2}\right)\right\}, \qquad \frac{2}{l} = \frac{2c\lambda}{bm^{\frac{1}{4}}}\left\{1 - \frac{b^2}{24c^2}\left(1 - \frac{1+m}{\lambda^2}\right)\right\},$$

whence (3) gives the approximation

(8) $\qquad \epsilon_I = \frac{SC_L}{48\pi c^2}\left(1 - \frac{1+m}{\lambda^2} \pm \frac{6m^{\frac{1}{2}}}{\lambda^2}\right),$

the upper sign for the closed working section. When $\alpha \to \beta$, we get the case of a circular section, and it will be found that (8) assumes the correct form for that type of section, see Ex. XIV, 18.

The case where the span is centred on the minor axis * presents no special difficulty, for the general method of 14·6 is of universal application.

EXAMPLES XIV

1. In a closed circular tunnel with the vortex pair replacing the aircraft centrally placed, show that the interference velocity at the centre of the span is the same as the velocity induced at that point by vortices placed at the inverse points (with respect to the circle) of the wing tips and of opposite circulations.

Verify that these image vortices make the circle a streamline.

2. Assuming uniform loading for an aircraft placed centrally in a closed tunnel of circular section, draw a graph to show how v_I varies across the span.

3. For a closed circular tunnel, show that the tunnel correction to the induced drag coefficient is the same as the change in that coefficient due to decreasing the aspect ratio from A to A' in free flight (assuming elliptic loading and the same lift coefficient) where

$$\frac{1}{A'} - \frac{1}{A} = \frac{\pi S}{8C}.$$

4. Obtain the formula 14·41 (1) for elliptic loading and complete the details of the integration.

Prove that, if $\eta^2 < 1$,

$$\frac{1}{\eta^2}\left(\frac{1}{\sqrt{(1-\eta^2)}} - 1\right) = \frac{1}{2} + \frac{3}{8}\eta^2 + \dots.$$

5. Prove that

$$\int_0^{\pi/2}\left(\cos^2\theta + \frac{3}{16}\left(\frac{b}{2R}\right)^4\sin^2 2\theta\right)d\theta = \frac{1}{4}\pi\left(1 + \frac{3}{16}\left(\frac{b}{2R}\right)^4\right).$$

---

* Rosenhead, *Proc. Roy. Soc.* (A), 140 (1933), has discussed the case of elliptic loading, with the aid of theta functions, when the span is along either axis.

6. Prove that for elliptic loading of a centrally placed aerofoil in a closed circular tunnel to the next order of approximation the correction to the induced drag coefficient is

$$\frac{SC_L{}^2}{8C}\left\{ 1 \; + \; \frac{3}{16}\left(\frac{b}{2R}\right)^4 + \; \frac{5}{64}\left(\frac{b}{2R}\right)^8 \right\}.$$

7. In the case of a closed circular tunnel, if the loading is uniform and the span is at distance h from the centre, prove, by means of the circle theorem, that

$$\epsilon_I = \frac{SC_L}{8C} \frac{R^4}{(R^2 - h^2)^2 + \frac{1}{4}b^2h^2}$$

at the centre of the span.

8. Perform the calculations which lead to 14·5 (4).

9. If there are two equal point vortices with circulations in the same sense, prove that on the perpendicular bisector of the line joining them the velocity potential is constant.

Prove also that at any point of the perpendicular bisector the velocity is normal to that line.

10. A wind tunnel has an open rectangular working section of height a and breadth h. Draw the image system for a small centrally placed vortex pair perpendicular to the height (use ex. 9). Hence show that the correction is the same (with the sign changed) as for a closed rectangle of height h and breadth a.

11. The origin is taken at the centre of a rectangular closed tunnel, the x-axis is parallel to the height and the y-axis to the breadth. Vortices of strengths κ, $-\kappa$ are at $(0, \frac{1}{2}b)$, $(0, -\frac{1}{2}b)$. Prove by the method of images, or otherwise, that

$$-u + iv = i\kappa\mu \; \text{sn} \; \lambda(z - \tfrac{1}{2}ib + \tfrac{1}{2}ia) \; - \; i\kappa\mu \; \text{sn} \; \lambda(z + \tfrac{1}{2}ib + \tfrac{1}{2}ia),$$

where K, K', λ are related by

$$\frac{K}{h} = \frac{K'}{a} = \frac{\lambda}{2},$$

and $\mu = \lambda m^{\frac{1}{2}}$.

12. The rectangular working section of a tunnel is open at one pair of opposite sides of the rectangle and closed at the other pair. Draw the image systems for a vortex pair centrally placed (i) parallel to the open sides, (ii) parallel to the closed sides.

13. Obtain the corrections for a closed elliptic section when the aerofoil is replaced by a symmetrical vortex pair on the minor axis.

14. Obtain the corrections for an open elliptic section when the aerofoil is replaced by a symmetrical vortex pair (i) on the major axis, (ii) on the minor axis.

15. Use the result of example 11 to show that

$$\epsilon_I = \frac{SC_L}{2\pi b^2}(1 \, - \, \theta \, \text{cs}(\theta \,|\, m_1)), \;\; \theta = b\lambda/2.$$

Hence obtain 14·8 (11) when θ is small.

16. If $\theta = \pi K'/4K$, use the approximations $K = \frac{1}{2}\pi(1 + \frac{1}{4}m)$, $K' = \log(4/\sqrt{m})$, which hold when m is small, to show that for small m

$$e^{2\theta} = 4/\sqrt{m}.$$

17. If m is determined from 14·91 (4), show that when m is small

$$\frac{\beta^2}{c^2} = \frac{e^{2\theta}}{4},$$

approximately, where $c^2 = \alpha^2 - \beta^2$ and θ is the same as in the preceding example.

18. Use the two preceding examples to show that as $\alpha \to \beta$, β remaining fixed, 14·91 (8) tends to the correction for a circular tunnel of radius β.

19. Show that wind tunnel walls will not interfere with the flow past an aircraft if they are so shaped as to coincide with streamlines in the flow of an unrestricted stream past the same aircraft.

20. An aircraft of volume V is suspended in a wind tunnel in which the pressure decreases downstream with the gradient $-\partial p/\partial z = P$. Show that the measured drag is approximately increased by the amount PV.

CHAPTER XV

SUBSONIC FLOW

15·0. Hitherto the compressibility of the air has been ignored. The Mach number (1·71) has been taken equal to zero.

In this chapter we shall consider *steady* motion at speeds for which the Mach number is not negligible.

The air will be treated as an ideal *compressible* fluid. The effect of viscosity is therefore ignored. The most important consequence of viscosity is probably the skin friction due to the drag in the boundary layer. The neglect of external forces (gravity) has already been explained (2·13) as implying merely that we are concerned only with aerodynamic pressure.

15·01. Thermodynamical considerations.

Consider a *unit mass* of gas, volume v, density ρ, so that

$$(1) \qquad\qquad v\rho = 1.$$

Let T be the absolute temperature (temperature measured from the absolute zero, about $-273°$ C.) of the gas. The gas is said to be *perfect* if it obeys the law (cf. 2·5 (1)).

$$(2) \qquad\qquad pv = RT, \quad \text{or} \quad p = R\rho T,$$

where p is the pressure and R is a constant. Thus of the four quantities p, v, ρ, T only two are independent.

Logarithmic differentiation of (2) gives the relations

$$(3) \qquad\qquad \frac{dp}{p} + \frac{dv}{v} = \frac{dT}{T}, \qquad \frac{dp}{p} = \frac{d\rho}{\rho} + \frac{dT}{T}.$$

We shall treat air as a perfect gas.

Let us imagine our unit mass of gas to receive a small quantity q of heat. *The first law of thermodynamics* asserts that heat is a form of energy.

Thus the small quantity of heat q, with a suitable choice of units of measurement, is equivalent to q units of mechanical energy. Hence communicating the heat q will supply energy to the gas; the gas expands so that its volume increases to $v + dv$, thus doing mechanical work $p\,dv$. We suppose the expansion to take place very slowly, so that no kinetic energy is developed.

Since no energy can disappear, we can write

$$(4) \qquad\qquad q = dE + p\,dv.$$

The quantity dE is the increase in the *internal energy* of the gas ; it is the excess of the energy supplied over the mechanical work done.

Hypothesis. In a perfect gas the internal energy E is a function of the absolute temperature T alone.

This hypothesis is a generalisation from the results of experiment. It is also known as *Joule's law.* It follows that

(5) $$dE = k\,dT$$

and (4) now becomes

(6) $$q = k\,dT + p\,dv.$$

If, in communicating the small quantity q of heat to the gas, the expansion is prevented ($dv = 0$), the temperature of the gas will rise, say dT, and we can write

$$q = c_v\,dT.$$

The quantity c_v is called the *specific heat at constant volume.* It is the quantity of heat required to raise the temperature one unit when the volume is kept constant. Putting $dv = 0$ in (6) gives

(7) $$k = c_v.$$

We similarly define c_p, the *specific heat at constant pressure*, as the quantity of heat required to raise the temperature one unit when the pressure is kept constant. Now, if p is constant, (3) gives $dv/v = dT/T$, and therefore from (6)

$$q = \left(k + \frac{pv}{T}\right)dT,$$

and therefore

$$c_p = k + R = c_v + R$$

from (7).

We therefore conclude that

(8) $$R = c_p - c_v.$$

Hypothesis. In a perfect gas c_p, c_v are constant. This is also based on experiment.

In the above we have denoted the small quantity of heat by q and not by what would seem the more natural notation dQ. The reason for this is that there is actually no function Q of which q is an exact differential. We can, however, write

(9) $$q = T\,dS,$$

where dS is the differential of a function S called the *entropy*.

To justify (9), observe that (6) and (7) give

$$\frac{q}{T} = c_v\frac{dT}{T} + \frac{p}{T}\,dv = c_v\frac{dp}{p} + c_p\frac{dv}{v} = c_v d \log\left(pv^\gamma\right)$$

where

(10) $$q = c_p/c_v,$$

which proves that γ/T is an exact differential, say, dS

$$dS = c_v \, d \log (pv^\gamma),$$

If the state changes from (p_1, v_1) to (p_2, v_2), the increase of entropy is therefore

(11) $$S_2 - S_1 = c_v \log (p_2 v_2{}^\gamma) - c_v \log (p_1 v_1{}^\gamma).$$

The *second law of thermodynamics* asserts that the entropy of an isolated system can never decrease, i.e. $dS \geqslant 0$.

If the entropy remains constant throughout the fluid, the flow is said to be *homentropic*. The condition for homentropic flow is therefore $dS = 0$. It follows from (11) that, if the flow is homentropic,

(12) $$pv^\gamma = \kappa, \quad \text{or} \quad p = \kappa \rho^\gamma,$$

where κ is a constant which depends on the entropy. This is the adiabatic law (cf. 2·32).

The steady flow of air is governed by the equations of motion and continuity (19·5) in the form

(13) $$-\frac{1}{\rho} \nabla p = (\mathbf{q} \nabla)\mathbf{q}, \quad \nabla (\rho \, \mathbf{q}) = 0,$$

and, as there are three unknowns p, ρ, $\mathbf{q}$, these equations are insufficient to determine the motion. In the case of homentropic flow, however, we can adjoin the adiabatic relation (12) and so obtain a determinate system of equations.

To calculate the internal energy we have

$$dE = c_v \, dT = \frac{c_v \, d(pv)}{R} = \frac{d(pv)}{\gamma - 1},$$

and thus, save for an added constant, we have the alternative forms

(14) $$E = \frac{pv}{\gamma - 1} = \frac{p}{(\gamma - 1)\, \rho} = c_v T.$$

15·1. Steady homentropic flow.

In this case we have Bernoulli's equation, 2·32, along a streamline

(1) $$\frac{\gamma}{\gamma - 1} \frac{p}{\rho} + \tfrac{1}{2} q^2 = C.$$

In 1·5 we found the speed of sound in air otherwise at rest, except for the small sonic disturbance, to be

(2) $$c_0{}^2 = \left(\frac{dp}{d\rho}\right)_0 = \frac{\gamma p_0}{\rho_0}.$$

We now define the *local speed of sound* or *sonic speed* by

$$(3) \qquad c^2 = \frac{dp}{d\rho} = \frac{\gamma p}{\rho} = \kappa \gamma \rho^{\gamma-1},$$

which gives the speed of propagation of small disturbances relative to air in the state (p, ρ). With this notation (1) becomes

$$(4) \qquad \frac{c^2}{\gamma - 1} + \tfrac{1}{2}q^2 = C = \frac{c_0{}^2}{\gamma - 1} = \tfrac{1}{2}q^2_{max},$$

where $c_0{}^2/(\gamma - 1)$ is what the left side would become, did the streamline to which (1) refers contain a stagnation point $q = 0$, and $\tfrac{1}{2}q^2_{max}$ is what the left side would become, were there a point at which $c = 0$. It is not asserted that such points actually occur on an arbitrary streamline but merely that each streamline has two such constants c_0, q_{max} associated with it.

Observe that (3) shows that when $c = 0$, $\rho = 0$, and therefore $p = 0$. It follows that q_{max} is the speed with which the air could flow into, or be in contact with, a vacuum, and this speed can be nowhere exceeded.

The graph of q^2 as a function of c^2 is the straight line AB in fig. 15·1. This clearly shows that along a streamline $c \leqslant c_0$, $q \leqslant q_{max}$. The straight line $q^2 - c^2 = 0$ cuts AB at the point C $(c^{\star 2}, q^{\star 2})$ where

$$(5) \qquad q^{\star 2} = c^{\star 2} = \frac{\gamma - 1}{\gamma + 1} q^2_{max} = \frac{2}{\gamma + 1} c_0{}^2.$$

The two portions AC, BC of this line correspond with physically different *régimes*.

At any point of AC we have $q < q^\star = c^\star < c$, so that the Mach number, $M = q/c$, is less than unity, provided that $q < c^\star$. Flow for which $M < 1$ is called *subsonic*.

At any point of BC we have $q > q^\star = c^\star > c$, so that $M > 1$, and the flow is then called *super-*

Fig. 15·1.

sonic. The condition for supersonic flow is therefore $q > c^\star$.

For these reasons the sound speed $c^\star$ is called the *critical speed*. For air

$$\gamma = 1 \cdot 405, \qquad \frac{q_{max}}{c_0} = 2 \cdot 20, \qquad \frac{q_{max}}{c^\star} = 2 \cdot 44.$$

As a simple illustration suppose that the motion is two-dimensional, that the streamlines are concentric circles, fig. 3·7 (iii), and that in each such circle

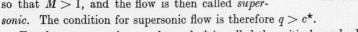

there is the same circulation $2\pi\kappa$. Then if q is the speed at radius r, we have the circulation $2\pi r q = 2\pi\kappa$ and $q = \kappa/r$, as in 4·1. Thus $M = q/c$ gives $c = q/M = \kappa/rM$. Substitution in (4) then yields

$$r^2 = \frac{\kappa^2(\gamma-1)}{2c_0{}^2}\left\{1 + \frac{2}{(\gamma-1)M^2}\right\} \geqslant \frac{\kappa^2(\gamma-1)}{2c_0{}^2} = a^2,$$

say. Thus

(6)
$$r^2 = a^2\left\{1 + \frac{2}{(\gamma-1)M^2}\right\}.$$

From this it appears that r has the minimum value a. This minimum value occurs when $M = \infty$, so that $q = q_{max}$, $c = 0$, and the pressure p is zero. Thus the motion cannot be continued into the region $r < a$. This region may, for example, be vacuous or occupied by a solid core (cf. 4·12).

As r increases, we see from (6) that M steadily decreases and attains the critical value unity when

$$r = r^\star = a\sqrt{\frac{\gamma+1}{\gamma-1}} = 2\cdot44a.$$

Thus in the region $a < r < r^\star$ the flow is supersonic, say between the two inner circles of fig. 3·7 (iii). When $r > r^\star$, $M < 1$ and the flow is subsonic. This is the analogue in a compressible fluid of the rectilinear vortex in an incompressible fluid.

15·2. Irrotational motion.

For irrotational motion there is a velocity potential ϕ such that $\mathbf{q} = -\nabla\phi$, and if the motion is steady, the equation of continuity (21·5) is $\nabla(\rho\,\mathbf{q}) = 0$, or $(\mathbf{q}\nabla)\rho + \rho\nabla\mathbf{q} = 0$, so that (see 21·5 (7))

(1)
$$-\rho\nabla^2\phi + u\frac{\partial\rho}{\partial x} + v\frac{\partial\rho}{\partial y} + w\frac{\partial\rho}{\partial z} = 0.$$

Now
$$d\rho = \frac{d\rho}{dp}\,dp = -\frac{1}{c^2}\rho d\left(\tfrac{1}{2}q^2\right) = -\frac{\rho}{2c^2}d(u^2 + v^2 + w^2),$$

using 15·1 (3) and Bernoulli's theorem in the form 2·31 (1). Since $u = -\partial\phi/\partial x$ we get

$$\frac{\partial\rho}{\partial x} = -\frac{\rho}{c^2}\left[u\frac{\partial u}{\partial x} + v\frac{\partial v}{\partial x} + w\frac{\partial w}{\partial x}\right] = \frac{\rho}{c^2}[u\phi_{xx} + v\phi_{yx} + w\phi_{zx}],$$

where $\phi_{xx} = \partial^2\phi/\partial x^2$, $\phi_{yx} = \partial^2\phi/\partial y\,\partial x$, and so on.

Substituting in (1), we have

(2)
$$\phi_{xx}\left(1 - \frac{u^2}{c^2}\right) + \phi_{yy}\left(1 - \frac{v^2}{c^2}\right) + \phi_{zz}\left(1 - \frac{w^2}{c^2}\right)$$
$$- \frac{2uv}{c^2}\phi_{xy} - \frac{2vw}{c^2}\phi_{yz} - \frac{2wu}{c^2}\phi_{zx} = 0,$$

where $u = - \partial\phi/\partial x$, $v = - \partial\phi/\partial y$, $w = - \partial\phi/\partial z$, and from 15·1 (4)

$$c^2 = \tfrac{1}{2}(\gamma - 1)(q^2_{max} - u^2 - v^2 - w^2).$$

Thus the velocity potential ϕ satisfies the non-linear differential equation (2). When $c = \infty$, (2) reduces to $\nabla^2\phi = 0$, the case of an incompressible fluid.

15·3. Linear perturbations.

In the case of a uniform wind V parallel to the x-axis, 15·2 (2) is identically satisfied by $\phi = Vx$. If we place an aerofoil in this stream, the motion will be perturbed and the velocity potential will become

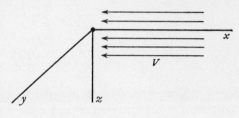

(1) $\Phi = Vx + \phi(x, y, z)$,

so that ϕ is the *perturbation potential*. The velocity at any point was $(- V, 0, 0)$ and is now $(- V + u, v, w)$.

The *perturbation is said to be linear* if u/V, v/V, w/V are small quantities of the first order whose squares and products are negligible.

Observe that there is no limitation of V which may be large or small.

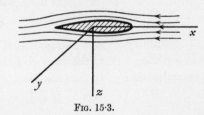

Fig. 15·3.

The approximation will fail near a stagnation point where $- V + u = 0$, so that $u/V = 1$.

The application of the linear theory will therefore be to *thin* aerofoils of *small camber* and at *small incidence*. Such aerofoils are suited to high speeds, so the linear theory should yield a good first approximation.*

To find the perturbations, let the suffix 0 refer to the undisturbed stream. Then $q^2 = (- V + u)^2 + v^2 + w^2 = V^2 - 2uV$ and therefore

(2) $$q = V\left(1 - \frac{u}{V}\right).$$

For the speed of sound, 15·1 (4) gives

$$c^2 = \tfrac{1}{2}(\gamma - 1)[q^2_{max} - (V^2 - 2uV)] = c_0^2 + (\gamma - 1)uV,$$

where c_0 is now the speed of sound in the undisturbed stream, whence we get

(3) $$c = c_0\left[1 + \frac{(\gamma - 1)uV}{2c_0^2}\right] = c_0\left[1 + \tfrac{1}{2}(\gamma - 1)M_0^2\frac{u}{V}\right],$$

---

* The application of the linear theory to aerofoils is particularly associated with the names of Glauert for subsonic, and Ackeret for supersonic flow.

where $M_0 = V/c_0$ is the Mach number of the undisturbed flow. The Mach number is therefore

(4)
$$M = \frac{q}{c} = M_0 \left[1 - \frac{u}{V} \{1 + \tfrac{1}{2}(\gamma - 1) M_0^2\} \right],$$

The pressure change is given by Bernoulli's theorem, 2·31 (1),

(5)
$$p - p_0 = dp = - \rho_0 q\, dq = u \rho_0 V.$$

Since
$$c^2 = \frac{dp}{d\rho} = \frac{p - p_0}{\rho - \rho_0} = \frac{u V \rho_0}{\rho - \rho_0},$$

we have

(6)
$$\rho - \rho_0 = d\rho = \frac{u V}{c^2} \rho_0 = \frac{u}{V} \rho_0 M_0^2.$$

In applying the above it is useful to observe that u is the incremental component of the disturbed velocity *opposite to* the velocity of the undisturbed flow.

With these approximations the equation 15·2 (2) satisfied by Φ becomes

$$(1 - M^2)\, \Phi_{xx} + \Phi_{yy} + \Phi_{zz} = 0,$$

and substituting (1) in this shows that the perturbation potential ϕ satisfies the same equation. Since the derivatives of ϕ are, on our hypothesis, small, we may replace M by M_0, thus obtaining

(7)
$$(1 - M_0^2)\, \phi_{xx} + \phi_{yy} + \phi_{zz} = 0.$$

This is the Prandtl-Glauert equation satisfied by the perturbation potential ϕ and therefore also by the complete potential (1).

15·4. Linearised subsonic flow.

Here $M_0 < 1$, and therefore the equation 15·3 (7), satisfied by the velocity potential, is

(1)
$$\beta^2 \frac{\partial^2 \phi (x, y, z)}{\partial x^2} + \frac{\partial^2 \phi (x, y, z)}{\partial y^2} + \frac{\partial^2 \phi (x, y, z)}{\partial z^2} = 0, \quad \beta^2 = 1 - M_0^2.$$

Let us make the *affine transformation*

(2)
$$x_i = \lambda x, \quad y_i = \lambda \beta y, \quad z_i = \lambda \beta z, \quad \lambda = \text{constant}$$

and write

(3)
$$\phi_i (x_i, y_i, z_i) = \nu \phi (x, y, z), \quad \nu = \text{constant}.$$

The affine transformation (2) is a mapping of the (x, y, z) space on the (x_i, y_i, z_i) space in such a way that all lines parallel to the x-axis are stretched by the factor λ, while lines parallel to the y-axis and the z-axis are stretched by the factor $\lambda \beta$. Clearly straight lines map into straight lines, planes into planes, and parallel lines or planes into parallel lines or planes. The mapping is *not* conformal unless $\beta = 1$, i.e. $M_0 = 0$.

From (2) we have

$$\frac{\partial}{\partial x} = \frac{\partial x_i}{\partial x} \frac{\partial}{\partial x_i} = \lambda \frac{\partial}{\partial x_i}, \quad \frac{\partial}{\partial y} = \lambda\beta \frac{\partial}{\partial y_i}, \quad \frac{\partial}{\partial z} = \lambda\beta \frac{\partial}{\partial z_i},$$

and therefore (1) becomes

$$(4) \qquad \frac{\partial^2 \phi_i(x_i, y_i, z_i)}{\partial x_i^2} + \frac{\partial^2 \phi_i(x_i, y_i, z_i)}{\partial y_i^2} + \frac{\partial^2 \phi_i(x_i, y_i, z_i)}{\partial z_i^2} = 0,$$

which is Laplace's equation.

Any solution $\phi_i(x_i, y_i, z_i)$ of (4) gives the flow of an incompressible fluid in the (x_i, y_i, z_i) space, and (2) and (3) then determine a corresponding linearised flow of compressible fluid in the (x, y, z) space, where the velocity potential is

$$(5) \qquad \phi(x, y, z) = \frac{1}{\nu} \phi_i(\lambda x, \lambda\beta y, \lambda\beta z).$$

If we regard ϕ and ϕ_i as perturbation potentials of the same uniform wind V in the negative direction of the x-axis in both spaces, we have for the complete velocity potentials

$$(6) \qquad Vx_i + \phi_i(x_i, y_i, z_i), \quad Vx + \phi(x, y, z) = Vx + \frac{1}{\nu} \phi_i(\lambda x, \lambda\beta y, \lambda\beta z).$$

If $(-V + u_i, v_i, w_i)$ and $(-V + u, v, w)$ are the perturbed velocities at corresponding points we have

$$(7) \qquad \begin{aligned} u_i &= -\frac{\partial \phi_i}{\partial x_i} = -\frac{\partial x}{\partial x_i} \frac{\partial}{\partial x} \nu\phi = \frac{\nu}{\lambda} u, \\ v_i &= -\frac{\partial \phi_i}{\partial y_i} = -\frac{\partial y}{\partial y_i} \frac{\partial}{\partial y} \nu\phi = \frac{\nu}{\lambda\beta} v, \\ w_i &= -\frac{\partial \phi_i}{\partial z_i} = -\frac{\partial z}{\partial z_i} \frac{\partial}{\partial z} \nu\phi = \frac{\nu}{\lambda\beta} w. \end{aligned}$$

15·41. Distortion of the streamlines.

In the unperturbed flow the streamlines are straight. Consider the streamline $y - h = 0$, $z - k = 0$. In the perturbed flow the quantities $y - h$, $z - k$ will differ from zero by quantities which by our linearisation hypothesis must be small. These small differences will be called the *distortions* of the streamlines. The differential equations of the streamlines are

$$\frac{dx}{-V+u} = \frac{dy}{v} = \frac{dz}{w},$$

and therefore to our order of approximation

$$\frac{dy}{dx} = -\frac{v}{V}, \quad \frac{dz}{dx} = -\frac{w}{V}.$$

Integrating, we get from 15·4 (7), (2)

$$y - h = -\int_{\infty}^{x} \frac{v}{V} dx = -\frac{\lambda\beta}{\nu} \int_{\infty}^{x_i} \frac{v_i}{V} \frac{dx_i}{\lambda} = \frac{\beta}{\nu}(y_i - h_i),$$

$$z - k = -\int_{\infty}^{x} \frac{w}{V} dx = -\frac{\lambda\beta}{\nu} \int_{\infty}^{x_i} \frac{w_i}{V} \frac{dx_i}{\lambda} = \frac{\beta}{\nu}(z_i - k_i),$$

where by 15·4 (2) $h_i = \lambda\beta h,$ $k_i = \lambda\beta k.$

Thus we can state that the distortions of the compressible flow at (x, h, k) are β/ν times the distortions of the incompressible flow at the point $(\lambda x, \lambda\beta h, \lambda\beta k)$ in the (x_i, y_i, z_i) space.

This result enables us to plot the streamlines of the compressible flow when the streamlines of the incompressible flow are known.

Consider the point $P_i(x, h, k)$ in the (x_i, y_i, z_i) space. In the undisturbed flow the streamline through P_i is straight and parallel to the x_i-axis. If we disturb the flow by introducing an aerofoil, this straight streamline will bend so as to pass through the adjacent point $P_i'(x, h + h', k + k')$ where h', k' are small distortions of the originally straight streamline. By what we have just proved, the point P_c in the compressible flow in the (x, y, z) space which corresponds with P_i is $P_c(x/\lambda, h/\lambda\beta, k/\lambda\beta)$, and the straight streamline, which passed through P_c in the undisturbed compressible flow, will bend so as to pass through the adjacent point P_c' whose coordinates are

$$\left(\frac{x}{\lambda},\ \frac{h}{\lambda\beta} + \frac{\beta h'}{\nu},\ \frac{k}{\lambda\beta} + \frac{\beta k'}{\nu}\right);$$

and as P_i' describes its streamline in the disturbed incompressible flow, so P_c' describes the corresponding streamline in the compressible flow. It is clear that the locus of P_c' can be plotted point by point from the locus of P_i'.

The reader's attention is explicitly directed to the observation that a streamline in one space does not map affinely into a streamline in the other space; that, for example, the relation $y_i = \lambda\beta y$ does not hold in general between the coordinates of P_i' and P_c'. In fact we are remarking that

$$h + h' \neq \lambda\beta\left(\frac{h}{\lambda\beta} + \frac{\beta h'}{\nu}\right) = h + \frac{\lambda\beta^2}{\nu} h'.$$

(But if we choose $\lambda\beta^2 = \nu$, the mapping will be affine.)

This conclusion might appear peculiar, if it were taken to imply that the compressible flow deduced from a given incompressible flow past an aerofoil A_i could lead to different compressible flows. But it must be remembered that the surface of the aerofoil A_i is itself a locus of streamlines and therefore the aerofoil A deduced from it will differ in shape according to our particular choice of λ and ν. No paradox is involved.

Conversely, we might start with the aerofoil A in the compressible flow and plot the form of A_i in the incompressible flow. From the behaviour of A_i the behaviour of A would then be deduced.

There is considerable advantage in locating the aerofoil A near the x-axis, $y = 0$, $z = 0$, for if we take $h = 0$, $k = 0$, we can now state that the distortions at $(x, 0, 0)$ in the compressible flow are β/ν times the distortions at $(\lambda x, 0, 0)$ in the incompressible flow.

In particular if we take $\lambda = 1$, $\nu = \beta$, the aerofoils A and A_i differ but slightly, provided the span is sufficiently small for the wing tips to be regarded as near to the x-axis.

15·42. Two-dimensional flow.

Consider two planes : the $x_i z_i$-plane occupied by incompressible fluid moving with velocity potential $V x_i$, and the xz-plane occupied by compressible fluid moving with velocity potential $V x$. Let a profile A_i which deviates but little from a part of the x_i-axis be placed in the $x_i z_i$-plane. Then the complete velocity potential will be

$$(1) \qquad\qquad V x_i + \phi_i(x_i, z_i).$$

If we make the affine transformation

$$(2) \qquad\qquad x_i = \lambda x, \quad z_i = \lambda \beta z.$$

the velocity potential

$$(3) \qquad\qquad V x + \frac{1}{\nu} \phi_i(\lambda x, \lambda \beta z)$$

will give a linearised subsonic compressible flow past a profile A in the xz-plane.

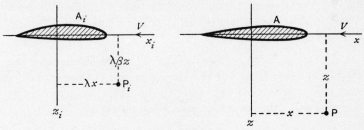

FIG. 15·42 (i).

The chords will be related by $c_i = \lambda c$, and the ordinate of the camber line at $(x, 0)$ in the compressible flow will be β/ν times the ordinate of the camber line at $(\lambda x, 0)$ in the incompressible flow.

From 15·4 (7) we have $w_i/V = \nu w/(\lambda \beta V)$ and therefore the incidences in the two planes are related by

$$(4) \qquad\qquad \alpha_i = \frac{\nu}{\lambda \beta} \alpha.$$

By Bernoulli's theorem the *pressure increase* due to the perturbation in the incompressible flow is

$$dp_i = p_i - p_0 = \tfrac{1}{2}\rho_0(V^2 - (V - u_i)^2) = \rho_0 u_i V,$$

and in the compressible flow from 15·3 (5) the pressure increase is $dp = \rho_0 u V$. Therefore

(5) $$dp_i = \nu\, dp/\lambda.$$

The lifts are given respectively by

$$L_i = \int p_i\, dx_i, \quad L = \int p\, dx,$$

and therefore

(6) $$L_i = \nu L.$$

For the circulations,

$$K_i = \int (u_i\, dx_i + w_i\, dz_i), \quad K = \int (u\, dx + w\, dz),$$

and so from 15·4 (7)

(7) $$K_i = \nu K.$$

Since $L_i = K_i \rho_0 V$ we have $L = K\rho_0 V$, and therefore the Kutta-Joukowski theorem holds also for linearised subsonic compressible flow.

We now consider some particular cases, recalling the result of 15·41 that the distortion of the streamlines in the compressible flow at $(x, 0)$ is β/ν times the distortion in the incompressible flow at $(\lambda x, 0)$.

Case I. $\lambda = 1, \nu = 1$. Here the chords are equal but the camber and thickness of A are those of A_i reduced in the ratio of $\beta : 1$. The lift, circulation, and pressure are the same at corresponding points in both flows. The incidence is reduced in the ratio $\beta : 1$.

Case II. $\lambda = 1/\beta, \nu = 1$. Now $c = \beta c_i$, so that the chord of A is less than the chord of A_i in the ratio $\beta : 1$ and the camber and thickness are reduced in the same ratio at corresponding points. The lift, shape, incidence, and circulation are unaltered ; the pressure is increased in the ratio $1 : \beta$.

Case III. $\lambda = 1, \nu = \beta$. In this case $\beta/\nu = 1$ and the distortions are the same in both flows. Thus A and A_i are identical profiles at the same incidence.

FIG. 15·42 (ii).

The circulation, lift, and pressure are increased in the ratio $1 : \beta$. Thus the effect of compressibility on a *given* profile is to increase the lift in the ratio $1 : \beta$ (Glauert's correction).* The disposition of corresponding streamlines is sketched on fig. 15·42 (ii) when $\beta = 0\cdot6$ so that $M_0 = 0\cdot8$.

15·43. Lifting line.

From 15·4 we see that the distortions at (x, h, k) in the compressible flow are β/ν times their values at the point $(\lambda x, \lambda\beta h, \lambda\beta k)$ in the incompressible.

If we take $\lambda = 1/\beta$, $\nu = 1$, it follows that the distortions at $(0, h, k)$ are in the ratio $\beta : 1$. If therefore we replace the aerofoil by a lifting line stretching from $(0, -b/2, 0)$ to $(0, b/2, 0)$, this line will occupy the same position in both flows. The lift, circulation and incidence (15·42) will be the same in both flows at any profile section but $c' = \beta c_i{}'$, and since from 11·3 we have

$$\Gamma_i{}' = \tfrac{1}{2}(a_0{}')_i\, c_i{}'\,(V\alpha_i{}' - w_i{}'),$$

we shall have in the compressible flow

$$\Gamma' = \tfrac{1}{2}\frac{(a_0{}')_i}{\beta}\, c'\,(V\alpha' - w'),$$

since, from 15·3 (7), $w_i{}' = w'$. It follows that the integral equation for the circulation (11·3) is the same for both flows except that $(a_0{}')_i$ is to be replaced by $(a_0{}')_i/\beta$. Therefore the method 11·5 still applies provided that we replace μ by μ/β. Thus, as far as the linearised theory applies, the only effect of compressibility on the lifting line theory is that the slope of the (C_L, α) graph must be increased in the ratio $1 : \beta$. For further details the reader is referred to a paper by Goldstein and Young.†

The foregoing theory has its application to aerofoils in conditions where shock waves do not develop. The flow past an aerofoil is such that the relative airspeed at certain places may be considerably greater than the forward speed. A simple instance is that of the circular cylinder in a stream V without circulation. Here the speed at the ends of a diameter perpendicular to the stream is $2V$. Should places occur on the aerofoil where the critical speed (15·1) is attained or passed, shock waves (16·4) may develop and part of the *régime* may become supersonic. The investigation of this mixed state is not yet on a satisfactory basis.

15·44. The hodograph method.

Consider two-dimensional steady motion. Let PQR be an arc of a curve in the plane of the flow, the (x, y) plane. From the points $P, Q, R, \dots$, draw vectors $\overrightarrow{PP_1}$, $\overrightarrow{QQ_1}$, $\overrightarrow{RR_1} \dots$, to represent the fluid velocity at these points. From a fixed point H draw vectors $\overrightarrow{HP'}$,

---

* The mere fact that $1/\beta \to \infty$ when $M_0 \to 1$, shows that the range of applicability of this correction is limited.

† *R. and M.*, No. 1909 (1943).

$\overrightarrow{HQ'}$, $\overrightarrow{HR'}$, ..., equal and parallel to these velocity vectors. The points P', Q', R', ..., describe the *hodograph* of the given curve PQR, and the plane of this curve is the *hodograph plane* of the given motion. If we take the axis Hu in the hodograph plane parallel to Ox in the plane of the flow, the velocity at P will be

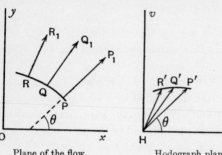

Plane of the flow.　　　　　Hodograph plane.

FIG. 15·44.

$$u + iv = qe^{i\theta},$$

and P' will have cartesian coordinates (u, v) or polar coordinates (q, θ). We shall assume, for the sake of simplicity of statement, that this mapping of one plane on the other is bi-uniform (3·6).*

We have seen in 15·2 that the velocity potential of an irrotational compressible flow satisfies a non-linear differential equation. We shall show that if (q, θ) or (u, v) are taken as variables, the equation becomes linear.

It is useful to introduce the stream function ψ. The equation of continuity (21·5) is, in the case of steady motion,

$$\frac{\partial(\rho u)}{\partial x} + \frac{\partial(\rho v)}{\partial y} = 0,$$

and we can satisfy this by taking

(1)
$$\rho u = - \rho_0 \frac{\partial \psi}{\partial y}, \qquad \rho v = \rho_0 \frac{\partial \psi}{\partial x},$$

where ρ_0 is any constant, which may be conveniently identified with the density, say, in the main stream, when we consider flow past an aerofoil. The function ψ is the *stream function*. Thus if ϕ is the velocity potential, we have

$$- d\phi = u\, dx + v\, dy, \qquad - \frac{\rho_0}{\rho} d\psi = - v\, dx + u\, dy,$$

and therefore, as is easily verified,

$$- \left(d\phi + \frac{i\rho_0}{\rho} d\psi \right) = (u - iv)dz = qe^{-i\theta}\, dz,$$

so that

$$dz = - \frac{e^{i\theta}}{q} \left(d\phi + \frac{i\rho_0}{\rho} d\psi \right).$$

---

* The locus of $u - iv$ is also a hodograph. It is the reflexion of the above in the real axis, and its use is frequently advantageous.

If suffixes denote partial differentiation, $z_q = \partial z/\partial q$, we have at once

$$z_q = -\frac{e^{i\theta}}{q}\left(\phi_q + \frac{i\rho_0}{\rho}\psi_q\right), \quad z_\theta = -\frac{e^{i\theta}}{q}\left(\phi_\theta + \frac{i\rho_0}{\rho}\psi_\theta\right),$$

and since $z_{q\theta} = z_{\theta q}$, we get

$$\frac{\partial}{\partial\theta}\left\{\frac{e^{i\theta}}{q}\left(\phi_q + \frac{i\rho_0}{\rho}\psi_q\right)\right\} = \frac{\partial}{\partial q}\left\{\frac{e^{i\theta}}{q}\left(\phi_\theta + \frac{i\rho_0}{\rho}\psi_\theta\right)\right\}.$$

Performing the differentiations and equating the real and imaginary parts, we get (see Ex. XV, 9), noting that ρ is independent of θ from 15·1 (1),

$$(2) \qquad \phi_q = q\psi_\theta\frac{\partial}{\partial q}\left(\frac{\rho_0}{q\rho}\right), \quad \phi_\theta = \frac{\rho_0 q}{\rho}\psi_q.$$

These are the hodograph equations. To get the equation satisfied by the stream function since $\phi_{q\theta} = \phi_{\theta q}$, we have

$$\frac{\partial}{\partial q}\left(\frac{\rho_0 q}{\rho}\psi_q\right) = \frac{\partial}{\partial\theta}\left\{q\psi_\theta\frac{\partial}{\partial q}\left(\frac{\rho_0}{q\rho}\right)\right\}, \text{ or}$$

$$(3) \qquad \frac{\rho_0 q}{\rho}\psi_{qq} + \psi_q\frac{\partial}{\partial q}\left(\frac{\rho_0 q}{\rho}\right) - q\psi_{\theta\theta}\frac{\partial}{\partial q}\left(\frac{\rho_0}{q\rho}\right) = 0,$$

since ρ is independent of θ.

Now
$$\frac{d}{dq}\left(\frac{\rho_0}{\rho}\right) = -\frac{\rho_0}{\rho^2}\frac{d\rho}{dp}\frac{dp}{dq} = \frac{\rho_0}{\rho^2}\frac{1}{c^2}q\rho = \frac{\rho_0}{\rho}\frac{q}{c^2},$$

using Bernoulli's theorem 2·31 (1) and $c^2 = dp/d\rho$.

We then find that (3) becomes finally

$$(4) \qquad q^2\frac{\partial^2\psi}{\partial q^2} + q(1 + M^2)\frac{\partial\psi}{\partial q} + (1 - M^2)\frac{\partial^2\psi}{\partial\theta^2} = 0, \quad M = \frac{q}{c}.$$

This is the linear equation satisfied by the stream function. The equation is due to Chaplygin.*

15·45. The hodograph equations for homentropic flow. Assuming the adiabatic relation $p/p_0 = (\rho/\rho_0)^\gamma$, put

$$(1) \qquad \tau = \frac{q^2}{q^2_{\max}}, \quad \beta = \frac{1}{\gamma - 1}\ (= 2·5).$$

The hodograph equations 15·44 (2) can then be written in the form (see Ex. XV, 10).

$$(2) \quad 2\tau(1 - \tau)^{\beta+1}\phi_\tau = -\{1 - (2\beta + 1)\tau\}\psi_\theta, \quad (1 - \tau)^\beta\phi_\theta = 2\tau\psi_\tau.$$

For these equations Chaplygin has given the formal solutions

$$(3) \qquad \psi = -B\theta - \Sigma B_m\psi_m(\tau)\sin(m\theta + \epsilon_m),$$
$$\phi = B\phi_0(\tau) + \Sigma B_m\phi_m(\tau)\cos(m\theta + \epsilon_m),$$

---

* See also R. Sauer, *Theoretische Einführung in die Gasdynamik*, Berlin (1943), p. 94.

where, in the notation for hypergeometric functions,[*]

$$\psi_m(\tau) = \frac{\tau^{\frac{1}{2}m} F(a_m, b_m ; c_m ; \tau)}{\tau_1^{\frac{1}{2}m} F(a_m, b_m ; c_m ; \tau_1)} = \frac{\psi_m^{(o)}(\tau)}{\psi_m^{(o)}(\tau_1)} \quad \text{say.}$$

Here B, B_m are constants and $\tau_1 = V^2/q^2_{max}$, where V is the speed of the main stream ; the parameters a_m, b_m, c_m are given by

$$a_m + b_m = m - \beta, \quad a_m b_m = -\tfrac{1}{2}m(m + 1)\beta, \quad c_m = m + 1 ;$$

and $\quad \phi_0(\tau) = \int \dfrac{1 - (2\beta + 1)\tau}{2\tau(1 - \tau)^{\beta+1}} d\tau, \quad \phi_m(\tau) = \dfrac{2\tau}{m(1 - \tau)^\beta} \dfrac{\partial \psi_m(\tau)}{\partial \tau}.$

Chaplygin has shown that when $q_{max} \to \infty$ (incompressible fluid), ϕ and ψ tend to the limiting form ϕ_i, ψ_i, given by

(4) $\qquad \phi_i + i\psi_i = B \log(q_i e^{-i\theta}) + \Sigma B_m [q_i e^{-i(\theta+\epsilon_m)}/V]^m.$

This is the expansion of the complex potential $\phi_i + i\psi_i$ of an incompressible flow in terms of the complex velocity $q_i e^{-i\theta} = - d(\phi_i + i\psi_i)/dz$. If, therefore, we start from an incompressible flow [†] as given by (4), assuming convergence, we can proceed to the compressible flow given by (3), for the constants B and B_m are now known. The boundaries of the field of the compressible flow will in general be different from those of the field of incompressible flow. In the applications this difference is small but difficult to evaluate precisely.

15·46. Velocity correction factor. [‡] The formal method of solution is quite unsuited to numerical calculation, for the hypergeometric functions have not been tabulated and the series do not appear to be reducible to a closed form. Some method of approximation is therefore required. In what follows suffix i will refer to incompressible flow.

In 15·45 (4), if V_i is the speed of the main stream, the speed q_i occurs in the forms §

$$\log \frac{q_i}{V_i} = \tfrac{1}{2} \log \frac{\tau_i}{\tau_{1i}}, \quad \text{and} \quad \left(\frac{q_i}{V_i}\right)^m = \left(\frac{\tau_i}{\tau_{1i}}\right)^{\frac{1}{2}m},$$

while in the compressible flow it occurs in the forms

$$\phi_0(\tau), \quad \frac{\psi_m^{(o)}(\tau)}{\psi_m^{(o)}(\tau_1)}, \quad \frac{\phi_m^{(o)}(\tau)}{\phi_m^{(o)}(\tau_1)}.$$

It is clear that the simplest and most useful approximations will therefore be of the types

$$\phi_0(\tau) = \log \xi(\tau), \quad \phi_m^{(o)}(\tau) = [\xi(\tau)]^m, \quad \psi_m^{(o)}(\tau) = [\eta(\tau)]^m,$$

---

[*] Milne-Thomson, *Calculus of Finite Differences* (1965), 9·8.

[†] The question has been examined by Bergman, *N.A.C.A. Technical Notes*, 972, 973 (Washington), 1945.

[‡] G. Temple and J. Yarwood, " The approximate solution of the hodograph equations for compressible flow ", *Rep. No. S.M.E.*, 3201, *R.A.E.*, 1942.

§ Observe that although $q_{max} = \infty$ for incompressible flow, $\tau/\tau_1 \to (q_i/V)^2$ as $q_{max} \to \infty$.

where the functions $\xi(\tau)$, $\eta(\tau)$ are independent of m, and are approximately equal to $\tau^{1/2}$ when τ is small. Temple and Yarwood proceed to find upper and lower approximating functions ξ_1, ξ_2, η_1, η_2 *independent of m, such that*

$$\xi_1{}^m \leqslant \phi_m{}^{(o)} \leqslant \xi_2{}^m, \qquad \eta_1{}^m \leqslant \psi_m{}^{(0)} \leqslant \eta_2{}^m,$$

and their functions are " best possible " approximations in the sense that no closer approximations will satisfy the above inequalities for all values of m. For details the reader is referred to the original paper. The applications of the approximations ξ and η are as follows. Let ϕ_i, ψ_i be the velocity potential and stream function of an incompressible flow, V_i being the speed of the main stream, and let

(1) $$\phi_i = \Phi(q_i/V_i, \theta), \qquad \psi_i = \Psi(q_i/V_i, \theta).$$

Then the velocity potential and stream function of the corresponding compressible flow are

(2) $$\phi = \Phi(\xi/\xi_1, \theta), \qquad \psi = \Psi(\eta/\eta_1, \theta),$$

where $\qquad \xi = \xi(\tau), \quad \eta = \eta(\tau), \quad \xi_1 = \xi(\tau_1), \quad \eta_1 = \eta(\tau_1),$

and, as before, $\tau = q^2/q^2{}_{max}$, $\tau_1 = V^2/q^2{}_{max}$, and V is the speed of the main stream. Let us arrange the constant so that $\psi_i = 0$ is the boundary in the incompressible flow. Then in the compressible flow the boundary is

$$\psi = \Psi(\eta/\eta_1, \theta) = 0,$$

and comparing this with (1) on the assumption that the boundaries are curvilinear, we have

(3) $$\frac{q_i}{V_i} = \frac{\eta(\tau)}{\eta(\tau_1)}$$

at the boundary.

The distortion of the boundary in the compressible flow can now be determined from the fact that the velocity potential has the same value at corresponding points in the compressible and incompressible flows. Hence, if ds, ds_i are corresponding elements of length of the boundaries, we have

(4) $$q\,ds = -\,d\Phi = q_i\,ds_i.$$

Let r, r_i be the radii of curvature of the boundaries at corresponding points. Then

(5) $$\frac{r}{r_i} = \frac{ds}{d\theta} \Big/ \frac{ds_i}{d\theta} = \frac{q_i}{q}.$$

From these results the distortion of curvilinear boundaries can be determined by geometrical construction.

The case of rectilinear boundaries is exceptional, for on such a boundary $\theta = $ constant and the shape is the same in both flows. Also

$$q = -\,d\Phi/ds = -\,(d\Phi/d\xi)(d\xi/dq)(dq/ds),$$

and therefore

(6)
$$s = -\int \frac{d\Phi}{d\xi} \frac{d\xi}{dq} \frac{dq}{q}.$$

Equation (4) still holds.

Again, writing

(7)
$$\frac{q}{V} = \frac{q_i}{V_i} f\left(\frac{q_i}{V_i}, M_0\right), \quad M_0 = \frac{V}{c_0},$$

where c_0 is the sound speed in the main stream, the function f is called the *velocity correction factor* at the Mach number M_0.

Write $\quad x = q/V, \quad y = q_i/V_i, \quad \mu = \frac{1}{2}\beta\tau_1 = \frac{1}{4}M_0^2/(1 + \frac{1}{5}M_0^2).$

Then it is shown that the approximation

(8)
$$\eta(\tau) = \tau^{1/2}(1 - \frac{1}{2}\beta\tau)$$

holds, with a proportional error of not more than 4 per cent., over the whole subsonic range $0 < \tau < \tau^\star$. From (3) and (8) we get $(1 - \mu)y = x(1 - \mu x^2)$ which, as is readily verified, is identically satisfied by

$$x = \frac{2\cos\frac{1}{3}(\pi + \theta)}{\surd(3\mu)}, \quad y = \frac{2\cos\theta}{(1 - \mu)\surd(27\mu)},$$

The second of these determines θ when y is given, and thus the velocity correction factor is, approximately,

$$\frac{x}{y} = 3(1 - \mu)\frac{\cos\frac{1}{3}(\pi + \theta)}{\cos\theta}.$$

EXAMPLES XV

1. If p, v, T, S are connected by

$$pv = RT, \quad S = c_v \log p + c_p \log v,$$

prove that

$$\frac{\partial T}{\partial p}\cdot\frac{\partial S}{\partial v} - \frac{\partial T}{\partial v}\cdot\frac{\partial S}{\partial p} = \frac{\partial p}{\partial T}\cdot\frac{\partial v}{\partial S} - \frac{\partial p}{\partial S}\cdot\frac{\partial v}{\partial T} = 1.$$

Explain the geometrical significance of this result.

2. Draw the graph $x = c^2/c^{\star 2}$ against $y = q^2/q^2_{max}$, showing the critical point dividing subsonic from supersonic flow.

3. If $c^2 = dp/d\rho$ gives the local speed of sound, obtain the following forms of Bernoulli's equation,

$$\frac{c^2\,d\rho}{\rho} + q\,dq = 0, \quad \frac{2c}{\gamma - 1}\,dc + q\,dq = 0, \quad \frac{1}{2}q^2 = \frac{1}{\gamma - 1}(c_0^2 - c^2).$$

4. Use the velocity potential, for incompressible flow,

$$\phi = -Vx + \frac{K\theta}{2\pi}$$

to prove that in compressible subsonic flow with the same circulation the radial and transverse components of velocity are

$$u_r = V \cos \theta, \quad u_\theta = -V \sin \theta - \frac{K}{2\pi r} \frac{\beta}{1 - M^2 \sin^2 \theta},$$

and hence prove the Kutta-Joukowski theorem for lift. Prove also that the drag is zero.

5. In subsonic flow past a lifting line, show that at a large distance l behind the aerofoil the downwash angle is

$$\epsilon = \frac{C_L}{2\pi A} \frac{b^2}{b'^2} [1 + (1 + \tfrac{1}{4}\beta^2 b'^2/l^2)^{\frac{1}{2}}],$$

where b' is the span of the rolled-up vortex wake (11·7).

6. Prove that, if compressible subsonic flow in a wind tunnel of width $2h$ is compared with incompressible flow in a tunnel of width $2\beta h$, the downwash in the compressible flow is β times the downwash in the incompressible flow.

(Goldstein and Young.)

7. Comparing compressible flow in a tunnel of breadth $2h$ with incompressible flow in a tunnel of breadth $2\beta h$, prove that the increase of longitudinal velocity at the working section is $1/\beta$ times the increase in the incompressible flow.

(Goldstein and Young.)

8. In a wind tunnel the stream is non-uniform owing to a pressure gradient. In the compressible flow the velocity components of the undisturbed stream are

$$u = V + \frac{1}{\beta} F_x(x, \beta y, \beta z), \quad v = F_y(x, \beta y, \beta z), \quad w = F_z(x, \beta y, \beta z)$$

and in the incompressible flow

$$u_i = V + F_x(x, y, z), \quad v_i = F_y(x, y, z), \quad w_i = F_z(x, y, z).$$

Prove that the corrections to be applied for non-uniformity of the stream in the compressible flow in which the velocity components are (u, v, w) are $1/\beta$ times the corrections in the incompressible flow with components (u_i, v_i, w_i).

(Goldstein and Young.)

9. With the notations of 15·44, prove that

$$\frac{\partial}{\partial \theta}\left(\frac{\partial z}{\partial q}\right) = -\frac{e^{i\theta}}{q}\left\{i\phi_q - \frac{\rho_0}{\rho}\psi_q + \phi_{\theta q} + i\frac{\rho_0}{\rho}\psi_{\theta q}\right\},$$

$$\frac{\partial}{\partial q}\left(\frac{\partial z}{\partial \theta}\right) = -e^{i\theta}\left\{-\frac{\phi_\theta}{q^2} + i\psi_\theta \frac{\partial}{\partial q}\left(\frac{\rho_0}{\rho q}\right) + \frac{\phi_{q\theta}}{q} + \frac{i\rho_0}{q\rho}\psi_{q\theta}\right\},$$

observing that ρ is a function of q but not of θ.

Hence prove that

$$\frac{i}{q}\left\{\phi_q + \frac{i\rho_0}{\rho}\psi_q\right\} = -\frac{\phi_\theta}{q^2} + i\psi_\theta \frac{\partial}{\partial q}\left(\frac{\rho_0}{\rho q}\right).$$

10. Obtain the hodograph equations for isentropic flow in the form 15·45 (2). Prove that

$$\frac{\partial}{\partial \tau}\left\{\frac{\tau}{(1-\tau)^\beta}\frac{\partial \psi}{\partial \tau}\right\} = -\frac{1}{4}\frac{1 - (2\beta + 1)\tau}{\tau(1-\tau)^{\beta+1}}\frac{\partial^2 \psi}{\partial \theta^2},$$

$$\frac{\partial}{\partial \tau}\left\{\frac{\tau(1-\tau)^{\beta+1}}{1 - (2\beta + 1)\tau}\frac{\partial \phi}{\partial \tau}\right\} = -\frac{1}{4}\frac{(1-\tau)^\beta}{\tau}\frac{\partial^2 \phi}{\partial \theta^2}.$$

11. Show that the change of variable given by

$$\frac{d\lambda}{\lambda} = \tfrac{1}{2}(1 - \tau)^\beta \frac{d\tau}{\tau}, \quad \frac{\lambda^2}{\tau} \to 1 \text{ when } \tau \to 0,$$

leads to the hodograph equations

$$\lambda \, \partial\phi/\partial\lambda = - F \, \partial\psi/\partial\theta, \quad \lambda \, \partial\psi/\partial\lambda = \partial\phi/\partial\theta,$$

where $\qquad\qquad F = [1 - (2\beta + 1)\tau](1 - \tau)^{-2\beta-1}.$

Hence show that the approximation $F = 1$ leads to

$$\frac{\lambda \, \partial(\phi + i\psi)}{\partial\lambda} = \frac{i \, \partial(\phi + i\psi)}{\partial\theta}.$$

12. If the adiabatic pressure density relation is replaced by

$$p - p_1 = C\left(\frac{1}{\rho_1} - \frac{1}{\rho}\right),$$

show that the graph of the above (p, ρ) relation will touch the adiabatic at the point (p_1, ρ_1) if $C = c_1{}^2\rho_1{}^2$, where c_1 is the speed of sound for the state (p_1, ρ_1).

13. Use 15·46 (8) and (3) to show that

$$y = \frac{q_i}{V_i} = x \frac{1 - \mu x^2}{1 - \mu},$$

and hence by writing $x = 2 \cos \tfrac{1}{3}(\pi + \theta)/\surd(3\mu)$, find the velocity correction factor x/y.

Draw a graph of $x/(y(1 - \mu))$ against $\cos \theta$.

14. With the notation of 15·46 if $C_{p, M_1} = (p - p_1)/\tfrac{1}{2}\rho_1 V^2$, where suffix $_1$ refers to the main stream, show that

$$C_{p, M_1} = \{[1 + \tfrac{1}{2}(\gamma - 1)M_1{}^2(1 - x^2)]^{\gamma/\gamma-1} - 1\}/\tfrac{1}{2}\gamma M_1{}^2.$$

Hence show that C_{p, M_1} can be calculated in terms of

$$C_{p, i} = (p - p_1)/\tfrac{1}{2}\rho_1 V_i{}^2$$

when the velocity correction factor is given.

CHAPTER XVI

SUPERSONIC FLOW

In this chapter we shall consider steady motion at speeds for which the Mach number (1·71, 15·1) exceeds unity.

The air will still be considered as an ideal compressible fluid so that the effect of viscosity and therefore of the boundary layer is ignored.

16·1. Moving disturbance.

Before considering supersonic flow let us examine a special problem. Let a feeble instantaneous disturbance such as a cry originate at a point P in air otherwise at rest.

Such a disturbance will spread in a spherical wave, with P as centre, with the speed of sound c, so that at times t, $2t$, $3t$, ... the disturbance will have reached points which lie on concentric spheres, centre P, radii ct, $2ct$, $3ct$, ... If, however, the air is moving with velocity V from right to left, the points reached by the disturbance at time nt will lie on a sphere of radius nct whose centre is at distance Vnt from P. If $V < c$ these spheres will not

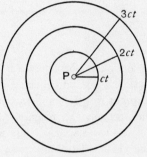

Fig. 16·1 (i).

intersect, and it is clear from fig. 16·1 (ii) that the disturbance will ultimately reach any pre-assigned point of space.

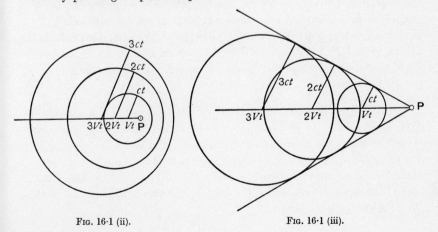

Fig. 16·1 (ii).   Fig. 16·1 (iii).

But when $V > c$ the state of affairs is different, fig. 16·1 (iii), for then the disturbance never reaches points which lie outside a cone whose vertex is P, whose axis is in th direction of V, and whose angle is 2μ, where $\sin \mu = c/V = 1/M$. The angle μ is called the *Mach angle* and the cone is the *Mach cone*.

In two-dimensional motion the Mach cone is replaced by a wedge and the lines in which the plane of the motion cuts the wedge are *Mach lines*.

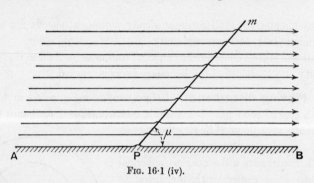

Fig. 16·1 (iv).

A similar phenomenon is observed when uniform flow $V (> c)$ takes place parallel to a wall which is smooth save for a small roughness (such as a projecting seam) at P. Here a disturbance originates at P and is continually renewed as the oncoming air reaches P. The waves continually generated at P give a noticeable disturbance only where they lie most densely, i.e. on m, the Mach line which issues from P. In the steady state the disturbance at every point of m is the same ; the disturbance is not damped, at least in theory, as we recede from the wall along m. If there are several such roughnesses, each will give rise to a Mach line. Along such a line there is air density slightly different from the density of the smooth flow (cf. 15·3 (6)), and this circumstance renders it possible to photograph the lines whose existence is thus well attested.

From this it appears that supersonic flow, in which the airspeed exceeds the critical value (15·1), is physically different from subsonic flow. This manifests itself mathematically by the change of the differential equations from the elliptic to the hyperbolic type (cf. 16·2).

16·15. Thrust due to a supersonic jet.

We consider flow from an ejector. Within the ejector the flow becomes supersonic immediately after passing the section at which the critical speed $c^\star$ is attained and thereafter emerges from the exit with velocity V, density ρ (cf. Ex. 2, 11, 12). Let ω be the area of the jet at exit. Then the mass flow is $m = \rho V \omega$. Let p be the pressure at exit and Π the external pressure. Then by the momentum theorem the reaction R of the gas upon the body is

$$R = mV + p\omega = mV_1 + \Pi\omega,$$

where V_1 is the velocity when the pressure is Π.

Thus

$$p - \Pi = \rho V(V_1 - V)$$

which gives V_1 when V, ρ, Π are known.

Now as the speed changes from V to V_1 there is, in general, an increase of entropy, so that V_1 will be smaller than in an isentropic transformation in which the pressure varies from ρ to Π.

Thus there is a loss of energy which will be the greater, the greater the difference between p and Π. The loss vanishes when the process remains isentropic and then

$$p = \Pi, \quad V = V_1$$

It follows that the ejector will be acting at its greatest efficiency if the exit pressure is equal to the external pressure.

16·2. Linearised supersonic flow.

We consider the case of two-dimensional steady motion in the xy-plane, the undisturbed wind being of speed V, where $V > c_0$, the sound-speed in the undisturbed flow, in the negative sense of the x-axis. The linearised equation satisfied by the velocity potential, 15·3 (7) is

$$(1) \qquad (1 - M_0{}^2)\frac{\partial^2 \phi}{\partial x^2} + \frac{\partial^2 \phi}{\partial y^2} = 0,$$

where now $M_0 = V/c_0$ exceeds unity. We write

$$(2) \qquad \sin \mu_0 = \frac{1}{M_0}, \; \cot \mu_0 = \surd(M_0{}^2 - 1),$$

so that μ_0 is the Mach angle (16·1) appropriate to the undisturbed flow. Then (1) becomes

$$(3) \qquad \frac{\partial^2 \phi}{\partial x^2} = \tan^2 \mu_0 \frac{\partial^2 \phi}{\partial y^2},$$

and it is easily verified by differentiation (cf. 1·5) that the general solution of (3) is

$$(4) \qquad \phi = f_1(y - x \tan \mu_0) + f_2(y + x \tan \mu_0),$$

where f_1 and f_2 are arbitrary functions.

We see here the essential mathematical difference between subsonic and supersonic flow. In subsonic flow we can reduce (1) to Laplace's equation by a real affine transformation (15·4). No such transformation exists for (3). In the terminology of the theory of partial differential equations the subsonic case leads to a differential equation of the *elliptic type*, the supersonic case to one of the *hyperbolic type*.

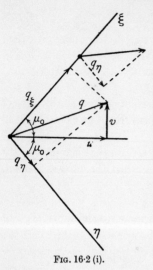

FIG. 16·2 (i).

The equations of the hyperbolic type (but not of the elliptic type) have an associated system of curves called *characteristics** along which discontinuities in the initial values are propagated. To elucidate this in the present case, we shall begin by stating, what will appear shortly, that the characteristics are the Mach lines $y - x \tan \mu_0 = \xi$, $y + x \tan \mu_0 = \eta$, where ξ and η are arbitrary constants. These lines form two systems of parallel lines. Along a line of the ξ-system ξ is constant, and along a line of the η-system η is constant. If the velocity at any point is $(-V + u, v)$ we have from (4), taking ϕ as the perturbation potential,

$$u = \tan \mu_0(f_1'(\xi) - f_2'(\eta)),$$
$$v = - f_1'(\xi) - f_2'(\eta).$$

Therefore the perturbation components parallel to the Mach lines ξ, η are respectively,

(5) $$q_\xi = u \cos \mu_0 + v \sin \mu_0 = - 2f_2'(\eta) \sin \mu_0,$$
$$q_\eta = u \cos \mu_0 - v \sin \mu_0 = 2f_1'(\xi) \sin \mu_0.$$

It follows that q_ξ is constant along a Mach line $\eta = $ constant, and that q_η is constant along a Mach line $\xi = $ constant. Also, when q_ξ, q_η are given, the velocity vector (u, v) is easily constructed, fig. 16·2 (i).

Now consider fig. 16·2 (ii) which shows a curve AC, at no point of which the tangent is a Mach line, and the parallelogram $ABCD$, formed by drawing the Mach lines through A and C.

We can prove the following fundamental theorem.

Theorem. If the perturbation velocity (u, v) is given at every point of AC, then the perturbation velocity at every point interior to the Mach line parallelogram $ABCD$ is uniquely determined.

Proof. Let P be any point inside $ABCD$. Let the

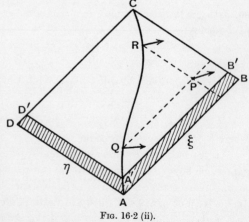

FIG. 16·2 (ii).

* See also 16·32.

Mach lines through P intersect AC in Q and R. Since the velocity at Q is known, q_η is known at Q and therefore at P where it has the same value. Similarly q_ξ is known at P, for it has the same value as at R where it is known. Therefore the velocity at P can be constructed.   Q.E.D.

Now consider a small arc AA' of the curve AC. If we alter the velocities which are given at points of AA' in any arbitrary manner, and draw the Mach lines $A'B'$, $A'D'$ through A', this alteration will only affect the calculated velocities at points within the shaded strips, and this is true however small AA' may be. It follows that a discontinuity arising in the velocity at A is propagated along the Mach lines AB, AD which pass through A. The Mach lines are therefore the characteristics mentioned above.

The linearised equation (1) permits this elementary demonstration owing to the simplicity of the Mach lines, but the above theorem is true in the general case of steady two-dimensional supersonic flow, when the Mach lines are not necessarily straight, provided that in the enunciation the parallelogram $ABCD$ is replaced by a curvilinear quadrilateral formed by the Mach lines through A and C. In all such cases the flow within the quadrilateral is determined solely by the velocities on the line AC, and is independent of the state of the flow outside.

As a particular important conclusion we see that a region of constant state (uniform wind) must be separated from a region where the state is different (constant or otherwise) by a Mach line, and this line is straight, for along it u and v are constant.

16·3. Flow round a corner.

Referring to 16·2 (4), consider the velocity potential

$$(1) \qquad \Phi = Vx + \theta V(y + x \tan \mu_0),$$

where θ is a *small* constant. When $\theta = 0$, Φ gives a uniform wind in the negative sense of the x-axis.

When $\theta \neq 0$, we get the velocity components

$$(2) \qquad - V - \theta V \tan \mu_0, \; - \theta V,$$

which are the components of a uniform flow at the angle

$$\tan^{-1} \frac{\theta V}{V + \theta V \tan \mu_0} = \theta$$

to the x-axis, in other words, a uniform wind of speed $V(1 + \theta \tan \mu_0)$ parallel to OB in fig. 16·3 (i).

Now the different states of uniform flow depicted must (16·2) be separated by a Mach line, and since O is a point of discontinuity for the flow parallel to AO, the Mach line must pass through O and be in the position of OD in the diagram. Alternatively, the line might conceivably be in the symmetrical position OD'. Let us, however, consider the case OD.

The flow depicted will then be given by the velocity potential

$$\Phi = Vx, \quad \text{when } y + x \tan \mu_0 > 0.$$

(3) $\qquad \Phi = Vx + \theta V(y + x \tan \mu_0), \text{ when } y + x \tan \mu_0 < 0.$

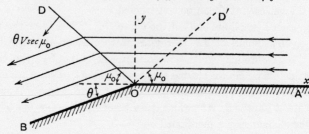

FIG. 16·3 (i).

This value of Φ satisfies 16·2 (3) everywhere, is continuous at the line OD, where $y + x \tan \mu_0 = 0$, and gives the undisturbed flow in the region AOD. It must therefore represent the type of flow depicted in fig. 16·3 (i) round the convex corner AOB.

The components of the incremental velocity in region BOD are, from (2),

(4) $\qquad u = -\theta V \tan \mu_0, \quad v = -\theta V,$

and therefore the incremental velocity is

(5) $\qquad q = \theta V \sec \mu_0, \text{ perpendicular to } OD.$

It therefore appears that there is an abrupt increase in velocity on crossing OD from the front (or side of the oncoming stream) to the back (or side of the deflected stream).

Also, from 15·3 (5), (6), in crossing OD from the front to the back, the pressure and density both decrease, in fact

(6) $\qquad p - p_0 = -\rho_0 \theta V^2 \tan \mu_0 = -c_0 \rho_0 q.$

(7) $\qquad \rho - \rho_0 = -\rho_0 \theta M_0{}^2 \tan \mu_0 = \dfrac{-2\rho_0 \theta}{\sin 2\mu_0}.$

This decrease of pressure and density is characteristic of *expansive* flow. The streamlines are wider apart after passing the bend, indicating expansion.

This is what the linearised theory has to say. This particular problem is, however, capable of exact solution,* which shows that the increase of velocity does not occur suddenly at a single Mach line m_1 (OD) but by a series of gradual transitions to a final Mach line m_2, the streamlines being curved between m_1 and m_2. This type of flow between m_1 and m_2 is known as a *Prandtl-Meyer expansion.*†

The linearised approximation is sufficiently exact for small deflections.

---

* See 16·35. † Th. Meyer, *Dissertation*, Göttingen, 1908.

The flow just described is irrotational and homentropic and is therefore reversible. If we reverse the direction of flow on all the streamlines the Mach

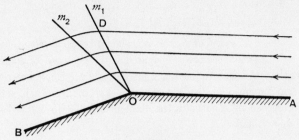

Fig. 16·3 (ii).

line OD will now " lean forward " against the oncoming flow. The flow will be *compressive*, that is to say, on crossing the Mach line from front to back the pressure and density will increase suddenly. It is not possible to predict in general terms which type of flow will arise unless all the boundaries are given, but as far as convex bends occur in the flow round an aerofoil, the flow is always expansive.

Compressive flow, however, occurs naturally when the bend is concave, fig. 16·3 (iii).

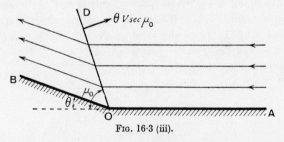

Fig. 16·3 (iii).

Here θ is negative, so that our equations give increases of pressure and density. The streamlines are nearer together behind the Mach line, indicating compression.

16·32. Characteristics.
Consider a geometrical (not a material) surface S conceived to be moving through fluid. Let the point P belonging to the surface have the velocity $\mathbf{q}_S$ and let $\mathbf{q}$ be the velocity of the fluid particle with which P instantaneously coincides. The velocity of the point P of the surface relative to the fluid is then $\mathbf{q}_S - \mathbf{q}$.

Def. A *characteristic* is a surface which moves through the fluid in such a way that the magnitude of the component of the velocity of each point P of the surface relative to the fluid in the direction of the normal to the surface at P is equal to the local speed of sound at P.

In symbols

$$(1) \qquad \mathbf{n}(\mathbf{q}_S - \mathbf{q}) = \pm c,$$

where c is the speed of sound at P, and $\mathbf{n}$ is the unit normal to the surface at P.

Since small disturbances are propagated with the speed of sound (1·5), it follows that the wave front of such a small disturbance is a characteristic. In the case of two-dimensional steady motion the characteristics will be cylindrical surfaces represented by a curve in the plane of the motion and will be *at rest*. Thus in (1) $\mathbf{q}_S = 0$ and

$$(2) \qquad \mathbf{nq} = \pm c \quad \text{or} \quad q_n = \pm c,$$

where q_n is the normal component of the fluid velocity.

Thus the projection of the fluid velocity vector at P on the normal to a characteristic at P is equal to c and if μ is the acute angle between the tangent to the characteristic and the fluid velocity

$$(3) \qquad \sin \mu = \frac{c}{q} = \frac{1}{M}$$

where M is the Mach number. The angle μ is the Mach angle (cf. 16·1).

It is clear from (3) that given the fluid velocity $\mathbf{q}$ at P there are exactly *two* characteristics through P whose tangents are equally inclined to $\mathbf{q}$.

The normal components of the velocity are both c while the tangential components are $q \cos \mu = t$, and $- q \cos \mu = - t$.

Therefore by Bernoulli's theorem (15·1)

$$(4) \qquad c^2 = k^2(q^2_{\max} - q^2 \cos^2 \mu) = k^2(q^2_{\max} - t^2),$$

$$(5) \qquad k^2 = \frac{\gamma - 1}{\gamma + 1} = \frac{c^{\star 2}}{q^2_{\max}}.$$

16·35. Prandtl-Meyer expansion.

To discuss flow round a corner consider fluid streaming with constant supersonic speed V_1 parallel to a straight wall AO which bends away from the stream into a second straight part OB at the corner O.

In the uniform stream of Mach number M_1 the Mach angle is given by $\sin \mu_1 = 1/M_1$ and is therefore known. Thus the flow will begin to turn the corner along a straight characteristic or Mach line m_1 in fig. 16·35. Assuming for the moment that the final state is uniform flow of speed V_2 parallel to OB, the turn will be completed on a second straight Mach line m_2. It then follows that all the Mach lines issuing from O are straight and that the velocity at each point of any one of them, say m, is the same. If ϕ is the velocity potential, we have

(1)
$$q_r = -\frac{\partial \phi}{\partial r}, \quad q_\theta = -\frac{1}{r}\frac{\partial \phi}{\partial \theta}$$

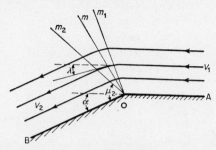

Fig. 16·35.

and q_r, q_θ are independent of r. Moreover since m is a characteristic, $q_\theta = c$ and therefore by Bernoulli's theorem, 15·1 (4) gives

(2)
$$\left(\frac{1}{r}\frac{\partial \phi}{\partial \theta}\right)^2 = c^2 = \tfrac{1}{2}(\gamma - 1)\left\{q^2_{max} - \left(\frac{\partial \phi}{\partial r}\right)^2 - \left(\frac{1}{r}\frac{\partial \phi}{\partial \theta}\right)^2\right\}.$$

Since q_r, q_θ are independent of r we must satisfy (1) and (2) by writing

(3)
$$\phi = rf(\theta),$$

where $f(\theta)$ is independent of r. Substitution in (2) then gives

$$\frac{1}{k^2}[f'(\theta)]^2 + [f(\theta)]^2 = q^2_{max},$$

where from 16·32 (5)

(4)
$$k^2 = \frac{\gamma - 1}{\gamma + 1} = \frac{c^{\star 2}}{q^2_{max}}.$$

The above equation has the obvious solution

$$f(\theta) = -q_{max}\sin(k\theta + \epsilon),$$

where ϵ is an arbitrary constant, and so

(5)
$$q_r = (kq_{max}\sin\theta + \epsilon), \quad q_\theta = c^{\star}\cos(k\theta + \epsilon).$$

Let us measure θ from the intial Mach line m_1. Then when $\theta = 0$

$$q_{max}\sin\epsilon = V_1\cos\mu_1, \quad c^{\star}\cos\epsilon = V_1\sin\mu_1,$$

so that

(6)
$$\tan\epsilon = k\cot\mu_1 = k\sqrt{(M_1{}^2 - 1)}.$$

To find the position of m_2 we have, on this line, $\theta = \theta_2 = \mu_1 + \alpha - \mu_2$, where α is the angle OB makes with AO i.e. the angle through which the oncoming stream has been deflected. Therefore

(7)
$$V_2\cos\mu_2 = q_{max}\sin(k\theta_2 + \epsilon), \quad V_2\sin\mu_2 = c^{\star}\cos(k\theta_2 + \epsilon).$$

By division, using the value of θ_2 given above,

$$\tan \mu_2 = \frac{c^\star}{q_{max}} \cot \{k(\mu_1 + \alpha - \mu_2)\},$$

an equation which determines μ_2, since μ_1 and α are known. Equation (7) then gives V_2 and so the Mach number, M_2, of the deflected stream.

To determine the pressure we have

$$\frac{\gamma p}{\rho} = c^2 = q_\theta{}^2 = c^{\star 2} \cos^2 (k\theta + \epsilon).$$

Now

$$\frac{p}{p_0} = \left(\frac{\rho}{\rho_0}\right)^\gamma \quad \text{and} \quad c_0{}^2 = \frac{\gamma p_0}{\rho_0} = \tfrac{1}{2}(\gamma + 1)c^{\star 2}.$$

Therefore

(8)
$$\left(\frac{p}{p_0}\right)^{(\gamma-1)/\gamma} = \frac{2 \cos^2 (k\theta + \epsilon)}{\gamma + 1}.$$

The maximum value of θ which is physically possible is that which makes $p = 0$, and so

(9)
$$k\theta_{max} + \epsilon = \tfrac{1}{2}\pi.$$

Thus if

(10)
$$\alpha + \mu_1 > \theta_{max},$$

i.e. if

(11)
$$\alpha > \frac{1}{k}(\tfrac{1}{2}\pi - \epsilon) - \mu_1,$$

the fluid will not be in contact with the wall OB but will be separated from it by a vacuum bounded by OB and the line $\theta = \theta_{max}$ which is simultaneously a characteristic and a streamline.

16·36. Complete Prandtl-Meyer expansion.

This corresponds to a situation similar to that in 16·35 but starting with a stream along AO whose Mach number is unity and arranging that the second wall OB does not interfere with the expansion which means that 16·35 (10) is satisfied.

If $M_1 = 1$, 16·35 (6) shows that $\epsilon = 0$, and therefore that

(1)
$$q_r = q_{max} \sin k\theta, \quad q_\theta = c^\star \cos k\theta.$$

When $\theta = 0$ it follows that $q_r = 0$, $q_\theta = c^\star$ and so the initial Mach line m_1 is perpendicular to AO.

Thus the intial stream has the velocity $V_1 = c^\star$ along AO.

Taking $\gamma = 1\cdot4$ for air we find $\theta_{max} = 220°$ nearly and so the acute angle between OA and m_2 is about $50°$.

Between OB and the final Mach line m_2 there is a region of cavitation.

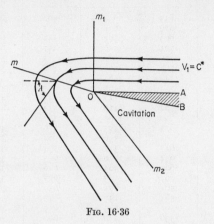

FIG. 16·36

Table* 16·36 shows corresponding values of M, p/p_0, T/T_0, ρ/ρ_0, q/q_{max}, μ, λ, in degrees, for the complete expansion.

Here p, T, ρ are pressure temperature and density. Suffix zero refers to stagnation conditions.

The angle λ in the last column is the angle through which the direction of motion is deflected from the original direction AO.

In the notation of fig. 16·36 the deflection of the stream on the Mach line m is

$$(2)\qquad \lambda = \theta - \left(\frac{\pi}{2} - \mu\right) = \frac{1}{k}\tan^{-1}[k\surd(M^2-1)] - \tan^{-1}\surd(M^2-1),$$

and entering the table with M we find λ or entering with λ we find M.

Therefore in the partial expansion of fig. 16·35, $\alpha = \lambda_2 - \lambda_1$ and

$$(3)\qquad \lambda_2 = \alpha + \lambda_1$$

where λ_1, λ_2 correspond to the initial and final Mach numbers M_1 and M_2. Thus from (2) and (3)

$$(4)\qquad \frac{1}{k}\tan^{-1}[k\surd(M_2{}^2 - 1)] - \tan^{-1}\surd(M_2{}^2 - 1) = \alpha + \beta$$

where

$$(5)\qquad \beta = \frac{1}{k}\tan^{-1}[k\surd(M_1{}^2 - 1)] - \tan^{-1}\surd(M_1{}^2 - 1).$$

---

* I am indebted to my friend Professor E. Carafoli for this table, which appears in his *High-Speed Aerodynamics*, Bucharest (1956).

TABLE 16·36

M	$\dfrac{p}{p_0}$	$\dfrac{T}{T_0}$	$\dfrac{\rho}{\rho_0}$	$\dfrac{q}{q_{max}}$	μ	λ
1·00	·528	·833	·634	·408	90·00	0·00
1·05	·498	·819	·608	·425	72·25	0·49
1·10	·468	·805	·582	·441	65·38	1·34
1·15	·440	·791	·556	·457	60·41	2·38
1·20	·412	·776	·531	·473	56·44	3·56
1·25	·386	·762	·507	·488	53·13	4·83
1·30	·361	·747	·483	·503	50·28	6·16
1·35	·337	·733	·460	·517	47·79	7·56
1·40	·314	·718	·437	·531	45·58	8·99
1·45	·293	·704	·416	·544	43·60	10·44
1·50	·272	·690	·395	·557	41·81	11·91
1·55	·253	·675	·375	·570	40·18	13·39
1·60	·235	·661	·356	·582	38·68	14·86
1·65	·218	·647	·337	·594	37·31	16·33
1·70	·203	·634	·320	·605	36·03	17·81
1·75	·188	·620	·303	·616	34·85	19·27
1·80	·174	·607	·287	·627	33·75	20·72
1·85	·161	·594	·271	·637	32·72	22·16
1·90	·149	·581	·257	·647	31·76	23·59
1·95	·138	·568	·243	·657	30·85	24·99
2·00	·128	·556	·230	·667	30·00	26·38
2·05	·118	·543	·218	·676	29·20	27·75
2·10	·109	·531	·206	·685	28·44	29·10
2·15	·101	·520	·195	·693	27·72	30·43
2·20	·0935	·508	·184	·701	27·04	31·73
2·25	·0865	·497	·174	·709	26·39	33·02
2·30	·0800	·486	·165	·717	25·77	34·28
2·35	·0740	·475	·156	·724	25·18	35·53
2·40	·0684	·465	·147	·732	24·62	36·75
2·45	·0633	·454	·139	·739	24·09	37·95
2·50	·0585	·444	·132	·745	23·58	39·12
2·55	·0542	·435	·125	·752	23·09	40·28
2·60	·0501	·425	·118	·758	22·62	41·41
2·65	·0464	·416	·111	·764	22·17	42·52
2·70	·0430	·407	·106	·770	21·74	43·62
2·75	·0398	·398	·0999	·776	21·32	44·69
2·80	·0368	·389	·0946	·781	20·92	45·74
2·85	·0341	·381	·0896	·787	20·54	46·79
2·90	·0317	·373	·0849	·792	20·17	47·79
2·95	·0293	·365	·0804	·797	19·82	48·78
3·00	·0272	·357	·0762	·802	19·47	49·75
3·05	·0252	·349	·0722	·806	19·14	50·71
3·10	·0234	·342	·0685	·811	18·82	51·65
3·15	·0217	·335	·0649	·815	18·51	52·57
3·20	·0202	·328	·0616	·820	18·21	53·47
3·25	·0188	·321	·0585	·824	17·92	54·35
3·30	·0175	·315	·0555	·828	17·64	55·22
3·35	·0162	·308	·0527	·832	17·37	56·07
3·40	·0151	·302	·0501	·835	17·10	56·91
3·45	·0141	·296	·0476	·839	16·85	57·73
3·50	·0131	·290	·0452	·843	16·60	58·53
3·55	·0122	·284	·0430	·846	16·36	59·32
3·60	·0114	·278	·0409	·849	16·13	60·09
3·65	·0106	·273	·0389	·853	15·90	60·85
3·70	·00990	·267	·0370	·856	15·68	61·60
3·75	·00924	·262	·0352	·859	15·47	62·33
3·80	·00863	·257	·0335	·862	15·26	63·04
3·85	·00806	·252	·0319	·865	15·05	63·75
3·90	·00753	·247	·0304	·868	14·86	64·44
3·95	·00704	·243	·0290	·870	14·66	65·12
4·00	·00659	·238	·0277	·873	14·48	65·78

Now M_1 is given. Therefore to determine M_2 we first find β from (5), and then enter the Table with $\alpha + \beta$ to obtain M_2 from (4).

16·37. Inversion of the deflection equation.

The equation 16·36 (2) which gives the angle λ through which the stream is deflected can be inverted, in some cases analytically, to give the Mach number M when λ is given.[*]

To this end write

$$(1) \qquad \sqrt{(M^2 - 1)} = \tan\left(\frac{\beta}{k}\right), \quad \lambda = \frac{\nu}{k}.$$

Then 16·36 (2) becomes

$$(2) \qquad \tan(\nu + \beta) = k \tan\left(\frac{\beta}{k}\right)$$

which is a form suited to finding β by iteration. If

$$(3) \qquad \frac{1}{k} = n, \quad \text{then} \quad \gamma = \frac{n^2 - 1}{n^2 + 1}$$

and (2) becomes

$$(4) \qquad n \tan(\nu + \beta) = \tan n\beta.$$

If n is an integer, the right hand side is a rational function of $\tan \beta$ and (4) leads to an equation in $\tan \beta$ of degree $n + 1$.

The resulting equation can be solved by numerical methods for any value of n, and analytically if $n = 1, 2, 3$.

If $n = 2$, then $\gamma = 5/3$ which corresponds to a monatomic gas, for example helium. Since a correlation exists between force coefficients measured in a helium tunnel and equivalent data in air this value of γ has practical applications. The solution is

$$(5) \qquad M = \frac{1 + (\tan \tfrac{1}{2}\lambda)^{2/3}}{1 - (\tan \tfrac{1}{2}\lambda)^{2/3}}.$$

If $n = 3$, then $\gamma = 5/4$. An example of the application of this case occurs in the nozzle section of a rocket motor.

The equation satisfied by $\cot \beta$ is

$$(6) \qquad \cot^4 \beta - 2 \cot^2 \beta - \tfrac{8}{3} \cot(\tfrac{1}{3}\lambda) \cot \beta - \tfrac{1}{3} = 0.$$

After some reduction (see Ex. XVI, 18) we get

$$M = \left(\frac{\psi + 3}{\psi + 1\cdot 5}\right)^{1/2} \frac{\psi + 3}{\psi - 3},$$

where

$$\psi = \left[\frac{\eta^3 + 1}{\eta + 1}\right]^{1/2} + \left\{(\eta + 1)\left[2 - \eta + \left(\frac{\eta^3 + 1}{\eta + 1}\right)^{1/2}\right]\right\}^{1/2},$$

$$\eta = [3 \cot^2(\tfrac{1}{3}\lambda) - 1]^{1/3}.$$

---

[*] R. Probstein *Journ. of Aeronautical Sci.* 24 (1957) 316, 317.

16·4. Shock waves. The state of affairs depicted in fig. 16·3 (iii) gives the linearised approximation to a stream turning a very obtuse angle.

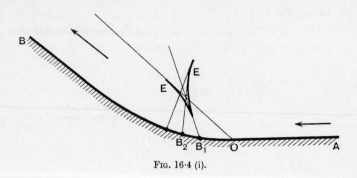

FIG. 16·4 (i).

If the corner is turned by a series of short straight pieces, $O B_1, B_1 B_2, \ldots$, each vertex would yield, on this theory, a Mach line, and these lines might have an envelope E. This would cause a mathematically ambiguous state behind the envelope, where (u, v) would not be uniquely determined. Such a state is not physically possible. Experimental observations indicate that this situation gives rise to a *shock line* S, which starts at the cusp of the envelope and runs between the two branches. In crossing this line the normal velocity decreases suddenly, the density, pressure, temperature and entropy suddenly increase.

Such shock waves can occur at the leading and trailing edges of an aerofoil moving at supersonic speed. They can also occur elsewhere on the aerofoil should the local relative speed exceed the critical speed of sound.

Consider the straight stationary shock wave occurring at an obtuse angle $\pi - \theta$ where θ is small.

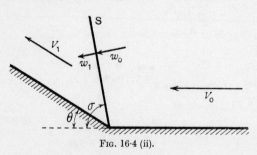

FIG. 16·4 (ii).

Let suffix 0 refer to conditions in front of the shock line S and suffix 1 to conditions behind that line, so that V_0 is the speed of the oncoming, V_1 that of the deflected flow. Let S make the angle σ with the direction of V_0 and let w_0, w_1 denote the components of V_0 and V_1 perpendicular to S. If we consider the condition in front and behind a small line element dl of S, the oncoming flux of matter must be the same as the departing flux (equation of continuity) so that

$$(1) \qquad\qquad \rho_0 w_0 = \rho_1 w_1.$$

Since the pressure thrust acts normally to dl there is no change of the momentum flux parallel to S, therefore

(2) $$\rho_0 w_0 V_0 \cos \sigma = \rho_1 w_1 V_1 \cos (\sigma - \theta).$$

The difference in pressure thrusts on dl must be equal to the normal flux of momentum through dl. Therefore

(3) $$p_1 - p_0 = \rho_0 w_0^2 - \rho_1 w_1^2.$$

These are the equations of ordinary mechanics. We obtain a fourth relation by applying the principle of conservation of energy including thermal energy.

If E is the internal energy per unit mass of air the total energy is $E + \frac{1}{2}V^2$ per unit mass. We equate the flux of energy to the rate at which work is done by the pressure thrusts. Thus

$$p_0 w_0 - p_1 w_1 = \rho_1 w_1 (E_1 + \tfrac{1}{2} V_1^2) - \rho_0 w_0 (E_0 + \tfrac{1}{2} V_0^2),$$

and therefore from (1)

$$\frac{p_0}{\rho_0} + E_0 + \tfrac{1}{2} V_0^2 = \frac{p_1}{\rho_1} + E_1 + \tfrac{1}{2} V_1^2.$$

Using 15·01 (14), we get

(4) $$\frac{\gamma p_0}{(\gamma - 1)\rho_0} + \tfrac{1}{2} V_0^2 = \frac{\gamma p_1}{(\gamma - 1)\rho_1} + \tfrac{1}{2} V_1^2.$$

This equation is of the same form as Bernoulli's equation, 2·32, but in fact the states (p_0, ρ_0) and (p_1, ρ_1) here correspond with different values of the entropy, so that (4) cannot be written down from the principles of isentropic flow on which Bernoulli's equation is based. The increase of entropy from 15·01 (11) is

(5) $$S_1 - S_0 = c_v \log \frac{p_1 \rho_0^{\gamma}}{p_0 \rho_1^{\gamma}}.$$

Observing that $w_0 = V_0 \sin \sigma$, $w_1 = V_1 \sin (\sigma - \theta)$, equations (2) to (4) reduce easily to the following set.

(6) $$V_0 \cos \sigma = V_1 \cos (\sigma - \theta),$$

(7) $$p_0 + \rho_0 V_0^2 \sin^2 \sigma = p_1 + \rho_1 V_1^2 \sin^2 (\sigma - \theta),$$

(8) $$\frac{\gamma p_0}{(\gamma - 1)\rho_0} + \tfrac{1}{2} V_0^2 \sin^2 \sigma = \frac{\gamma p_1}{(\gamma - 1)\rho_1} + \tfrac{1}{2} V_1^2 \sin^2 (\sigma - \theta),$$

the last being got by squaring both sides of (6) and subtracting half the result from (4).

Putting $\quad \Delta \rho_0 = \rho_1 - \rho_0, \quad \Delta p_0 = p_1 - p_0,$

we get, after some easy reductions,

(9) $$1 + \frac{\Delta \rho_0}{\rho_0} = \frac{\tan \sigma}{\tan (\sigma - \theta)} \quad \text{from (1) and (6)},$$

(10) $$\Delta p_0 = \rho_0 V_0^2 \sin^2 \sigma \frac{\Delta \rho_0}{\rho_0 + \Delta \rho_0} \quad \text{from (1) and (3)},$$

(11) $$\frac{\Delta p_0}{\Delta \rho_0} = \gamma \frac{2p_0 + \Delta p_0}{2\rho_0 + \Delta \rho_0} \quad \text{from (8) and (10).}$$

From these equations $\Delta p_0, \Delta \rho_0$ and σ may be calculated when θ, p_0, ρ_0, V_0 are given.

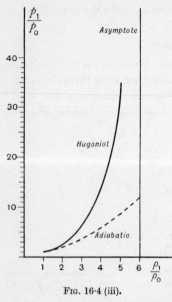

FIG. 16·4 (iii).

When θ is infinitesimal, (9) gives

$$1 + \frac{\Delta \rho_0}{\rho_0} = \frac{\tan \sigma}{\tan \sigma - \theta \sec^2 \sigma}$$
$$= 1 + \frac{\theta}{\sin \sigma \cos \sigma}.$$

Comparing this with 16·3 (7), we see that, when θ is small and negative, $\sigma = \mu_0$, which justifies the use of the Mach line instead of the shock line in this case.

Returning to the general case, we get from (11)

$$\frac{p_1}{p_0} - \frac{\rho_1}{\rho_0} = \frac{\gamma - 1}{2}\left(1 + \frac{p_1}{p_0}\right)\left(\frac{\rho_1}{\rho_0} - 1\right)$$

which determines the Hugoniot curve of p_1/p_0 against ρ_1/ρ_0. When $p_1/p_0 \to \infty$, we get

$$\frac{\rho_1}{\rho_0} \to \frac{\gamma + 1}{\gamma - 1} = 6 \text{ approximately}$$

for air. Thus a shock wave can compress air at most to six times its original density. The dotted curve is the adiabatic $p_1/p_0 = (\rho_1/\rho_0)^\gamma$. When $\Delta p \to 0$, $\Delta \rho \to 0$, (11) goes over into the differential equation $dp/d\rho = \gamma p/\rho$ of the adiabatic. The two curves therefore touch at their starting point $\rho_1/\rho_0 = 1$. The ratio p/ρ and therefore the temperature rise more steeply in the Hugoniot curve than in the adiabatic.

Finally, we may note that the conditions in front of the shock line here discussed must be supersonic. The conditions behind may be either supersonic or subsonic. It is the normal component of velocity which is reduced, the component tangential to the shock front is unaltered. Thus the velocity is refracted towards the shock front in passing from front to back. If the shock front is sufficiently oblique to the oncoming air, the conditions behind may still be supersonic.

16·45. The shock polar. In the hodograph plane represent the velocity $\mathbf{V}_0$ of the oncoming flow by the segment OA of the u-axis. From O draw the vector OP to represent the velocity $\mathbf{V}_1$ (components $u, v,$) of the stream deflected by the shock through the angle θ. The locus of P is the *shock polar* belonging to $\mathbf{V}_0$.

With the notations of fig. 16·45 (i) we have

(1) $w = V_0 \cos \sigma, \quad w_0 = V_0 \sin \sigma, \quad w_1 = V_0 \sin \sigma - \dfrac{v}{\cos \sigma}.$

Now from 16·4 (1) and (2) we see that the tangential component w of the velocity parallel to the shock front is unaltered and then from 16·4 (4)

$$\frac{2\gamma}{\gamma - 1} \frac{p_0}{\rho_0} = q^2_{max} - w^2 - w_0{}^2, \quad \frac{2\gamma}{\gamma - 1} \frac{p_1}{\rho_1} = q^2_{max} - w^2 - w_1{}^2.$$

Substitute for p_0, p_1 in 16·4 (3) and eliminate ρ_0, ρ_1 by means of 16·4 (1). We then get Prandtl's relation

(2) $w_0 w_1 = k^2(q^2_{max} - w^2), \quad k^2 = (\gamma - 1)/(\gamma + 1)$

Substitution in (2) from (1) then gives

(3) $V_0{}^2 \sin^2 \sigma - V_0 v \tan \sigma = k^2(q^2_{max} - V_0{}^2 \cos^2 \sigma)$

which together with

(4) $\tan \sigma = \dfrac{V_0 - u}{v}$

determines the locus of $P(u, v)$ i.e. the shock polar.

Elimination of σ between (3) and (4) leads directly to the equation of the shock polar namely

(5) $v^2\{k^2(q^2_{max} - V_0{}^2) + V_0(V_0 - u)\} = (V_0 - u)^2(V_0 u - k^2 q^2_{max}).$

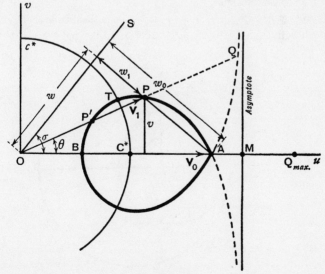

FIG. 16·45 (i).

The shock polar is, therefore a cubic curve (in fact the Folium of Descartes) symmetrical with respect to the u-axis which it meets at the points A, B in fig. 16·45 (i),

(6) $$u = V_0, \quad u = k^2 q^2{}_{max}/V_0, \quad v = 0$$

so that

(7) $$OA \cdot OB = k^2 q^2{}_{max} = c^{\star 2} = OC^{\star 2}.$$

Thus A and B are inverse points with respect to the sonic circle $u^2 + v^2 = c^{\star 2}$. Points on the polar within this circle correspond with a subsonic regime after the shock. When

(8) $$OM = u = \frac{k^2(q^2{}_{max} - V_0{}^2) + V_0{}^2}{V_0},$$

v is infinite and the real asymptote is therefore $u = OM$. The points on the infinite branch of the Folium, shown dotted in fig. 16·45 (i), have also a physical interpretation. In fact if we produce OP to meet the infinite branch at Q, an initial velocity represented by OQ will, after the shock, be reduced to the velocity represented by OA.

The shock polar corresponding with given values V_0 and q_{max} can be constructed point by point as follows.

Mark the points A, B, M given by (4) and (8), and on AB, MB as diameters draw circles C_1, C_2.

Join any point Q on C_2 to B and let QB intersect C_1 in R. Then the point P where AR meets QN, the perpendicular to AB, is a point on the shock polar. The proof is left as an exercise.

In the shock polar as $\theta \to 0$, i.e. as P approaches the double point A of the polar, the shock becomes weaker and weaker and the conditions of no shock

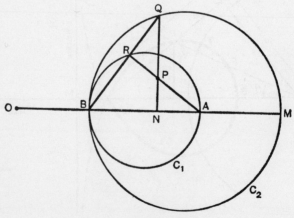

FIG. 16·45 (ii).

are being approached, so that the direction of the shock line must tend to coincide
with a Mach line. Therefore the angle between the tangents to the shock polar
at the double point A must be $\pi - 2\mu$ where μ is the Mach angle.

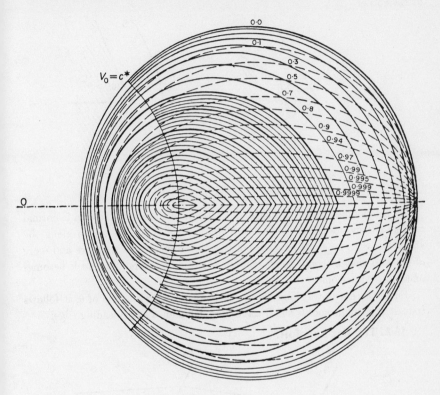

FIG. 16·45 (iii).

Fig. 16·45 (iii), due to Busemann, shows a family of shock polars for $c^\star < V_0$
$\leqslant q_{max}$. They all enclose the point $c^\star$, and lie within the circle to which they
tend when $V_0 \to q_{max}$. On the dotted curves the ratio of the stagnation pressure behind the shock to that before it has the constant value shown.

16·46 Critical angle. Consider fig. 16·45 (i). The direction of the shock
line which deflects the stream through the angle θ is obtained by drawing a
normal to AP, where P the point where the line through O, which makes the
angle θ with the oncoming stream, cuts the polar. Also we get from this
construction $V_1 = OP$. Since OP cuts the polar at a second point P' there is a
second possible shock line perpendicular to AP'; but experimental results seem
to indicate that, for compressive flow at a bend, the one corresponding with P
is that which actually occurs.

The tangent OT from O (the point T lies on the circle $u^2+v^2=c^{\star 2}$) gives the *critical angle* $\theta^\star$ where the two possible shock lines coincide.

If $\theta > \theta^\star$ the above construction fails and there is a curved shock line in front of the corner.

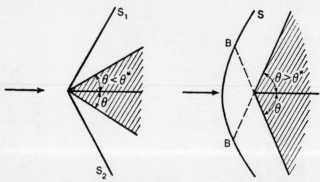

FIG. 16·46.

At the point A of the detached shock the incidence of the stream is normal to S and therefore the flow behind A becomes subsonic. As we go along the shock line S from A the incidence of the stream on S becomes more and more oblique so that at some point such as B the flow behind the shock becomes once more supersonic.

Since the entropy behind the shock increases with increase of θ, it follows that at supersonic speeds small wave drag demands a sharp leading edge.

16·6. The flat aerofoil.

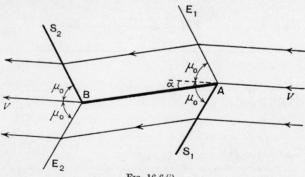

FIG. 16·6 (i).

The simplest supersonic aerofoil, and the one with the least drag, is the flat plate. We shall investigate the results of applying the linearised theory,* and we therefore assume small incidence α. For delta wings, see Chapter XVII.

* For a more general treatment and investigation of the conditions in the wake see M. J. Lighthill, *R. and M.* 1930 (1944).

Referring to fig. 16·6, the air encounters the leading edge at A as a parallel stream. We have thus a Prandtl-Meyer expansion at A, Mach line E_1 and a compression shock S_1; the flow is then parallel to the plate and undergoes a shock along S_2 at B, and an expansion along E_2; and finally leaves parallel to the original stream.

For small incidence the linearised theory allows us to replace S_1 and S_2 by the Mach lines appropriate to the speed V, and the results of 16·3 are immediately applicable.

On the upper surface the pressure is   16·3 (6) $p_U = p_0 - \rho_0 \alpha V^2 \tan \mu_0$, and on the lower surface $p_L = p_0 + \rho_0 \alpha V^2 \tan \mu_0$.

Thus, if c is the chord we have a force, upwards, perpendicular to the plate

$$(1) \qquad F = c(p_L - p_U) = 2\rho_0 \alpha V^2 c \tan \mu_0.$$

The lift is $L = F \cos \alpha = F$, and the drag is $D = F \sin \alpha = F\alpha$. From these we get the coefficients

$$(2) \qquad C_L = 4\alpha \tan \mu_0, \qquad C_D = 4\alpha^2 \tan \mu_0.$$

Since the loading is uniformly distributed the centre of pressure coefficient is $C_p = \frac{1}{2}$.

If we set aside the linearised theory the expansions at A and B are schematically shown in fig. 16·6 (ii).

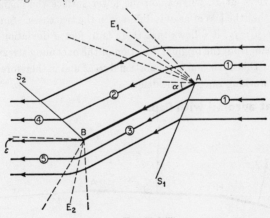

FIG. 16·6 (ii).

Lines through A, B parallel to the main stream determine upper regions 1, 2, 4 and lower regions 1, 3, 5 as shown. The force due to pressure is normal to the aerofoil and of magnitude

$$(3) \qquad F = c(p_3 - p_2) = cp_1\left(\frac{p_3}{p_1} - \frac{p_2}{p_1}\right),$$

where c is the chord and suffixes refer to the above named regions. Since the pressures are constant over the aerofoil, the resultant for F acts at the mid-point of the chord. The pressure ratios are determined from Table 16·36.

The lift coefficient is

$$\text{(4)} \qquad C_L = \frac{F \cos \alpha}{\frac{1}{2}\rho V^2 c} = \frac{2}{\gamma M^2}\left(\frac{p_3}{p_1} - \frac{p_2}{p_1}\right)\cos \alpha$$

where ρ is the density, V the speed and M the Mach number of the oncoming stream.

The wave drag coefficient is

$$\text{(5)} \qquad C_D = \frac{F \sin \alpha}{\frac{1}{2}\rho V^2 c} = \tan \alpha . C_L$$

If the incidence is small we can write α for $\tan \alpha$

The wave drag here given is due to the air mass crossing shock waves with a corresponding increase of entropy. Thus the wave drag and the entropy would become infinite if the shock waves were to propagate to infinity. Fig. 16·6 (ii) shows, however, that this does not happen, since the shock waves are met by expansion waves so that their intensity is finally damped to zero. Thus the dissipation of energy is restricted to a finite region; the increase of entropy is finite and so therefore is the drag.

We note also that along the upper surface the entropy changes on crossing the shock wave S_2 at B, whereas on the lower surface the change occurs on crossing S_1 so that the loss of head is different in the two cases. Since physically we must have $p_4 = p_5$, it follows that there is in fact a deviation say ϵ of the departing stream from the original direction of the oncoming stream. Lighthill has shown that this deviation is proportional to α^2 and is therefore negligible in any practical working range of incidence.

16·61. Flat aerofoil with flap.

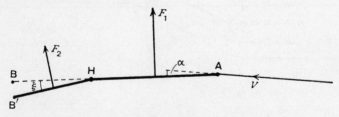

FIG. 16·61.

If the aerofoil has a flap hinged at the distance Ec from the trailing edge,[*] lowered through the angle ξ, the forces on aerofoil and flap are, by precisely as before,

$$F_1 = (1 - E) c . 2\rho\alpha V^2 \tan \mu_0, \qquad F_2 = Ec . 2\rho(\alpha + \xi)V^2 \tan \mu_0,$$

---

[*] This case is considered by A. R. Collar, $R.$ and $M.$ 2004 (1943).

acting at the mid-points of AH, BH respectively. The lift coefficient is therefore

$$C_L = (4\alpha + 4E\xi)\tan\mu_0,$$

so that lowering the flap increases the lift. Similarly, raising the flap will decrease it, but we may observe that in either case there may be a shock wave at the hinge.

The pitching moment is easily calculated about any point from the disposition of the forces shown in fig. 16·61.

16·7. Flow past a polygonal profile.

Consider the slender polygonal profile sketched in fig. 16·7, at incidence α.

Replacing possible shock lines and Prandtl-Meyer expansions by Mach lines as in 16·6, the linearised theory gives a system of parallel Mach lines m, m_1, ..., at each vertex on the upper side and m', m_1', ..., at each vertex on the lower side. Between consecutive Mach lines there is uniform flow parallel to a side of the polygon, and in each such field of uniform flow there exists a pressure in excess, or in defect, of the pressure of the main stream. The pressure thrusts on each side of the polygon can then be replaced by a normal force at the centre of that side exactly as in 16·6, and the calculation of the lift, drag, and moment presents no particular difficulty for a given polygon.

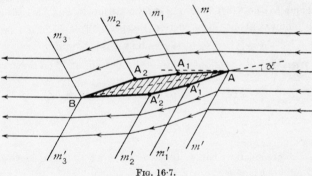

FIG. 16·7.

16·71. Flow past a general profile.

The general profile can of course be considered as the limiting case of the polygonal profile of 16·7, but it is simpler to deal with it directly. The type of profile which we shall consider will be slender, to allow the application of the linearised theory, and will have a sharp point at the leading and trailing edges.

From each point of the surface there will spring a Mach line. We take the chord AB as x-axis and the y-axis to be directed through A. Let $P(x, y)$ be a point on the upper surface, draw PM perpendicular to AB to meet the lower surface at $P'(x, y')$. The tangent at P meets the x-axis at T. Then the angle

PTB is dy/dx and the flow at P is along TP, so that it has been deflected through the angle

(1)
$$\theta = \alpha - \frac{dy}{dx},$$

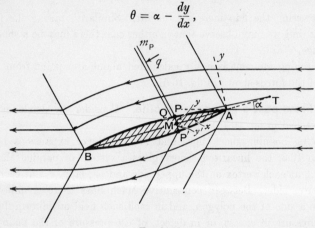

Fig. 16·71.

from the direction of the oncoming wind V. Now consider an element $PQ = ds$ of the upper surface. Since all the gradients are small we can write $ds = dx$. If dp_U is the incremental pressure $p - p_0$ at P the thrust on PQ is $dp_U\,dx$, perpendicular to the tangent at P. The contributions of this thrust to lift, drag, and moment about A are therefore respectively

$$- dp_U\,dx\cos\theta, \qquad - dp_U\,dx\sin\theta, \qquad + x\,dp_U\,dx.$$

The corresponding contributions from the element at P' on the lower surface are

$$dp_L\,dx\cos\theta', \qquad dp_L\,dx\sin\theta', \qquad - x\,dp_L\,dx,$$

where

(2)
$$\theta' = \alpha - \frac{dy'}{dx}.$$

Since θ, θ' are small we can write $\cos\theta = 1$, $\sin\theta = \theta$, and so we get for lift drag, and pitching moment,

(3)
$$L = \int_0^c (dp_L - dp_U)\,dx, \qquad D = \int_0^c (dp_L\,\theta' - dp_U\,\theta)\,dx,$$

$$M_A = \int_0^c (dp_U - dp_L)\,x\,dx,$$

where $c = AB$ is the chord of the profile.

Now the incremental velocity q at the Mach line m is perpendicular to that line (15·61). Also the flow at P must be along the tangent. Therefore

(4)
$$\tan\theta = \frac{q\cos\mu_0}{V + q\sin\mu_0}, \qquad \text{or,} \quad q = V\theta\sec\mu_0,$$

since θ and q are small. Therefore * from 16·3 (6),

(5) $dp_U = p_U - p_0 = -c_0\rho_0 V\theta \sec \mu_0 = -c_0\rho_0\left(\alpha - \dfrac{dy}{dx}\right) V \sec \mu_0.$

Similarly,

$$dp_L = p_L - p_0 = c_0\rho_0 V\theta' \sec \mu_0 = c_0\rho_0\left(\alpha - \dfrac{dy'}{dx}\right) V \sec \mu_0.$$

Therefore

$$dp_L - dp_U = c_0\rho_0\left(2\alpha - \dfrac{dy'}{dx} - \dfrac{dy}{dx}\right) V \sec \mu_0,$$

$$\theta'\,dp_L - \theta\,dp_U = c_0\rho_0 V \sec \mu_0\left[\left(\alpha - \dfrac{dy}{dx}\right)^2 + \left(\alpha - \dfrac{dy'}{dx}\right)^2\right].$$

Now $\displaystyle\int_0^c \frac{dy}{dx}\,dx = 0 = \int_0^c \frac{dy'}{dx}dx.$

Therefore, from (3),

$$L = 2\alpha c\, c_0\rho_0 V \sec \mu_0, \qquad D = c\, c_0\rho_0 V \sec \mu_0\,[2\alpha^2 + 2A_1{}^2 + 2A_2{}^2].$$

(6) $M_A = -c_0\rho_0 V \sec \mu_0 \displaystyle\int_0^c \left(2\alpha - \frac{dy}{dx} - \frac{dy'}{dx}\right) x\,dx,$

where

$$2A_1{}^2 = \frac{1}{c}\int_0^c \left(\frac{dy}{dx}\right)^2 dx, \qquad 2A_2{}^2 = \frac{1}{c}\int_0^c \left(\frac{dy'}{dx}\right)^2 dx,$$

so that $A_1{}^2$ and $A_2{}^2$ are never negative. To get the lift and drag coefficients we divide by $\tfrac{1}{2}\rho_0 V^2 c = \tfrac{1}{2}\rho_0 c\, c_0 V M_0 = \tfrac{1}{2}\rho_0 c\, c_0 V \operatorname{cosec} \mu_0.$
Thus

$$C_L = 4\alpha \tan \mu_0, \qquad C_D = 4(\alpha^2 + A_1{}^2 + A_2{}^2)\tan \mu_0,$$

where $\tan \mu_0 = 1/\sqrt{(M_0{}^2 - 1)}$.   Thus it appears that, to the linearised approximation, C_L is independent of the form of the profile, and decreases as M_0 increases. On the other hand, C_D depends on the shape, unless $A_1 = A_2 = 0$, which is the case of the flat aerofoil, and then C_D is least.

The efficiency C_L/C_D is a function of the incidence, in fact

$$\frac{C_L}{C_D} = \frac{\alpha}{\alpha^2 + A_1{}^2 + A_2{}^2},$$

which is independent of the Mach number.

---

* Note that here c_0 is the speed of sound in the undisturbed flow, and c is the chord of the aerofoil.

For a given aerofoil the efficiency is greatest when $\alpha = \sqrt{(A_1{}^2 + A_2{}^2)}$. Thus for any supersonic aerofoil, which is not a flat plate, there is an optimum incidence independent of the Mach number.*

Before discussing the moment M_A, consider

$$(7) \quad \int_0^c \left(\frac{dy}{dx} + \frac{dy'}{dx}\right) x\, dx = -\int_0^c (y + y')\, dx = -\int_0^c 2h\, dx = -2c^2 A,$$

where we have integrated by parts and then put $2h = y + y'$, so that h is the ordinate of the camber line (1·14), and $c^2 A$ is the area enclosed between the camber line and the chord.

With this notation the pitching moment coefficient about the leading edge is

$$(8) \quad C_m = -2(\alpha + 2A)\tan\mu_0,$$

and therefore the centre of pressure coefficient is

$$(9) \quad C_p = -\frac{C_m}{C_L} = \tfrac{1}{2} + \frac{A}{\alpha}.$$

Thus for all supersonic aerofoils symmetrical with respect to the chord the centre of pressure is at the mid-point of the chord, since then $A = 0$.

EXAMPLES XVI

1. If a uniform wind is deflected expansively through the angle θ, show that the new Mach angle μ is given approximately by

$$\mu - \mu_0 = -\theta[\tfrac{1}{2}(\gamma - 1) + \tfrac{1}{2}(\gamma + 1)\tan^2\mu_0].$$

2. In expansive supersonic flow round a polygonal bend the air stream is deflected through the small angle θ_n at the nth corner $(n = 1, 2, 3, \ldots,)$. If p_n is the pressure and μ_n the local Mach angle after the nth corner is passed, prove that, approximately,

$$\frac{p_n}{p_{n-1}} = 1 - 2\gamma\theta_n \operatorname{cosec} 2\mu_n, \quad \mu_n = \mu_{n-1} - \tfrac{1}{2}\theta_n[(\gamma + 1)\sec^2\mu_{n-1} - 2].$$

(Lighthill.)

3. In the preceding example, show that, if the bend is continuous,

$$\frac{d\mu}{d\theta} = -\tfrac{1}{2}(\gamma + 1)\sec^2\mu + 1,$$

and hence prove, or verify, that

$$\theta = f(\mu_0) - f(\mu)$$

where

$$f(\mu) = \sqrt{\left(\frac{\gamma + 1}{\gamma - 1}\right)}\tan^{-1}\sqrt{\left(\frac{\gamma + 1}{\gamma - 1}\right)}\tan\mu - \mu.$$

(Lighthill.)

* G. I. Taylor, *R. and M.* 1467 (1932), states that a symmetrical aerofoil with thickness ratio 1/20 has a maximum efficiency of 8·8, neglecting skin friction.

4. In Ex. 3, prove that if the bend is continuous

$$\frac{1}{p}\frac{dp}{d\theta} = -2\gamma \operatorname{cosec} 2\mu,$$

and hence prove, or verify, that

$$\frac{p}{p_0} = \frac{g(\mu)}{g(\mu_0)}, \text{ where}$$

$$g(\mu) = \left(\frac{\sin^2\mu}{\gamma - \cos 2\mu}\right)^{\gamma/(\gamma-1)}. \qquad \text{(Lighthill.)}$$

5. In Ex. 4, prove that the velocity V at the deflection θ is given by

$$\frac{V}{V_0} = \sqrt{\left(\frac{\gamma - \cos 2\mu_0}{\gamma - \cos 2\mu}\right)},$$

where V_0 is the velocity of the undeflected stream. (Lighthill.)

6. Show that the pressure behind a plane shock front which deflects through the angle θ is, approximately,

$$p_1 = p_0[1 - 2\gamma\theta \operatorname{cosec} 2\mu_0].$$

7. Show that 16·4 (5) does in fact yield an increase of entropy in the case of a shock wave, and that the increase is approximately

$$c_v \frac{\gamma^3 - \gamma}{12}\left(\frac{\Delta\rho}{\rho_0}\right)^3,$$

where $\Delta\rho/\rho_0 = 2\theta \operatorname{cosec} 2\mu_0$ nearly.

8. Calculate the drag coefficient for a flat aerofoil with the flap lowered (16·61)

9. With the diagram and notations of 16·61, prove that

$$\frac{\partial C_L}{\partial \alpha} = \frac{4}{\sqrt{(M^2 - 1)}}, \quad \frac{\partial C_L}{\partial \xi} = \frac{4E}{\sqrt{(M^2 - 1)}},$$

and that for moments about the point rc of the chord from the leading edge

$$C_m = \frac{4(r - \frac{1}{2})}{\sqrt{(M^2 - 1)}}\alpha + \frac{4E(r - 1 + \frac{1}{2}E)}{\sqrt{(M^2 - 1)}}\xi.$$

(Collar.)

10. With the diagram and notations of 16·61, prove that the moment about the hinge of the force on the flap is

$$H = -\frac{\rho_0 V^2 E^2 c^2}{\sqrt{(M^2 - 1)}}(\alpha + \xi)$$

and that

$$\frac{\partial C_H}{\partial \alpha} = \frac{\partial C_H}{\partial \xi} = -\frac{2}{\sqrt{(M^2 - 1)}}.$$

(Collar.)

11. Calculate the lift and drag coefficients for a supersonic profile of six equal sides symmetrical about one diagonal, the sides which meet that diagonal each making an angle $\pi/90$ with it.

12. Show that in a supersonic aerofoil the lift coefficient is proportional to the height, measured perpendicularly to the direction of motion, of the leading edge above the trailing edge.

13. A biconvex aerofoil consisting of two circular arcs is such that the arcs make angles ϵ and ζ with the chord at the leading edge. Prove that in supersonic flow

$$C_D = (4\alpha^2 + \tfrac{2}{3}\epsilon^2 + \tfrac{2}{3}\zeta^2)\tan\mu_0, \quad C_m = -2(\alpha + \tfrac{1}{6}\epsilon - \tfrac{1}{6}\zeta)\tan\mu_0,$$

where ζ is the angle at the lower arc.

Evaluate these when $M_0 = 1.7$, $\epsilon = 0.279$, $\zeta = 0.120$.   (Taylor.)

14. If a biconvex supersonic aerofoil bounded by two circular arcs is placed at zero incidence, show that the (p, x) curve where p is the pressure excess and x is measured along the chord is antisymmetrical with respect to the centre of the chord.

15. With the other notations of 16·4 let w denote the component of velocity at the shock front parallel to the shock line. Prove that

$$\frac{2\gamma}{\gamma - 1}\frac{p_0}{\rho_0} = q^2_{max} - w^2 - w_0^2, \quad \frac{2\gamma}{\gamma - 1}\frac{p_1}{\rho_1} = q^2_{max} - w^2 - w_1^2.$$

16. Use Ex. 15 to obtain Prandtl's relation

$$w_0 w_1 = c^{\star 2} - k^2 w^2, \quad k^2 = (\gamma - 1)/(\gamma + 1).$$

17. If, with the notations of 15, $\rho_0 w_0 = \rho_1 w_1 = m$, prove that

$$p_1 - p_0 = m(w_0 - w_1), \quad \gamma(p_1 + p_0) = m(w_0 + w_1).$$

18. In 16·37 (6) show that the appropriate value of $\cot\beta$ is determined from the fact that the maximum value of the deflection λ, associated with $M = \infty$, is π and therefore that the coefficient of $\cot\beta$ in the equation is always negative since

$$\cot\tfrac{1}{3}\lambda \geqslant \frac{1}{\sqrt{3}}.$$

CHAPTER XVII

SUPERSONIC SWEPTBACK AND DELTA WINGS

17·0. The principle of two-dimensional linearised supersonic flow was treated in Chapter XVI. In this chapter we shall be concerned with a special type of linearised supersonic flow known as *conical*. The theory is illustrated mainly in its application to thin aerofoils whose boundary is a plane rectilinear polygon. Throughout we shall consider the air to be inviscid and the motion to be steady, irrotational, and homentropic.

17·01. Conical flow. Consider the steady supersonic flow of air round a corner depicted in fig. 16·3 (ii). It is characteristic of this particular flow pattern that the *state* of the fluid, that is to say, its velocity, density, temperature and pressure, is the same at every point of a half-line or ray drawn from a fixed point on the corner into the fluid. Although each such ray is a line of constant state, the state may be different on different rays.

Def. A flow in which there exists a fixed point O such that each ray from O is a line of constant state, is called a *conical flow*, vertex O.

Thus in a conical flow the state may depend upon the direction of the ray but not upon the distance along the ray.

Take O as origin and let P (x, y, z) be any point on a ray through O. Then Q (xt, yt, zt) is equally a point on this ray. Consider any one of the variables which define the state, say the pressure $p = p(x, y, z)$. In conical flow, vertex O, the pressure is the same at P and Q. Therefore

$$p(x, y, z) = p(xt, yt, zt).$$

This means that p is a homogeneous function of degree zero in the variables x, y, z.* Another way of stating this fundamental fact is to say that p depends solely on the ratios $x : y : z$ and not on the individual values x, y, z. In other words we can say, for example, that p depends only on the pair of ratios $y/x, z/x$ and write

$$p = p\left(\frac{y}{x}, \frac{z}{x}\right).$$

This function is clearly unaltered when we replace x, y, z by xt, yt, zt.

17·02. Rectangular aerofoil. To illustrate the idea of conical flow in three-dimensions consider a rectangular flat aerofoil of finite span and negligible

---

* A function $f(x, y, z)$ is said to be homogeneous of degree n in the variables x, y, z if
$$f(xt, yt, zt) = t^n f(x, y, z).$$

thickness placed with its leading edge OO' perpendicular to an oncoming supersonic wind which meets the aerofoil at zero incidence. We can think of this
aerofoil as having been obtained by cutting off two (infinite) ends from an aerofoil of infinite span (two-dimensional aerofoil).

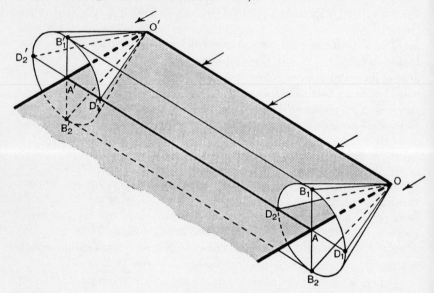

FIG. 17·02 (i).

Each point of the leading edge is the vertex of a Mach cone (61·1) and all
these cones touch an envelope, or Mach wedge, consisting of two planes indicated
by $O'B_1'B_1O$, $O'B_2'B_2O$ in fig. (i). In particular the cone whose vertex is O
touches the planes along the generators OB_1, OB_2 and the cone whose vertex is
O' touches them along the generators $O'B_1'$, $O'B_2'$. Like the Mach wedge, the
Mach cones vertices O and O' extend downstream without limit. In fig. (i) only
parts of these cones and the wedge are shown. Take a section by a plane perpendicular to the wind and not so far downstream that the cones of vertices O
and O' intersect. We then obtain a diagram like fig. (ii), wherein the cones are

FIG. 17 02 (ii).

represented by circles centres A and A'. In the region outside the figure, $B_1'D_2'B_2'B_2D_1B_1$ the flow is that of the undisturbed wind. In the inner region between the circles and the planes $B_1'B_1$, B_2B_2', the flow is the same as that over an aerofoil of infinite span, for no part of this region comes within the Mach cones from the wing tips O, O' of fig. (i). The flows in the regions interior to these cones, shown as the interiors of the circles in fig. (ii) are conical flows with vertices O, O' respectively and it is the object of this chapter to discuss such flows. Before doing so, however, let us dispose of the two-dimensional part of the total flow.

17·03. Two-dimensional sweep-back.

The problem of the flat aerofoil of infinite span at small incidence α in a supersonic wind V perpendicular to the leading edge was solved in 16·6. In flowing over the upper surface of the aerofoil the velocity component in the direction of the undisturbed flow. will be increased to $V + u_\infty$, where, from 16·3 (4), on the upper surface of the aerofoil

$$(1) \qquad u_\infty = \alpha V \tan\mu_0, \qquad \sin\mu_0 = c_0/V.$$

In the flow over the lower surface the velocity component in the direction of the undisturbed flow is $V - u_\infty$. The resultant *lifting pressure*, that is to say the difference of pressures just above and just below the same point of the aerofoil, is, from 16·6,

$$(2) \qquad 2\rho_0 u_\infty V = 2\rho_0\alpha V^2 \tan\mu_0 = 2\rho_0\alpha V^2(V^2/c_0^2 - 1)^{-1/2},$$

In the above, ρ_0, c_0, μ_0 refer to the density, sonic speed, and Mach angle of the undisturbed wind.

To find the corresponding lifting pressure for a swept-back aerofoil of infinite span consider fig. 17·03 which shows an aerofoil of infinite span with the leading edge inclined at the angle δ, where $\delta > \mu_0$, to the oncoming wind V.

At a point O of the leading edge, we distinguish two lines, l_e the leading edge and l_v the line of the wind vector, and two planes OAB, the plane of the aerofoil, and OCD, the plane which

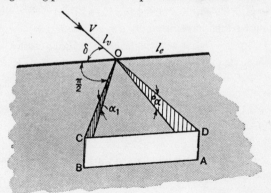

Fig. 17·03.

contains l_e and l_v. We draw three more planes, (i) OAD through l_v perpendicular to the plane of the aerofoil, (ii) OBC perpendicular to the leading edge l_e,

(iii) $ABCD$ perpendicular to the plane OCD, and parallel to the leading edge l_e. By their intersections these planes determine, in fig. 17·03, the angle $AOD = \alpha$, the incidence, and the angle $BOC = \alpha_1$. Then

$$\tan \alpha / \tan \alpha_1 = OC/OD = \sin \delta,$$

and therefore since α is small,

(3) $\alpha = \alpha_1 \sin \delta.$

The velocity V can be resolved into $V \sin \delta$ along OC and $V \cos \delta$ parallel to the leading edge. If the component $V \cos \delta$ were absent, the wind would have velocity $V \sin \delta$ perpendicular to the leading edge and the incidence would be α_1, and from (1) the perturbation velocity of $V \sin \delta$ would be

(4) $u_{1\infty} = \alpha_1 V \sin \delta \tan \mu_1, \qquad \sin \mu_1 = c_0/(V \sin \delta),$

in the direction OC. If we now restore the component $V \cos \delta$ by imposing this constant velocity, the dynamical circumstances are unaltered and the perturbation velocity of V for the swept-back wing is still given by (4) and is still in the direction OC. This perturbation velocity has the components

(5) $u_0 = u_{1\infty} \sin \delta, \qquad v_0 = - u_{1\infty} \cos \delta,$

in the direction of the undisturbed wind and to starboard respectively.

From the second equations of (1) and (4) we have $\sin \mu_1 = \sin \mu_0/\sin \delta$, from which it follows easily that $\tan \mu_1 = \operatorname{cosec} \delta(\operatorname{cosec}^2 \mu_0 - \operatorname{cosec}^2 \delta)^{-1/2}$. Combining this with (3) and (4), we get

(6) $u_{1\infty} = \alpha V \operatorname{cosec} \delta(\cot^2 \mu_0 - \cot^2 \delta)^{-1/2}$

$$= u_\infty \operatorname{cosec} \delta \cot \mu_0 (\cot^2 \mu_0 - \cot^2 \delta)^{-1/2}.$$

Now introduce the angle λ defined by

(7) $\cos \lambda = \tan \mu_0 \cot \delta = \cot \delta/\cot \mu_0,$

and we then have, from (6),

(8) $u_{1\infty} = u_\infty \operatorname{cosec} \delta \operatorname{cosec} \lambda.$

So that, (5) gives

(9) $u_0 = u_\infty \operatorname{cosec} \lambda, \qquad v_0 = - u_\infty \cot \delta \operatorname{cosec} \lambda.$

These give the perturbation velocities for a swept-back aerofoil where u_∞ is given by (1). The lifting pressure is

(10) $2\rho_0 u_0 V = 2\alpha\rho_0 V^2 \tan \mu_0 \operatorname{cosec} \lambda,$

and thus the lift on any part of the aerofoil may be calculated by multiplication of (10) by the area of the part.

17·1. Linearisation for the flat aerofoil.

Take rectangular axes Ox, Oy, Oz and suppose there is a supersonic wind $(V, 0, \alpha V)$ so that its velocity potential is $- Vx - \alpha Vz$. This wind may be described otherwise as a wind of speed $V(1 + \alpha^2)^{1/2}$ at incidence $\tan^{-1} \alpha$ to the plane $z = 0$, fig. 17·1.

Let a portion of the plane $z = 0$ be occupied by a flat aerofoil, to fix our ideas say a triangular aerofoil or *delta wing*, placed with a vertex at the origin O. The

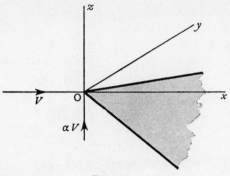

FIG. 17·1.

presence of this aerofoil will disturb the wind stream and the velocity potential will become

$$(1) \qquad \Phi = -Vx - \alpha Vz + \phi,$$

where ϕ is a velocity potential whose negative gradient, $-\nabla\phi$, gives the perturbation velocity (u, v, w) so that the velocity at any point will be

$$(2) \qquad (V + u, v, \alpha V + w).$$

Since the air cannot flow through the aerofoil we have the boundary condition $\alpha V + w = 0$ or

$$(3) \qquad w = -\alpha V \quad \text{on the aerofoil.}$$

Observe that statements (1), (2) and (3) involve no approximation.

We now introduce the assumption that α and consequently the perturbation is small, so that (u, v, w) are small quantities whose squares and products may be neglected. Also the speed of the wind stream is $V(1 + \alpha^2)^{1/2}$ which to this order of approximation is V and the perturbation velocity potential satisfies 15·3 (7). In the present case

$$M_0 = \operatorname{cosec} \mu_0 = c_0/V(1 + \alpha^2)^{1/2} = c_0/V$$

and therefore $M_0^2 - 1 = \cot^2 \mu_0$. We can now write 15·3 (7) in the form

$$(4) \qquad \frac{\partial^2\phi}{\partial y^2} + \frac{\partial^2\phi}{\partial z^2} = \cot^2 \mu_0 \frac{\partial^2\phi}{\partial x^2}.$$

Equation (4) is the linearised equation satisfied by the perturbation velocity potential in the case of a steady supersonic wind stream slightly disturbed by a flat aerofoil. The equation has more extended applications, the only essential condition being that the disturbance shall be small. For example, the same equation is satisfied by the perturbation velocity potential due to an aerofoil of small thickness whose camber surface lies nearly in the plane $z=0$ and whose slope referred to this plane is everywhere small, see 17·19. Similarly the equation may be used to discuss the perturbation due to a cone of infinitesimal angle whose generators lie near to the x-axis or to a slender body of revolution whose axis of revolution lies near to the x-axis.

Since $u = -\phi_x, v = -\phi_y, w = -\phi_z$, whose suffixes denote partial differentiation, it follows that u, v, w themselves all satisfy (4).

It might be well to remark here that the axes just introduced differ slightly from those employed in 15·3, where the wind is in the negative x-direction. The perturbation quantities there found are converted to use with the present axes by changing the sign of V. Thus, with the axes of this section,

$$(5) \qquad p - p_0 = -u\rho_0 V, \qquad \rho - \rho_0 = -u\rho_0 M_0{}^2/V$$

and are, naturally, still linear functions of u.

If S denotes any of the quantities ϕ, u, v, w, p, ρ, it follows that in the linearised case S satisfies (4) and therefore

$$(6) \qquad \frac{\partial^2 S}{\partial y^2} + \frac{\partial^2 S}{\partial z^2} = \cot^2 \mu \, \frac{\partial^2 S}{\partial x^2},$$

where we have written μ instead of μ_0, since in the sequel the Mach angle μ will always be taken to be that of the undisturbed wind stream.

On the upper surface of the aerofoil u is positive and $p - p_0$ is therefore negative. Therefore, from (5), the lifting pressure is

$$(7) \qquad 2\rho u V,$$

where ρ is the density of the undisturbed stream.

Since the lifting pressure determines the force on the aerofoil (see 17·2), it follows from (7) that our problem is essentially to determine u.

17·12. Solution of the linearised equation for conical flow.

As shown in 17·01 the pressure, and the perturbation velocity component u on which the pressure depends, are independent of the distance from the vertex in the case of conical flow. If the vertex is taken as origin, we require therefore those solutions of

$$(1) \qquad - \cot^2 \mu \, \frac{\partial^2 S}{\partial x^2} + \frac{\partial^2 S}{\partial y^2} + \frac{\partial^2 S}{\partial z^2} = 0,$$

which are homogeneous functions of degree zero in x, y, z.

We now show that the required solutions can be inferred from the homogeneous solutions of zero degree of Laplace's equation,

$$(2) \qquad \frac{\partial^2 S}{\partial X^2} + \frac{\partial^2 S}{\partial y^2} + \frac{\partial^2 S}{\partial z^2} = 0,$$

which reduces to (1) by the substitution,

$$(3) \qquad X = ix \tan \mu.$$

In polar coordinates

$$X = R_1 \cos \omega, \quad y = R_1 \cos \theta \sin \omega, \quad z = R_1 \sin \theta \sin \omega.$$

Laplace's equation becomes *

$$(4) \qquad \frac{\partial}{\partial R_1}\left(R_1{}^2 \frac{\partial S}{\partial R_1}\right) + \frac{1}{\sin \omega}\frac{\partial}{\partial \omega}\left(\sin \omega \frac{\partial S}{\partial \omega}\right) + \frac{1}{\sin^2 \omega}\frac{\partial^2 S}{\partial \theta^2} = 0,$$

and when S is independent of R_1 (the distance from the origin) the first term is identically zero. The substitution

$$(5) \qquad\qquad \chi = \log \tan \tfrac{1}{2}\omega$$

gives $\partial S/\partial \omega = \operatorname{cosec} \omega\, \partial S/\partial \chi$ and so (4) becomes

$$(6) \qquad\qquad \frac{\partial^2 S}{\partial \chi^2} + \frac{\partial^2 S}{\partial \theta^2} = 0,$$

which, as is easily verified by substitution, has the general real-valued solution

$$(7) \qquad\qquad S = F(\chi + i\theta) + \bar{F}(\chi - i\theta),$$

where F is an arbitrary, twice differentiable, function. Now from (5), observing that

$$\tan \tfrac{1}{2}\omega = \sin \omega/(1 + \cos \omega),$$

$$e^{\chi + i\theta} = e^{i\theta} \tan \tfrac{1}{2}\omega = \frac{\sin \omega \cos \theta + i \sin \omega \sin \theta}{1 + \cos \omega} = \frac{y + iz}{X + R_1}.$$

Thus, $\chi + i\theta$ is a function of $(y + iz)/(X + R_1)$ and therefore from (7) the general real-valued solution of Laplace's equation of zero degree is S_0, where

$$(8) \qquad\qquad S_0 = f_1\left(\frac{y + iz}{X + R_1}\right) + \bar{f}_1\left(\frac{y - iz}{X + R_1}\right),$$

and where f_1 denotes an arbitrary function.

From (8) we can deduce the general solution of zero degree of (1) by using the substitution (3). In fact, since $R_1{}^2 = X^2 + y^2 + z^2$, the substitution (3) gives

$$X + R_1 = ix \tan \mu + \sqrt{(y^2 + z^2 - x^2 \tan^2 \mu)} = i(x \tan \mu + R_2),$$

where

$$(9) \qquad\qquad R_2{}^2 = x^2 \tan^2 \mu - y^2 - z^2$$

and so

$$\frac{y + iz}{X + R_1} = (-i)\frac{y + iz}{x \tan \mu + R_2}.$$

Therefore, the general real-valued solution of zero degree of (1) is

$$(10) \qquad\qquad S_0 = f\left(\frac{y + iz}{x \tan \mu + R_2}\right) + \bar{f}\left(\frac{y - iz}{x \tan \mu + R_2}\right),$$

where f is an arbitrary, twice differentiable, function.

---

* Milne-Thomson, *Theoretical Hydrodynamics*, 16·1.

This solution is appropriate to all points (x, y, z) such that R_2^2 is positive, that is to say, to all points within the Mach cone

(11) $$y^2 + z^2 = x^2 \tan^2 \mu.$$

If we now introduce Busemann's complex variable ϵ which is defined by

(12) $$\epsilon = \frac{y + iz}{x \tan \mu + R_2}, \quad R_2 = (x^2 \tan^2 \mu - y^2 - z^2)^{1/2},$$

the general solution (11) assumes the convenient form

(13) $$S_0 = f(\epsilon) + \bar{f}(\bar{\epsilon}).$$

The properties of the variable ϵ will be discussed in the next section.

17·13. The complex variable ϵ.

This variable is defined by 17·12 (12). Introduce dimensionless variables r, s, $s^\star$ which satisfy the relations

(1) $$s = r \cos \theta = \frac{y}{x \tan \mu}, \quad s^\star = r \sin \theta = \frac{z}{x \tan \mu}.$$

It then follows that

(2) $$\epsilon = \frac{s + is^\star}{1 + (1 - s^2 - s^{\star 2})^{1/2}} = \frac{r(\cos \theta + i \sin \theta)}{1 + (1 - r^2)^{1/2}} = Re^{i\theta},$$

where

(3) $$R = \frac{r}{1 + (1 - r^2)^{1/2}} = \frac{1 - (1 - r^2)^{1/2}}{r}.$$

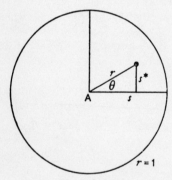

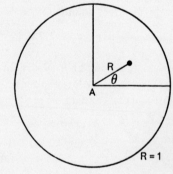

Physical plane. ϵ-plane.

FIG. 17·13.

If we mark the point $re^{i\theta}$ in one Argand diagram, which we shall call the *physical plane*, and the point $Re^{i\theta}$ in another, the ϵ-plane, we see that these points correspond in a mapping in which the points of a ray from the origin which makes an angle θ with the real axis in the physical plane, are mapped on the points of a ray from the origin which makes an angle θ with the real axis in the ϵ-plane. This mapping is not conformal.

It also appears from (3) that in the mapping the origins correspond, for

$R = 0$ when $r = 0$, and so do the circumferences $r = 1$ and $R = 1$. In the case of other corresponding points within these circumferences we have $R < r$. Moreover, R is complex when $r > 1$. Thus we regard the correspondence as a mapping of the circle $r \leqslant 1$ in the physical plane on the circle $R \leqslant 1$ in the ϵ-plane.

Further, since $re^{i\theta}$ takes the same value at all points of a ray which lies within the Mach cone of the point $(0, 0, 0)$, it follows that the points in the physical plane within the circumference $r = 1$ give a complete representation so far as conical flow is concerned of all points within the Mach cone.

Thus the state of a conical flow will be completely known if the variables which define the state are found as functions of (r, θ) or what amounts to the same as functions of ϵ. In this sense the unit circles of fig. 17·13 represent the Mach cone of the actual flow. We shall refer to either as the *Mach circle*.

An intuitive physical interpretation of s, $s^\star$ is obtained from the observation that $(\cot \mu, s, s^\star)$ are the cartesian coordinates of a point within the circular section of the Mach cone by the plane $x = \cot \mu$. It is in this sense that we may reasonably call the plane of $re^{i\theta}$ the physical plane.

Notation. In diagrams, corresponding points in the physical and ϵ-planes will always be marked with the same letter. For flat thin aerofoils which lie in the plane $z = 0$ the section by the plane $x = \cot \mu$ will be a straight line lying along the real axis of the physical plane, and the points of section of the leading and trailing edges (see 17·16) will be marked L and T with or without suffix. In so far as these points lie within the Mach circle in the physical plane we shall write (with due regard to sign) $s = l$ at a leading and $s = t$ at a trailing edge. The corresponding points in the ϵ-plane will be denoted by $\epsilon = L$, $\epsilon = T$ (i.e. with capital letters) as for example in fig. 17·6. In fig. 17·62 the leading edges will be $s = l_1$, $s = -l_2$ in the physical and $\epsilon = L_1$, $\epsilon = -L_2$ in the ϵ-plane.

The following formulae will be found useful. The proofs are left as simple exercises.

(5) $$R_2{}^2 = x^2 \tan^2 \mu - y^2 - z^2 = x^2 \tan^2 \mu (1 - r^2).$$

(6) $$\surd(1 - r^2) = \frac{1 - R^2}{1 + R^2}.$$

(7) $$\frac{2\surd(1 - r^2)}{1 + \surd(1 - r^2)} = 1 - R^2 = 1 - \epsilon\bar{\epsilon}.$$

(8) $$r = \frac{2R}{1 + R^2}.$$

(9) $$\frac{r}{1 - r} = \frac{2R}{(1 - R)^2}, \quad \frac{r}{1 + r} = \frac{2R}{(1 + R)^2}.$$

(10) $$\epsilon + \frac{1}{\epsilon} = \frac{2}{s} \quad \text{when } \epsilon \text{ is real.}$$

17·14. Complex velocity components. In 17·12 (13) the general real valued solution of zero degree of the linearised equation was found to be of the form $f(\epsilon) + \bar{f}(\epsilon)$. Since the perturbation velocity components u, v, w are solutions of this type, it follows that they are the real parts of functions of the complex variable ϵ. Thus we can write

$$(1) \qquad u + iu^\star = \mathcal{U}(\epsilon), \quad v + iv^\star = \mathcal{V}(\epsilon), \quad w + iw^\star = \mathcal{W}(\epsilon),$$

where $u^\star$, $v^\star$, $w^\star$ are also real. To these latter quantities we attribute no physical significance. We note, however, that

$$(2) \qquad u^\star - iu = -i\mathcal{U}(\epsilon), \quad v^\star - iv = -i\mathcal{V}(\epsilon), \quad w^\star - iw = -i\mathcal{W}(\epsilon),$$

so that $u^\star$, $v^\star$, $w^\star$ are perturbation velocity components of some other conical flow.

The functions $\mathcal{U}$, $\mathcal{V}$, $\mathcal{W}$, will be called the *complex velocity components*.

Just as we may regard a two-dimensional flow problem as solved when the complex potential $w(z)$ has been found, so *we regard a problem concerning conical flow as solved when the complex velocity component $\mathcal{U}$ has been found*, for then u and therefore the pressure and so the forces are determinable. Moreover, v and w can also be found (17·15) and so the whole flow is determinable.

17·15. Compatibility equations. The complex velocity components $\mathcal{U}$, $\mathcal{V}$, $\mathcal{W}$ of 17·14 are not independent, for their real parts u, v, w are velocity components of an irrotational flow. Since u, v, w are homogeneous functions of degree zero in x, y, z, the velocity potential from which they are derived must be homogeneous of degree unity. Let $f(x, y, z)$ be a homogeneous function of degree n. Then, if suffixes denote partial differentiation, a well-known theorem of Euler states that

$$(1) \qquad xf_x + yf_y + zf_z = nf.$$

In the case of the perturbation velocity potential ϕ, we have $n = 1$ and so

$$\phi = x\phi_x + y\phi_y + z\phi_z = -ux - vy - wz.$$

Therefore $d\phi = -u\,dx - v\,dy - w\,dz - x\,du - y\,dv - z\,dw$

But $d\phi = \phi_x\,dx + \phi_y\,dy + \phi_z\,dz = -u\,dx - v\,dy - w\,dz.$

Therefore

$$(2) \qquad x\,du + y\,dv + z\,dw = 0.$$

Since (17·14), $u^\star$, $v^\star$, $w^\star$ are also perturbation velocities of some conical flow, it follows similarly that

$$(3) \qquad x\,du^\star + y\,dv^\star + z\,dw^\star = 0.$$

Now, $du + i\,du^\star = d\mathcal{U} = \mathcal{U}'\,d\epsilon$, if a dash denotes differentiation with respect to ϵ. If, therefore, we multiply (3) by i and add to (2), we get

$$(4) \qquad x\mathcal{U}' + y\mathcal{V}' + z\mathcal{W}' = 0.$$

Partial differentiation of (4) with respect to x gives

$$\mathcal{U}' + (\mathcal{U}x'' + y\mathcal{V}'' + z\mathcal{W}'')\epsilon_x = 0.$$

It follows from this and the two similar equations obtained by differentiating (4) with respect to y and z that

(5)
$$\frac{\mathcal{U}'}{\epsilon_x} = \frac{\mathcal{V}'}{\epsilon_y} = \frac{\mathcal{W}'}{\epsilon_z}.$$

These are the *compatibility equations* expressing that $\mathcal{U}$, $\mathcal{V}$, $\mathcal{W}$ are compatible with irrotational flow.

We now proceed to eliminate x, y, z. From 17·12 (12) by logarithmic differentiation

$$\frac{d\epsilon}{\epsilon} = \frac{dy + i\,dz}{y + iz} - \frac{\tan \mu\,dx + dR_2}{x \tan \mu + R_2}.$$

Now
$$R_2\,dR_2 = x \tan^2 \mu\,dx - y\,dy - z\,dz.$$

Therefore
$$\frac{\epsilon_x}{-2\epsilon \tan \mu} = \frac{\epsilon_y}{\dfrac{2R_2 + 2y\epsilon}{x \tan \mu + R_2}} = \frac{\epsilon_z}{\dfrac{2iR_2 + 2z\epsilon}{x \tan \mu + R_2}}.$$

Using 17·13 (5) to (9), we obtain, after reduction,

(6)
$$\frac{\mathcal{U}'}{2\epsilon \tan \mu} = \frac{\mathcal{V}'}{-(1 + \epsilon^2)} = \frac{\mathcal{W}'}{-i(1 - \epsilon^2)}$$

as the final form of the compatibility equations.

The problem of linearised conical flow is essentially that of finding functions $\mathcal{U}(\epsilon)$, $\mathcal{V}(\epsilon)$, $\mathcal{W}(\epsilon)$ holomorphic within the Mach circle $|\epsilon| = 1$ except at certain singularities defined by the boundary conditions (see 17·17). It follows from (6) that if one of these functions can be found, the other two can be determined by integration.

On account of the presence of the factor ϵ in the denominator of the first member of (6) it follows that $\mathcal{V}'$ and $\mathcal{W}'$ will have a singularity at $\epsilon = 0$ unless $\mathcal{U}'(0) = 0$. Such a singularity cannot occur on the surface of the aerofoil (except possibly at an edge), so that if $\epsilon = 0$ is a point on the aerofoil and not at an edge,

(7)
$$\frac{d\mathcal{U}(\epsilon)}{d\epsilon} = 0 \quad \text{when } \epsilon = 0.$$

We now prove an important property of $u^\star$. From the compatibility equations (6), we have

(8)
$$d(w + iw^\star) = \frac{-i(1 - \epsilon^2)}{2\epsilon \tan \mu}(du + i\,du^\star).$$

Now, on the surface of the aerofoil ϵ is real and w is constant, 17·1 (3). Therefore, $dw = 0$ and equating the real parts of (8), we find that $du^\star = 0$ on the aerofoil. Thus $u^\star =$ constant on the surface of the aerofoil. Without loss of generality we can take the constant to be zero so that

(9) $u^\star = 0$ on the surface of the aerofoil.

This result has the very useful consequence that

(10) $u = \mathcal{U}(\epsilon)$ on the surface of the aerofoil,

since on that surface ϵ is real. In other words, there is no necessity to separate the real part in order to find u. Moreover, the lifting pressure is $2\rho V \mathcal{U}(\epsilon)$.

17·16. Polygonal aerofoils. As explained in 17·1 we assume the aerofoil to occupy a portion of the plane $z = 0$ and to be exposed to a wind $(V, 0, \alpha V)$. Consider a flat polygonal aerofoil. Such an aerofoil can be rotated in the plane $z = 0$ to take various orientations with respect to the wind stream. Let us consider one of these orientations and one vertex O of the polygon through which there necessarily pass two edges.

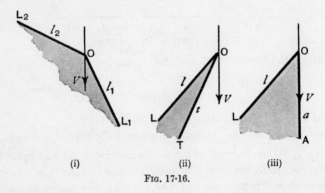

(i)   (ii)   (iii)

FIG. 17·16.

Fig. 17·16 shows three cases. In each figure a line is drawn through O to indicate the direction of V.

Def. If a line drawn parallel to the direction of V leaves the aerofoil when crossing an edge, that edge is said to be *trailing*; if it enters the aerofoil when crossing an edge, that edge is said to be *leading*. An edge drawn from the vertex parallel to the direction of V will be called *axial*.

Thus, in fig. (ii), OT is a trailing edge; in figs. (ii) and (iii) OL is a leading edge. In fig. (i) both edges OL_1, OL_2 are leading. In fig. (iii) OA is an axial edge.

We shall use the letters L, T, A to distinguish leading, trailing and axial edges, so that the edge character at O will be denoted by LL (two leading edges) in fig. (i), by LT (one leading, one trailing edge) in fig. (ii), by LA (one leading, one axial edge) in fig. (iii). We shall consider only those cases in which neither edge through O lies within the upstream part of the Mach cone of O.

The edges at O, when the aerofoil is presented in a certain orientation to a supersonic stream, are further characterised by lying inside or outside the Mach cone whose vertex is O, which will be called the Mach cone of O, or simply the Mach cone. Since the axis of the Mach cone is, to the linear approximation, in the direction of V, an axial edge necessarily lies within the Mach cone. In the case of edges other than axial the fact that they lie within the Mach cone will be indicated by suffix i. Thus, $L_i T_i$ indicates an aerofoil with one leading and one trailing edge at O both inside the Mach cone, as in fig. 17·6 (i). Similarly $L_i A$ has both edges inside the Mach cone as in fig. 17·65.

Def. An edge through O leading, trailing or axial which lies inside the Mach cone of O is termed a *subsonic edge* ; if it lies outside the Mach cone, it is termed a *supersonic edge* ; if it lies on the Mach cone, it is termed a *sonic edge*.

That an edge lies outside the Mach cone will be indicated by suffix o ; that the edge lies on the Mach cone will be indicated by suffix s. Thus $L_o L_o$, fig. 17·3 (i), has two leading edges, both supersonic ; $L_o A$, fig. 17·5 (i), has one supersonic leading edge and one axial and therefore subsonic edge ; $L_s A$, fig. 17·51, has one sonic leading edge and one axial edge.

The case of a sonic edge can always be regarded as a limiting case of either a supersonic or subsonic edge. This circumstance gives point to the observation that the sub- or supersonic character of an edge, for a given orientation of the aerofoil depends in general on the Mach number. Consider the LL delta wing illustrated in fig. 17·3 (i). When $M = 1$, the Mach angle is $\pi/2$ and the Mach cone a half-space so that both edges are subsonic, form $L_i L_i$. As M increases, the Mach angle decreases and we pass through the form $L_s L_i$ to $L_o L_i$ and with further increase of M to the form $L_o L_o$. In discussing various cases it is advantageous to use a uniform notation in the diagrams. We shall label leading and trailing edges by l and t, and the points where they meet the plane of the

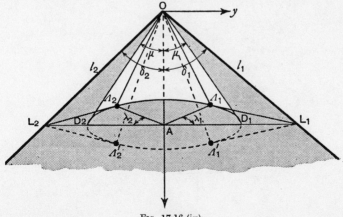

FIG. 17·16 (iv).

Mach circle by L and T, adding suffixes if necessary. An axial edge will be labelled by a and the centre of the Mach circle by A.

When the leading edge l is supersonic two planes can be drawn through l to touch the Mach cone. The points in which these planes touch the Mach circle will be denoted by $\varLambda$ and $\varLambda'$.

The ends of the diameter of the Mach circle which lie in the plane of the

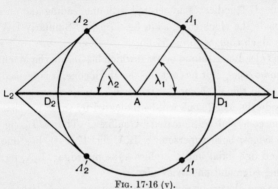

FIG. 17·16 (v).

aerofoil are D_1, D_2, the acute angle between $A\varLambda$ and D_1D_2 will be denoted by λ with suffix where necessary.

Thus, fig. 17·16 (iv) shows a case L_oL_o. There are two leading edges l_1, l_2 and four points $\varLambda_1, \varLambda_1'$ and $\varLambda_2, \varLambda_2'$ such that the planes through $l_1, \varLambda_1$

and $l_1, \varLambda_1'$ touch the Mach cone as do also the planes $l_2, \varLambda_2$ and $l_2, \varLambda_2'$. The Mach circle is shown separately in fig. (v). The angles which the edges make with the axis of the Mach cone are denoted by δ_1 and δ_2. Note that

$$(1) \qquad \cos \lambda_1 = \frac{A\varLambda_1}{AL_1} = \frac{A\varLambda_1}{OA} \cdot \frac{OA}{AL_1} = \tan \mu \cot \delta_1.$$

Similarly, $\cos \lambda_2 = \tan \mu \cot \delta_2$.

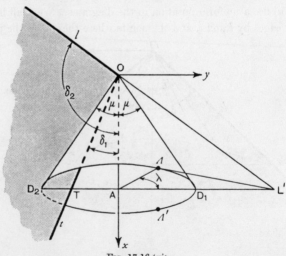

FIG. 17·16 (vi).

Fig. 17·16 (vi) shows a case $L_0 T_i$. Observe that here the ray l which lies along the swept-forward supersonic leading edge does not meet the plane of the Mach circle, but that the opposite ray meets the plane in the point labelled by L' in the figure. Observe also that a plane through l which touches the Mach

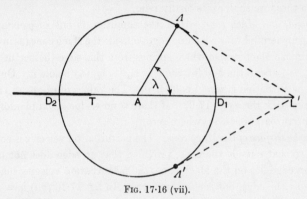

FIG. 17·16 (vii).

cone also passes through OL'. Fig. (vii) shows the Mach circle. In this diagram (cf. (1) above),

(2) $\cos \lambda = \tan \mu \cot AOL' = - \tan \mu \cot \delta_2$.

There is no point in multiplying further these illustrations, but a strict adherence to the notations here established will render the succeeding diagrams self-explanatory, and obviate the necessity of a detailed explanation in each individual case.

From the physical standpoint, in fig. 17·16 (iv) the region between the tangent planes l_1, Λ_1 ; l_1, Λ_1' and the Mach cone is a region of two-dimensional flow, so that if $\pm u_1$ is the appropriate perturbation velocity obtained from 17·03 the perturbation velocity will be u_1 on the arc $D_1 \Lambda_1$ of the Mach circle in fig. (v) and $- u_1$ on the arc $D_1 \Lambda_1'$. Similarly, it will be u_2 on the arc $D_2 \Lambda_2$ and $- u_2$ on the arc $D_2 \Lambda_2'$ where $\pm u_2$ is the two-dimensional perturbation velocity in the region between the tangent planes through l_2 and the Mach cone. On the arcs $\Lambda_1 \Lambda_2$ and $\Lambda_1' \Lambda_2'$ we shall have $u = 0$ for these arcs are adjacent to the undisturbed region outside the Mach cone. Similarly, in fig. 17·16 (vi) if $\pm u_0$ is the perturbation velocity in the region between the tangent planes through l and the Mach cone, in fig. (vii) we shall have $u = u_0$ on the arc $D_2 \Lambda$ and $u = - u_0$ on the arc $D_2 \Lambda'$, while $u = 0$ on the arc $\Lambda' D_1 \Lambda$.

17·17. Boundary conditions. We now consider conditions which the perturbation velocity components (u, v, w) must satisfy in the case of a flat polygonal aerofoil at small incidence α in a supersonic stream V, the aerofoil lying as usual in the plane $z = 0$. We are primarily concerned with the conditions immediately outside the Mach cone of the intersection of two edges; on the Mach cone, and inside it. To fix our ideas and for simplicity of statement we shall suppose that the Mach cone in question is not intersected by the Mach cone of any other vertex of the polygon. Figs. 17·16 (vi), (vii) show a typical case and to these we shall occasionally refer.

(A) *Outside the Mach cone.* Here the condition is one of constant state, Considering an aerofoil with a leading edge l outside the Mach cone (as in fig. 17·16 (vi)), within the Mach wedge which springs from this edge the perturbation is. say, (u_0, v_0, w_0) above the aerofoil and $(- u_0, - v_0, w_0)$ below it. Outside the wedge the perturbation is zero, i.e. $(0, 0, 0)$. The values of (u_0, v_0, w_0) are given by the sweep-back theory of 17·03. If there is no wedge, as in planforms $L_i T_i$, $L_i L_i$, the perturbation is everywhere zero.

(B) *On the surface of the Mach cone.* The perturbation velocity is continuous and equal to that outside the cone, provided that an edge does not lie on the cone. If an edge lies on the Mach cone, the perturbation velocity may be discontinuous at the edge, but is elsewhere zero. In fig. 17·16 (vii) $u = u_0$ on the arc ΛD_2, $u = - u_0$ on the arc $\Lambda' D_2$ and $u = 0$ on the arc $\Lambda' D_1 \Lambda$ of the Mach circle. The conditions on the Mach cone are of fundamental importance in determining the flow within it.

(C) *In the plane of the aerofoil, off the wing.* Here Bernoulli's theorem shows that we must have $u = 0$ to ensure continuity of the pressure which is a linear function of u (15·3 (7)). This condition holds both inside and outside the Mach cone, see e.g. fig. 17·41 (i).

The component v can have a discontinuity of sign but not of magnitude. Such a discontinuity of sign implies a vortex sheet and will occur behind a trailing edge but will not extend beyond the axis of the Mach cone, see figs. 17·16 (vi), 17·41 (i).

The component w is continuous.

(D) *On the aerofoil.* We consider points inside the Mach cone and *not* at an edge. Here u has opposite signs but the same magnitude just above and just below a given point of the aerofoil (condition of lift).

We have also the important conditions (17·15 (8))

(1) $$u^\star = 0, \quad u = \mathcal{U}(\epsilon).$$

Moreover, u, v, w are all finite and continuous on the surface of the aerofoil, and if the point $\epsilon = 0$ is on the aerofoil, we have from 17·15 (7)

(2) $$d\mathcal{U}(\epsilon)/d\epsilon = 0, \quad \text{when } \epsilon = 0.$$

This condition is not obligatory unless $\epsilon = 0$ is on the aerofoil.

The component w satisfies the condition (17·1 (3))

(3) $$w = - \alpha V.$$

(E) *Edge conditions.* The basic assumption is that at a subsonic edge, leading or trailing, the flow is qualitatively the same as in incompressible flow.

Consider a point of the aerofoil at a small distance d from an edge. Then

(i) At a *subsonic trailing edge* the velocity remains finite as $d \to 0$ (Kutta-Joukowski condition). It then follows from (C) above that

(4) $$u = 0 \text{ at a subsonic trailing edge.}$$

(ii) At a *subsonic leading edge* (cf. 7·5)

(5) $$u = kd^{-\frac{1}{2}},$$

where k is a constant, and $u \to \infty$ in this manner when $d \to 0$.

To examine the implications of (5) consider fig. 17·52 (i). Let $AL_1 = l_1$. Then, if P is a point on the aerofoil near to L_1, $AP = l_1 - d$ and the corresponding points in the ϵ-plane, fig. 17·52 (ii), are

$$L_1 = \frac{1 - \sqrt{(1 - l_1^2)}}{l_1}, \quad \epsilon = \frac{1 - \sqrt{[1 - (l_1 - d)^2]}}{l_1 - d}.$$

Expanding by the binomial theorem, we find that

$$\epsilon - L_1 = a_0 d + a_1 d^2 + \dots$$

and therefore (5) gives, to the lowest order,

(6) $$u = C(\epsilon - L_1)^{-\frac{1}{2}},$$

where C is a constant, in this case purely imaginary, since $\epsilon < L_1$. Thus, using (1) we see that near L_1, $\mathcal{U}(\epsilon)$ behaves like $C(\epsilon - L_1)^{-\frac{1}{2}}$ therefore in the neighbourhood of L_1

(7) $$\mathcal{U}(\epsilon) = \frac{f(\epsilon)}{(\epsilon - L_1)^{\frac{1}{2}}},$$

where $f(\epsilon)$ is holomorphic and therefore finite at $\epsilon = L_1$. This result exposes the nature of the singularity at a subsonic leading edge.

We also observe that, on the pattern of 7·5, the existence of the singularity (6) at a leading edge should lead to a suction force in the plane of the wing at this edge. G. N. Ward * obtains a force of magnitude, per unit length,

(8) $$\frac{\pi \rho k^2}{(1 - M^2 \sin^2 \delta)^{\frac{1}{2}}},$$

where δ is the inclination of the edge to the x-axis and k is a constant.

17·18. Circular boundary value problems.

To find a function $f(\epsilon)$, holomorphic within the Mach circle γ, that is $|\epsilon| \leqslant 1$, whose real part takes given values on the circumference.

Since a point on the circumference can be specified by the angle θ measured

---

* *Quarterly Journal of Mechanics and Applied Math.*, 2 (1949), p. 136.

from the real axis, we can suppose that the real part of $f(\epsilon)$ is to take the value $F(\theta)$ on the circumference, where $F(\theta)$ is given. Now write

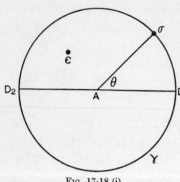

FIG. 17·18 (i).

$$(1) \qquad \sigma = e^{i\theta},$$

so that σ is a point on the circumference and $\theta = (\log \sigma)/i$. Then, if we write

$$(2) \quad F(\theta) = F\{(\log \sigma)/i\} = g(\sigma),$$

the function $g(\sigma)$ is likewise known.

The real part of $f(\epsilon)$ is $\tfrac{1}{2}f(\epsilon) + \tfrac{1}{2}\bar{f}(\bar{\epsilon})$, and on the circumference

$$\epsilon = \sigma, \quad \bar{\epsilon} = e^{-i\theta} = 1/\sigma.$$

Therefore

$$(3) \quad f(\sigma) + \bar{f}(1/\sigma) = 2g(\sigma).$$

Multiply this equation by $d\sigma/[2\pi i(\sigma - \epsilon)]$ and integrate right round γ. Then

$$(4) \qquad \frac{1}{2\pi i} \int_{(\gamma)} \frac{f(\sigma)\, d\sigma}{\sigma - \epsilon} + \frac{1}{2\pi i} \int_{(\gamma)} \frac{\bar{f}(1/\sigma)\, d\sigma}{\sigma - \epsilon} = \frac{1}{\pi i} \int_{(\gamma)} \frac{g(\sigma)\, d\sigma}{\sigma - \epsilon}.$$

By Cauchy's formula * the first integral is $f(\epsilon)$ and by the residue theorem (3·53) the second integral (see below) is $\bar{f}(0)$. Therefore

$$(5) \qquad f(\epsilon) + \bar{f}(0) = \frac{1}{\pi i} \int_{(\gamma)} \frac{g(\sigma)\, d\sigma}{\sigma - \epsilon}$$

which solves the problem.

In many applications $g(\sigma)$ is constant over arcs of the circumference γ and the calculation of the integral in (5) is then very simple. Suppose for example, that $g(\sigma)$ is equal to u_0 from $\theta = -\lambda$ to $\theta = \lambda$ and is zero on the rest of the circumference. Writing for brevity, $\epsilon_1 = e^{i\lambda}$, $\bar{\epsilon}_1 = e^{-i\lambda}$,

$$\frac{1}{\pi i} \int_{(\gamma)} \frac{g(\sigma)\, d\sigma}{\sigma - \epsilon} = \frac{u_0}{\pi i} \Big[\log(\sigma - \epsilon) \Big]_{\sigma = \bar{\epsilon}_1}^{\sigma = \epsilon_1},$$

and therefore

$$(6) \qquad f(\epsilon) = -\bar{f}(0) + \frac{u_0}{\pi i} [\log(\epsilon_1 - \epsilon) - \log(\bar{\epsilon}_1 - \epsilon)].$$

We observe that that $\bar{f}(0)$ is a constant whose value can be determined only by further information. If, for example, we know that $f(\epsilon)$ is real when ϵ is real, the fact that $f(\epsilon) = u_0$ when $\epsilon = 1$ (i.e. $\theta = 0$) determines $\bar{f}(0)$ from the equation

$$(7) \qquad u_0 = -\bar{f}(0) + \frac{u_0}{\pi i} [\log(\epsilon_1 - 1) - \log(\bar{\epsilon}_1 - 1)],$$

which gives $\bar{f}(0) = u_0 \lambda/\pi$.

The same simple principle can be used to determine a function $f(\epsilon)$ such that

---

* Milne-Thomson, *Theoretical Hydrodynamics*, 5·59.

(i) $f(\epsilon)$ is holomorphic within a *semi-circle* of unit radius, (ii) $f(\epsilon)$ is real-valued on the diameter, (iii) the real part of $f(\epsilon)$ takes given values on the semi-circle.

To solve this problem we use the following artifice. Complete the unit circle γ. The lower semicircle in fig. 17·18 (ii) is then the reflection of the upper in the given diameter $D_1 D_2$. We then add to the three conditions above the condition (iv) that $f(\epsilon)$ shall be deter-mined in the whole circle in such a way that the functional values at every point and its reflection in the diameter $D_1 D_2$ shall be conjugate complex quantities.* This reduces the problem to the one just solved.

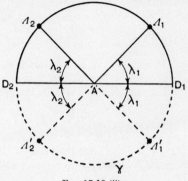

FIG. 17·18 (ii).

To take a specific example, suppose that in fig. 17·18 (ii) the diameter $D_1 D_2$ is the real axis of the ϵ-plane and that on the arc $D_1 \Lambda_1$ the real part of $f(\epsilon) = u_1$, on the arc $\Lambda_1 \Lambda_2$ the real part of $f(\epsilon) = 0$, and on the arc $\Lambda_2 D_2$ the real part of $f(\epsilon) = u_2$. Then condition (iv) demands that the real part of $f(\epsilon)$ shall be $u_2, 0, u_1$ on the respective arcs $D_2 \Lambda_2', \Lambda_2' \Lambda_1', \Lambda' D_1$ where Λ_1', Λ_2' are the reflections of Λ_1, Λ_2 in $D_1 D_2$. In this and other cases, we shall suppose that $-\pi \leqslant \theta \leqslant \pi$ so that with the notation of the figure we can write for the values of σ at the respec-tive points $\Lambda_1, \Lambda_2, \Lambda_2', \Lambda_1'$,

$$\epsilon_1 = e^{i\lambda_1}, \quad \epsilon_2 = e^{i(\pi - \lambda_2)} = -e^{-i\lambda_2}, \quad \bar{\epsilon}_2 = e^{i(-\pi + \lambda_2)} = -e^{i\lambda_2}, \quad \bar{\epsilon}_1 = e^{-i\lambda_1}.$$

Then (5) gives

$$(8) \qquad f(\epsilon) + \bar{f}(0) = \frac{1}{\pi i} \int_{\bar{\epsilon}_1}^{\epsilon_1} \frac{u_1 \, d\sigma}{\sigma - \epsilon} + \frac{1}{\pi i} \int_{\epsilon_2}^{\bar{\epsilon}_2} \frac{u_2 \, d\sigma}{\sigma - \epsilon}$$

$$= \frac{u_1}{\pi i} \log \frac{\epsilon_1 - \epsilon}{\bar{\epsilon}_1 - \epsilon} - \frac{u_2}{\pi i} \log \frac{\epsilon - \epsilon_2}{\epsilon - \bar{\epsilon}_2},$$

and we note in passing that $\epsilon_1 - \epsilon$ and $\epsilon - \epsilon_2$ have been written rather than $\epsilon - \epsilon_1$ and $\epsilon_2 - \epsilon$ in order that a diagram may readily show the arguments of these numbers in conformity with their lying between $-\pi$ and π.

Since by hypothesis $f(\epsilon)$ is real when $\epsilon = 0$, we have $\bar{f}(0) = f(0)$ and so, putting $\epsilon = 0$ in (8) we get

$$2\bar{f}(0) = \frac{2u_1}{\pi} \lambda_1 + \frac{2u_2}{\pi} \lambda_2,$$

and therefore finally

$$(9) \qquad f(\epsilon) = \frac{u_1}{\pi} \left\{ -i \log \frac{\epsilon_1 - \epsilon}{\bar{\epsilon}_1 - \epsilon} - \lambda_1 \right\} + \frac{u_2}{\pi} \left\{ i \log \frac{\epsilon - \epsilon_2}{\epsilon - \bar{\epsilon}_2} - \lambda_2 \right\}.$$

---

* This is an application of the principle of reflection. See, e.g. Milne-Thomson, *Theoretical Hydrodynamics*, 5·53.

It should be verified that this function satisfies all the conditions. There are no arbitrary elements.

Returning to (4) to show that the second integral is $\bar{f}(0)$ we note that $f(\epsilon)$ is by hypothesis holomorphic within γ and therefore can be developed in an infinite series, convergent within γ,

$$f(\epsilon) = a_0 + a_1\epsilon + a_2\epsilon^2 + \dots$$

and so

$$\bar{f}(1/\sigma) = \bar{a}_0 + \frac{\bar{a}_1}{\sigma} + \frac{\bar{a}_2}{\sigma^2} + \dots .$$

Now, by Cauchy's residue theorem (3·53),

$$(10) \qquad \frac{1}{2\pi i}\int_{(\gamma)} \frac{\bar{a}_0\, d\sigma}{\sigma - \epsilon} = \bar{a}_0 = \bar{f}(0) \quad \text{and} \quad \frac{1}{2\pi i}\int_{(\gamma)} \frac{d\sigma}{\sigma^n(\sigma - \epsilon)} = 0,$$

if $n > 0$, by the same theorem. This proves the statement.

We calculate here an integral which will be used in 17·4

$$(11) \qquad J = \frac{1}{2\pi i}\int_{(\gamma)} \frac{\log \sigma\, d\sigma}{\sigma - \epsilon} = \log(\sigma_0 - \epsilon) + a,$$

where a is an arbitrary constant and σ_0 is an arbitrary point of γ.

To prove this, we have

$$\frac{\partial J}{\partial \epsilon} = \frac{1}{2\pi i}\int_{(\gamma)} \frac{\log \sigma\, d\sigma}{(\sigma - \epsilon)^2} = -\frac{1}{2\pi i}\int_{(\gamma)} \log \sigma\, d\left(\frac{1}{(\sigma - \epsilon)}\right)$$

$$= -\frac{1}{2\pi i}\left[\frac{\log \sigma}{\sigma - \epsilon}\right]_{(\gamma)} + \frac{1}{2\pi i}\int_{(\gamma)} \frac{d\sigma}{\sigma(\sigma - \epsilon)},$$

on integration by parts. The last integral vanishes by the residue theorem. If we start at an arbitrary point of σ_0 of γ and go round, $\log \sigma$ increases by $2\pi i$ while $\sigma - \epsilon$ returns to its initial value. Therefore

$$\frac{\partial J}{\partial \epsilon} = -\frac{1}{\sigma_0 - \epsilon} \quad \text{and} \quad J = \log(\sigma_0 - \epsilon) + a$$

on integration.

17·19. Non-lifting aerofoils.

The boundary conditions of 17·17 were formulated for flat aerofoils of zero thickness lying in the plane $z = 0$.

Consider a more general aerofoil whose upper and lower surfaces have the respective equations

$$(1) \qquad z = f_U(x, y), \quad z = f_L(x, y).$$

To keep within the linear theory z must be small and so must $\partial z/\partial x$. The equations (1) may be written in the equivalent forms

$$(2) \qquad z = \tfrac{1}{2}(f_U + f_L) + \tfrac{1}{2}(f_U - f_L),$$

$$(3) \qquad z = \tfrac{1}{2}(f_U + f_L) - \tfrac{1}{2}(f_U - f_L).$$

We call

(4) $z = \frac{1}{2}(f_U + f_L)$

the *camber surface* (cf. 8·0). This corresponds to a nearly flat aerofoil of zero thickness.

The equation

(5) $z^2 = \frac{1}{4}(f_U - f_L)^2$

represents an aerofoil of small thickness $f_U - f_L$ symmetrical with respect to the plane $z = 0$, for if the point (x, y, z) lies on (5), so does the point $(x, y, -z)$.

When such a symmetrical aerofoil is placed at zero incidence in a wind which blows in the x-direction no lift is experienced. For this reason, it is called a *non-lifting aerofoil*. The determination of the perturbation velocity components for the non-lifting aerofoil (5) allows us by superposition to solve the problem for the aerofoil (1).

As a simple and manageable example consider an aerofoil whose section by any plane $y = $ constant is two straight lines inclined to the plane $z = 0$ at the small angle β and whose planform is that of a delta wing. The boundary conditions are (1) $w = \beta V$ on the upper surface, (2) $w = -\beta V$ on the lower surface, (3) $w = 0$ in the plane $z = 0$, off the aerofoil. These conditions are sufficient to determine $\mathcal{W}$ and therefore $\mathcal{U}$ and $\mathcal{V}$ from the compatibility equations.

It is not our intention to investigate the various cases at length. They are in fact easily discussed by the methods which follow later. To take one illustration, consider the planform L_oA of fig. 17·5 (i).

The boundary conditions are shown in fig. 17·19 where $w_0 = \beta V$.

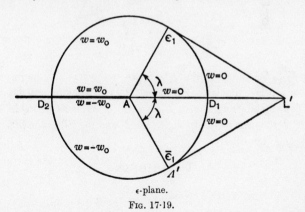

ϵ-plane.

Fig. 17·19.

These are precisely the conditions satisfied by $\mathcal{U}(\epsilon)$ in 17·4, and we take over the solution of that section by replacing $\mathcal{U}$ by $\mathcal{W}$, u_0 by βV and ϵ_2 by $\epsilon_1 = e^{i\lambda}$, where $\cos \lambda = -\tan \mu \cot \delta$. This gives

(6) $$\mathcal{W}(\epsilon) = \frac{i\beta V}{\pi} \{\log (\epsilon_1 - \epsilon)(\bar{\epsilon}_1 - \epsilon) - \log \epsilon\}.$$

There is a logarithmic singularity at $\epsilon = 0$.

17·2. Force on a delta wing. The pressure on the part of the wing, if any, which lies outside the Mach cone of its vertex can be dealt with by the sweep-back theory of 17·03. We shall therefore restrict attention to that part of the wing which lies inside the Mach cone. For simplicity of exposition we shall suppose that no part of the wing OPQ lies outside the Mach cone of O. We shall further suppose that the acute angle between PQ and the direction of the undisturbed wind exceeds the Mach angle μ so that neither vertex P or Q lies within the Mach cone of the other.

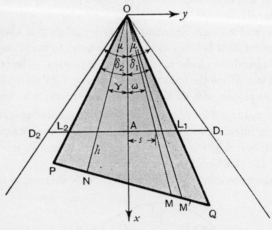

FIG. 17·2.

In fig. 17·2, $OA = \cot \mu$ and $D_1 L_1 A L_2 D_2$ is the trace of the physical plane. Draw $ON = h$ perpendicular to PQ and let ON make angle γ with Ox. Draw the rays OM, OM' making angles $\omega, \omega + d\omega$ with Ox and meeting PQ in M and M'. Then, if dS is the area of the triangle OMM' and if $p(\omega)$ is the lifting pressure on the ray OM, the lift is

(1) $$L = \int p(\omega) \, dS,$$

taken over the triangle OPQ. Now

(2) $dS = \frac{1}{2}hMM' = \frac{1}{2}h^2 \, d \, [\tan (\gamma + \omega)] = \frac{1}{2}h^2 \sec^2 (\gamma + \omega) \, d\omega.$

Therefore

(2) $$L = \frac{1}{2}h^2 \int_{-\delta_2}^{\delta_1} p(\omega) \sec^2 (\gamma + \omega) \, d\omega.$$

The resultant thrust on the triangle OMM' is $p(\omega)\,dS$ at its centroid whose distance from O is $\frac{2}{3}OM = \frac{2}{3}h\sec(\gamma+\omega)$. Thus the pitching moment M_y (about Oy) and the rolling moment M_x (about Ox) are given by

$$(3) \qquad M_y = -\tfrac{1}{3}h^3 \int_{-\delta_2}^{\delta_1} p(\omega)\sec^3(\gamma+\omega)\cos\omega\,d\omega,$$

$$(4) \qquad M_x = \tfrac{1}{3}h^3 \int_{-\delta_2}^{\delta_1} p(\omega)\sec^3(\gamma+\omega)\sin\omega\,d\omega.$$

Calculations of drag and yawing moment must take into account the suction force, 17·17 (8), at the leading edges. Into this we shall not enter.

The lift L and the moments M_x, M_y can be made to depend upon the single integral

$$(5) \qquad J = \int_{-\delta_2}^{\delta_1} p(\omega)\tan(\gamma+\omega)\,d\omega.$$

We have

$$\frac{\partial}{\partial\gamma}\tan(\gamma+\omega) = \sec^2(\gamma+\omega)$$

and therefore

$$(6) \qquad L = \tfrac{1}{2}h^2 \frac{\partial J}{\partial\gamma}.$$

Now, let

$$(7) \qquad M_1 = M_x\cos\gamma - M_y\sin\gamma, \quad M_2 = M_x\sin\gamma + M_y\cos\gamma$$

denote the rolling moment about ON and the pitching moment about the line through O perpendicular to ON in the plane of the wing. Then from (3) and (4),

$$(8) \qquad M_1 = \tfrac{1}{3}h^3 \int_{-\delta_2}^{\delta_1} p(\omega)\sec^3(\gamma+\omega)\sin(\gamma+\omega)\,d\omega = \tfrac{1}{6}h^3 \frac{\partial^2 J}{\partial\gamma^2}.$$

$$(9) \qquad M_2 = -\tfrac{1}{3}h^2 \int_{-\delta_2}^{\delta_1} p(\omega)\sec^2(\gamma+\omega)\,d\omega = -\tfrac{1}{3}h^3 \frac{\partial J}{\partial\gamma}.$$

Thus, with the aid of (7) M_x and M_y are expressible in terms of $\partial J/\partial\gamma$ and $\partial^2 J/\partial\gamma^2$.

The above formulae are susceptible of various manipulations. In particular

$$(10) \qquad M_1 = \tfrac{1}{3}h \frac{\partial L}{\partial\gamma}, \quad M_2 = -\tfrac{2}{3}hL,$$

the second result being obvious from statical considerations.

It is often convenient to express the lift and moments in terms of ϵ. In the ϵ-plane, on the wing, ϵ is real and $u = \mathcal{U}(\epsilon)$. Therefore $p(\omega) = 2\rho V \mathcal{U}(\epsilon)$, while

$$\frac{2}{s} = \epsilon + \frac{1}{\epsilon}, \quad s = \cot\mu\tan\omega.$$

Elimination of s gives

(11) $\tan(\gamma + \omega) - \tan\gamma = \dfrac{\tan\omega \sec^2\gamma}{1 - \tan\omega \tan\gamma} = \dfrac{2\epsilon \sec^2\gamma \tan\mu}{\epsilon^2 - 2\epsilon \tan\gamma \tan\mu + 1}$.

For brevity, write

(12) $\qquad g = g(\epsilon) = \dfrac{\epsilon}{\epsilon^2 - 2\beta\epsilon + 1}, \quad \beta = \tan\gamma \tan\mu.$

Then, from (1), (2) and (11),

(13) $\qquad L = 2\rho V h^2 \sec^2\gamma \tan\mu \displaystyle\int_{(L_2 L_1)} \mathcal{U}\, dg.$

Integration by parts gives

(14) $\qquad \dfrac{L}{2\rho V h^2 \sec^2\gamma \tan\mu} = \Big[\mathcal{U}g\Big]_{(L_2 L_1)} - \displaystyle\int_{(L_2 L_1)} g\, d\mathcal{U},$

where g is given by (12). The form (14) is appropriate only when $\mathcal{U}$ is not infinite at the limits.

The lift coefficient is got by dividing the lift by $\tfrac{1}{2}\rho V^2 S$, where the area S is given by

(15) $\qquad S = \tfrac{1}{2}h^2[\tan(\delta_1 + \gamma) + \tan(\delta_2 - \gamma)].$

17·3. Planform $L_o L_o$. Here both edges are leading and both supersonic, i.e. outside the Mach cone of their point of intersection.

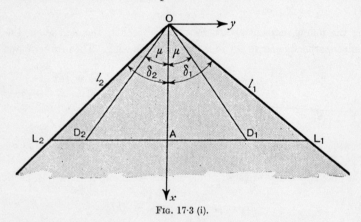

Fig. 17·3 (i).

Fig. (i) shows the planform and fig. (ii) the ϵ-plane in the standard notation of 17·16,

(1) $\qquad \cos\lambda_1 = \tan\mu \cot\delta_1, \quad \cos\lambda_2 = \tan\mu \cot\delta_2.$

(2) $\qquad \epsilon_1 = e^{i\lambda_1}, \quad \epsilon_2 = e^{i(\pi - \lambda_2)} = -e^{-i\lambda_2}.$

Outside the Mach cone in the region $L_2 D_2 \Lambda_2$ the flow is two-dimensional with

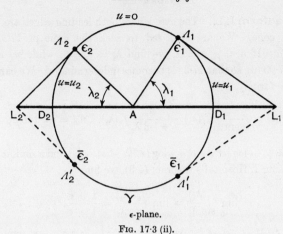

ϵ-plane.

FIG. 17·3 (ii).

perturbation velocity component u_2 parallel to Ox and similarly in the region $L_1 \varLambda_1 D_1$ the component is u_1 parallel to Ox, where from 17·03,

(3) $u_1 = u_\infty \operatorname{cosec} \lambda_1, \quad u_2 = u_\infty \operatorname{cosec} \lambda_2, \quad u_\infty = \alpha V \tan \mu.$

In the regions $L_2 D_2 \varLambda_2', \varLambda_1' D_1 L_1$ the components are $-u_2, -u_1$ respectively.

Thus, on the semicircle $D_1 \varLambda_1 D_2$, we have $u = u_1$ on the arc $D_1 \varLambda_1$, $u = 0$ on the arc $\varLambda_1 \varLambda_2$, $u = u_2$ on the arc $\varLambda_2 D_2$ and $u^\star = 0$ on the diameter $D_1 D_2$.

These data combined with the principle of reflection explained in 17·18 determine the function $\mathcal{U}(\epsilon)$ whose real part takes the above values on the boundary of the semicircle. This particular boundary value problem has already been solved in 17·18 and in fact from 17·18 (9), we have

(4) $\mathcal{U}(\epsilon) = \dfrac{u_1}{\pi} \left\{ - i \log \dfrac{\epsilon_1 - \epsilon}{\bar{\epsilon}_1 - \epsilon} - \lambda_1 \right\} + \dfrac{u_2}{\pi} \left\{ i \log \dfrac{\epsilon - \epsilon_2}{\epsilon - \bar{\epsilon}_2} - \lambda_2 \right\}.$

The complex velocity component $\mathcal{U}$ in the lower semicircle is then the same as (4) with the sign changed.

The lift on the part within the Mach cone can be calculated by means of 17·2 (14), the limits of integration being $\epsilon = -1, \epsilon = 1$. To get the total lift we add the lift on the triangular areas outside the Mach cone which is got by multiplying them by $2\rho u_1 V, 2\rho u_2 V$. Division by $\frac{1}{2}\rho V^2 S$, where

$$S = \tfrac{1}{2} h^2 \left[\tan(\delta_1 + \gamma) + \tan(\delta_2 - \gamma) \right]$$

gives the lift coefficient

$$C_L = \frac{4\alpha \tan \mu}{(1 - \tan^2 \mu \tan^2 \gamma)^{\frac{1}{2}}}.$$

Note that this is independent of the angle of the delta wing but depends on the angle of yaw γ of the perpendicular ON in fig. 17·2. See Ex. XVII, 5, 6.

17·31. Planform L_sL_s. The case where both leading edges are generators of the Mach cone. We are thus led to seek the limit of 17·3 (4) when $\delta_1 \to \mu, \delta_2 \to \mu$. If $\delta_1 = \mu, \delta_2 = \mu$, we find $\lambda_1 = \lambda_2 = 0$, while u_1 and u_2 are infinite, see 17·03 (9), so that 17·3 (4) becomes indeterminate. We can, however, write it in the form

$$\mathcal{U}(\epsilon) = \frac{u_\infty}{\pi \sin \lambda_1} f_1(\lambda_1) + \frac{u_\infty}{\pi \sin \lambda_2} f_2(\lambda_2), \quad u_\infty = \alpha V \tan \mu,$$

where $f_1(\lambda_1) = \{- i \log (e^{i\lambda_1} - \epsilon) + i \log (e^{-i\lambda_1} - \epsilon) - \lambda_1\}$ and a similar expression for $f_2(\lambda_2)$. Using L'Hospital's theorem (1·9), we find

$$\lim_{\lambda_1 \to 0} \frac{f_1(\lambda_1)}{\sin \lambda_1} = \lim_{\lambda_1 \to 0} \frac{f_1'(\lambda_1)}{\cos \lambda_1} = \frac{1 + \epsilon}{1 - \epsilon},$$

and similarly for the limit of $f_2(\lambda_2)/\sin \lambda_2$ we get $(1 - \epsilon)/(1 + \epsilon)$. Thus, for planform L_sL_s,

$$(1) \qquad \mathcal{U}(\epsilon) = \frac{2u_\infty}{\pi} \frac{1 + \epsilon^2}{1 - \epsilon^2}, \quad u_\infty = \alpha V \tan \mu.$$

We observe that at the sonic edges $\epsilon = \pm 1$, $\mathcal{U}$ has a simple pole. Let us calculate the lift on an isosceles delta wing of this form. For an isosceles wing, $\gamma = 0$ and therefore 17·2 (12) gives $g = \epsilon/(\epsilon^2 + 1)$ and so 17·2 (13) gives

$$L = 2\rho V h^2 \tan \mu \int_{-1}^{1} \frac{2u_\infty}{\pi} \frac{1 + \epsilon^2}{1 - \epsilon^2} \cdot \frac{1 - \epsilon^2}{(1 + \epsilon^2)^2} \, d\epsilon$$

$$= 2\alpha\rho V^2 h^2 \tan^2 \mu.$$

From this it appears that in spite of the poles at the leading edges there is finite lift.

In the present case the area is $h^2 \tan \mu$ and so the lift coefficient is

$$C_L = L/\tfrac{1}{2}\rho V^2 S = 4\alpha \tan \mu.$$

17·4. Delta wing of constant lift distribution.

In this type of aerofoil the perturbation velocity in the x-direction is assumed to be constant, equal to u_0 say on the upper surface and to $- u_0$ on the lower. In so far as part of the aerofoil lies outside the Mach cone this part will be flat with sweep-back perturbation velocity u_0. In the Mach cone the aerofoil must be slightly cambered in the manner appropriate to make $u = u_0$ on its upper and $u = - u_0$ on its lower surface. The lifting pressure, and so the lift, will then be uniform over the aerofoil.

We illustrate this principle by taking the planform L_oA.

Figs. 17·4 (i) and (ii) show the planform and boundary conditions in our standard notation.

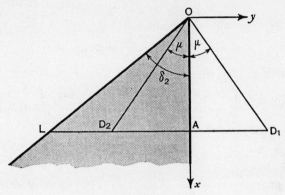

FIG. 17·4 (i).

On making a circuit round A from a point near A on the lower surface to a point near A on the upper surface u changes from $-u_0$ to u_0, which indicates that at $\epsilon = 0$, $\mathcal{U}(\epsilon)$ has a logarithmic singularity of the form

$$-\frac{iu_0}{2\pi}\log \epsilon,$$

for on the lower surface $\epsilon = Re^{-i\pi}$ and on the upper $\epsilon = Re^{i\pi}$.

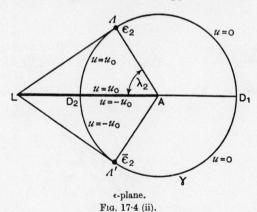

ϵ-plane.
FIG. 17·4 (ii).

Within the Mach circle γ there is no other singularity and therefore

(1) $$\mathcal{U}(\epsilon) + \frac{iu_0}{\pi}\log \epsilon = f(\epsilon),$$

where $f(\epsilon)$ is holomorphic everywhere inside γ.

Since $\mathcal{U}(\epsilon) + \overline{\mathcal{U}}(\bar{\epsilon}) = 2u$, we have on γ (see 17·18),

$$f(\sigma) + \bar{f}(1/\sigma) = 2u + \frac{iu_0}{\pi} \log \sigma - \frac{iu_0}{\pi} \log \frac{1}{\sigma}$$

$$= 2u + \frac{2iu_0}{\pi} \log \sigma,$$

and therefore from 17·18 (5)

$$f(\epsilon) + \bar{f}(0) = \frac{1}{\pi i} \int_{\epsilon_2}^{-1} \frac{u_0 \, d\sigma}{\sigma - \epsilon} - \frac{1}{\pi i} \int_{-1}^{\bar{\epsilon}_2} \frac{u_0 \, d\sigma}{\sigma - \epsilon} + \frac{2iu_0}{\pi} \cdot \frac{1}{2\pi i} \int_{(\gamma)} \frac{\log \sigma \, d\sigma}{\sigma - \epsilon}.$$

Using 17·18 (11) we get $f(\epsilon) + \bar{f}(0)$

$$= \frac{u_0}{\pi i} \{2 \log (-1 - \epsilon) - \log (\epsilon_2 - \epsilon) - \log (\bar{\epsilon}_2 - \epsilon)\} + \frac{2iu_0}{\pi} \log (\sigma_0 - \epsilon) + a,$$

where a is an arbitrary constant. Since there can be no singularity at $\epsilon = -1$ which is on the aerofoil we must take $\sigma_0 = -1$ in order to remove the term $2 \log (-1 - \epsilon)$ and then

$$f(\epsilon) = \frac{iu_0}{\pi} \log \{(\epsilon - \epsilon_2)(\epsilon - \bar{\epsilon}_2)\} + \text{constant}$$

and therefore from (1)

$$(1) \qquad \mathcal{U}(\epsilon) = \frac{iu_0}{\pi} \{\log [(\epsilon - \epsilon_2)(\epsilon - \bar{\epsilon}_2)] - \log \epsilon\},$$

no constant being necessary since $u = u_0$ when $\arg \epsilon = \pi$.

Here

$$\epsilon_2 = -e^{-i\lambda_2}, \quad \cos \lambda_2 = \tan \mu \cot \delta_2.$$

Since there is no pressure interference between aerofoils of constant lift distribution the solutions appropriate to them can be superposed by addition.

17·41. Planform L_oT_o. Here both leading and trailing edges are outside the Mach cone of their intersection. Thus on the surface of the aerofoil the

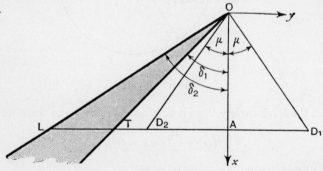

Fig. 17·41 (i).

motion is two-dimensional, in other words the lift distribution is constant and
equal, say, to $2\rho u_0 V$, where u_0 is the x-component of the perturbation velocity
on the upper surface. In fig. 17·41 (i) the wing is OLT and OA is the axis. Thus

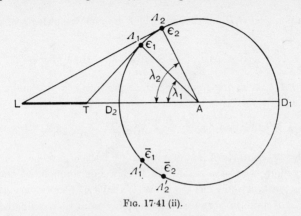

FIG. 17·41 (ii).

the wing OLT can be regarded as the superposition of a wing OLA with lifting
pressure $2\rho u_0 V$ and a wing OTA with lifting pressure $-2\rho u_0 V$. Thus, from
17·4 (1), we have for the velocity inside the Mach cone

$$(1) \qquad \mathcal{U}(\epsilon) = \frac{iu_0}{\pi}[\log\{(\epsilon - \epsilon_2)(\epsilon - \bar\epsilon_2)\} - \log\{(\epsilon - \epsilon_1)(\epsilon - \bar\epsilon_1)\}],$$

where $\epsilon_1 = -e^{-i\lambda_1}$, $\epsilon_2 = -e^{-i\lambda_2}$, $\cos\lambda_1 = \tan\mu\cot\delta_1$, $\cos\lambda_2 = \tan\mu\cot\delta_2$ in
the notation of figs. (i) and (ii).

Since $(\epsilon - \epsilon_2)(\epsilon - \bar\epsilon_2)$ and $(\epsilon - \epsilon_1)(\epsilon - \bar\epsilon_1)$ are real and positive when ϵ is real,
for they are products of complex conjugates, it follows from (1) that $u = 0$ on the
diameter D_1D_2 as should be the case, 17·17 (C). Let P be a point on the circum-
ference of the upper semicircle in fig. (ii) and let the arcs $\Lambda_1\Lambda_2$ and $\Lambda_1'\Lambda_2'$ sub-
tend the angles ϕ, ϕ' respectively at P. Then, equating the real parts of (1), we
have $u = u_0(\phi + \phi')/\pi$ or $u_0(\phi - \phi')/\pi$ according as P lies on the arc $\Lambda_1\Lambda_2$ or not.
In the former case $\phi' = \pi - \phi$ and in the latter $\phi' = \phi$. Thus $u = u_0$ on the arc
$\Lambda_1\Lambda_2$ and $u = 0$ on the remaining arcs of the semicircle. Similarly, on the lower
semicircle $u = -u_0$ on the arc $\Lambda_1'\Lambda_2'$ and $u = 0$ on the remaining arcs. These
statements determine u on the Mach circle and therefore in the regions outside it.

The physically important property of the solution, however, lies in the
determination of the *side-wash* v. From the compatibility equations, 17·15 (6),
we have

$$2\tan\mu\,\frac{d\mathcal{V}}{d\epsilon} = -\frac{1 + \epsilon^2}{\epsilon}\frac{d\mathcal{U}}{d\epsilon}.$$

Substitution from (1) and a simple integration lead to

(2) $\mathcal{V}(\epsilon)$

$$= \frac{iu_0}{\pi} \cot \mu \left\{ \cos \lambda_2 \log \frac{(\epsilon - \epsilon_2)(\epsilon - \bar{\epsilon}_2)}{\epsilon} - \cos \lambda_1 \log \frac{(\epsilon - \epsilon_1)(\epsilon - \bar{\epsilon}_1)}{\epsilon} \right\}.$$

If ϵ is a point on D_1D_2 all the logarithms are real except $\log \epsilon$, and equating real parts

(3) $v = \text{Real part of } \dfrac{iu_0}{\pi} \cot \mu (\cos \lambda_1 - \cos \lambda_2) \log \epsilon.$

If we regard AD_2 as having an upper surface on which $\epsilon = Re^{i\pi}$ and a lower surface on which $\epsilon = Re^{-i\pi}$, we have on the upper surface

(4) $v = - \dfrac{u_0}{\pi} \cot \mu (\cos \lambda_1 - \cos \lambda_2)\pi = - u_0(\cot \delta_1 - \cot \delta_2),$

while on the lower surface of AD_2 the value of v is the negative of this. Outside the Mach cone in the region within the Mach wedge through the trailing edge the state is constant and v has the value (4) above AT, and its negative below. Thus there is a uniform vortex sheet extending from the trailing edge to the axis of the Mach cone.

If ϵ is a point on AD_1, ϵ is real and positive and so from (3) $v = 0$, so that the vortex sheet does not extend beyond the axis.

On the other hand, the axis, at which $\epsilon = 0$, is a logarithmic singularity of v which is infinite. The existence of this singularity is a consequence of the linearisation of the problem, not a physical necessity. Thus the general conclusion is that behind the trailing edge there is a vortex sheet extending to a region near the axis. The vortex lines trail downstream for the trailing edge.

17·5. Planform L_0A. Consider a flat aerofoil of the planform shown in

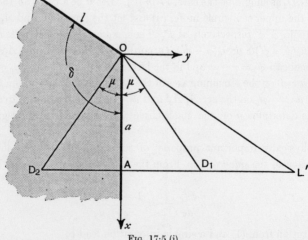

Fig. 17·5 (i).

fig. 17·5 (i). The ϵ-plane is shown in fig. (ii) together with the boundary values of u on the unit circle γ. Here

(1) $$\epsilon_1 = e^{i\lambda}, \quad \cos\lambda = -\tan\mu\cot\delta,$$

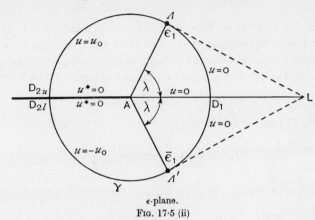

ϵ-plane.

FIG. 17·5 (ii)

the negative sign because in this figure λ is acute and δ is obtuse. We assume that $\delta < \pi - \mu$, otherwise O would be inside the Mach cones of the points of l. We have

(2) $$u_0 = u_\infty \operatorname{cosec}\lambda = \alpha V \tan\mu \operatorname{cosec}\lambda$$

and $u^\star = 0$ on AD_2. To solve the boundary value problem we begin by writing

(3) $$\nu^2 = \epsilon,$$

thus mapping the interior of the circle γ in the ϵ-plane on the interior of the *semicircle* $D_{2l}\Lambda'\Lambda D_{2u}$ of the unit circle γ' in the ν-plane shown in fig. (iii).

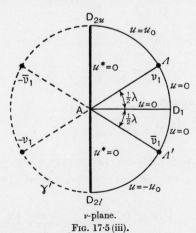

ν-plane.

FIG. 17·5 (iii).

Here D_{2u} denotes the point D_2 regarded as on the upper surface of the aerofoil and D_{2l} denotes the point D_2 regarded as on the lower surface, so that in fig. (iii), AD_{2u} is the map of the upper and AD_{2l} is the map of the lower surface on both of which $u^\star = 0$.

Thus, in the ν-plane, we have, as in the case of planform L$_o$L$_o$ (in the ϵ-plane), the semicircle boundary problem of 17·18 which is solved by ascribing to u the same values on reflection in the diameter $D_{2l}D_{2u}$ as on the semi-circumference $D_{2l}D_1D_{2u}$. If σ is a point on the circumference γ', we have

$$(4) \qquad \mathcal{U}(\sigma) + \overline{\mathcal{U}}(1/\sigma) = 2u,$$

leading as in 17·18 to, if $\nu_1{}^2 = e^{i\lambda}$,

$$(5) \qquad \mathcal{U}(\nu) + \overline{\mathcal{U}}(0) = \frac{2u_0}{2\pi i}\int_{\nu_1}^{-\bar{\nu}_1} \frac{d\sigma}{\sigma - \nu} - \frac{2u_0}{2\pi i}\int_{-\nu_1}^{\bar{\nu}_1} \frac{d\sigma}{\sigma - \nu}.$$

Since $u = u^\star = 0$ at $\nu = 0$, $\mathcal{U}(0) = 0$ and therefore (5) gives

$$(6) \qquad \mathcal{U}(\nu) = \frac{-iu_0}{\pi} \log \frac{(\nu_1 + \nu)(\bar{\nu}_1 + \nu)}{(\nu_1 - \nu)(\bar{\nu}_1 - \nu)}, \quad \nu^2 = \epsilon,$$

and this gives the value of u on the aerofoil since there $u^\star = 0$. To calculate u in the physical plane, put, for points on the upper surface, $\nu = ik$, $k > 0$, then from (6),

$$(7) \qquad u = \frac{iu_0}{\pi} \log \frac{1 - k^2 - 2ik\cos\frac{1}{2}\lambda}{1 - k^2 + 2ik\cos\frac{1}{2}\lambda} = \frac{2u_0}{\pi} \tan^{-1} \frac{2k\cos\frac{1}{2}\lambda}{1 - k^2},$$

where we have used the fact that $\nu_1{}^2 = e^{i\lambda}$ and therefore $\nu_1 + \bar{\nu}_1 = 2\cos\frac{1}{2}\lambda$

Now on the upper surface of the aerofoil $\epsilon = Re^{i\pi} = -R$, and so

$$k^2 = R = \frac{1 - \sqrt{(1 - r^2)}}{r},$$

and therefore, from 17·13 (9),

$$(8) \qquad \frac{2k}{1 - k^2} = \frac{2\sqrt{R}}{1 - R} = \sqrt{\frac{2r}{1 - r}}.$$

Thus, the lifting pressure is

$$(9) \qquad \frac{4\rho u_0 V}{\pi} \tan^{-1}\left\{\cos\tfrac{1}{2}\lambda\sqrt{\frac{2r}{1 - r}}\right\}.$$

It follows that the ratio of the lift coefficient C_L on the part within the Mach cone to the lift coefficient C_{L_0} on that part operating as a two-dimensional aerofoil is

$$(10) \qquad \frac{C_L}{C_{L_0}} = \int_0^1 \frac{u}{u_0} dr = \frac{2}{\pi}\int_0^1 \tan^{-1}\sqrt{\frac{2r\cos^2\frac{1}{2}\lambda}{1 - r}}\, dr.$$

Put

$$\tan^2 \theta = \frac{2r \cos^2 \frac{1}{2}\lambda}{1 - r} = t^2.$$

Then

$$r = 1 - \frac{2 \cos^2 \frac{1}{2}\lambda}{2 \cos^2 \frac{1}{2}\lambda + \tan^2 \theta},$$

and so integrating by parts

$$\frac{\pi C_L}{2 C_{L_0}} = \left[r\theta \right]_{r=0}^{r=1} - \int_{r=0}^{r=1} r \, d\theta = 2 \cos^2 \frac{1}{2}\lambda \int_0^\infty \frac{dt}{(1 + t^2)(2 \cos^2 \frac{1}{2}\lambda + t^2)}.$$

(11) $$\frac{C_L}{C_{L_0}} = \frac{(\sqrt 2) \cos \frac{1}{2}\lambda}{1 + (\sqrt 2) \cos \frac{1}{2}\lambda},$$

which gives the *lift reduction ratio* due to the conical flow. In the case of the rectangular aerofoil of fig. 17·02 (i) at each tip we have a flow of the type just described. Here $\delta = \frac{1}{2}\pi$, $\cos \lambda = 0$, $\lambda = \frac{1}{2}\pi$ and the reduction ratio is $\frac{1}{2}$.

17·51. Planform L_sA. We can regard this as a limiting case of L_0A, when $\delta = \mu$, $\cos \lambda = -1$, $\lambda = \pi$. As in 17·31 the insertion of these values in 17·5 (6) leads to an indeterminate form so we again have recourse to L'Hospital's theorem (1·9).

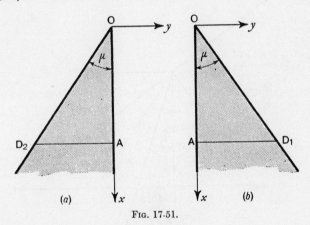

Fig. 17·51.

We begin by writing 17·5 (6) in the form

$$\mathcal{U}(\nu) = \frac{-iu_\infty}{\pi \sin \lambda} f(\lambda),$$

where

$$f(\lambda) = \log (\nu^2 + 2\nu \cos \tfrac{1}{2}\lambda + 1) - \log (\nu^2 - 2\nu \cos \tfrac{1}{2}\lambda + 1),$$

$$f'(\lambda) = \frac{-\nu \sin \frac{1}{2}\lambda}{\nu^2 + 2\nu \cos \frac{1}{2}\lambda + 1} - \frac{\nu \sin \frac{1}{2}\lambda}{\nu^2 - 2\nu \cos \frac{1}{2}\lambda + 1}.$$

Thus the required value is

$$\mathscr{U}(\nu) = \lim_{\lambda \to \pi} \frac{-iu_\infty f(\lambda)}{\pi \sin \lambda} = \lim_{\lambda \to \pi} \frac{-iu_\infty f'(\lambda)}{\pi \cos \lambda} = \frac{-2iu_\infty \nu}{\pi(\nu^2 + 1)}.$$

This value is correct, for on the upper surface of the aerofoil, where $\nu = ik$, $k > 0$, u is positive. Since $\nu^2 = \epsilon$, we have

(1) $$\mathscr{U}(\epsilon) = \frac{2u_\infty(-\epsilon)^{\frac{1}{2}}}{\pi(1 + \epsilon)}, \quad u_\infty = \alpha V \tan \mu,$$

when the aerofoil is disposed as in fig. 17·51 (a), for there $-\epsilon$ is positive on the surface, u is real, and we take the positive value of the square root for the upper surface and the negative for the lower. Comparing with 17·31 we see that again there is a pole at $\epsilon = -1$, i.e. on the leading edge.

If the aerofoil is disposed as in fig. 17·51 (b) we change the sign of ϵ thus obtaining

(2) $$\mathscr{U}(\epsilon) = \frac{2u_\infty \epsilon^{\frac{1}{2}}}{\pi(1 - \epsilon)}$$

with a pole at the leading edge.

17·52. Planform L_oL_i.

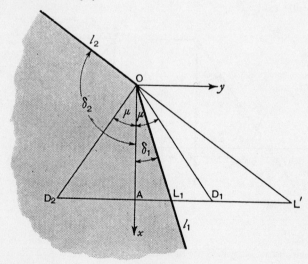

Fig. 17·52 (i).

Fig. 17·52 (i) shows the plan of the aerofoil in which we shall suppose that

(1) $$AL_1 = \cot \mu \tan \delta_1 = l.$$

Fig. (ii) shows the ϵ-plane in which at L_1 we put

(2) $$\epsilon = L = \frac{1 - \sqrt{(1 - l^2)}}{l}.$$

If we write

(3) $$\nu^2 = \frac{\epsilon - L}{1 - \epsilon L},$$

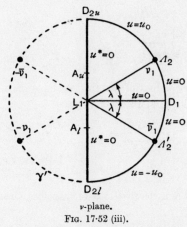

ϵ-plane.

Fig. 17·52 (ii).

the region within the Mach circle γ of fig. (ii) is mapped upon the region within
the semicircle of fig. (iii) the centre of this semicircle being now the map of the
edge L_1.

ν-plane.

Fig. 17·52 (iii).

The point corresponding to Λ_2 in the ν-plane is ν_1 where,

(4) $\qquad \nu_1{}^2 = (\epsilon_2 - L)/(1 - \epsilon_2 L) = e^{2i\lambda}, \ \cos 2\lambda = \tfrac{1}{2}(\nu_1{}^2 + \bar{\nu}_1{}^2),$

where $\epsilon_2 = e^{i\lambda_2}, \cos\lambda_2 = -\tan\mu\cot\delta_2$, and since, from 17·13 (8), $l = 2L/(1 + L^2)$,
we find, after a simple reduction,

(5) $\qquad \cos 2\lambda = \dfrac{\cos\lambda_2 - l}{1 - l\cos\lambda_2} = \dfrac{-\tan\mu\cot\delta_2 - \cot\mu\tan\delta_1}{1 + \tan\delta_1\cot\delta_2}.$

Since l_1 is a subsonic leading edge, 17·17 (7) gives

(6)
$$\mathcal{U}(\epsilon) = \frac{f_1(\epsilon)}{(\epsilon - L)^{\frac{1}{2}}},$$

and therefore

(7)
$$\mathcal{U}(\nu) = \frac{f_2(\nu)}{\nu},$$

where $f_2(\nu)$ is holomorphic at $\nu = 0$ and so admits an expansion in series of the form

$$f_2(\nu) = a_0 + a_1\nu + a_2\nu^2 + \ldots ,$$

so that

$$\frac{f_2(\nu)}{\nu} = \frac{a_0}{\nu} + a_1 + a_2\nu + \ldots = \frac{a_0}{\nu} + f(\nu),$$

where $f(\nu)$ is holomorphic in the semicircle. For subsequent convenience, write

$$a_0 = iCu_0/\pi.$$

Then

(8)
$$\mathcal{U}(\nu) = \frac{iCu_0}{\pi\nu} + f(\nu).$$

Here C is a constant which will turn out to be real.

The boundary condition is

$$\mathcal{U}(\sigma) + \overline{\mathcal{U}}(1/\sigma) = 2u,$$

which from (8) is equivalent to

$$f(\sigma) + \bar{f}(1/\sigma) = -\frac{iCu_0}{\pi\sigma} + \frac{iCu_0\sigma}{\pi} + 2u.$$

Using the boundary values of fig. (iii) we have once more the semicircle boundary value problem of 17·18 which gives

$$f(\nu) + \bar{f}(0) = \frac{1}{2\pi i}\int_{(\gamma')}\frac{-iCu_0\,d\sigma}{\pi\sigma(\sigma - \nu)} + \frac{1}{2\pi i}\int_{(\gamma')}\frac{iCu_0\sigma\,d\sigma}{\pi(\sigma - \nu)}$$
$$- \frac{1}{\pi i}\int_{-\nu_1}^{\bar{\nu}_1}\frac{u_0\,d\sigma}{\sigma - \nu} + \frac{1}{\pi i}\int_{\nu_1}^{-\bar{\nu}_1}\frac{u_0\,d\sigma}{\sigma - \nu}.$$

By the residue theorem the first integral vanishes, the second gives $iCu_0\nu/\pi$ and we have

$$f(\nu) + \bar{f}(0) = \frac{iCu_0\nu}{\pi} - \frac{iu_0}{\pi}\log\frac{(\nu_1 + \nu)(\bar{\nu}_1 + \nu)}{(\nu_1 - \nu)(\bar{\nu}_1 - \nu)}$$

and therefore from (8)

(9)
$$\mathcal{U}(\nu) = \frac{iCu_0}{\pi}\left(\nu + \frac{1}{\nu}\right) - \frac{iu_0}{\pi}\log\frac{(\nu_1 + \nu)(\bar{\nu}_1 + \nu)}{(\nu_1 - \nu)(\bar{\nu}_1 - \nu)},$$

the constant $\bar{f}(0)$ turning out to be zero since when $\nu = i$, $\mathcal{U}(\nu) = u_0$ and the two sides of (9) agree.

To determine the constant C we note, 17·15 (7), that since $\epsilon = 0$ is on the aerofoil $d\mathcal{U}/d\epsilon = 0$ when $\epsilon = 0$. Now

$$\frac{d\mathcal{U}}{d\epsilon} = \frac{d\mathcal{U}}{dv} \cdot \frac{dv}{d\epsilon}$$

and from (3) we see that $dv/d\epsilon \neq 0$ when $\epsilon = 0$, i.e. when $v^2 = -L$. Therefore

$$(10) \qquad \frac{d\mathcal{U}}{dv} = 0 \quad \text{when } v^2 = -L.$$

Differentiate (9), put $v^2 = -L$, and equate to zero. This gives

$$C = \frac{2L(\nu_1 + \bar{\nu}_1)}{(1 - L)^2 + L(\nu_1 + \bar{\nu}_1)^2}.$$

Now, from (4),

$$\nu_1 + \bar{\nu}_1 = 2 \cos \lambda.$$

Also from 17·13 (9)

$$2L/(1 - L)^2 = l/(1 - l),$$

and therefore

$$(11) \qquad C = \frac{2l \cos \lambda}{1 - l + 2l \cos^2 \lambda} = \frac{2l \cos \lambda}{1 + l \cos 2\lambda}.$$

This completes the solution. To bring it to a usable form we calculate u on the upper surface of the aerofoil where $v = ik$, $k > 0$, so that (9) gives

$$(12) \qquad \frac{\pi u}{u_0} = C\left(\frac{1}{k} - k\right) + 2 \tan^{-1} \frac{2k \cos \lambda}{1 - k^2}.$$

On the portion AL_1 of the aerofoil

$$\epsilon = R = [1 - \sqrt{(1 - r^2)}]/r,$$

and from (3),

$$k^2 = (L - R)/(1 - LR).$$

Therefore

$$\left(\frac{1}{k} - k\right)^2 = \frac{1}{k^2} + k^2 - 2 = \frac{(1 - L)^2(1 + R)^2}{L(1 + R^2) - R(1 + L^2)} = \frac{2(1 - l)(1 + r)}{l - r}.$$

on the use of 17·13 (8) and (9). Put

$$(13) \qquad x = \left(\cos \lambda \sqrt{\frac{2}{1 - l}}\right) \sqrt{\frac{l - r}{1 + r}}.$$

Then

$$(14) \qquad \frac{\pi u}{u_0} = 2 \tan^{-1} x + 2C \cos \lambda \left(\frac{1}{x}\right).$$

Here l and r are measured in the physical plane, $\cos \lambda$ is given by (5) and so (14) determines u in terms of the geometry of the aerofoil.

To determine the lift reduction ratio C_L/C_{L_0}, we have

(15) $$\frac{C_L}{C_{L_0}} = \int_{-1}^{l} \frac{u}{u_0}\,dr = \frac{2}{\pi}\int_{-1}^{l}\left[\tan^{-1}x + \frac{C\cos\lambda}{x}\right]dr,$$

where from (13)

(16) $$x = f\sqrt{\frac{l-r}{1+r}}, \quad f = \sqrt{\frac{2\cos^2\lambda}{1-l}}\,.$$

To integrate (15) we observe that integration by parts gives

$$\int_{-1}^{l}\tan^{-1}x\,dr = \frac{\pi}{2} + \int_{0}^{\infty}\frac{r\,dx}{1+x^2},$$

where from (16) $r = (lf^2 - x^2)/(x^2 + f^2)$. The remaining steps are somewhat lengthy but straightforward, the final result being

(17) $$\frac{C_L}{C_{L_0}} = (1+l)\left\{\frac{(\sqrt{2})\cos\lambda}{\sqrt{(1-l)}+(\sqrt{2})\cos\lambda} + \frac{l\cos\lambda\,\sqrt{(2-2l)}}{1+l\cos 2\lambda}\right\},$$

where λ is given by (5).

17·53. Planform L_oT_i.

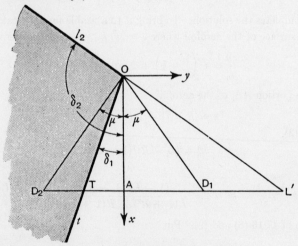

Fig. 17·53 (i).

The planform is shown in fig. 17·53 (i) where we put

(1) $$t = AT = \cot\mu\tan\delta_1.$$

In the ϵ-plane, T is the point $\epsilon = -T$ and

(2) $$\cos\lambda_2 = -\tan\mu\cot\delta_2.$$

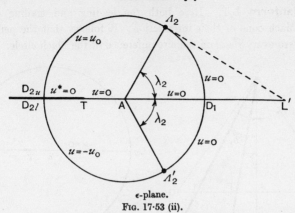

ϵ-plane.

FIG. 17·53 (ii).

We proceed as in 17·52 by writing

(3)
$$\nu^2 = \frac{\epsilon + T}{1 + \epsilon T},$$

which maps the circle in the ϵ-plane on a semicircle in the ν-plane, cf. fig. 17·52 (iii). Since we have here a subsonic trailing edge there is no singularity at T where $u = 0$. Thus the method of 17·52 gives at once

(4)
$$\mathcal{U}(\nu) = - \frac{iu_0}{\pi} \log \frac{(\nu + \nu_1)(\nu + \bar{\nu}_1)}{(\nu - \nu_1)(\nu - \bar{\nu}_1)},$$

where

(5)
$$\nu_1 = e^{i\lambda}, \quad \cos 2\lambda = \frac{\cos \lambda_2 + t}{1 + t \cos \lambda_2}.$$

On the upper surface of the aerofoil $\nu = ik$, $k > 0$ and $\epsilon = -R$. Proceeding exactly as before, we get

(6)
$$u = \frac{2u_0}{\pi} \tan^{-1} \frac{2k \cos \lambda}{1 - k^2}, \quad k^2 = \frac{R - T}{1 - RT}$$

and we find that

(7)
$$\frac{k}{1 - k^2} = \frac{1}{\sqrt{[2(1 + t)]}} \sqrt{\frac{r - t}{1 - r}}.$$

Putting, cf. 17·52 (13),

(8)
$$x = \left(\cos \lambda \sqrt{\frac{2}{1 + t}} \right) \sqrt{\frac{r - t}{1 - r}}.$$

(9)
$$\frac{\pi u}{u_0} = 2 \tan^{-1} x,$$

and the lift reduction ratio is

(10)
$$\frac{C_L}{C_{L_0}} = \frac{b(1 - t)}{1 + b}, \quad b^2 = \frac{2 \cos^2 \lambda}{1 + t}.$$

17·6. Planform L_iT_i. Here both the leading and trailing edges are within the Mach cone of their intersection. It follows that the perturbation velocity is zero everywhere on the circumference of the Mach circle.

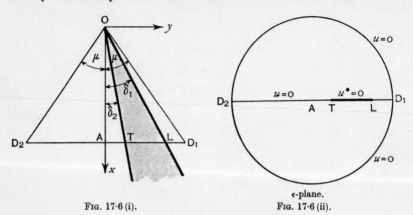

FIG. 17·6 (i).

ϵ-plane.
FIG. 17·6 (ii).

If $\sigma = e^{i\theta}$ is a point of the circumference of the Mach circle,

$$\mathcal{U}(\sigma) + \overline{\mathcal{U}}(1/\sigma) = 0$$

and so

(1)
$$\mathcal{U}(\epsilon) = - \overline{\mathcal{U}}\left(\frac{1}{\epsilon}\right).$$

Now, at the point $\epsilon = L$, subsonic leading edge, $\mathcal{U}(\epsilon)$ behaves like $(\epsilon - L)^{-\frac{1}{2}}$, while at $\epsilon = T$, subsonic trailing edge, $u = 0$.

Consider the function

$$f(\epsilon) = \frac{\epsilon - T}{\epsilon - L},$$

then

$$f\left(\frac{1}{\epsilon}\right) = \frac{1 - T\epsilon}{1 - L\epsilon}.$$

It follows that

(2)
$$\mathcal{U}(\epsilon) = iB_1\left[\frac{(\epsilon - T)(1 - T\epsilon)}{(\epsilon - L)(1 - L\epsilon)}\right]^{\frac{1}{2}}$$

satisfies all the conditions if B_1 is a real constant, for it satisfies (1) and has the appropriate behaviour at $\epsilon = L$ and $\epsilon = T$. Moreover, the quantity within the brackets is positive for real values of ϵ unless $T < \epsilon < L$ when it is negative. Thus the value of u given by (2) vanishes on D_2T and LD_1, while $u^\star$ vanishes on TL. Lastly, when $\epsilon = e^{i\theta}$ is on the circumference of the Mach circle

$$\frac{(\epsilon - T)(1 - T\epsilon)}{(\epsilon - L)(1 - L\epsilon)} = \frac{(e^{i\theta} - T)(e^{-i\theta} - T)}{(e^{i\theta} - L)(e^{-i\theta} - L)}$$

and both numerator and denominator are the products of conjugate complex numbers so that the expression is real and positive. Therefore $\mathcal{U}(\epsilon)$ is purely imaginary on the Mach circle and $u = 0$.

Thus the solution is complete except for the determination of the constant B_1 (see 17·61).

To express the perturbation velocity component u in terms of the physical plane variable s, let $AT = t$ and $AL = l$ in fig. (i) so that

(3) $$t = \cot \mu \tan \delta_2, \quad l = \cot \mu \tan \delta_1.$$

Then

(4) $$T = \frac{1 - \sqrt{(1 - t^2)}}{t}, \quad L = \frac{1 - \sqrt{(1 - l^2)}}{l},$$

while (2) can be written

(5) $$\mathcal{U}(\epsilon) = iB_1 \left(\frac{T}{L}\right)^{\frac{1}{2}} \left[\frac{\epsilon + \dfrac{1}{\epsilon} - \left(T + \dfrac{1}{T}\right)}{\epsilon + \dfrac{1}{\epsilon} - \left(L + \dfrac{1}{L}\right)}\right]^{\frac{1}{2}}.$$

Now on the aerofoil, from 17·13 (10),

$$\epsilon + \frac{1}{\epsilon} = \frac{2}{s}, \quad T + \frac{1}{T} = \frac{2}{t}, \quad L + \frac{1}{L} = \frac{2}{l},$$

and therefore on the aerofoil

(6) $$u = \pm B \left[\frac{s - t}{l - s}\right]^{\frac{1}{2}},$$

where

(7) $$B = B_1 \frac{l}{t} \left[\frac{1 - \sqrt{(1 - t^2)}}{1 - \sqrt{(1 - l^2)}}\right]^{\frac{1}{2}}.$$

In (6) the sign which makes u positive is to be taken on the upper or suction surface of the aerofoil.

The method of 17·2 can now be applied to find the lift and moments for any portion of the aerofoil. These results will all involve the constant B. The determination of B necessarily involves the use of elliptic integrals. The details are given in 17·61.

17·61. Determination of the constant B. The determination is based on the fact that the real part of the complex velocity component $\mathcal{W}$ vanishes on the Mach circle where there is no perturbation and that the real part is $-\alpha V$ on the surface of the aerofoil.

From the compatibility equations 17·15 (6),

(1) $$\frac{d\mathcal{W}}{d\epsilon} = \frac{1}{2} i \cot \mu \left(\epsilon - \frac{1}{\epsilon}\right) \frac{d\mathcal{U}}{d\epsilon}.$$

We begin with the identity

(2) $4(\epsilon - L)(1 - L\epsilon) = (1 - L)^2(1 + \epsilon)^2 - (1 + L)^2(1 - \epsilon)^2.$

Now write

(3) $$\eta = \frac{1 + \epsilon}{1 - \epsilon}, \quad a = \frac{1 + L}{1 - L}, \quad b = \frac{1 + T}{1 - T}.$$

Then 17·6 (2) gives

(4) $$\mathcal{U} = iB_1 \frac{1 - T}{1 - L} \left[\frac{\eta^2 - b^2}{\eta^2 - a^2}\right]^{\frac{1}{2}}.$$

Now take as squared modulus of elliptic functions (14·7),

(5) $$m = \frac{b^2}{a^2} = \frac{(1 + t)(1 - l)}{(1 - t)(1 + l)}.$$

The substitution

(6) $$\eta = a \operatorname{ns} \zeta,$$

then gives

(7) $\eta^2 - b^2 = a^2(\operatorname{ns}^2 \zeta - m) = a^2 \operatorname{ds}^2 \zeta, \quad \eta^2 - a^2 = a^2 \operatorname{cs}^2 \zeta,$

so that

(8) $$\mathcal{U} = iB_2 \operatorname{dc} \zeta, \quad B_2 = \frac{1 - T}{1 - L} B_1.$$

From (3) we have

$$\frac{1}{2}\left(\epsilon - \frac{1}{\epsilon}\right) = \frac{- 2\eta}{\eta^2 - 1} = \frac{- 2a \operatorname{ns} \zeta}{a^2 \operatorname{ns}^2 \zeta - 1},$$

while $d\mathcal{U}/d\zeta = iB_2 m_1 \operatorname{sc} \zeta \operatorname{nc} \zeta$, where $m_1 = 1 - m$ is the complementary squared modulus. Combining these results with (1), we get

$$\frac{d\mathcal{W}}{d\zeta} = 2am_1 \cot \mu \, B_2 \frac{\operatorname{sn}^2 \zeta}{(1 - \operatorname{sn}^2 \zeta)(a^2 - \operatorname{sn}^2 \zeta)}$$

(9) $$= \frac{2am_1}{1 - a^2} \cot \mu \, B_2 \left[- \operatorname{sc}^2 \zeta + \frac{1}{a^2} \frac{\operatorname{sn}^2 \zeta}{1 - (\operatorname{sn}^2 \zeta)/a^2}\right].$$

Integration will lead to elliptic integrals of the second and third kinds.* To bring the second term on the right to the standard form, we put

(10) $m \operatorname{sn}^2 \beta = 1/a^2, \quad \operatorname{sn}^2 \beta = 1/b^2.$

The standard form and notation for the elliptic integral of the third kind is then

(11) $$\Pi(\zeta, \beta) = \int_0^\zeta \frac{m \operatorname{sn} \beta \operatorname{cn} \beta \operatorname{dn} \beta \operatorname{sn}^2 \zeta \, d\zeta}{1 - m \operatorname{sn}^2 \beta \operatorname{sn}^2 \zeta}.$$

---

* The reader may consult Whittaker and Watson, *Modern Analysis*, 4th edition, Cambridge (1927), Chapter 22, for the details which cannot be included in this summary sketch of the method. See also Milne-Thomson in *Handbook of Mathematical Functions* National Bureau of Standards, 1964.

Observing that at D_1, $w = 0$, $\epsilon = 1$, $\eta = \infty$, $\zeta = 0$, we get on integrating (9) from 0 to ζ,

$$(12) \quad \mathcal{W} = \frac{2a \cot \mu\, B_2}{1 - a^2}\left[E(\zeta) - \mathrm{sc}\,\zeta\, \mathrm{dn}\,\zeta + \frac{m_1 \mathrm{sn}\,\beta}{\mathrm{cn}\,\beta\, \mathrm{dn}\,\beta}\, \Pi(\zeta, \beta)\right],$$

where $E(\zeta)$ is the elliptic integral of the second kind.

Now, on the aerofoil,

$$a\, \mathrm{ns}\,\zeta = \eta = \frac{1 + \epsilon}{1 - \epsilon}, \quad T < \epsilon < L,$$

so that $b < \eta < a$. Also* when $\eta = a$, $\zeta = K$ and when $\eta = b$, $\mathrm{sn}\,\zeta = m^{-\frac{1}{2}}$, $\zeta = K + iK'$. Thus on the surface of the aerofoil ζ varies from K to $K + iK'$. If in (12) we put $\zeta = K + i\xi$, where $0 < \xi < K'$, and equate the real parts, we get

$$(13) \quad \alpha V = \frac{2a \cot \mu\, B_2}{(a^2 - 1)}\left[E + \frac{m_1 \mathrm{sn}\,\beta}{\mathrm{cn}\,\beta\, \mathrm{dn}\,\beta}\, KZ(\beta)\right],$$

where E is the complete elliptic integral of the second kind and $Z(\beta)$ is the Jacobian zeta function. All the quantities involved in (13) are known in terms of the geometry of the aerofoil by the preceding equations. In particular, m is given by (5) and $m_1 = 1 - m$. Thus the constant B_2 and therefore B_1 from (8) and so B from 17·6 (7) can be found from tables.† Making the reductions we get

$$(14) \quad B = \frac{\alpha V \tan \mu}{D}\, \frac{l}{[(1 + l)(1 - t)]^{\frac{1}{2}}}, \quad D = E + \frac{m_1 \mathrm{sn}\,\beta}{\mathrm{cn}\,\beta\, \mathrm{dn}\,\beta}\, KZ(\beta).$$

17·62. Planform L_iL_i. Both leading edges are within the Mach cone of their intersection.

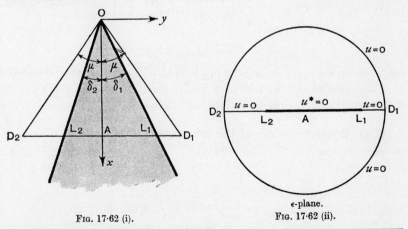

Fig. 17·62 (i).

ϵ-plane.
Fig. 17·62 (ii).

---

* For K and K', see 14·7.

† Milne-Thomson, *Jacobian elliptic function tables*, Dover Publications, New York (1960).

Since $u = 0$ on the Mach cone we have as before

(1) $$\mathcal{U}(\epsilon) = -\overline{\mathcal{U}}\left(\frac{1}{\epsilon}\right),$$

and the same reasoning as in 17·6 shows that the required solution is of the form

(2) $$\mathcal{U}(\epsilon) = iB_1\left[\frac{(\epsilon + L_2)(1 + L_2\epsilon)}{(\epsilon - L_1)(1 - L_1\epsilon)}\right]^{\frac{1}{2}} + iB_2\left[\frac{(\epsilon - L_1)(1 - L_1\epsilon)}{(\epsilon + L_2)(1 + L_1\epsilon)}\right]^{\frac{1}{2}},$$

where B_1 and B_2 are real constants, for (2) satisfies (1), has the appropriate singularities at the leading edges $\epsilon = L_1, \epsilon = -L_2$, makes $u = 0$ on the Mach circle and on D_2L_2, L_1D_1, and gives $u^\star = 0$ when $-L_2 < \epsilon < L_1$.

Since $\epsilon = 0$ is now a point on the aerofoil, we must have $d\mathcal{U}/d\epsilon = 0$ when $\epsilon = 0$ from 17·15 (7). This gives at once

(3) $$\frac{B_1}{L_1} = \frac{B_2}{-L_2} = C_1 \quad \text{say},$$

and (2) and (3) solve the problem in terms of the real constant C_1.

To express u on the aerofoil in terms of the variable s of the physical plane, let, in fig. (i), $l_1 = AL_1, l_2 = AL_2$, so that

(4) $$l_1 = \cot\mu \tan\delta_1, \quad l_2 = \cot\mu \tan\delta_2.$$

Then

(5) $$L_1 = \frac{1 - \sqrt{(1 - l_1{}^2)}}{l_1}, \quad L_2 = \frac{1 - \sqrt{(1 - l_2{}^2)}}{l_2},$$

and therefore on the model of 17·6, we find that

(6) $$u = \pm C\left\{l_1\left(\frac{l_2 + s}{l_1 - s}\right)^{\frac{1}{2}} + l_2\left(\frac{l_1 - s}{l_2 + s}\right)^{\frac{1}{2}}\right\},$$

where

(7) $$C = C_1[L_1L_2/(l_1l_2)]^{\frac{1}{2}},$$

and where on the upper surface the sign is to be taken which makes u positive.

The method of 17·2 can now be applied to find lift and moment in terms of C. The constant C is evaluated in the next section.

17·63. Determination of the constant C. The procedure here is on the same lines as in 17·61. We write

(1) $$\eta = \frac{1 + \epsilon}{1 - \epsilon}, \quad a = \frac{1 + L_1}{1 - L_1}, \quad b = \frac{1 - L_2}{1 + L_2}.$$

With the aid of identities of the type 17·61 (2), and the substitution

(2) $$\eta = a \operatorname{ns} \zeta, \quad m = \frac{b^2}{a^2} = \frac{(1 - l_1)(1 - l_2)}{(1 + l_1)(1 + l_2)},$$

we reduce 17·62 (2) and (3) to

(3) $$\mathscr{U} = iB_3\{(a^2 - 1)\,\mathrm{dc}\,\zeta - (1 - b^2)\,\mathrm{cd}\,\zeta\},$$

where

(4) $$B_3 = \tfrac{1}{4}C_1(1 - L_1)(1 + L_2).$$

The compatibility equation 17·61 (1) then gives, after some reduction,

$$\frac{d\mathscr{W}}{d\zeta} = 2am_1 \cot \mu\, B_3\,\{\mathrm{nc}^2\,\zeta - \mathrm{nd}^2\,\zeta\}.$$

Integrate this from 0 to ζ, observing that $w = 0$ at D_1 where $\zeta = 0$. Then

(5) $$\frac{\mathscr{W}}{2a \cot \mu\, B_3} = -\,2E(\zeta) + m_1\zeta + \mathrm{sc}\,\zeta\,\mathrm{dn}\,\zeta + m\,\mathrm{sd}\,\zeta\,\mathrm{cn}\,\zeta.$$

Now, on the aerofoil $\zeta = K + i\xi,\ 0 < \xi < K'$, while the real part of $\mathscr{W}$ is $-\alpha V$. Therefore, equating the real parts of (5), we get

(6) $$\alpha V = 2a \cot \mu\, B_3(2E - m_1K).$$

This determines B_3 in terms of the geometry of the aerofoil and therefore C_1 from (4), and so C from 17·62 (7). We get

(7) $$C = \frac{\alpha V \tan \mu}{F}\,\frac{1}{[(1 + l_1)(1 + l_2)]^{\frac{1}{2}}}, \quad F = 2E - m_1K.$$

17·64. A limit case of planform L_iL_i. If, in fig. 17·62 (i), we let δ_1 and δ_2 tend to μ the limiting case so obtained is that considered in 17·31. In the present case, 17·62 (2), (3) give when $L_1 = L_2 = 1$,

$$\mathscr{U}(\epsilon) = iC_1\left[\frac{(1 + \epsilon)^2}{-(1 - \epsilon)^2}\right]^{\frac{1}{2}} - iC_1\left[\frac{-(1 - \epsilon)^2}{(1 + \epsilon)^2}\right]^{\frac{1}{2}}$$

$$= 2C_1\frac{1 + \epsilon^2}{1 - \epsilon^2}.$$

Now, from 17·63 (4), (6),

$$C_1 = \frac{4 \tan \mu}{2(1 + L_2)^2}\,\frac{\alpha V}{(2E - m_1K)}.$$

When $L_2 = 1$, $m = 0$, $m_1 = 1$, $E = K = \tfrac{1}{2}\pi$ and therefore

$$C_1 = \frac{\alpha V \tan \mu}{\pi}$$

in agreement with 17·31.

17·65. Planform L_iA. This may be regarded as a limit case of L_iT_i when $T \to 0$ or L_iL_i when $L_2 \to 0$. Taking the former, we have from 17·6 (2) when $T = 0$.

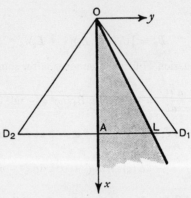

Fig. 17·65.

$$(1) \qquad \mathcal{U}(\epsilon) = iB_1 \left[\frac{\epsilon}{(\epsilon - L)(1 - L\epsilon)} \right]^{\frac{1}{4}}$$

and from 17·61 (8) and (9)

$$(2) \qquad \frac{d\mathcal{W}}{d\zeta} = \frac{2am_1 \cot \mu}{1 - L} B_1 \frac{\operatorname{sn}^2 \zeta}{(1 - \operatorname{sn}^2 \zeta)(a^2 - \operatorname{sn}^2 \zeta)}.$$

Now, when $T = 0$, 17·61 (3) and (5) give

$$(3) \qquad b = 1, \quad m = \frac{1}{a^2} = \frac{(1 - L)^2}{(1 + L)^2} = \frac{1 - l}{1 + l}$$

and therefore from (2)

$$(4) \qquad \frac{d\mathcal{W}}{d\zeta} = \frac{2am \cot \mu}{1 - L} B_1 (\operatorname{nc}^2 \zeta - \operatorname{nd}^2 \zeta).$$

If we integrate this and equate the real parts at a point of the aerofoil, we get, as in 17·63,

$$(5) \qquad B_1 = \frac{\alpha V(1 - L) \tan \mu}{2a(2E - m_1K)} \frac{m_1}{m} = \frac{2L}{(1 + L)} \cdot \frac{\alpha V \tan \mu}{(2E - m_1K)}$$

from (3), where

$$(6) \qquad L = [1 - \sqrt{(1 - l^2)}]/l, \quad \text{and} \quad l = \cot \mu \tan \delta.$$

At a point of the aerofoil $\epsilon + 1/\epsilon = 2/s$ and (1) gives

$$(7) \qquad u = \pm B_1 \left[\frac{ls}{2L(l - s)} \right]^{\frac{1}{4}}.$$

This can also be obtained as the limit of 17·6 (6) when $t \to 0$.

If we take instead the limit case of L$_i$L$_i$ when $L_2 \to 0$, we have from 17·62 (3), $B_2 = 0$. Thus 17·62 (2) coincides with (1) above and therefore the same result is obtained.

In the case where the leading edge becomes a generator of the Mach cone $L \to 1$ and (1) gives

$$\mathcal{U} = \frac{B_1 \epsilon^{\frac{1}{2}}}{1 - \epsilon}.$$

Now, when $L = 1$, $m = 0$, $m_1 = 1$, $E = K = \pi/2$ and so $B_1 = 2\alpha V \tan \mu/\pi$ which agrees with the result found in 17·51.

17·7. Interference of conical flows.

So far we have considered only the flow within the Mach cone of a single vertex. When a point lies within the Mach cones of two or more vertices the flows interfere. Consider, for example, a flat rectangular aerofoil at small incidence α with its leading edge OO' at right angles to the oncoming wind as in fig. 17·02 (i). If the chord of the aerofoil is sufficiently great each axial edge will intersect the Mach cone of O or O'.

This case is illustrated in fig. 17·7 where the axial edges OM, $O'M'$ are temporarily assumed to continue indefinitely downstream and to intersect the generators of the Mach cones of O and O' at A

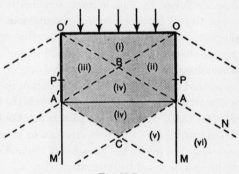

FIG. 17·7.

and A'; the generators of the Mach cones of O, O', A, A', which lie in the plane of the aerofoil, are shown by dotted lines. By their intersections these generators determine on the aerofoil four regions, marked (i) to (iv) in the diagram, namely the triangles OBO', OBA, $O'BA'$ and the parallelogram $A'CAB$. We denote by u_1, u_2, u_3, u_4 the x-component of the perturbation velocity at a general point within these respective regions. In region (i) the motion is two-dimensional and if we restrict attention to the upper surface of the aerofoil, u_1 has the constant value $\alpha V \tan \mu$. In region (ii), u_2 can be obtained from 17·5 (6) and similarly u_3 in region (iii). At a point within region (iv) u_4 will be obtained by combining u_2 and u_3 in a manner to be determined. Since $u_2 = u_1 + (u_2 - u_1)$ we can regard the effect of the Mach cone of O at a point in region (ii) to be a perturbation $(u_2 - u_1)$, superposed by addition, of the constant two-dimensional component u_1. Similarly, the Mach cone of O' causes in region (iii) a perturbation $u_3 - u_1$ of u_1. The perturbation of u_1, which both Mach cones cause at a point of region (iv) is therefore the sum of these, so that

$$u_4 - u_1 = (u_2 - u_1) + (u_3 - u_1)$$

and

(1) $u_4 = u_2 + u_3 - u_1.$

To verify this, observe that u_1, u_2, u_3 satisfy the linearised equation 17·1 (6) and so therefore does u_4. Also, u_4 is continuous on AB where $u_3 = u_1$, so that $u_4 = u_2$. Similarly, $u_4 = u_3$ on $A'B$. At B $u_1 = u_2 = u_3$ and therefore $u_4 = u_1$.

Similarly, for the w-component we arrive at the equation

(2) $w_4 = w_2 + w_3 - w_1$

and since $w_1 = w_2 = w_3 = -\alpha V$ on the surface of the aerofoil, $w_4 = -\alpha V$ and all the boundary conditions are satisfied in region (iv).

Now, consider a point within the Mach cone of A. Here u has to satisfy certain boundary conditions within the region (v) between AC and AM and the condition $u = 0$ in the region (vi) between AM and AN. We cannot satisfy all these conditions by a finite linear combination of the type (1). Such regions can be treated by superposition of continuous fields of conical flow, that is by integrating conical flow solutions. Into this we cannot enter, but the inference remains that the superposition by addition of a finite number of conical flows cannot describe the flow on the aerofoil outside the regions (i) to (iv). Thus, for the method of (1) to be applicable no part of the trailing edge of the aerofoil can lie downstream of AC, $A'C$, in other words the trailing edge must lie within the area $O'A'CAO$. Subject to this limitation the trailing edge may be straight or curved provided only that its direction or tangent at no point makes an acute angle less than μ with the direction of the undisturbed wind. In all such cases the lift can be found by integration of u_1, u_2, u_3, u_4 over the appropriate areas.

As a simple illustration, consider the rectangular aerofoil $OO'P'P$ where P lies between O and A and PP' is parallel to OO'. Let S denote the total area and let S_1, S_2, S_3, S_4 be areas of regions (i), (ii), (iii), (iv) respectively which lie on the aerofoil. If we denote mean values by a bar, the lift L is given by

$$L/(2\rho V) = S_1 u_1 + S_2 \bar{u}_2 + S_3 \bar{u}_3 + S_4 \bar{u}_4.$$

Now, from 17·5 (11), when $\delta = \frac{1}{2}\pi$, $\bar{u}_2 = \bar{u}_3 = \frac{1}{2}u_1$ and therefore

$$L = \rho V u_1 (2S_1 + S_2 + S_3) = \rho V u_1 (S + S_1 - S_4).$$

This result holds for all parallel positions of PP' between OO' and AA'.

EXAMPLES XVII

1. Perform the reduction which leads to the compatibility equations, 17·15 (6).

2. Show that the velocity potential

$$\phi = \epsilon[x^2 - \cot^2 \mu (y^2 + z^2)]^{\frac{1}{2}} f'(\epsilon) + xf(\epsilon),$$

where $f(\epsilon)$ is an arbitrary function, is a general homogeneous solution of degree unity of 17·1 (4). (Poritsky.)

3. Determine, in terms of λ, the value of $\bar{f}(0)$ in 17·18 (7).

4. With the notations of 17·2, prove that

$$M_x = \tfrac{1}{3}h \sec \gamma \frac{\partial}{\partial \gamma}(L \cos^2 \gamma), \quad M_y = -\tfrac{1}{3}h \operatorname{cosec} \gamma \frac{\partial}{\partial \gamma}(L \sin^2 \gamma).$$

5. In the case of planform $L_0 L_o$, in the notations of 17·2 and 17·3, show that, on integration by parts,

$$2 \int_{\epsilon=-1}^{\epsilon=1} \mathcal{U}(\epsilon)\, dg = u_1 A_1 + u_2 A_2 + u_1 B_1 + u_2 B_2,$$

where

$$A_1 = \frac{1}{1-\beta} - \frac{1}{\cos \lambda_1 - \beta}, \quad A_2 = \frac{1}{1+\beta} - \frac{1}{\cos \lambda_2 + \beta},$$

$$B_1(1-\beta^2)^{\frac{1}{2}} = \frac{\sin \lambda_1}{\cos \lambda_1 - \beta}, \quad B_2(1-\beta^2)^{\frac{1}{2}} = \frac{\sin \lambda_2}{\cos \lambda_2 + \beta}.$$

6. In Ex. 5, show that the terms $u_1 A_1$, $u_2 A_2$ when multiplied by $\rho V h^2 \sec^2 \gamma \tan \mu$ cancel the lift on the triangular parts of the delta wing which lie outside the Mach cone.

7. Calculate the perturbation velocity component w inside and outside the Mach cone in the case of planform $L_o T_o$.

8. Find the perturbation velocity components for planform $L_o T_s$ regarded as a limit case of $L_o T_o$.

9. Solve the case of planform $L_o T_s$ regarded as a limit of $L_o T_i$ and show that the result agrees with that of Ex. 8.

10. Perform the integration which leads to 17·5 (11).

11. Explain why the solution for planform $L_s L_s$ at incidence α cannot be obtained by adding the solutions for two planforms $L_s A$ which are mirror images in the axis. Verify by comparing the actual solutions.

12. Perform the integrations which lead to the lift reduction ratio for planform $L_o L_i$.

13. Work out in detail the solution for planform $L_o T_i$.

14. The wind direction is along the bisector of the angle 2δ of a delta wing, where $\delta < \mu$. Obtain the formulae

$$u = \alpha V C \tan^2 \delta (\tan^2 \delta - s^2 \tan^2 \mu)^{-\frac{1}{2}},$$
$$v = -\alpha V C s \tan \mu (\tan^2 \delta - s^2 \tan^2 \mu)^{-\frac{1}{2}},$$

where in the notation of 17·6, $C(2E - m_1 K) = 1 + m^{\frac{1}{2}}$, and explain to what portions of the surface of the wing they apply.

15. Show that the lift coefficient for the part of the delta wing of Ex. 14 in the form of an isosceles triangle of height h is $2\pi\alpha C \tan \delta$.

16. The wind direction is that of the bisector of the angle 2δ of a delta wing, where $\delta > \mu$. Prove that on the upper surface of the wing, inside the Mach cone,

$$u = \frac{2\alpha V \tan \mu}{\pi \sin \lambda} \sin^{-1} \frac{\sin \lambda}{\sqrt{(1 - s^2 \cos^2 \lambda)}}, \quad \cos \lambda = \tan \mu \cot \delta.$$

17. In the delta wing of the previous example, show that on the wing

$$v = \frac{-2\alpha V \cot \lambda}{\pi} \cos^{-1} x,$$

where $x^2 = (1 - s^2)/(1 - s^2 \cos^2 \lambda)$, and explain to what portions of the surface of the wing this result applies. Examine the value of v on the remaining parts of the wing.

18. Show that the lift on the part of the delta wing of Ex. 16 cut off by a line perpendicular to the wind in the plane of the wing and at distance h from the vertex is

$$2\alpha\rho V^2 h^2 \tan^2 \mu \sec \lambda.$$

Find the induced drag and the lift and induced drag coefficients.

19. A symmetrical flat delta wing is defined as having for planform an isosceles triangle and for vertex the intersection of the equal sides. If 2δ is the angle between the equal sides, prove, with the notations of fig. 17·2, that

$$\delta = \tfrac{1}{2}(\delta_1 + \delta_2), \quad \gamma = \tfrac{1}{2}(\delta_2 - \delta_1).$$

Explain why the direct application of the cone field theory of this chapter necessitates the assumption that the angle of yaw γ shall not exceed $\tfrac{1}{2}\pi - \mu$.

20. A symmetrical delta wing whose vertical angle is 2δ is yawed through the angle γ, and has both leading edges within the Mach cone. Prove that the lift coefficient is

$$\pi\alpha C \cos \gamma \sec \mu \sec \delta \left[\cos 2\gamma - \cos 2\mu \cos 2\delta - 2A\right]^{\frac{1}{2}},$$

where $A^2 = \sin (\mu + \delta + \gamma) \sin (\mu + \delta - \gamma) \sin (\mu - \delta + \gamma) \sin (\mu - \delta - \gamma)$, and C has the value defined in Ex. 14. (Roper.)

21. A symmetrical delta wing has one leading edge on the Mach cone and one leading edge within it. Prove that the lift coefficient is

$$4\alpha \cos \gamma \left[\tan (\mu - \gamma) \tan \mu\right]^{\frac{1}{2}},$$

where γ is the angle of yaw. (Roper.)

22. For a symmetrical delta wing of planform $L_o L_o$ yawed through the angle γ, prove that the lift coefficient is

$$4\alpha \sin \mu \cos \gamma \left[\sec (\mu + \gamma) \sec (\mu - \gamma)\right]^{\frac{1}{2}}.$$

Discuss the limiting case of this result when one edge lies on the Mach cone.

23. A symmetrical delta wing of planform $L_o L_i$ is yawed through the angle γ. Prove that the lift coefficient is

$$\alpha(\sqrt{32}) \cos \gamma \sin \mu \left[\sin \delta \operatorname{cosec} (\mu + \delta + \gamma) \sec (\mu - \gamma)\right]^{\frac{1}{2}},$$

where 2δ is the angle of the wing.

Obtain the result for planform $L_s L_i$ as a limiting case. (Roper.)

24. Show that $\mathcal{U}(\epsilon) = V\beta^2 \log \epsilon$ gives the complex velocity component for flow past a cone of infinitesimal angle 2β placed with its axis along the undisturbed wind direction. Examine the components $\mathcal{V}(\epsilon)$, $\mathcal{W}(\epsilon)$.

25. Calculate the lift coefficient for the aerofoil $OO'A'CA$ of fig. 17·7.

26. A delta wing of planform $L_i A$ is symmetrical with respect to the plane $z = 0$, consisting of two flat triangular plates each inclined at the small angle β to the plane $z = 0$. Prove that, at zero incidence,

$$\mathcal{U}_{(\epsilon)} = \frac{\beta V \tan \mu}{\pi} \frac{2L}{1 - L^2} \log \frac{\epsilon - L}{L\epsilon - 1},$$

and find the components $\mathcal{V}$ and $\mathcal{W}$.

27. Obtain the solution for a delta wing of planform $L_i L_i$ symmetrical with respect to the plane $z = 0$, at zero incidence by superposition of two wings of planform $L_i A$, the wings consisting of triangular plates each inclined at the angle β to the plane $z = 0$.

28. Show that in the case of planform $L_o L_o$ the flow within the Mach cone of the air above the aerofoil is completely independent of the flow below it. Use this principle to deduce the flow past a thin flat aerofoil $L_o L_o$ at incidence α from the solution for the symmetrical aerofoil consisting of two planes inclined at the angle 2α and placed at zero incidence.

CHAPTER XVIII

SIMPLE FLIGHT PROBLEMS

18·0. Hitherto we have been concerned with the main lifting system of an aircraft, the wings. We now consider the complete machine.

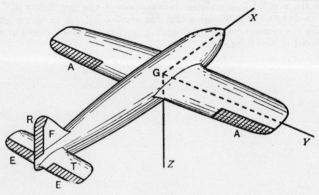

FIG. 18·0.

Fig. 18·0 shows in outline a monoplane aircraft with the control surfaces shaded.

When the control surfaces are in their neutral positions the aircraft, like the aerofoil, has a median plane of symmetry, and when properly loaded the centre of gravity G lies in this plane.

There is in the diagram a fixed fin F in the plane of symmetry.

For simplicity of description, suppose the aircraft to be in straight horizontal flight, with all the controls in their neutral positions.

The lift is given by the wings and tail, T (we ignore the effect of the body, which should be allowed for), and the propeller thrust overcomes the drag, if we suppose this thrust to be horizontal.

The disposition of the axes of reference is shown in the diagram, the lateral axis, GY, being perpendicular to the plane of symmetry and positive to starboard.

The symmetry will not be disturbed if the elevators E are moved. Raising the elevators will decrease the lift on the tail (see 8·37), and will cause a *pitching* moment, positive when the nose tends to be lifted.

Moving the rudder R to starboard will cause a *yawing* moment tending to deflect the nose to starboard, the positive sense.

The ailerons, A, move in opposite senses, one up, one down, by a single motion of the control column. If we depress the port aileron and therefore simultaneously raise the starboard one, the lift on the port wing will increase and that on the starboard wing will decrease so that a *rolling* moment will be caused tending to dip the starboard wing, and this sense will be positive.

This movement will also cause a yawing moment, for the drag on the two wings will likewise be altered. To minimise this the ailerons are generally geared to move differentially so that one moves through a greater angle than the other.

Motion of ailerons or rudder will disturb the symmetry of the machine. A single-engined aircraft also has dynamical asymmetry (tendency to list).

For the present we shall consider some flight situations in which the controls are supposed always to be operated so as to maintain the motions we require.

18·1. Linear flight. When the velocity **V** of the aircraft is in a fixed straight line the flight is said to be *linear*.

When **V** is in the plane of symmetry the flight is said to be *symmetric*.

There are three types of linear symmetric flight : gliding, horizontal, and climbing.

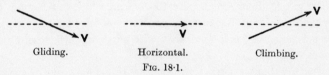

Gliding.   Horizontal.   Climbing.

FIG. 18·1.

Of these the only flight possible without use of the engines is gliding.

These flights can, of course, be steady (**V** constant) or accelerated.

In the case of steady flight the resultant force on the aircraft must be zero.

The forces are : (1) propeller thrust (under control of the pilot), (2) weight (not under control of the pilot), (3) aerodynamic force (in some measure under the control of the pilot by use of ailerons, rudder, and elevators).

18·2. Stalling. We have seen that the lift coefficient C_L is a function of the absolute incidence α, and the considerations of 1·71 show that it is also a function of the Reynolds' number $R = Vl/\nu$. For a given aircraft we could therefore draw a surface which is the locus of the point (C_L, α, R) which is the *characteristic lift surface* for that aircraft. Since the aircraft is given, l is given, and for a given state of the air ν is given, so that in this case C_L is a function of incidence α and of the forward speed V. From this point of view the characteristic surface may then be regarded as the locus of the point (C_L, α, V). Now consider three points of this surface (C_{L1}, α_1, V_1), (C_{L2}, α_2, V_2), (C_{L3}, α_3, V_3), where we suppose $V_1 < V_2 < V_3$ and $\alpha_1 < \alpha_2 < \alpha_3$.

Fig. 18·2 (i) shows the (C_L, α) graphs corresponding with the values V_1, V_2, V_3 of V, the straight portions being practically in the same line for all. This diagram may be thought of as showing sections of the characteristic surface by

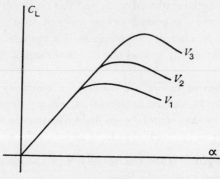

FIG. 18·2 (i).

planes $V = V_1$, $V = V_2$, $V = V_3$. In all our previous work we have considered C_L to be directly proportional to α, i.e. we have restricted ourselves to the linear part of the graph about which pure theory can make statements. Graphs of the type shown in fig. (i) must necessarily be obtained from experimental measurements, and the graph shows that, with increasing incidence, C_L rises to a maximum value $C_{L\,\mathrm{max}}$ and then decreases.

It is generally, but not always, the case that $C_{L\,\mathrm{max}}$ for a given V increases as V increases. If we draw the sections of the characteristic surface by the planes

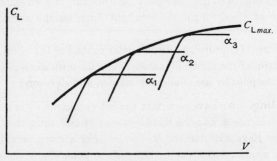

FIG. 18·2 (ii).

$\alpha = \alpha_1$, $\alpha = \alpha_2$, $\alpha = \alpha_3$, we get (C_L, V) graphs of the type shown in fig. 18·2 (ii), where the heavy line shows the graph of $C_{L\,\mathrm{max}}$ as a function of V.

The *stalled state* is that in which the airflow on the suction side of the aerofoil is turbulent. It is found that, just before the stalled state sets in, the lift coefficient attains its maximum value, and the corresponding speed is called the *stalling speed* V_S. Thus the stalling speed corresponding with a given $C_{L\max}$ can be read from the heavy graph of fig. (ii).

Stalling speed is a function of incidence. Let us draw the sections of the characteristic surface by the planes $C_L = C_{L1}$, $C_L = C_{L2}$, $C_L = C_{L3}$, fig. 18·2 (iii), giving (α, V) curves.

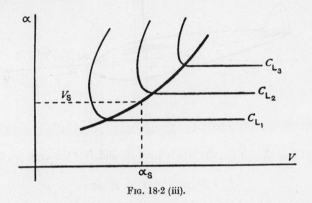

FIG. 18·2 (iii).

Here the heavy curve shows the stalling speed as a function of the *stalling incidence* α_S. Any point (α, V) above this curve corresponds with stalled flight, any point below it with *normal flight*. In the sequel we shall be concerned only with normal flight.

It should be observed that the foregoing discussion only applies to speeds V such that the airflow speed over the aerofoil nowhere approaches the speed of sound, i.e. we neglect variation of the Mach number.

The graphs of the above type are in all cases deduced from experiments, generally in wind tunnels in conditions corresponding with linear flight at constant speed. When the aircraft flies in a curved path the graphs will differ slightly from the above, but investigations made by Wieselsberger show that the changes are of the order of the square of the ratio of the span to the radius of curvature of the path and may therefore, in general, be neglected.

Moreover, since the danger of stalling is generally greatest when the aircraft is about to land and is therefore flying near to the stalling incidence and at a low speed, it is for most calculations sufficient to substitute one of the C_L graphs shown in fig. (i) for the whole group, namely the one which corresponds with the landing speed.

If, in the foregoing, we substitute R for V throughout, the conclusions may be held to apply to a family of geometrically similar aircraft. For such a family there will be one characteristic lift surface.

The results of this section apply, without modification, if α is the geometrical incidence.

18·3. Gliding. When an aircraft glides steadily with the engine off, the resultant aerodynamic force **A** balances the weight (including buoyancy) **W**, i.e. **A** + **W** = 0.

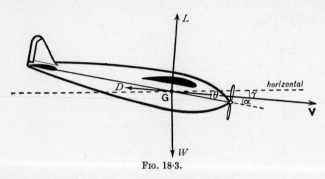

FIG. 18·3.

Thus, if L is the lift (on the whole aircraft) and D is the drag (on the whole aircraft),

$$L = W \cos \gamma, \quad D = W \sin \gamma,$$

where γ is the angle which the direction of motion makes with the horizontal, called the *gliding angle*. It follows that

$$\tan \gamma = \frac{D}{L} = \frac{C_D}{C_L}.$$

This equation determines γ in terms of C_D and C_L. It should be observed that C_L and C_D are here the lift and drag coefficients for the whole aircraft. They could not be inferred solely from the corresponding coefficients for the wings alone. They are, of course, functions of the incidence α and the speed V (or more properly of the Reynolds' number). It follows that, with the limitations described in 18·2, the gliding angle is a function of the incidence, and therefore the gliding speed is a function of the gliding angle.

The *attitude* * of the machine is the angle which a line fixed in the machine makes with the horizontal. If we measure incidence α and attitude θ from the same line we have $\theta = \alpha - \gamma$. Observe that θ and α can be negative, as in fig. 18·3 ; γ is necessarily positive.

Note also that the direction of the glide does not, in general, coincide with any fixed direction in the machine, in other words the attitude is a function of the incidence.

The extreme attitude is that assumed when the machine is diving vertically, the *terminal velocity dive*. In this case the lift vanishes, the incidence is that of zero lift, and if the dive is undertaken from a sufficiently great height, the weight just balances the drag, the speed being then the terminal speed, which

---

* Cf. Lewis Carroll, *Through the Looking Glass*, Ch. VII.

may be five or six times the stalling speed (18·2). The attitude will then be about $-90°$.

18·31. Straight horizontal flight. Here the engine is required. There are now three forces ; engine thrust T, weight, and aerodynamic force.

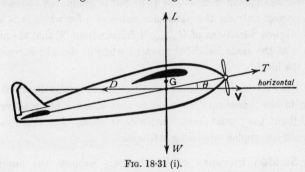

FIG. 18·31 (i).

By proper choice of chord the incidence may be taken equal to the attitude and

$$L = W - T \sin \theta, \quad D = T \cos \theta.$$

In practice θ is small, so that $T = D, L = W$, and

$$(1) \qquad C_L = \frac{L}{\frac{1}{2}\rho V^2 S} = \frac{W}{\frac{1}{2}\rho V^2 S} = \frac{w}{\frac{1}{2}\rho V^2},$$

where $w = W/S$ is called the *wing loading*, i.e. the average load per unit area of wing plan. When w, ρ (i.e. height), and V are given, C_L is determined, and therefore incidence from the (C_L, α) graph. At the stalling speed V_S, (1) becomes

$$(2) \qquad C_{L\max} = \frac{w}{\frac{1}{2}\rho V_S^2}.$$

To determine the stalling speed, we find where the graph of $C_L V^2 = 2w/\rho$ cuts the $(C_{L\max}, V)$ graph (18·2).

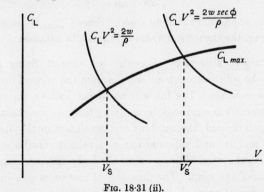

FIG. 18·31 (ii).

We see that V_S increases with height (i.e. with decrease of ρ). From (1) and (2)

(3) $C_L(\tfrac{1}{2}\rho V^2) = C_{L\max}(\tfrac{1}{2}\rho V_S^2) = w.$

The air-speed indicator measures $\tfrac{1}{2}\rho V^2$ but is graduated to read V. It is, therefore, correct only for the particular value of ρ for which it is graduated, but, if we neglect variations of $C_{L\max}$, it follows from (3) that the aircraft will stall always at the same indicated airspeed when in straight horizontal flight, whatever the height.

Note. For convenience of explanation fig. (i) and similar diagrams elsewhere are drawn in the " nose-up " attitude so that θ and α are positive. In fast horizontal flight the " nose-down " attitude is usual. For this θ is negative. The same remark applies to gliding attitudes.

18·32. Sudden increase of incidence. Suppose the aircraft to be flying steadily and horizontally, so that if C_L' is the lift coefficient, $C_L' \cdot \tfrac{1}{2}\rho V^2 S = W$. A sudden increase of incidence will increase the lift to $C_L \cdot \tfrac{1}{2}\rho V^2 S$ and the aircraft will acquire an upward acceleration f given by

$$(C_L - C_L')\tfrac{1}{2}\rho V^2 S = \frac{W}{g}f,$$

so that it will begin to describe a curved path of radius of curvature r given by $f = V^2/r$, where

$$r = \frac{2w}{g\rho}\,\frac{1}{C_L - C_L'}.$$

In this analysis we ignore the change in drag. If the speed is high, C_L' is small and C_L cannot exceed $C_{L\max}$ for speed V. Thus the absolute minimum value of r is given by

(1) $r_{\min} = \dfrac{2w}{g\rho C_{L\max}} = \dfrac{V_S^2}{g},$

where V_S is the appropriate stalling speed. Since $C_{L\max}$ is accompanied by a rather large drag, the theoretical value (1) cannot be attained.

18·33. Straight side-slip. Consider an aircraft flying steadily and horizontally to be rolled through an angle ϕ from the vertical and held in this position by the controls. The lift will no longer balance the weight. If in fig. 18·33 the aircraft is supposed to be flying towards the reader so that the starboard wing is dipped, the machine will accelerate in the direction of the resultant of L and W, and will continue to accelerate until a steady state is reached owing to the wind blowing across the body and producing a *side force* in the direction of the span. The direction of motion is now inclined to the plane of symmetry at angle β, say, measured positively when the direction of

motion is to starboard. The aircraft is now moving crab-wise in a straight
path and is said to be *side-slipping*. If V is the speed, the component $V \sin \beta$
perpendicular to the plane of symmetry is called the *velocity of side-slip*, or
simply side-slip.

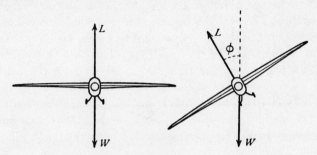

Fig. 18·33.

Side-slip will neither diminish the drag nor increase the lift as compared
with symmetric flight at the same speed. If D' and L' are the drag and lift in
the steady side-slip induced by the above manœuvre then $L' < L$, and $D' > D$.
Thus the gliding angle γ' will be given by

$$\tan \gamma' = \frac{D'}{L'} > \frac{D}{L} = \tan \gamma,$$

or $\gamma' > \gamma$. The effect of side-slip is therefore to increase the gliding angle
without reducing the speed. This fact is of use to pilots in landing.

18·4. Banked turn. This is steady motion in a horizontal circle with
the plane of symmetry inclined to the vertical. The direction of motion is
longitudinal and there is no side-force.

If ϕ is the angle of bank and r the radius of
the turn, we have

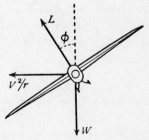

$$L \cos \phi = W, \quad L \sin \phi = \frac{W}{g} \cdot \frac{V^2}{r},$$

$$\tan \phi = \frac{V^2}{gr}.$$

If we ignore differences due to the difference in
speed at the outer and inner wing tips, it follows
from $L = W \sec \phi$ that $C_L = w \sec \phi/(\tfrac{1}{2}\rho V^2)$, and

Fig. 18·4.

therefore, as in 18·31, the stalling speed V_S' is
determined by the intersection of the curve $C_L V^2 = 2w \sec \phi/\rho$ with the
$(C_{L\max}, V)$ graph ; see fig. 18·31 (ii). From the relation

$$C_{L\max}(\tfrac{1}{2}\rho V_S'^2) = w \sec \phi,$$

it appears that banking increases the stalling speed, and if we treat $C_{L\,\text{max}}$ as effectively constant, the increase is in the ratio $\sqrt{\sec \phi} : 1$. This explains the danger of banking at too low a speed.

18·5. Lanchester's phugoids.

A *phugoid* * is the path of a particle which moves under gravity in a vertical plane and which is acted upon by a force L normal to the path and proportional to V^2, the square of the speed.

Since no work is done by the force L, it follows that $\frac{1}{2}V^2 - gz$, the total energy of the particle (per unit mass), is constant, z being the depth of the

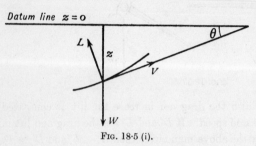

FIG. 18·5 (i).

particle below a fixed horizontal line when the speed is V. We can choose the position of this line so that the constant energy is zero, and then we shall have

(1) $V^2 = 2gz.$

If θ is the inclination of the path to the horizontal,

we have, by resolving along the normal,

(2) $$L - W \cos \theta = \frac{W}{g} \frac{V^2}{R},$$

where R is the radius of curvature.

If we could imagine an aircraft to fly at constant incidence, and so arrange that the propeller thrust exactly balances the drag, the centre of gravity of the aircraft would describe a phugoid, for then $L = \frac{1}{2}\rho V^2 S \,.\, C_L$ and C_L is constant for constant incidence (if we neglect the effect of curvature of the path on C_L; see 18·2).

Now suppose that V_1 is the speed at which the aircraft would fly in steady straight horizontal flight at the same incidence as in the phugoid. Then $W = \frac{1}{2}\rho V_1^2 S \,.\, C_L$, so that (2) will give

$$\frac{V^2}{V_1^2} - \cos \theta = \frac{V^2}{gR}.$$

Putting $V_1^2 = 2gz_1$ we get

(3) $$\frac{z}{z_1} - \cos \theta = \frac{2z}{R}.$$

Now if ds is an element of the arc of the path in fig. 18·5 (i) we see that

(4) $$\frac{1}{R} = \frac{d\theta}{ds}, \qquad \sin \theta = -\frac{dz}{ds},$$

---

* Greek $\phi\epsilon\nu\gamma\omega$,*take flight.*

and therefore (3) can be written in the equivalent form

$$\frac{d}{dz}(z^{\frac{1}{2}}\cos\theta) = \frac{z^{\frac{1}{2}}}{2z_1}.$$

If we integrate this equation, we add an arbitrary constant which we shall denote by $C\sqrt{z_1}$, and we then get, after division by $\sqrt{z}$,

(5) $$\cos\theta = \tfrac{1}{3}\frac{z}{z_1} + C\sqrt{\frac{z_1}{z}},$$

and using (4),

(6) $$\frac{z_1}{R} = \tfrac{1}{3} - \frac{C}{2}\sqrt{\frac{z_1{}^3}{z^3}}.$$

Let us examine the consequences of taking the positive square root in (5) and (6). It is then easy to show that $\cos\theta > 1$ if $C > 2/3$, so that no phugoid is possible. If $C = 2/3$, (5) gives $\cos\theta = 1$, so that $\theta = 0$ and $R = \infty$. We then get the horizontal straight-line phugoid, at depth z_1 below the datum line.

FIG. 18·5 (ii).

If $C = 0$, (6) gives $R = 3z_1$ and the phugoid reduces to a set of semicircles of radius $3z_1$. The cusps are on the datum line and the paths correspond with unsuccessful attempts at "looping the loop".

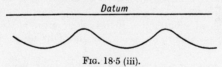

FIG. 18·5 (iii).

For $0 < C < 2/3$, we get trochoidal-like paths.

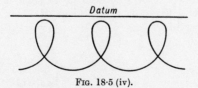

FIG. 18·5 (iv).

If $C < 0$, the paths are looped.

If z_1 and the initial values of z and θ are prescribed, (6) shows that for a given value of C there are two possible radii of curvature owing to the ambiguity of sign of the square root. Thus, for example, an aircraft describing a trochoidal-like path as in fig. 18·5 (iii) might, owing to a sudden gust, find itself describing a loop as in fig. 18·5 (iv).

18·51. The phugoid oscillation. Let an aircraft describing the straight-line phugoid, corresponding with $C = 2/3$, $z = z_1$, $\cos \theta = 1$, have its path slightly disturbed, for example, by a temporary gust. It may then begin to describe a sinuous path of small slope (as in fig. 18·5 (iii)) having the straight line as mean. This motion is called the *phugoid oscillation*. Since the vertical upward acceleration is $- d^2z/dt^2$, and since $\cos \theta = 1$ to the first order, we have, for the vertical motion

$$L - W = - \frac{W}{g} \frac{d^2z}{dt^2},$$

and therefore, from 18·5,

$$\frac{d^2z}{dt^2} + \frac{gz}{z_1} = g.$$

This is simple harmonic motion whose period is $2\pi \sqrt{(z_1/g)}$, showing that the disturbed motion is stable. In terms of the speed V_1 the period is $\pi \sqrt{2} \, V_1/g$s If V_1 is measured in knots (6080 ft./hr.) we get for the period $0\cdot234 \, V_1$ sec. Thus for a speed of 200 knots the period is about 47 sec.

EXAMPLES XVIII

1. If the (C_L, α) graph is given by

$$C_L = b \sin \frac{a\alpha}{b},$$

find the maximum lift coefficient and the stalling angle. Draw the graph carefully for $b = 2$, $a = 2\pi$.

2. Assuming the values of C_L, C_D given in Ex. I, 21 to apply to a complete aircraft, draw graphs to show the gliding angle, and the attitude angle in gliding, as functions of the incidence.

3. Calculate the minimum initial radius of curvature of the path when an aircraft whose stalling speed is 70 mi./hr. in straight flight has the incidence suddenly increased.

4. Show that, if an aircraft executes a flat turn, i.e. moves in a horizontal circle with the plane of symmetry vertical, the aircraft must side-slip.

5. For small angles of bank ϕ, show that the angle of side-slip is given by $\beta = - 2\cdot5C_L\phi$, assuming the coefficient of side-force is $C_Y = - 0\cdot4\beta$.
Prove also that in a flat turn

$$\beta = - \frac{5w \cos \beta}{g\rho r},$$

where r is the radius of the turn and w is the wing loading.

6. Show that the minimum radius of a true banked turn, for a given angle of bank ϕ, is

$$\frac{2w \operatorname{cosec} \phi}{g\rho \, C_{L \max}}.$$

7. Plot $\cos \theta$ as a function of z/z_1 from 18·5 (5), taking different values of C, and positive square roots. Show how to infer the general form of the phugoids as C varies.

8. Let x be the horizontal displacement in a phugoid motion. Express $dz/dx = \tan \theta$ as a function of z and hence show how a phugoid may be plotted point by point with the aid of numerical integration

CHAPTER XIX

MOMENTS

19·0. Pitching moment. Referring to fig. 18·0, the moment M of the aerodynamic forces about the lateral axis GY is called pitching moment.

Pitching moment does not induce yaw or roll. It is the only *symmetric* moment and it is controlled by the use of the elevators.

Pitching moment is positive when it tends to raise the nose (sense z to x).

The coefficient of pitching moment is

$$C_m = \frac{M}{\frac{1}{2}\rho V^2 Sc},$$

where c is the mean chord and S is the plan area.

The angular velocity of pitching is denoted by q and is positive in the sense z to x.

19·1. Static stability. Imagine an aircraft to be free to turn about a fixed horizontal axis coinciding with the lateral axis through its centre of gravity G.

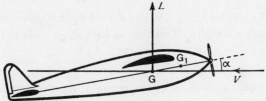

Fig. 19·1.

If a given horizontal wind stream V impinges on the aircraft, there will be equilibrium at a certain incidence α, the pitching moment about G will be zero, and the lift L will act vertically through G.

If the incidence α is changed by $d\alpha$, a pitching moment dM will be generated about G.

The equilibrium is said to be *statically stable* if dM and $d\alpha$ have opposite signs, i.e. if the pitching moment tends to restore the original incidence. The condition for static stability is therefore $\partial M/\partial \alpha < 0$.

The quantity $\partial M/\partial \alpha$ is therefore a measure of the static stability. Since $\partial M/\partial \alpha$ is proportional to $\partial C_m/\partial \alpha$, this latter quantity is also a measure of the static stability.

If $\partial M/\partial \alpha > 0$, the equilibrium is said to be *statically unstable*; a change of incidence will tend to increase.

If $\partial M/\partial \alpha = 0$, the equilibrium is said to be *statically neutral*; a change of incidence will tend neither to increase nor to decrease.

Observe that it is the pitching moment with respect to the centre of gravity which is here in question, and when the aircraft is in free flight, it is the sign of $\partial M/\partial \alpha$ which determines the nature, stable or unstable, of the static stability. Thus, if in fig. 19·1 the centre of gravity is moved forward to G_1, we shall have a new pitching moment M_1 about G_1 such that

$$M_1 = M - L \,.\, GG_1 \cos \alpha = M - L \,.\, GG_1,$$

ignoring the drag and treating α as a small angle. Therefore

$$(1) \qquad \frac{\partial M_1}{\partial \alpha} = \frac{\partial M}{\partial \alpha} - GG_1 \frac{\partial L}{\partial \alpha},$$

to that moving the centre of gravity forward decreases the former measure $\partial M/\partial \alpha$ of the static stability, and therefore increases the static stability of the aircraft. Similarly, moving the centre of gravity aft decreases the static stability. In the foregoing it is assumed that the aircraft is below the stalling incidence so that $\partial L/\partial \alpha$ is positive.

19·11. Metacentric ratio. In 19·1 (1) dividing by $\frac{1}{2}\rho V^2 Sc$ and writing x_1 for GG_1 we get

$$\left(\frac{\partial C_m}{\partial \alpha} \right)_1 = \frac{\partial C_m}{\partial \alpha} - \frac{x_1}{c} \frac{\partial C_L}{\partial \alpha},$$

where the suffix 1 refers to G_1. Now in normal flight $\partial C_L/\partial \alpha$ is constant, since the (C_L, α) graph is straight. Therefore we have $dC_L = a \, d\alpha$ and

$$(1) \qquad \left(\frac{\partial C_m}{\partial C_L} \right)_1 = \frac{\partial C_m}{\partial C_L} - \frac{x_1}{c}.$$

Now let G_0, distant x_0 from G, be the centre of gravity position which gives neutral equilibrium, i.e. such that $(\partial C_m/\partial C_L)_0 = 0$. It follows from (1) that

$$\frac{\partial C_m}{\partial C_L} - \frac{x_0}{c} = 0,$$

and therefore

$$(2) \qquad \left(\frac{\partial C_m}{\partial C_L} \right)_1 = - \frac{(x_1 - x_0)}{c} = - H, \text{ say.}$$

The dimensionless number H is called * the *metacentric ratio*; it is the ratio of the distance between the actual centre of gravity G_1 and the position G_0 of the centre of gravity when the static stability is neutral, to the mean chord.

When H is positive $(\partial C_m/\partial C_L)_1$ is negative, so that there is static stability; when H is negative there is static instability, and these cases correspond respectively with G_1 being forward or aft of G_0.

---

* The alternative name *static margin* is also used. It is the aerodynamic analogue of the metacentric height in naval architecture. In a ship the metacentre is the centre of gravity position for flotation in neutral equilibrium.

19·12. Neutral centre of gravity positions. Consider a monoplane flying at incidence α. Let O be the leading edge of the mean chord of the wings, O' the leading edge of the mean chord of the tail. We shall take as chord of the tail the line through O' parallel to OA the mean chord of the wings. For

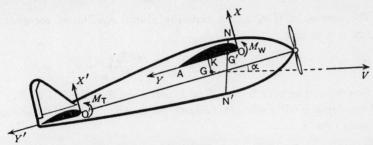

FIG. 19·12.

temporary convenience we specify the aerodynamic force on the wings by X, Y perpendicular and parallel to the chord acting at O and the moment M_W about O. The corresponding system for the tail at O' is X', Y', M_T. Let G be the centre of gravity and draw GK perpendicular to OA. The moment M of all the air forces about G is then

(1) $M = M_W + OK . X + GK . Y + M_T - (l - OK) X' - (f - GK) Y'$,

where l, the *tail lever arm*,* is the distance which O is ahead of O' measured parallel to OA, and f is the height of O above O' measured perpendicularly to OA.

If L, D are the lift and drag on the wings, and L', D' the lift and drag on the tail, we have by resolution, observing that α is a small angle,

(2) $X = L + \alpha D, \quad Y = D - \alpha L, \quad X' = L' + \alpha D', \quad Y' = D' - \alpha L'$.

Since, in general, l is large compared with OK and $f - GK$ is small, we neglect $OK . X'$ and the last term of (1).

We now introduce the dimensionless number

(3) $\kappa = \dfrac{lS'}{cS}$,

where S, S' are the plan areas of the wings and tail and c is the mean chord of the wings. The number κ, which is the ratio of two volumes, is known as the *tail volume ratio*. Dividing (1) by $\frac{1}{2}\rho V^2 Sc$ we get

(4) $C_m = C_{mW} + xC_X + yC_Y + \dfrac{\kappa c'}{l} C_{mT} - \kappa C_{X'}$,

where c' is the chord of the tail and

(5) $OK = cx, \quad GK = cy.$

---

* This term is also applied to the distance of G from the centre of pressure of the tail.

Since c' is small compared with l we may conveniently neglect the penultimate term of (4). Differentiate (4) with respect to α. Then

$$\frac{\partial C_m}{\partial \alpha} = x\frac{\partial C_X}{\partial \alpha} + y\frac{\partial C_Y}{\partial \alpha} - \kappa\frac{\partial C_X'}{\partial \alpha} + \left(\frac{\partial C_m}{\partial \alpha}\right)_W.$$

The position of $G(x_0, y_0)$ for statically neutral equilibrium occurs when $\partial C_m/\partial \alpha = 0$, so that

$$(6) \qquad x_0\frac{\partial C_X}{\partial \alpha} + y_0\frac{\partial C_Y}{\partial \alpha} - \kappa\frac{\partial C_X'}{\partial \alpha} + \left(\frac{\partial C_m}{\partial \alpha}\right)_W = 0.$$

Now from (2) we have $X = L, X' = L'$ nearly and therefore $\partial C_X/\partial \alpha, \partial C_X'/\partial \alpha$ are the slopes of lift coefficient-incidence graphs which will be taken to have the same value a. Also

$$\left(\frac{\partial C_m}{\partial \alpha}\right)_W = \left(\frac{\partial C_m}{\partial C_L}\right)_W\frac{\partial C_L}{\partial \alpha} = -\tfrac{1}{4}a,$$

if we assume the two-dimensional result found in 8·34, that the moment about the quarter point is independent of the incidence so that $C_{mW} = \text{constant} - \tfrac{1}{4}C_L$. Lastly, from (2), we have

$$C_Y = C_D - \alpha C_L = C_{D_0} + C_{D_i} - \alpha C_L.$$

Let $-\alpha_0$ be the value of α for which the lift vanishes. Then the absolute incidence is

$$\alpha + \alpha_0 = \alpha_e + \epsilon,$$

where α_e is the effective incidence and ϵ is the angle of downwash. Also, from 11·24, $C_{D_i} = \epsilon C_L$, and $C_L = a\alpha_e$, so that

$$C_Y = C_{D_0} + \alpha_0 C_L - \frac{C_L{}^2}{a},$$

and therefore

$$\frac{\partial C_Y}{\partial \alpha} = a\frac{\partial C_Y}{\partial C_L} = a\left(\alpha_0 - \frac{2C_L}{a}\right) = a(\alpha_0 - 2\alpha_e).$$

Thus (6) becomes

$$(7) \qquad x_0 + y_0(\alpha_0 - 2\alpha_e) - \kappa - \tfrac{1}{4} = 0.$$

Therefore the locus of neutral positions of G is a line such as NN' in fig. 19·12, inclined to the perpendicular to the direction of motion at the small angle $2\alpha_e - \alpha_0 - \alpha = \alpha_e - \epsilon$.

When G is forward of this line there is static stability. The line NN' cuts the chord OA at G' where $OG' = c(\kappa + \tfrac{1}{4})$, a point whose position depends, to the order of approximation here used, only on the tail volume ratio κ. The practical conclusion is that there is static stability when G is forward of the point G'.

We may also observe that (1), being linear in x and y, would still lead to a relation of the form (7), had the terms which we have neglected been retained.

19·2. Asymmetric moments.   These are rolling and yawing moments. They are asymmetric in the sense that they tend to deviate the plane of symmetry of the aircraft.   We refer to fig. 18·0.

Rolling moment, L, or L_R when confusion with lift may arise, is the moment of the aerodynamic forces about the longitudinal axis GX.   It is controlled by the use of the ailerons and is positive when it tends to depress the starboard wing tip (sense y to z).

The coefficient of rolling moment is

$$C_l = \frac{L_R}{\frac{1}{2}\rho V^2 Sb}.$$

Yawing moment, N, is the moment of the aerodynamic forces about the normal axis GZ.   It is controlled by the use of the rudder, and is positive when it tends to deviate the nose to starboard (sense x to y).

The coefficient of yawing moment is

$$C_n = \frac{N}{\frac{1}{2}\rho V^2 Sb}.$$

Note the use of b the span, instead of c the chord, in the definition of C_l and C_n

The angular velocity of rolling is denoted by p, positive in the sense y to z, and the angular velocity of yawing is denoted by r, positive in the sense x to y.

19·3. The strip hypothesis.   Consider an aircraft rolling with angular velocity p.   A point distance y from the plane of symmetry on the starboard wing will have a velocity yp parallel to the normal axis.   Thus the strip or section of the starboard wing between the planes y and $y + dy$ will move through the air with a velocity compounded of V the forward velocity of the aircraft as a whole and the velocity yp.   Similarly, a strip of the port wing at the same distance y will have a velocity compounded of V and $- yp$.   Thus each such strip has a different resultant velocity.   Similar considerations apply to a yawing aircraft.

The *strip hypothesis* asserts that we may calculate the aerodynamic force on each strip as if it were an isolated aerofoil moving with the resultant velocity which it has on account of its local position on the aircraft.

The strip hypothesis is manifestly a first approximation to a state of affairs in which the trailing vortices from each strip possibly interfere with the others, but this first approximation is confirmed by experiment to give results which are in excellent agreement with the facts as observed, even at incidences above the stall.

19·4. Moments due to rolling.

We consider strips at distance y from the plane of symmetry.

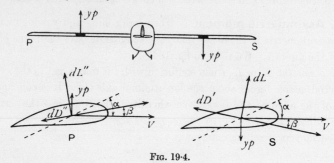

FIG. 19·4.

The resultant velocity of the starboard strip S makes an angle β with the direction of forward motion of the aircraft where

$$\beta = \tan \beta = \frac{py}{V}.$$

Thus, by the strip hypothesis, the incidence is $\alpha + \beta$, where α is the incidence when $p = 0$. We are here treating py as small compared with V so that the resultant forward speed $\sqrt{(V^2 + p^2 y^2)}$ can be replaced by V. If C_L', C_D' are the lift and drag coefficients at incidence α, at incidence $\alpha + \beta$ they will be, by Taylor's theorem,

$$C_L' + \beta \frac{\partial C_L'}{\partial \alpha}, \quad C_D' + \beta \frac{\partial C_D'}{\partial \alpha}.$$

The components of the lift dL' and drag dD', resolved perpendicularly to and against the direction of motion, will be $dL' + \beta dD'$, $dD' - \beta dL'$. The corresponding coefficients for the port strip will be got by writing $-\beta$ for β in the above and the forces will have components $dL'' - \beta dD''$, $dD'' + \beta dL''$.

The two strips will therefore contribute to the rolling and yawing moments (using wind axes).

(1) $$dL_R = -y(dL' - dL'') - \beta y(dD' + dD'').$$

(2) $$dN = y(dD' - dD'') - \beta y(dL' + dL'').$$

Now $$dL' = \tfrac{1}{2}\rho V^2 c' dy \left(C_L' + \beta \frac{\partial C_L'}{\partial \alpha} \right),$$

$$dD' = \tfrac{1}{2}\rho V^2 c' dy \left(C_D' + \beta \frac{\partial C_D'}{\partial \alpha} \right),$$

where c' is the chord of the strip. Also dL'', dD'' are obtained from the above by changing the sign of β. Thus substituting and integrating over half the span, since we have taken an element from each wing, we get

(3) $$L_R = -\rho V p \int_0^{b/2} c' \left(C_D' + \frac{\partial C_L'}{\partial \alpha} \right) y^2 \, dy.$$

$$(4) \qquad N = - \rho V p \int_0^{b/2} c' \left(C_L' - \frac{\partial C_D'}{\partial \alpha} \right) y^2 \, dy.$$

These are the rolling moment and yawing moment due to rolling.

If we draw the graph of $c'(C_D' + \partial C_L'/\partial \alpha)$ against y, the area under the graph for the whole span $(- \frac{1}{2}b, \frac{1}{2}b)$ will be $S(C_D + \partial C_L/\partial \alpha)$, where S is the plan area of the aerofoil and C_D, C_L are the coefficients for the aerofoil as a whole. The integral on the right of (3) will be the second moment of half this area with respect to y and may therefore be written in the form

$$\tfrac{1}{2}S(C_D + \partial C_L/\partial \alpha) \, J_1{}^2 \frac{b^2}{4},$$

where J_1 is dimensionless and $J_1 b/2$ represents the radius of gyration of the area considered. Thus dividing by $\frac{1}{2}\rho V^2 Sb$ we have

$$(5) \qquad C_l = - \tfrac{1}{2} \left(C_D + \frac{\partial C_L}{\partial \alpha} \right) J_1{}^2 \frac{pb}{2V}.$$

Here $pb/(2V)$ is equal to the ratio of the wing tip speed due to rolling to the forward speed.

Similarly from (4) we get

$$(6) \qquad C_n = - \tfrac{1}{2} \left(C_L - \frac{\partial C_D}{\partial \alpha} \right) J_2{}^2 \frac{pb}{2V},$$

where $J_2 b/2$ is the radius of gyration of the area under the $[c'(C_L' - \partial C_D'/\partial \alpha), y]$ curve on the semi-span.

In discussing dynamic stability the important quantities are $\partial L_R/\partial p$ and $\partial N/\partial p$, for these give the changes in L_R and N due to changes in p. They are best obtained from *derivative coefficients* of rolling and yawing due to rolling with the notations

$$C_{lp} = \frac{\partial C_l}{\partial \left(\dfrac{pb}{2V} \right)} = - \tfrac{1}{2}J_1{}^2 \left(C_D + \frac{\partial C_L}{\partial \alpha} \right),$$

$$C_{np} = \frac{\partial C_n}{\partial \left(\dfrac{pb}{2V} \right)} = - \tfrac{1}{2}J_2{}^2 \left(C_L - \frac{\partial C_D}{\partial \alpha} \right).$$

Thus, for example, the increment in rolling moment due to a small angular velocity of roll ω will be

$$dL_R = \tfrac{1}{2}\rho V^2 Sb C_{lp} \omega b/2V.$$

19·5. Moments due to yawing. Unlike rolling, yawing does not alter the incidence of the strips, so that if dL', dD' refer to the starboard strip at distance y, we have as in 19·4 (1), (2)

$$dL_R = - y(dL' - dL''), \quad dN = y(dD' - dD'').$$

If r is the angular velocity of yaw,

$$dL' = C_L' c' \, dy \, . \, \tfrac{1}{2}\rho (V - ry)^2, \quad dD' = C_D' c' \, dy \, . \, \tfrac{1}{2}\rho (V - ry)^2,$$

and dL'', dD'' are got by changing the sign of r. Integrating over half the span, since we have taken an element from each wing, we get

$$L_R = 2\rho V r \int_0^{b/2} c' C_L' y^2 \, dy, \qquad N = -2\rho V r \int_0^{b/2} c' C_D' y^2 \, dy.$$

As in 19·4, we draw the curves $(c'C_L', y)$, the lift grading curve, and $(c'C_D', y)$, the drag grading curve, on the span. Their areas will be SC_L and SC_D respectively, and if their radii of gyration are $J_3 b/2$, $J_4 b/2$, we shall get

$$C_l = C_L . J_3{}^2 . \frac{br}{2V}, \qquad C_n = -C_D . J_4{}^2 . \frac{br}{2V}$$

for the coefficients of rolling and yawing moment due to yawing. We note that $br/2V$ is the ratio of the wing tip speed due to yaw to the forward speed. The derivative coefficients will be obtained by partial differentiation with respect to $br/2V$ in the forms

$$C_{lr} = J_3{}^2 C_L, \quad C_{nr} = -J_4{}^2 C_D.$$

19·6. Rolling moment due to side-slip.
Consider an aircraft whose wings are inclined upwards, at the small angle γ, to the plane through the

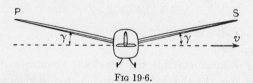

FIG 19·6.

mean chord perpendicular to the plane of symmetry. This angle γ is known as the *dihedral angle*. Let V be the speed of the aircraft, v the velocity component perpendicular to the plane of symmetry, i.e. the side-slip. We shall suppose v/V to be so small that velocity component in the plane of symmetry is V. The side-slip velocity v can be resolved on the starboard wing S into a component $v \cos \gamma = v$ along the wing, and a component $v \sin \gamma = v\gamma$ perpendicular to the wing. The component along the wing gives rise to a cross-wind which we shall not consider further, while the component $v\gamma$ has the effect of altering the incidence of the strip of wing at distance y along the wing. In fact the starboard strip has its incidence increased by $\beta = v\gamma/V$, while the port strip has its incidence decreased by the same amount (see fig. 19·4, reading $v\gamma$ for yp in that figure).

The investigation of 19·4 holds with this difference that β is now constant instead of being a function of y. Thus from 19·4 (3) we get for the rolling moment

$$L_R = -\rho V^2 \beta \int_0^{b/2} c' \left(C_D' + \frac{\partial C_L'}{\partial \alpha} \right) y \, dy = -\tfrac{1}{2} S \left(C_D + \frac{\partial C_L}{\partial \alpha} \right) J \left(\frac{b}{2} \right) \rho V^2 \beta,$$

where $bJ/2$ is the y-coordinate of the centroid of the graph of $c'(C_D' + \partial C_L'/\partial \alpha)$ against y over the semi-span $(0, b/2)$. Thus the coefficient of rolling moment due to side-slip v is

$$C_l = -\tfrac{1}{2}J\left(C_D + \frac{\partial C_L}{\partial \alpha}\right)\frac{v\gamma}{V}.$$

In normal flight this is negative. The tendency of this rolling moment is therefore to raise the wing towards which the aircraft is side-slipping and so to promote lateral stability. This stabilising effect is absent if $\gamma = 0$, i.e. when there is no dihedral angle.

The derivative coefficient of rolling moment due to side-slip is

$$C_{lv} = \frac{\partial C_l}{\partial(v/V)} = -\tfrac{1}{2}J\gamma\left(C_D + \frac{\partial C_L}{\partial \alpha}\right).$$

19·7. Change of axes. The asymmetric moments have been calculated for simplicity of exposition with respect to wind axes. They can, of course, be calculated directly by the foregoing principles with respect to chord axes, or we can deduce the results for chord axes by the following method.

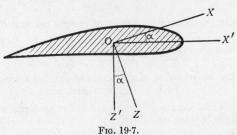

FIG. 19·7.

Let d___es denote the moment coefficients, etc. with respect to wind axes ; suppose we want to calculate C_l, for example. If α is the incidence, we get by simple resolution $C_l = C_l' \cos \alpha - C_n' \sin \alpha$. To calculate, say, C_{lp}, we first note that angular velocity p about OX is equivalent to angular velocities $p' = p \cos \alpha$ about OX' and $r' = -p \sin \alpha$ about OZ'. Then

$$C_{lp} = \frac{\partial C_l}{\partial(bp/2V)} = \frac{2V}{b}\frac{\partial C_l}{\partial p}, \qquad C_{lp}' = \frac{2V}{b}\frac{\partial C_l'}{\partial p'},$$

$$\frac{\partial C_l}{\partial p} = \frac{\partial(C_l'\cos\alpha - C_n'\sin\alpha)}{\partial p}$$

$$= \left(\frac{\partial C_l'}{\partial p'}\frac{\partial p'}{\partial p} + \frac{\partial C_l'}{\partial r'}\frac{\partial r'}{\partial p}\right)\cos\alpha - \left(\frac{\partial C_n'}{\partial p'}\frac{\partial p'}{\partial p} + \frac{\partial C_n'}{\partial r'}\frac{\partial r'}{\partial p}\right)\sin\alpha.$$

Thus $C_{lp} = C_{lp}'\cos^2\alpha - (C_{lr}' + C_{np}')\sin\alpha\cos\alpha + C_{nr}'\sin^2\alpha.$

If α is small, the usual case, this simplifies to

$$C_{lp} = C_{lp}' - \alpha(C_{lr}' + C_{np}').$$

EXAMPLES XIX

1. Show that the increase in pitching moment about the centre of gravity due to an increase $d\alpha$ in incidence is

$$- \tfrac{1}{2}H\rho V^2 Sca\, d\alpha,$$

where H is the metacentric ratio.

2. If the pitching moment about the centre of gravity is $k\alpha + k_0$, where k and k_0 are constants, find how the metacentric ratio depends on k.

3. Draw a graph to show how the neutral equilibrium position (on the mean chord) of the centre of gravity, measured from the leading edge, depends on the tail volume ratio κ.

Explain the interpretation of this graph for negative values of κ.

4. If the coefficients of rolling and yawing moment were defined with respect to chord instead of span (as for the pitching moment), show that the usual values would have to be multiplied by the aspect ratio.

5. Discuss the relation between the strip hypothesis and loading law.

6. Use the data for Clarke YH (Ex. I, 21) to show that in normal flight $\partial C_D/\partial\alpha$ may reasonably be neglected in calculating C_n and C_{np} due to rolling, in normal flight.

Deduce the corresponding simplified results, indicating how J_2 is related to the load grading curve.

7. Show that the derivative coefficient of yawing moment due to side-slip is given by

$$C_{nv} = - \tfrac{1}{2}J\gamma\left(C_L - \frac{\partial C_D}{\partial\alpha}\right)$$

where J refers to the line of action of the resultant drag on a wing.

Discuss the order of magnitude of C_{nv} in normal flight.

8. If dashes refer to wind axes, prove the following formulae for coefficients referred to chord axes, α being small.

$$C_{lr} = C_{lr}' + (C_{lp}' - C_{nr}')\alpha,$$
$$C_{np} = C_{np}' + (C_{lp}' - C_{nr}')\alpha,$$
$$C_{nr} = C_{nr}' + (C_{lr}' + C_{np}')\alpha.$$

9. Referring to 19·4, show that, if p is so large that the approximation $\beta = \tan\beta$ is not permissible, then for the starboard strip

$$dL' = \tfrac{1}{2}\rho V^2 \sec^2\beta\, c'\, dy\, [C_L' \cos\beta + C_D' \sin\beta],$$

and that $p\, dy = V \sec^2\beta\, d\beta$, where C_L', C_D' correspond with the incidence $\alpha + \beta$.

Obtain a similar expression for the port strip and hence obtain an integral to give C_l.

10. If dashes refer to wind axes and the wind axis Ox' makes the angle α with the chord axis Ox, prove that in normal flight

$$C_{lv} = C_{lv}', \quad C_{nv}' = C_{nv} + \alpha C_{lv},$$

approximately.

CHAPTER XX

STABILITY

20·0. An aircraft in steady motion is always liable to disturbance, as for example, when a sudden gust is encountered, or if the pilot alters the controls. In the present chapter we propose to examine how an aircraft will respond to such a disturbance, which will be supposed small. We shall, in particular, obtain the equations of motion due to a disturbance of steady straight flight, gliding, horizontal, or climbing. The equations will then be further particularised by considering horizontal flight. By restricting the discussion to disturbances in the plane of symmetry, we reach the problem of longitudinal stability. The method, which is of general application, should thus be sufficiently exemplified.

20·01. Equations of motion. If G is the centre of gravity of a rigid body, all the forces acting on the body can be replaced by a single force acting at G together with a couple.

The forces acting on an aircraft are (1) the air force due to pressure, (2) the propeller thrust, (3) the weight. Let us take rectangular axes Gx, Gy, Gz (fig. 18·0) passing through G and fixed in the aircraft (chord axes). Referred to these axes the resultant of (1) and (2) will be a force

$$(1) \qquad \mathbf{F} = \mathbf{i}X + \mathbf{j}Y + \mathbf{k}Z,$$

and the weight will be a force

$$m\mathbf{g} = \mathbf{i}mg_1 + \mathbf{j}mg_2 + \mathbf{k}mg_3,$$

where m is the mass of the aircraft, $\mathbf{g}$ the (vector) acceleration due to gravity and $\mathbf{i}, \mathbf{j}, \mathbf{k}$ are unit vectors along the axes of reference.

The force $\mathbf{F}$ will be called the aerodynamic force although it includes the propeller thrust and the buoyancy due to hydrostatic pressure. This slight departure from our previous use of the term will cause no subsequent difficulty, for we shall be concerned only with variations in $\mathbf{F}$.

With the centre of gravity G as base-point, let the velocity of G, the angular velocity of the aircraft, and the angular momentum * about G be

$$(2) \qquad \mathbf{v} = \mathbf{i}u' + \mathbf{j}v' + \mathbf{k}w'.$$

$$(3) \qquad \boldsymbol{\Omega} = \mathbf{i}p' + \mathbf{j}q' + \mathbf{k}r'.$$

$$(4) \qquad \mathbf{h} = \mathbf{i}h_1 + \mathbf{j}h_2 + \mathbf{k}h_3.$$

---

* Or, moment of momentum. The terms are interchangeable.

Then the equations of rate of change * of linear and angular momentum are

(5) $$\frac{d}{dt}(m\mathbf{v}) = m\dot{\mathbf{v}} + \mathbf{\Omega}_{\wedge} m\mathbf{v} = \mathbf{F} + m\mathbf{g}.$$

(6) $$\frac{d\mathbf{h}}{dt} = \dot{\mathbf{h}} + \mathbf{\Omega}_{\wedge}\mathbf{h} = \mathbf{L},$$

where the couple, referred to above, is

(7) $$\mathbf{L} = \mathbf{i}L + \mathbf{j}M + \mathbf{k}N.$$

Written in full the equations (5) and (6) are equivalent to the six equations

(8) $$m(\dot{u}' - v'r' + w'q') = X + mg_1,$$

(9) $$m(\dot{v}' - w'p' + u'r') = Y + mg_2,$$

(10) $$m(\dot{w}' - u'q' + v'p') = Z + mg_3,$$

(11) $$\dot{h}_1 - h_2 r' + h_3 q' = L,$$

(12) $$\dot{h}_2 - h_3 p' + h_1 r' = M,$$

(13) $$\dot{h}_3 - h_1 q' + h_2 p' = N.$$

Also if A, B, C, D, E, F are the moments and products of inertia

$$(A = \Sigma m(y^2 + z^2), \quad D = \Sigma m\, yz, \quad \text{etc.}),$$

then

(14) $$h_1 = Ap' - Fq' - Er', \quad h_2 = -Fp' + Bq' - Dr',$$
$$h_3 = -Ep' - Dq' + Cr'.$$

For the justification of the above statements the reader is referred to works on Dynamics.

In what follows we shall use the vector formulation, which is simpler, but the same equations can, of course, be obtained with the use of equations (8) to (13) instead of (5) and (6.)

20·1. Straight flight. Consider an aircraft in steady straight flight, at incidence α, with velocity $\mathbf{V}$ in the plane of symmetry, so that $\mathbf{v} = \mathbf{V}, \mathbf{\Omega} = 0$. We seek the equations of motion when the steady state is slightly disturbed by giving the aircraft a *small* additional velocity and a *small* angular velocity.

In the steady state there is no rotation and

$$\mathbf{F} + m\mathbf{g} = 0.$$

Let the motion be slightly disturbed, at time $t = 0$, so that at time t the velocity is $\mathbf{V}_1 + \mathbf{u}$, where the components of $\mathbf{V}_1$ along the axes are the constants $V\cos\alpha, V\sin\alpha, 0$ and the angular velocity is $\boldsymbol{\omega}$, where

(1) $$\mathbf{u} = \mathbf{i}u + \mathbf{j}v + \mathbf{k}w, \quad \boldsymbol{\omega} = \mathbf{i}p + \mathbf{j}q + \mathbf{k}r.$$

---

* The terms $\mathbf{\Omega} \wedge m\mathbf{v}$ and $\mathbf{\Omega} \wedge \mathbf{h}$ arise as the result of the application of the usual rule for writing down the rate of change of a vector (21·14) when the axes are rotating. Time rate of change with respect to axes fixed in the aircraft is indicated by the dot, while d/dt indicates differentiation with respect to axes fixed in space.

The effect of the angular velocity is that in a short time dt the aircraft will undergo a small rotation $d\boldsymbol{\chi} = \boldsymbol{\omega}\, dt$. If therefore we write for the rotation at time t,*

$$\boldsymbol{\chi} = \mathbf{i}\beta + \mathbf{j}\theta + \mathbf{k}\gamma, \tag{2}$$

we shall have

$$\boldsymbol{\omega} = \dot{\boldsymbol{\chi}} = \mathbf{i}\dot{\beta} + \mathbf{j}\dot{\theta} + \mathbf{k}\dot{\gamma}. \tag{3}$$

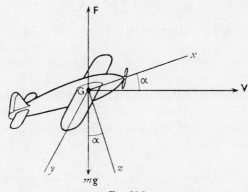

FIG. 20·1.

It will be assumed throughout that $\boldsymbol{\chi}$, $\boldsymbol{\omega}$, $\mathbf{u}$ are small quantities of the first order and we shall therefore neglect products of these quantities. Another result of the rotation will be that the aircraft will acquire angular momentum $\mathbf{h}$ which is also small.

On account of the disturbance the force $\mathbf{F}$ will become $\mathbf{F} + d\mathbf{F}$, and owing to the rotation of the axes the weight will be described by the vector $m(\mathbf{g} + d\mathbf{g})$ instead of by $m\mathbf{g}$. Since $\mathbf{g}$, the acceleration due to gravity, is fixed in magnitude and in direction in space, the change of $\mathbf{g}$ in time t will be zero, in other words (see 21·14),

$$d\mathbf{g} + \boldsymbol{\chi}_{\wedge}\mathbf{g} = 0. \tag{4}$$

The equations of motion (see 20·01 (5), (6)) will now be

$$m(\dot{\mathbf{V}}_1 + \dot{\mathbf{u}}) + \boldsymbol{\omega}_{\wedge} m(\mathbf{V}_1 + \mathbf{u}) = \mathbf{F} + d\mathbf{F} + m(\mathbf{g} + d\mathbf{g}),$$

$$\dot{\mathbf{h}} + \boldsymbol{\omega}_{\wedge}\mathbf{h} = d\mathbf{L},$$

where $d\mathbf{L}$ is the moment of the forces about G.

Now $\mathbf{F} + m\mathbf{g} = 0$, $\dot{\mathbf{V}}_1 = 0$ and $\boldsymbol{\omega}_{\wedge}\mathbf{u}$, $\boldsymbol{\omega}_{\wedge}\mathbf{h}$ are negligible, therefore using (4) we get

$$m\dot{\mathbf{u}} + m(\dot{\boldsymbol{\chi}}_{\wedge}\mathbf{V}_1 + \boldsymbol{\chi}_{\wedge}\mathbf{g}) = d\mathbf{F}. \tag{5}$$

$$\dot{\mathbf{h}} = d\mathbf{L}. \tag{6}$$

---

* The rotation χ will be assumed small. A small rotation can be represented to the first order by a unique vector. A finite rotation cannot be so represented, for the final result of given finite rotations in roll, pitch, and yaw depends upon the order in which they take place.

These are the general equations of straight flight slightly disturbed from the steady state. They apply to horizontal, gliding or climbing flight.

As to **h**, we have from 20·01 (14),

$$h_1 = A\dot\beta - F\dot\theta - E\dot\gamma, \quad h_2 = -F\dot\beta + B\dot\theta - D\dot\gamma, \quad h_3 = -E\dot\beta - D\dot\theta + C\dot\gamma.$$

When the ailerons and rudder are in their neutral positions, Gy is a principal dynamical axis and $D = F = 0$. Thus

$$(7) \qquad \mathbf{h} = \mathbf{i}\,(A\dot\beta - E\dot\gamma) + \mathbf{j}\,B\dot\theta + \mathbf{k}\,(C\dot\gamma - E\dot\beta).$$

20·11. Simplifying assumption.

It appears from 20·01 (5), (6) that **F** and **L** depend not only on **v**, Ω but on their time rates of change, and therefore $d\mathbf{F}$ and $d\mathbf{L}$ will, in general, depend on $\mathbf{v} + d\mathbf{v}$, $\Omega + d\Omega$, and the corresponding changed time rates. In the case of 20·1 we have $\mathbf{v} = \mathbf{V}$, $d\mathbf{v} = \mathbf{u}$, $\Omega = 0$, $p = \boldsymbol{\omega}$, and therefore $d\mathbf{F}$ and $d\mathbf{L}$ will depend on **u**, $\boldsymbol{\omega}$ and their time rates.

We shall assume that $d\mathbf{F}$ and $d\mathbf{L}$ depend solely on **u**, $\boldsymbol{\omega}$ and not on their time rates, with the single exception that the pitching moment component of $d\mathbf{L}$ will be taken to depend also on $\dot w$, the rate of change of the downwash velocity w.

The reason for this exception is that the tail plane is, in general, traversing the downwash created by the wings at an earlier time and may therefore be expected to be influenced by $\dot w$.

With this assumption we shall have

$$d\mathbf{F} = \mathbf{i}\,dX + \mathbf{j}\,dY + \mathbf{k}\,dZ, \quad d\mathbf{L} = \mathbf{i}\,dL + \mathbf{j}\,dM + \mathbf{k}\,dN,$$

where, for example,

$$dX = \left(\frac{\partial X}{\partial u'}\right)_0 du' + \left(\frac{\partial X}{\partial v'}\right)_0 dv' + \left(\frac{\partial X}{\partial w'}\right)_0 dw'$$
$$+ \left(\frac{\partial X}{\partial p'}\right)_0 dp' + \left(\frac{\partial X}{\partial q'}\right)_0 dq' + \left(\frac{\partial X}{\partial r'}\right)_0 dr'$$

the zero suffix indicating that after the differentiation the steady values of the velocity are to be inserted, namely $u' = V$, $v' = w' = 0$, $p' = q' = r' = 0$. Also

$$d\mathbf{v} = \mathbf{u} = \mathbf{i}u + \mathbf{j}v + \mathbf{k}w, \quad d\Omega = \boldsymbol{\omega} = \mathbf{i}p + \mathbf{j}q + \mathbf{k}r,$$

and therefore $du' = u, dp' = p$, etc.

If we write $(\partial X/\partial u')_0 = X_u$, and so on, we get

$$dX = uX_u + vX_v + wX_w + pX_p + qX_q + rX_r,$$

and two similar expressions for dY and dZ. In like manner we have analogous expressions for dL, dM, dN. Writing dM in full to show the term in $\dot w$, we have

$$dM = uM_u + vM_v + wM_w + \dot wM_{\dot w} + pM_p + qM_q + rM_r.$$

Thus it appears that $d\mathbf{F}$ consists of 18 terms and $d\mathbf{L}$ of 19 on account of the extra term $\dot wM_{\dot w}$. Of the 37 derivatives $X_u, \ldots, L_u, \ldots$, thus arising, 18 are zero.

Proof. No symmetric disturbance can cause an asymmetric reaction, e.g. pitching will not induce yaw or roll. Thus

$$Y_u, Y_w, Y_q; \; L_u, L_w, L_q; \; N_u, N_w, N_q$$

are all zero.

Again the symmetric reaction due to an asymmetric disturbance must by symmetry be independent of the sign of the disturbance, e.g. if an angular velocity of roll could cause a pitching moment pM_p, this pitching moment would necessarily have the same sign if p were replaced by $-p$. Since M_p is independent of p it must therefore vanish. Thus

$$X_p, X_r, X_v; \; Z_p, Z_r, Z_v; \; M_p, M_r, M_v$$

are all zero.

Therefore using 20·1 (3) we have

(1) $\quad d\mathbf{F} = \mathbf{i}\,(uX_u + wX_w + \dot\theta X_q) + \mathbf{j}\,(vY_v + \dot\beta Y_p + \dot\gamma Y_r)$
$$+ \mathbf{k}\,(uZ_u + wZ_w + \dot\theta Z_q).$$

(2) $\quad d\mathbf{L} = \mathbf{i}\,(\dot\beta L_p + \dot\gamma L_r + vL_v) + \mathbf{j}\,(\dot\theta M_q + uM_u + wM_w + \dot w M_{\dot w})$
$$+ \mathbf{k}\,(\dot\beta N_p + \dot\gamma N_r + vN_v).$$

We can also take account of small moments exerted by the controls by adding to $d\mathbf{L}$ the vector $\mathbf{i}L_0 + \mathbf{j}M_0 + \mathbf{k}N_0$. Application of L_0 and N_0 would of course introduce products of inertia D and F which have been taken to be zero, but as these will necessarily be small and multiplied in the equations of motion by small factors we shall neglect them.

The derivatives in (1) and (2) fall into five classes : *force velocity derivatives* X_u, X_w, Y_v, Z_u, Z_w ; *force rotary derivatives* X_p, Y_q, Y_r, Z_q ; *moment velocity derivatives* L_v, M_u, M_w, N_v ; *moment rotary derivatives* L_p, L_r, M_q, N_p, N_r ; *moment acceleration derivatives* $M_{\dot w}$, the only representative.

20·12. The equations of disturbed horizontal flight.

At this stage we shall suppose that $\mathbf{V}$ is horizontal so that initially Gz makes an angle α with the vertical, fig. 20·1. Had the flight contemplated been other than horizontal this angle would have had some different initial value. Referring to 20·1 (5), (6), and evaluating the vector products (21·13) $\dot{\boldsymbol\chi}_\wedge \mathbf{V}_1$ and $\boldsymbol\chi_\wedge \mathbf{g}$, where *

$$\mathbf{g} = -\,\mathbf{i}g\sin\alpha + \mathbf{k}g\cos\alpha, \qquad \mathbf{V}_1 = \mathbf{i}V\cos\alpha + \mathbf{k}V\sin\alpha$$

we get the following six equations of motion :

(1) $\quad m\dot u - uX_u - wX_w + (mV\sin\alpha - X_q)\,\dot\theta + mg\,\theta\cos\alpha = 0,$

(2) $\quad -uZ_u + m\dot w - wZ_w - (mV\cos\alpha + Z_q)\,\dot\theta + mg\,\theta\sin\alpha = 0,$

(3) $\quad -uM_u - \dot w M_{\dot w} - wM_w + B\ddot\theta - \dot\theta M_q = M_0,$

---

* When the steady flight is not horizontal the above expression for $\mathbf{g}$ is replaced by

$$\mathbf{g} = -\,\mathbf{i}g\sin\Theta + \mathbf{k}g\cos\Theta,$$

where Θ is the attitude angle (18·3), i.e. the angle between Gx and the horizontal, positive when Gx is above the horizontal.

which contain the variables u, w, θ and form the *symmetric group*, and

(4) $m\dot{v} - vY_v - (mV \sin \alpha + Y_p)\dot{\beta} - (mg \cos \alpha)\beta$
$$+ (mV \cos \alpha - Y_r)\dot{\gamma} - (mg \sin \alpha)\gamma = 0,$$

(5) $-vL_v + A\ddot{\beta} - L_p\dot{\beta} - E\ddot{\gamma} - L_r\dot{\gamma} = L_0,$

(6) $-vN_v - E\ddot{\beta} - N_p\dot{\beta} + C\ddot{\gamma} - N_r\dot{\gamma} = N_0,$

which contain the variables v, β, γ, and form the *asymmetric group*.

Equations (1), (2), (4) are derived from 20·1 (5) and (3), (5), (6) are derived from 20·1 (6).

The above six equations are ordinary linear differential equations with constant coefficients. When the derivatives X_u, L_v, etc., are known they can be solved by various standard methods and the response of the aircraft to a given disturbance can be calculated. The equations, however, are best expressed in a particular system of units which we shall proceed to consider.

20·2. The parameter μ.

We define the *relative aircraft density* by the dimensionless * number

$$\mu = \frac{m}{\rho lS} = \frac{\text{mass of the aircraft}}{\text{air density} \times \text{typical length} \times \text{plan area}}.$$

Here the typical length l may be conveniently taken either as the chord c or the tail lever arm in discussing the symmetric equations, and as the span b when discussing the asymmetric equations. Observing that lS is a measure of volume, we see that μ is the ratio of the average aircraft density, taking lS as the volume, to the air density. Since air density decreases as height increases, μ increases with height.

It is also convenient to introduce a dimensionless parameter k defined by $k\rho V^2 S = mg$. In straight horizontal flight $k = \frac{1}{2}C_L$.

If the flight path is inclined to the horizontal at the angle Θ we have $k = \frac{1}{2}C_L \sec \Theta$.

20·21. Units.

The quantities appearing in the equations of motion derived in 20·12 are all measured in absolute units, that is to say, if the units of mass, length and time are given, the unit of velocity is the unit of length divided by the unit of time, the unit of acceleration is the unit of length divided by the square of the unit of time, the unit of force is the unit of mass multiplied by the unit of acceleration, etc. The equations of motion will preserve the *same form* whether the ft.-lb.-sec., c.g.s., or any other system of absolute units is used. For the equations of motion of the aircraft it is convenient to use the following system :

---

* See 1·7.

Unit of mass m, the mass of the aircraft.

Unit of length l, the typical length.

Unit of time $\tau = \dfrac{m}{\rho V S} = \dfrac{\mu l}{V}.$

Observe that τ is the time required to travel the distance μl.

In terms of these units we can draw up a list of derived units for the quantities which occur in our equations.

Quantity	Unit	Quantity	Unit
Velocity	$\dfrac{l}{\tau} = \dfrac{V}{\mu}$	Moment	$\dfrac{ml^2}{\tau^2} = \dfrac{\rho V^2 S l}{\mu}$
Angular velocity	$\dfrac{1}{\tau} = \dfrac{V}{\mu l}$	Force velocity derivative	$\dfrac{m}{\tau} = \rho V S$
Acceleration	$\dfrac{l}{\tau^2} = \dfrac{V^2}{\mu^2 l}$	Force rotary derivative	$\dfrac{ml}{\tau} = \rho V S l$
Angular acceleration	$\dfrac{1}{\tau^2} = \dfrac{V^2}{\mu^2 l^2}$	Moment velocity derivative	$\dfrac{ml}{\tau} = \rho V S l$
Moment of inertia	ml^2	Moment rotary derivative	$\dfrac{ml^2}{\tau} = \rho V S l^2$
Force	$\dfrac{ml}{\tau^2} = \dfrac{\rho V^2 S}{\mu}$	Moment acceleration derivative	$ml = \rho S l^2 \mu$

The equations of motion, when expressed with the above system of units, are said to be in *non-dimensional form*.[*]

With this system of units we have

(1) $$V = \frac{\mu l}{\tau} = \mu,$$

since $l = 1$, $\tau = 1$. Also from 20·2

$$\frac{k}{\mu} = \frac{gl}{V^2} = \frac{g}{\mu^2},$$

from (1) so that $g = k\mu$, and in horizontal flight

(2) $$g = \tfrac{1}{2}\mu C_L.$$

20·3. Expression in non-dimensional form. Consider equation 20·12 (1).

(1) $$m\dot{u} - uX_u - wX_w + (mV \sin \alpha - X_q)\dot{\theta} + mg\theta \cos \alpha = 0.$$

As we stated in 20·21 this equation will take exactly the same form in the units m, l, τ of 20·21. The mass m is therefore to be replaced by unity and

---

[*] Glauert, *R. and M.* 1093 (1927).

from 20·21 (1) and (2) we may replace V by μ and g by $\frac{1}{2}\mu C_L$. Since in practice the derivatives are usually negative, it is convenient to write $-x_u$ for X_u, $-x_q$ for X_q and so on. Equation (1) then becomes

(2)　　　$\dot{u} + ux_u + wx_w + (\mu \sin \alpha + x_q)\dot{\theta} + \frac{1}{2}\mu C_L \theta \cos \alpha = 0,$

which is the so-called non-dimensional form of (1).

Now consider equation 20·12 (3).

(3)　　　$-uM_u - \dot{w}M_{\dot{w}} - wM_w + B\ddot{\theta} - \dot{\theta}M_q = M_0.$

With the new system of units, we first divide every term by B and then write $-m_u$ for M_u/B, $-m_{\dot{w}}$ for $M_{\dot{w}}/B$, $-m_q$ for M_q/B, m_0 for M_0/B.

The above equation then becomes

(4)　　　$um_u + \dot{w}m_{\dot{w}} + wm_w + \ddot{\theta} + \dot{\theta}m_q = m_0.$

Observe that in the non-dimensional equations $\dot{u} = du/d\tau$, etc. We can always revert to the original units by using the table of 20·21. If, for example, we find $u = 6$ from the non-dimensional equations, the corresponding component in ft./sec. is $6V/\mu$, where V ft./sec. is the forward speed.

The whole set of non-dimensional equations is as follows :

The *symmetric group*, which contains only the symmetric variables u, w, θ,

$$\dot{u} + ux_u + wx_w + (\mu \sin \alpha + x_q)\dot{\theta} + \frac{1}{2}C_L \mu\theta \cos \alpha = 0,$$
$$uz_u + \dot{w} + wz_w - (\mu \cos \alpha - z_q)\dot{\theta} + \frac{1}{2}C_L \mu\theta \sin \alpha = 0,$$
$$um_u + \dot{w}m_{\dot{w}} + wm_w + \ddot{\theta} + \dot{\theta}m_q = m_0.$$

The *asymmetric group*, which contains only the asymmetric variables β, γ, v

$$\dot{v} + vy_v - (\mu \sin \alpha - y_p)\dot{\beta} - \frac{1}{2}C_L\mu\beta \cos \alpha + (\mu \cos \alpha + y_r)\dot{\gamma} - \frac{1}{2}C_L\mu\gamma \sin \alpha = 0.$$

$$vl_v + \ddot{\beta} + l_p\dot{\beta} - \frac{E}{A}\ddot{\gamma} + l_r\dot{\gamma} = l_0,$$

$$vn_v - \frac{E}{C}\ddot{\beta} + n_p\dot{\beta} + \ddot{\gamma} + n_r\dot{\gamma} = n_0.$$

Notes. (i) The term " non-dimensional " should not mislead. Change of units cannot alter the physical character of the equations. The equations still depend on the mass m, the speed V, and the length l, but only through the dimensionless number μ.

(ii) In the above equations the only terms which depend on height explicitly are those which contain μ.

(iii) Since incidence is usually small, we may then write $\cos \alpha = 1$, $\sin \alpha = \alpha$.

(iv) All the derivatives $x_u, \ldots, n_r$ must be known before numerical calculations can be performed. Theoretical values are available for some, others must be obtained from experiment. We shall assume their values to be known.

(v) Some writers * put $m_{\dot{w}}/\mu$ and μl_0, μm_0, μn_0 for what are here written as $m_{\dot{w}}$, l_0, m_0, n_0.

(vi) When the attitude angle is Θ we replace in the above equations $\frac{1}{2}C_L \cos \alpha$ by $\frac{1}{2}C_L$ and $\frac{1}{2}C_L \sin \alpha$ by $\frac{1}{2}C_L \tan \Theta$.

(vii) In forming the coefficients l_v, l_p, l_r we divide by A instead of B. Similarly, for the coefficients n_v, n_p, n_r we divide by C.

(viii) Coefficients have not been introduced for the ratios E/A, E/C, which are anyway pure numbers.

20·4. Dynamical stability.

The problem of the stability of an aircraft is essentially that of finding how the machine will respond to a disturbance which is initially small.

(1) If a small disturbance remains small, or dies away after it is applied, the aircraft is *inherently stable*.

(2) A small disturbance may go on increasing. This indicates *inherent instability*.

(3) If the initial disturbance, while increasing, increases so slowly that the pilot has time to adopt countermeasures before a dangerous increase has arisen, inherent instability may be tolerated, especially in fighter aircraft, for a certain degree of instability is found to make the aircraft more immediately responsive to the controls. Inherent instability cannot be tolerated in an aircraft which is designed for long passages, for the continual vigilance might cause too great a strain on the pilot.

Our first problem is therefore to ascertain whether an aircraft is inherently stable.

Our second problem, in the case of an aircraft not inherently stable, is to discuss the rate of growth of an initial disturbance.

It may be observed that a statically unstable aircraft may still be dynamically stable.

The discussion of stability is simply a discussion of the solutions of the differential equations. We shall illustrate the method, which is the same for both symmetric and asymmetric groups, by considering the symmetric group, i.e. we shall discuss longitudinal or pitching stability.

20·5. Longitudinal stability.

We take the non-dimensional form of the equations of the symmetric group given in 20·3, and to investigate the inherent stability we take $m_0 = 0$. If we put $u = A_1 e^{\lambda t}$, $w = B_1 e^{\lambda t}$, $\theta = C_1 e^{\lambda t}$, where A_1, B_1, C_1 are arbitrary constants, and substitute in the equations, we get

(1)
$$(\lambda + x_u)A_1 + x_w B_1 + [(\mu \sin \alpha + x_q)\lambda + \tfrac{1}{2}C_L \mu \cos \alpha]C_1 = 0,$$
$$z_u A_1 + (\lambda + z_w)B_1 + [(-\mu \cos \alpha + z_q)\lambda + \tfrac{1}{2}C_L \mu \sin \alpha]C_1 = 0,$$
$$m_u A_1 + (\lambda m_{\dot{w}} + m_w)B_1 + (\lambda^2 + \lambda m_q)C_1 = 0.$$

---

* See B. M. Jones in *Aerodynamic Theory*, edited by W. F. Durand, Berlin (1935), Vol. V.

Eliminating A_1, B_1, C_1 we get the determinantal equation

$$(2) \quad \begin{vmatrix} \lambda + x_u & x_w & (\mu \sin \alpha + x_q)\lambda + \tfrac{1}{2}C_L \mu \cos \alpha \\ z_u & \lambda + z_w & (-\mu \cos \alpha + z_q)\lambda + \tfrac{1}{2}C_L \mu \sin \alpha \\ m_u & \lambda m_{\dot{w}} + m_w & \lambda^2 + \lambda m_q \end{vmatrix} = 0,$$

which on reduction is an equation of the fourth degree in λ, say,

$$(3) \qquad \lambda^4 + B\lambda^3 + C\lambda^2 + D\lambda + E = 0.$$

This equation has four roots, and since the coefficients are real, complex roots will occur in conjugate pairs. A real root λ will give rise to a term, in the solution, of the type $\kappa_1 e^{\lambda t}$, and a pair of complex roots $\mu \pm i\nu$ will give rise to a term of the type $\kappa_2 e^{\mu t} \cos (\nu t + \epsilon)$.

As t tends to infinity $\kappa_1 e^{\lambda t}$ tends to infinity or zero according as λ is positive or negative. In the former case the term represents a *divergence*, and in the latter a *subsidence*.

A term like $K_2 e^{\mu t} \cos (\nu t + \epsilon)$ oscillates with an amplitude which tends either to infinity, *increasing oscillation*, or to zero, *damped oscillation*, according as μ is positive or negative.

It follows therefore that for stability all the roots of (3) must, if real, be negative or, if complex, have a negative real part.

This conclusion continues to hold if two or more of the roots of (3) are coincident, in which case terms of the type $t e^{\lambda t}$, etc., may appear in the solution.

Thus it appears that there will certainly be inherent stability if the roots of (3) are such that their *real parts* are all negative.

The necessary and sufficient condition that the real parts of the roots of (3) shall all be negative is that the terms of the set

$$(5) \qquad\qquad B, C, D, E, BCD - D^2 - EB^2$$

shall all be positive, as will be proved in 20·51.

The number $R = BCD - D^2 - EB^2$ is called *Routh's discriminant*.

20·51. Routh's discriminant.

It was stated in 20·5 that the real parts of the roots of the quartic equation

$$(1) \qquad f(\lambda) = \lambda^4 + B\lambda^3 + C\lambda^2 + D\lambda + E = 0$$

are negative, if, and only if, each term of the set 20·5 (5) is positive.

*Proof.** The necessary and sufficient condition that a quadratic equation $\lambda^2 + a\lambda + b = 0$ shall have the real parts of its roots negative is $a > 0$, $b > 0$.

Any quartic such as (1) can be factorised in the form

$$(\lambda^2 + a\lambda + b)(\lambda^2 + c\lambda + d) = 0,$$

---

* I am indebted to Professor B. M. Brown for this elegant proof.

where a, b, c, d are all real. This follows from the fact that complex roots can only occur in conjugate pairs and any such pair satisfies a quadratic equation with real coefficients.

If the roots of (1) are to have negative real parts, a, b, c, d are each > 0. These conditions are necessary and sufficient.

Equating coefficients, we have

(2) $B = a + c$.

(3) $C = b + d + ac$.

(4) $D = ad + bc$.

(5) $E = bd$.

It follows that it is necessary for B, C, D, E to be all positive.

The critical case arises when, say, $a = 0$, so that (1) has two purely imaginary conjugate roots. In this case $B = c$, $C = b + d$, $D = bc$, $E = bd$. Eliminating b, c, d, we obtain the discriminant

$$R = BCD - B^2E - D^2 = 0.$$

In the general case, substituting for B, C, D, E, we obtain

$$R = (a + c)(b + d + ac)(ad + bc) - (ad + bc)^2 - bd(a + c)^2$$
$$= ac\,BD + (ab + cd)(ad + bc) - bd(a + c)^2,$$

(6) $R = ac\,[(b - d)^2 + BD]$,

so that if a, b, c, $d > 0$, then $R > 0$.

We now have as necessary conditions B, C, D, E, $R > 0$. We can show that they are sufficient, for if B, $D > 0$ we see from (6) that $ac > 0$. Since $ac > 0$, it follows from (2) that if $B > 0$, then a, $c > 0$. If D, $E > 0$, and a, $c > 0$, it follows from (4) and (5) that b, $d > 0$.   Q.E.D.

20·52. Method of Gates.

An elegant graphical method of solving the problem of dynamical stability has been given by Gates,* who introduces two non-dimensional co-ordinates

$$X = \frac{\bar{h}}{k_B}, \qquad Y = \frac{a'S'}{2k_BS}.$$

Here $\bar{h}c$ is the distance of the centre of gravity of the aircraft aft of the focus of the wings, k_Bml^2 is the moment of inertia about the lateral axis, l is the tail-lever arm, S' is the plan area of the tail-plane, S is the plan area of the wings, and a' is the slope of the (C_L, α) graph for the tail-plane. These quantities have been chosen as the most easily adjustable in the designing stage. Routh's discriminant R and the coefficient E of the quartic are then expressed in terms of X, Y, and the curves $R = 0$, $E = 0$ are plotted for different

---

* R. and M. 1118 (1928).

values of C_L and fixed μ, or for different values of μ for fixed C_L. Fig. 20·52 shows the relevant part of such curves for $C_L = 1$, $\mu = 5$, for a gliding aircraft.

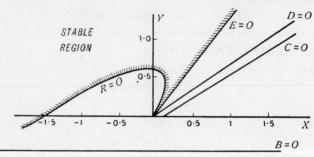

FIG. 20·52.

If the point (X, Y) is on the shaded side of these curves we have $R > 0$ and $E > 0$ so that there is stability, as is visible from the fact that B, C, D are then also positive.

20·6. Laplace transform.

The solution of the set of three simultaneous differential equations of disturbed motion with given initial values of the variables, while straightforward, can be very laborious. In particular, for the symmetric group we might be interested only in one variable, namely θ. We shall explain an operational method based on the Laplace transform which enables the solution for any particular variable to be written down in terms of initial conditions without evaluating any arbitrary constants, or separating the solution into the sum of a complementary function and a particular integral.

Definition. The Laplace transform of a given function of t, say $f(t)$, is denoted by $(f(t))^\star$ where

$$(f(t))^\star = \int_0^\infty e^{-pt} f(t)\, dt.$$

The Laplace transform is thus a function of p, not of t, and we shall mainly be concerned with inferring $f(t)$ when $(f(t))^\star$ is given. So long as l_0, m_0, n_0 are constants the only type of solution which can arise from the differential equations is of the form $t^n e^{at}$, where n may be zero, and a may be complex. We have

$$(1) \qquad (e^{at})^\star = \int_0^\infty e^{(a-p)t}\, dt = \frac{1}{p - a}.$$

Differentiate partially n times with respect to a. Then

$$(2) \qquad (t^n e^{at})^\star = \frac{n!}{(p - a)^{n+1}}.$$

This last result includes all we want. In particular, we can put $a = \lambda + i\mu$; or $n = 0$; or $a = 0$. In this way we deduce from (2) a table of Laplace transforms as follows.

$(1)^\star$	$\dfrac{1}{p}$	$(t^n e^{at})^\star$	$\dfrac{n!}{(p-a)^{n+1}}$
$(t^n)^\star$	$\dfrac{n!}{p^{n+1}}$	$(e^{\lambda t} \cos \mu t)^\star$	$\dfrac{p-\lambda}{(p-\lambda)^2 + \mu^2}$
$(e^{at})^\star$	$\dfrac{1}{p-a}$	$(e^{\lambda t} \sin \mu t)^\star$	$\dfrac{\mu}{(p-\lambda)^2 + \mu^2}$
$(\cos \mu t)^\star$	$\dfrac{p}{p^2 + \mu^2}$	$(t e^{\lambda t} \cos \mu t)^\star$	$\dfrac{(p-\lambda)^2 - \mu^2}{[(p-\lambda)^2 + \mu^2]^2}$
$(\sin \mu t)^\star$	$\dfrac{\mu}{p^2 + \mu^2}$	$(t e^{\lambda t} \sin \mu t)^\star$	$\dfrac{2\mu(p-\lambda)}{[(p-\lambda)^2 + \mu^2]^2}$

This table contains all the transforms likely to be required. The table can be used either to find the transform of a given function or to find the function whose transform is given. The reversibility of the table is here assumed, but a rigorous proof is possible.

20·61. Method of solution. Consider, for example, the equations of the symmetric group given in 20·3. We multiply each equation by e^{-pt} and integrate from 0 to ∞. Now if u_0 is the value of u when $t = 0$, and if

$$(u)^\star = \int_0^\infty e^{-pt} u \, dt, \text{ we have } \int_0^\infty \dot{u} e^{-pt} \, dt = -u_0 + p(u)^\star.$$

Again, if θ_0, $\dot\theta_0$ are the values of θ and $\dot\theta$ when $t = 0$, we get, by successive partial integrations,

$$\int_0^\infty \ddot\theta e^{-pt} \, dt = -\dot\theta_0 - p\theta_0 + p^2(\theta)^\star.$$

Thus the equations of the symmetric group yield

$$(u)^\star(p + x_u) + (w)^\star x_w + (\theta)^\star(p\mu \sin\alpha + px_q + \tfrac{1}{2}\mu\, C_L \cos\alpha)$$
$$= u_0 + (\mu \sin\alpha + x_q)\theta_0,$$

$$(u)^\star z_u + (w)^\star(p + z_w) + (\theta)^\star(-p\mu\cos\alpha + pz_q + \tfrac{1}{2}\mu\, C_L \sin\alpha)$$
$$= w_0 - (\mu\cos\alpha - z_q)\theta_0,$$

$$(u)^\star m_u + (w)^\star(pm_{\dot w} + m_w) + (\theta)^\star(p^2 + pm_q)$$
$$= m_{\dot w} w_0 + \dot\theta_0 + \theta_0(p + m_q) + m_0/p.$$

Thus, we have three linear simultaneous equations from which $(\theta)^\star$, $(u)^\star$, $(w)^\star$ may be determined, or any one of them separately. Thus using the determinantal method of solution we get

$$(\theta)^{\star}$$

$$= \frac{\begin{vmatrix} p + x_u & x_w & u_0 + (\mu \sin \alpha + x_q)\,\theta_0 \\[2mm] z_u & p + z_w & w_0 - (\mu \cos \alpha - z_q)\,\theta_0 \\[2mm] m_u & pm_{\dot{w}} + m_w & m_{\dot{w}}w_0 + \dot{\theta}_0 + (p + m_q)\,\theta_0 + \dfrac{m_0}{p} \end{vmatrix}}{\begin{vmatrix} p + x_u & x_w & p(\mu \sin \alpha + x_q) + \tfrac{1}{2}\mu C_L \cos \alpha \\[2mm] z_u & p + z_w & p(-\mu \cos \alpha + z_q) + \tfrac{1}{2}\mu C_L \sin \alpha \\[2mm] m_u & pm_{\dot{w}} + m_w & p^2 + pm_q \end{vmatrix}}$$

This gives $(\theta)^{\star}$ as the ratio of two polynomials in p and we proceed to express $(\theta)^{\star}$ as the sum of partial fractions. We then infer θ as a function of t from the table of Laplace transforms given in 20·6.

The values of u_0 and w_0 are not without interest for they represent the velocities imparted to the aircraft when it encounters a horizontal or vertical gust of wind.

20·62. Sudden application of elevators. The aircraft is flying steadily on a straight horizontal course when a small movement of the elevators is made, causing a small couple whose non-dimensional measure is m_0. In this case we have $u_0 = w_0 = \theta_0 = \dot{\theta}_0 = 0$, and we get from 18·61 a result of the form

$$(\theta)^{\star} = \frac{m_0(p^2 + px_u + pz_w + x_u z_w - x_w z_u)}{p(p^4 + Bp^3 + Cp^2 + Dp + E)},$$

where the polynomial in the denominator is exactly the same as that in 20·5.

To discuss a simple numerical case, let us suppose that the values of the derivative coefficients are such that the above expression reduces to

$$(\theta)^{\star} = \frac{m_0(p^2 + 6\cdot24p + 2\cdot04)}{p(p^4 + 12p^3 + 52p^2 + 92p + 51)}.$$

Expressing this in partial fractions by any of the usual algebraic methods, we get

$$\frac{1}{m_0}(\theta)^{\star} = \frac{\cdot04}{p} + \frac{\cdot16}{p+1} - \frac{\cdot64}{p+3} + \frac{\cdot44(p+4)}{(p+4)^2 + 1}.$$

Then from the table of transforms in 18·6 we get

$$\theta = m_0\,[\cdot04 + \cdot16e^{-t} - \cdot64e^{-3t} + \cdot44e^{-4t}\cos t],$$

where, of course, t is " non-dimensional " time τ. Thus it appears, that in this case, θ settles down, as t increases, to the steady value $\cdot04m_0$, and there is therefore stability for this manœuvre.

The above numbers were chosen to afford simple algebra and arithmetic. In the practical application it will be necessary to solve the quartic equation which is obtained by equating the denominator to zero. Any real roots can be approximated by graphical methods or by Horner's method. To obtain complex roots the somewhat laborious but practicable method of Ferrari, which is explained in books on Algebra, can be used (see Ex. XX, 6).

20·63. Approximate solution of the quartic. In discussing *longitudinal stability* the quartic in λ 20·51 (1) is usually such that D and E are small compared with B and C. This circumstance leads to a method whereby the quartic may be resolved into two quadratics by a method of successive approximations. To see this, suppose the quartic to be replaceable by

$$(1) \qquad (\lambda^2 + b\lambda + c)(\lambda^2 + \gamma\lambda + \delta) = 0,$$

where b and c are large compared with γ and δ. Comparing the above product with 20·51 (1) we get at once

$$(2) \qquad b + \gamma = B, \quad c + b\gamma + \delta = C, \quad b\delta + c\gamma = D, \quad c\delta = E.$$

On the assumption that γ, δ are small we get the first approximation,

$$(3) \qquad b_1 = B, \quad c_1 = C, \quad \delta_1 = \frac{E}{C}, \quad \gamma_1 = \frac{D}{C} - \frac{BE}{C^2}.$$

Inserting these values in (2) we get the second approximation,

$$(4) \quad b_2 = B - \gamma_1, \quad c_2 = C - b_1\gamma_1 - \delta_1, \quad \delta_2 = \frac{E}{c_2}, \quad \gamma_2 = \frac{D}{c_2} - \frac{b_2}{c_2}\delta_2,$$

and the process may be continued as often as is needed to attain any required accuracy.

In practice the first approximation is usually sufficient so that the quartic is replaced by the two quadratics

$$(5) \qquad \lambda^2 + B\lambda + C = 0, \quad \lambda^2 + \frac{CD - BE}{C^2}\lambda + \frac{E}{C} = 0.$$

Thus in the typical case cited by Gates (see p. 399, footnote), the quartic,

$$(6) \qquad \lambda^4 + 5\lambda^3 + 7\lambda^2 + \tfrac{1}{2}\lambda + \tfrac{1}{5} = 0,$$

is replaced by the two quadratics got from (5), which on multiplication will be found to be equivalent to considering the quartic

$$\lambda^4 + 5{\cdot}05\lambda^3 + 7{\cdot}28\lambda^2 + \tfrac{1}{2}\lambda + \tfrac{1}{5} = 0$$

instead of (6).

With regard to (5) the first quadratic corresponds with a short period oscillation which is usually heavily damped and therefore stable. The second quadratic gives the long period or phugoid oscillation (18·51) and, if we replace

this quadratic by the approximation $\lambda^2 + E/C = 0$, it appears that the period *
of the phugoid oscillation is $2\pi\sqrt{(C/E)}$ nearly, and this will be a damped oscillation if $CD - BE$ is positive, a criterion which is tantamount to neglecting
the term D^2 in Routh's discriminant.

In the case of the quartic which arises from consideration of *lateral stability*
the above circumstance of relative magnitude of the coefficients is absent.
However, in this case it will be found that the quartic has a real root nearly
equal to $- l_p$, and such a real root can always be approximated by graphical
or other methods.

Moreover, the existence of one real root implies that there is a second real
root, for imaginary roots occur in conjugate pairs, which can also be found.
Removal of these real roots by long division will then reduce the quartic to a
quadratic.

EXAMPLES XX

1. If $m = 12,000$ lb., $S = 500$ ft.^{2}, $l = 8$ ft., find the value of μ at sea-level,
and draw a graph to show how μ varies up to 20,000 ft.

Express numerically the units of 20·21 at sea-level in terms of the above data.

2. Show that the unit of time τ is given approximately by

$$\tau = 0·45\sqrt{\frac{wC_L}{\sigma}} \text{ sec.,}$$

where w is the wing loading, and σ is the relative air density ρ/ρ_0, where ρ_0 refers to
sea-level.

3. Derive in detail the whole set of 6 non-dimensional equations of disturbed
straight flight.

Write out these equations in full when the aircraft is climbing on a path at
inclination Θ to the horizontal.

4. Find the non-dimensional equations of a steady glide at gliding angle γ when
slightly disturbed.

5. Calculate Routh's discriminant for the equation
$$100\lambda^4 + 505\lambda^3 + 728\lambda^2 + 50\lambda + 20 = 0.$$

6. Prove the identity
$$\lambda^4 + B\lambda^3 + C\lambda^2 + D\lambda + E$$
$$= (\lambda^2 + \tfrac{1}{2}B\lambda + \theta)^2 - \{(\tfrac{1}{4}B^2 + 2\theta - C)\lambda^2 + (B\theta - D)\lambda + \theta^2 - E\}$$
and show that the term in curled brackets is a perfect square if θ is a root of the
cubic equation
$$(B\theta - D)^2 - (B^2 + 8\theta - 4C)(\theta^2 - E) = 0.$$
[Ferrari's method consists of determining θ and thus factorising the quartic as
the difference of two squares.]

7. Show that the quartic
$$\lambda^4 + 14·67\lambda^3 + 62·16\lambda^2 + 8·65\lambda + 2·15 = 0,$$
has the roots $\quad\quad -7·27 \pm 2·71i, \; -0·07 \pm 0·17i.$

---

* Remember that the unit of time is τ (20·21).

8. In the preceding example show that the periods of oscillations are 37·0 and 2·32 units.

Convert these to seconds if $m = 2000$ lb., $V = 150$ ft./sec., $S = 400$ ft.², taking standard air at 5,000 ft. altitude.

Find the time taken by these oscillations to reduce the amplitude to half its initial value.

9. Solve the quartic 20·63 (6) by using the first approximation. Proceed to the second approximation and hence estimate the percentage error in the modulus of the roots given by the first approximation.

10. Assuming that $z_u = C_L$ and $m_{\dot{w}} = 0$ and neglecting x_u, x_w, prove that the period of the phugoid oscillation is nearly

$$\frac{\pi V \sqrt{2}}{g} \sqrt{\frac{\mu m_w + z_w m_q}{\mu m_w - \mu m_u z_w / C_L}}.$$

What assumption is necessary to deduce the result of 18·51?

11. Discuss the longitudinal stability associated with the equation

$$\lambda^4 + 4·53\lambda^3 + 5·09\lambda^2 + 0·407\lambda + 0·122 = 0,$$

and determine the periods of the phugoid oscillation.

12. Use a typical diagram of Gates's method to show that longitudinal stability is largely determined by $E > 0$.

Write out in full the condition $E > 0$.

13. Find the quartic for λ arising from the consideration of lateral stability.

Show that in this equation the relative aircraft density appears only as a factor of l_v and n_v. Deduce that an increase of load or an increase of height are equivalent to an increase in these coefficients.

14. Show that the lateral stability quartic

$$\lambda^4 + 12·6\lambda^3 + 11·5\lambda^2 + 10·7\lambda + 0·76 = 0$$

implies inherent stability.

Solve the equation, showing that two roots are real, and find the period of the motion arising from the complex roots.

CHAPTER XXI

VECTORS

21·1. Scalars and vectors. Pure numbers and physical quantities which do not require direction in space for their complete specification are called *scalar quantities*, or simply *scalars*. Volume, density, mass and energy are familiar examples. Air pressure is also a scalar. The thrust on an infinitesimal plane area due to air pressure is, however, not a scalar, for to describe the thrust completely, the direction in which it acts must also be known.

A *vector quantity*, or simply a *vector*, is a quantity which needs for its complete specification both magnitude and direction, and which obeys the parallelogram law of composition (addition), and certain laws of multiplication which will be formulated later. Examples of vectors are readily furnished by velocity, linear momentum and force. Angular velocity and angular momentum are also vectors, as is proved in books on Mechanics.

A vector can be represented completely by a straight line drawn in the direction of the vector and of appropriate magnitude to some chosen scale. The sense of the vector in this straight line can be indicated by an arrow.

In some cases a vector must be considered as *localised* in a line. For instance, in finding the position of the centre of pressure of an aerofoil the actual line of action of the aerodynamic force is required.

We represent a vector by a single letter in clarendon (heavy) type, and its magnitude by the corresponding letter in italic type. Thus if **q** is the velocity vector, its magnitude is q, the airspeed. Similarly, angular velocity (or surface vorticity) **ω** has the magnitude ω. In manuscript work, a vector can be indicated by underlining it with a wavy line. The vector whose end points are P and Q can be conveniently indicated by using an arrow, thus :

$$\vec{PQ}, \quad \vec{QP},$$

according as the sense is from P to Q or from Q to P. Such vectors have opposite signs

$$\vec{PQ} = -\vec{QP}.$$

A *unit vector* is a vector whose magnitude is unity. Any vector can be represented by a numerical (scalar) multiple of a unit vector parallel to it. Thus if $\mathbf{a}_1$ is a unit vector parallel to the vector **a**, we have $\mathbf{a} = a\mathbf{a}_1$.

If with each point of space there corresponds a scalar we have a *scalar field*. or example, the air pressures in the air round an aerofoil constitute a scalar field; so do the values of the velocity potential in the same region. Similarly, if with each point there corresponds a vector, that is to say a scalar and a direction, so that a vector is, as it were, tied to each point of space, then a *vector field* is defined. Air velocity in the air round an aerofoil is an obvious example.

21·11. Resolution of vectors.

Consider three mutually perpendicular axes of reference (cartesian axes) OX, OY, OZ. If P is any point, we call $\overrightarrow{OP}$ the *position vector* of P (with respect to O). Complete the rectangular parallelepiped shown in fig. 21·11. Then with the notation of that figure

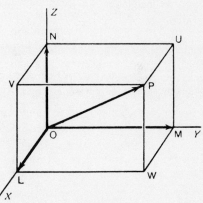

$$\overrightarrow{OM} + \overrightarrow{OL} = \overrightarrow{OL} + \overrightarrow{OM} = \overrightarrow{OW},$$

$$\overrightarrow{ON} + \overrightarrow{OW} = \overrightarrow{OP},$$

$$\overrightarrow{OP} = \overrightarrow{OL} + \overrightarrow{OM} + \overrightarrow{ON},$$

where we have applied the parallelogram law of addition, and have assumed that the order in which the additions are performed is irrelevant. We have thus re-

Fig. 21·11.

solved the vector $\overrightarrow{OP}$ into the sum of three vectors along the axes. Now let **i** be a unit vector along OX, **j** a unit vector along OY, and **k** a unit vector along OZ. Then $\overrightarrow{OL} = OL \cdot \mathbf{i}$, $\overrightarrow{OM} = OM \cdot \mathbf{j}$, $\overrightarrow{ON} = ON \cdot \mathbf{k}$, so that

$$\overrightarrow{OP} = OL \cdot \mathbf{i} + OM \cdot \mathbf{j} + ON \cdot \mathbf{k}.$$

Here OL, OM, ON are called the resolved parts, or components, of the vector $\overrightarrow{OP}$ along the axes. They are in fact the coordinates (x, y, z) of P, so that if we write $\mathbf{r} = \overrightarrow{OP}$, we can express the state of affairs in the form

(1) $\mathbf{r} = \mathbf{i}x + \mathbf{j}y + \mathbf{k}z.$

More generally let $\overrightarrow{OP}$ represent some other vector such as **q**, the fluid velocity. Then, if (u, v, w) are the components of **q**, we have

(2) $\mathbf{q} = \mathbf{i}u + \mathbf{j}v + \mathbf{k}w.$

In a similar way any vector can be expressed in terms of its cartesian components and the cartesian unit vectors **i**, **j**, **k**. We shall use these unit vectors conventionally without further reference to their definition.

It is often convenient to denote the components of any vector, say, **a** by a_x, a_y, a_z so that

$$\mathbf{a} = \mathbf{i}a_x + \mathbf{j}a_y + \mathbf{k}a_z.$$

Now let Q be the point $(x + dx, y + dy, z + dz)$. Then we can write

$$\overrightarrow{OQ} = \mathbf{r} + d\mathbf{r} = \mathbf{i}(x + dx) + \mathbf{j}(y + dy) + \mathbf{k}(z + dz),$$

so that

(3) $$d\mathbf{r} = \mathbf{i}\,dx + \mathbf{j}\,dy + \mathbf{k}\,dz.$$

If we imagine an air particle to move from P to Q in the time dt we can write for its velocity

(4) $$\mathbf{q} = \frac{d\mathbf{r}}{dt} = \mathbf{i}\frac{dx}{dt} + \mathbf{j}\frac{dy}{dt} + \mathbf{k}\frac{dz}{dt}.$$

21·12. The scalar product of two vectors.

Let **a**, **b**, be two vectors of magnitudes a, b, represented by the lines OA, OB issuing from the point O.

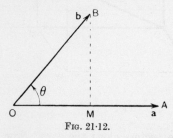

FIG. 21·12.

Let θ be the angle between the vectors, i.e. the angle AOB measured positively in the sense of rotation from **a** to **b**.

The scalar product of the vectors **a**, **b** is denoted by **ab** and defined by the equation

(1) $$\mathbf{ab} = ab\cos\theta.$$

The scalar product of two vectors is thus a scalar and is measured by the product $OA \,.\, OM$, where M is the projection of B on OA.

It is clear from the definition that

$$\mathbf{ba} = ba\cos(-\theta) = ab\cos\theta = \mathbf{ab},$$

so that the order of the two factors is irrelevant.

Observe that if θ is obtuse the scalar product is negative.

If θ is a right angle, that is, *if the vectors are perpendicular, their scalar product is zero*, for $\cos\theta = 0$.

As an example, suppose $\boldsymbol{\omega}_S{}^P$ is the surface vorticity at a point P of a vortex sheet S, and **n** is the unit normal vector to the sheet at P. Then $\boldsymbol{\omega}_S{}^P$ and **n** are perpendicular, therefore $\boldsymbol{\omega}_S{}^P\mathbf{n} = 0$.

Again, if $\mathbf{ab} = 0$, we have $ab\cos\theta = 0$ so that either $\cos\theta = 0$ or $ab = 0$. The second alternative means that one at least of the vectors is of zero magnitude. If neither vector is zero, $\mathbf{ab} = 0$ implies that **a** and **b** are perpendicular.

As particular applications of the definition (1) we have, for the scalar products of the unit vectors **i**, **j**, **k**, the following important consequences:

(2) $$\mathbf{ii} = \mathbf{i}^2 = 1 = \mathbf{j}^2 = \mathbf{k}^2,$$
$$\mathbf{ij} = \mathbf{jk} = \mathbf{ki} = 0.$$

Thus if $\mathbf{a} = \mathbf{i}a_x + \mathbf{j}a_y + \mathbf{k}a_z$, $\mathbf{b} = \mathbf{i}b_x + \mathbf{j}b_y + \mathbf{k}b_z$, we have *

(3) $$\mathbf{ab} = a_x b_x + a_y b_y + a_z b_z,$$

on developing the product and using (2).

Taking the velocity vector $\mathbf{q} = \mathbf{i}u + \mathbf{j}v + \mathbf{k}w$ and the infinitesimal position vector $d\mathbf{r} = \mathbf{i}\,dx + \mathbf{j}\,dy + \mathbf{k}\,dz$ of 19·11, we have from (3)

$$\mathbf{q}\,d\mathbf{r} = u\,dx + v\,dy + w\,dz,$$

and if we integrate by moving $\mathbf{r}$ round a closed curve C we have (cf. 3·2, 9·21)

$$\int_{(C)} \mathbf{q}\,d\mathbf{r} = \int_{(C)} (u\,dx + v\,dy + w\,dz) = \text{circ } C.$$

21·13. The vector product of two vectors.

Let $\mathbf{a}, \mathbf{b}$ be two vectors of magnitudes a, b inclined (as in 21·12) at the angle θ measured positively from $\mathbf{a}$ to $\mathbf{b}$. We define the *vector product* $\mathbf{a}_{\wedge}\mathbf{b}$ as the vector whose *magnitude* is $ab \sin \theta$, and whose *direction* is perpendicular both to $\mathbf{a}$ and $\mathbf{b}$, and whose *sense* is such that rotation from $\mathbf{a}$ to $\mathbf{b}$ is related to the sense of $\mathbf{a}_{\wedge}\mathbf{b}$ by the right-handed screw rule, fig. 21·13 (i).

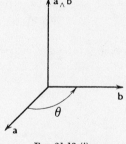

FIG. 21·13 (i).

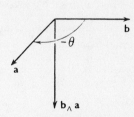

FIG. 21·13 (ii).

It follows from the definition that this vector multiplication is not commutative, for since $ba \sin(-\theta) = -ab \sin \theta$, we have

(1) $$\mathbf{b}_{\wedge}\mathbf{a} = -(\mathbf{a}_{\wedge}\mathbf{b}) = -\mathbf{a}_{\wedge}\mathbf{b},$$

see fig. 21·13 (ii).

When two vectors are parallel their vector product is zero, for then $\theta = 0$ or π.

Conversely, $\mathbf{a}_{\wedge}\mathbf{b} = 0$ implies that either $a = 0$ or $b = 0$ or that $\mathbf{a}$ is parallel to $\mathbf{b}$. Note that two vectors are parallel, irrespectively of sense, if their representative lines are parallel. Thus the opposite vectors $\mathbf{V}$ and $-\mathbf{V}$ are parallel.

---

* We shall assume that scalar (and also vector) multiplication is distributive, i.e.
$\mathbf{a}(\mathbf{b} + \mathbf{c}) = \mathbf{ab} + \mathbf{ac}$. See Ex. XXI, 26, 27.

We can now form the vector products of the unit vectors $\mathbf{i}$, $\mathbf{j}$, $\mathbf{k}$, in fact

$$\mathbf{i}_\wedge \mathbf{i} = \mathbf{j}_\wedge \mathbf{j} = \mathbf{k}_\wedge \mathbf{k} = 0.$$

(2) $\mathbf{i}_\wedge \mathbf{j} = \mathbf{k} = -\mathbf{j}_\wedge \mathbf{i}, \quad \mathbf{j}_\wedge \mathbf{k} = \mathbf{i} = -\mathbf{k}_\wedge \mathbf{j}, \quad \mathbf{k}_\wedge \mathbf{i} = \mathbf{j} = -\mathbf{i}_\wedge \mathbf{k}.$

Thus for the vectors $\mathbf{a}$, $\mathbf{b}$ of 21·12 we have *

$$\mathbf{a}_\wedge \mathbf{b} = (\mathbf{i}a_x + \mathbf{j}a_y + \mathbf{k}a_z)_\wedge (\mathbf{i}b_x + \mathbf{j}b_y + \mathbf{k}b_z)$$

$$= a_x \mathbf{i}_\wedge (\mathbf{i}b_x + \mathbf{j}b_y + \mathbf{k}b_z) + a_y \mathbf{j}_\wedge (\mathbf{i}b_x + \mathbf{j}b_y + \mathbf{k}b_z) + a_z \mathbf{k}_\wedge (\mathbf{i}b_x + \mathbf{j}b_y + \mathbf{k}b_z),$$

(3) $= \mathbf{i}(a_y b_z - a_z b_y) + \mathbf{j}(a_z b_x - a_x b_z) + \mathbf{k}(a_x b_y - a_y b_x),$

from (2). The determinantal form of this result is convenient, namely,

(4)
$$\mathbf{a}_\wedge \mathbf{b} = \begin{vmatrix} \mathbf{i} & \mathbf{j} & \mathbf{k} \\ a_x & a_y & a_z \\ b_x & b_y & b_z \end{vmatrix}.$$

The last two rows are in the order of the terms of the product. For $\mathbf{b}_\wedge \mathbf{a}$ these rows would therefore be interchanged, which changes the sign but not the absolute magnitude of the product (cf. (1) above).

Consider the vectors $\boldsymbol{\chi}$ of 20·1 (2) and $\mathbf{g}$ of 20·12. For these

$$\boldsymbol{\chi}_\wedge \mathbf{g} = \begin{vmatrix} \mathbf{i} & \mathbf{j} & \mathbf{k} \\ \beta & \theta & \gamma \\ -g \sin \alpha & 0 & g \cos \alpha \end{vmatrix} = \mathbf{i}\theta g \cos \alpha + \mathbf{j}g(-\gamma \sin \alpha - \beta \cos \alpha) + \mathbf{k}\theta g \sin \alpha.$$

The vector product has a direct application to angular velocity, small rotations, and moments.

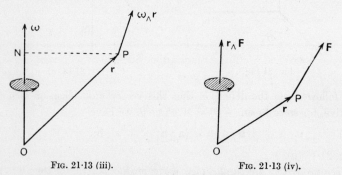

FIG. 21·13 (iii).   FIG. 21·13 (iv).

Thus, if in fig. (iii) P is a point of a rigid body which is moving about the fixed point O with angular velocity $\boldsymbol{\omega}$, and if PN is perpendicular to the axis of rotation, the velocity of P is, into the paper and of magnitude, $\omega \cdot PN = \omega \cdot OP \sin \theta$. Thus the velocity of P is $\boldsymbol{\omega}_\wedge \mathbf{r}$, where $\mathbf{r}$ is the position vector of P as shown.

---

* See footnote on p. 342.

Similarly, in time dt the body undergoes a small rotation $\boldsymbol{\omega}\, dt$, and P therefore undergoes a displacement of translation $(\boldsymbol{\omega}\, dt)_\wedge \mathbf{r}$.

Again, the vector moment with respect to O of a force $\mathbf{F}$ acting at P, fig. (iv), is $\mathbf{r}_\wedge \mathbf{F}$. The torque about any line through O is the component of this vector along that line.

Referring to fig. (i), we note that $ab \sin \theta$ measures the area of the parallelogram of which $\mathbf{a}$, $\mathbf{b}$ are two adjacent sides. The vector product $\mathbf{a}_\wedge \mathbf{b}$ can therefore be regarded as a directed measure of area. It is a vector whose magnitude measures the area and whose direction is normal to the area.

21·14. Equations of motion of an aircraft.

We now derive the equations 20·01 (5), (6). Referring to 20·01, the linear momentum of the aircraft is $m\mathbf{v}$, the moment of momentum referred to axes fixed in the aircraft at G, the centre of gravity, is $\mathbf{h}$ and the machine is rotating with angular velocity $\boldsymbol{\Omega}$. We consider the changes in $m\mathbf{v}$ and $\mathbf{h}$ during a small time dt. In this time G undergoes a translation $\mathbf{v}\, dt$ and the frame of reference a rotation $\boldsymbol{\Omega}\, dt$. We consider separately the effects of the translation and rotation, which, being infinitesimal, are additive. Lastly, we consider the rates of change when the axes remain at rest.

Fig. 21·14 (a).

To consider the effect of the translation we ignore the rotation. Since $m\mathbf{v}$ is merely moved parallel to itself it undergoes no change.

The moment of momentum $\mathbf{h}$ at time t is $\mathbf{h}$ with respect to G. At time $t + dt$ when G has moved to G' the momentum of momentum, still with respect to G, is increased by the moment about G of the linear momentum $m\mathbf{v}$ at G',

i.e. by $\overrightarrow{GG'}_\wedge m\mathbf{v} = \mathbf{v}\, dt_\wedge m\mathbf{v} = 0$, since $\mathbf{v}$ and $\mathbf{v}$ are parallel. Thus no change arises from the translation.

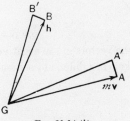

Fig. 21·14 (b).

To find the effect of the rotation, we ignore the translation, G remains fixed and the frame undergoes a rotation $\boldsymbol{\Omega}\, dt$.

To an observer moving with the frame, $m\mathbf{v}$ and $\mathbf{h}$ appear to be unchanged.

Let GA, GB represent $m\mathbf{v}$, $\mathbf{h}$ at time t. At time $t + dt$ they will be represented by GA', GB' where

$$\overrightarrow{AA'} = \boldsymbol{\Omega}\, dt_\wedge \overrightarrow{GA}, \quad \overrightarrow{BB'} = \boldsymbol{\Omega}\, dt_\wedge \overrightarrow{GB}$$

and therefore with regard to the original position of the frame the rates of change, got by dividing by dt are $\boldsymbol{\Omega}_\wedge m\mathbf{v}$, $\boldsymbol{\Omega}_\wedge \mathbf{h}$ respectively.

Lastly, the axes remaining at rest, $m\mathbf{v}$ and $\mathbf{h}$ increase at the rates $m\dot{\mathbf{v}}$, $\dot{\mathbf{h}}$, where we use a dot to indicate time rate of change with respect to axes fixed in the aircraft.

Thus the total rates of change are respectively

$$m\dot{\mathbf{v}} + \Omega_\wedge m\mathbf{v}, \quad \dot{\mathbf{h}} + \Omega_\wedge \mathbf{h},$$

whence equations 20·01 (5) and (6) follow from Newton's laws of motion.

21·2. Triple products. Let $\mathbf{a}, \mathbf{b}, \mathbf{c}$ be three vectors. Then $\mathbf{a}_\wedge \mathbf{b}$ is a vector and can be combined with $\mathbf{c}$ by scalar multiplication to form the *triple scalar product* $(\mathbf{a}_\wedge \mathbf{b})\mathbf{c}$.

The triple scalar product is a scalar which measures the volume of the parallelepiped with concurrent edges $\mathbf{a}, \mathbf{b}, \mathbf{c}$, so that

$$(\mathbf{a}_\wedge \mathbf{b})\mathbf{c} = \mathbf{b}(\mathbf{c}_\wedge \mathbf{a}) = \mathbf{c}(\mathbf{a}_\wedge \mathbf{b}),$$

preserving the cyclic order $\mathbf{a}, \mathbf{b}, \mathbf{c}$. On the other hand,

$$(\mathbf{a}_\wedge \mathbf{b})\mathbf{c} = -(\mathbf{b}_\wedge \mathbf{a})\mathbf{c}.$$

The proof is left to the reader. Note also that $(\mathbf{a}_\wedge \mathbf{b})\mathbf{c} = \mathbf{a}(\mathbf{b}_\wedge \mathbf{c})$.
The *triple vector product* is the vector product of $(\mathbf{a}_\wedge \mathbf{b})$ and $\mathbf{c}$, i.e.,

$$(\mathbf{a}_\wedge \mathbf{b})_\wedge \mathbf{c}.$$

Similarly, with those vectors we can form other triple vector products such as $\mathbf{a}_\wedge (\mathbf{b}_\wedge \mathbf{c})$, $(\mathbf{c}_\wedge \mathbf{a})_\wedge \mathbf{b}$.

Theorem. $\qquad \mathbf{a}_\wedge (\mathbf{b}_\wedge \mathbf{c}) = -(\mathbf{ab})\mathbf{c} + (\mathbf{ca})\mathbf{b}.$

Proof. Using 21·13 (3) we have

$$\mathbf{a}_\wedge (\mathbf{b}_\wedge \mathbf{c}) = (\mathbf{i}a_x + \mathbf{j}a_y + \mathbf{k}a_z)_\wedge \{\mathbf{i}(b_y c_z - b_z c_y) + \mathbf{j}(b_z c_x - b_x c_z) + \mathbf{k}(b_x c_y - b_y c_x)\}$$

$$= \mathbf{i}P_x + \mathbf{j}P_y + \mathbf{k}P_z, \text{ where}$$

$$P_x = a_y(b_x c_y - b_y c_x) - a_z(b_z c_x - b_x c_z)$$

$$= -c_x(a_x b_x + a_y b_y + a_z b_z) + b_x(a_x c_x + a_y c_y + a_z c_z)$$

$$= -(\mathbf{ab})c_x + (\mathbf{ca})b_x,$$

and two similar results, whence the theorem follows. $\qquad$ Q.E.D.

As a useful mnemonic, observe that the order of the vectors in the negative term is the same as the order on the left with the brackets in a different position. Thus an alternative result is

$$(\mathbf{a}_\wedge \mathbf{b})_\wedge \mathbf{c} = -\mathbf{a}(\mathbf{bc}) + \mathbf{b}(\mathbf{ac}),$$

but note that $(\mathbf{a}_\wedge \mathbf{b})_\wedge \mathbf{c}$ and $\mathbf{a}_\wedge (\mathbf{b}_\wedge \mathbf{c})$ are *different* vectors.

Similarly, $(\mathbf{b}_\wedge \mathbf{c})_\wedge \mathbf{a} = -\mathbf{b}(\mathbf{ca}) + \mathbf{c}(\mathbf{ab})$, $(\mathbf{c}_\wedge \mathbf{a})_\wedge \mathbf{b} = -\mathbf{c}(\mathbf{ab}) + \mathbf{a}(\mathbf{bc})$.
Thus by addition (cf. 10·61),

$$(\mathbf{a}_\wedge \mathbf{b})_\wedge \mathbf{c} + (\mathbf{b}_\wedge \mathbf{c})_\wedge \mathbf{a} + (\mathbf{c}_\wedge \mathbf{a})_\wedge \mathbf{b} = 0.$$

Again, $\qquad \mathbf{q}_\wedge (\mathbf{n}_\wedge \mathbf{q}) = -(\mathbf{qn})\mathbf{q} + (\mathbf{qq})\mathbf{n} = \mathbf{n}q^2$

as in 10·6, $\mathbf{n}$ and $\mathbf{q}$ are perpendicular.

21·3. Vector differentiation ; the operator nabla.

In 1·41 we proved that the thrust due to air pressure on an infinitesimal volume of air $d\tau$ is $-\,d\tau\,\partial p/\partial s$ in the direction of a line element ds (fig. 1·41). It follows that the thrusts in the directions of the axes are

$$- d\tau \frac{\partial p}{\partial x}, \quad - d\tau \frac{\partial p}{\partial y}, \quad - d\tau \frac{\partial p}{\partial z},$$

and therefore the total thrust is a vector $d\mathbf{T}$ with the above components, i.e.,

$$d\mathbf{T} = -\,d\tau \left(\mathbf{i}\,\frac{\partial p}{\partial x} + \mathbf{j}\,\frac{\partial p}{\partial y} + \mathbf{k}\,\frac{\partial p}{\partial z} \right) = -\,d\tau \left(\mathbf{i}\,\frac{\partial}{\partial x} + \mathbf{j}\,\frac{\partial}{\partial y} + \mathbf{k}\,\frac{\partial}{\partial z} \right) p.$$

We now define the operator *nabla*,* written ∇, by

$$(1) \qquad\qquad \nabla = \mathbf{i}\,\frac{\partial}{\partial x} + \mathbf{j}\,\frac{\partial}{\partial y} + \mathbf{k}\,\frac{\partial}{\partial z}.$$

We see that ∇ is a vector differentiation operator analogous to a scalar differentiation operator such as $\partial/\partial x$. Thus $d\mathbf{T} = -\,d\tau\,\nabla\,p$.

There is another way of regarding the above result. Consider the infinitesimal volume of air in fig. 21·3. We can suppose its surface to be divided into elementary patches of area such as dS. Let $\mathbf{n}$ be the *outward* unit normal vector to dS. Then the thrust due to pressure on dS is the vector $-\,p\mathbf{n}\,dS$, minus since the thrust is in the opposite sense to $\mathbf{n}$.

Fig. 21·3.

The total thrust on the whole volume is the vector sum of the thrusts on the elements of area such as dS and therefore, approximately,

$$d\mathbf{T} = \Sigma(\,- p\mathbf{n}\,dS) = -\int_{(S)} p\mathbf{n}\,dS,$$

where we have written an integral over the whole surface S of the volume instead of the sum. The equation thus written is, of course, approximate in the usual sense of the infinitesimal calculus, in that it is exact only to the lowest order of infinitesimals retained. Thus we have, to this order,

$$(2) \qquad\qquad - d\tau\,\nabla\,p = -\int_{(S)} \mathbf{n}p\,dS, \text{ approximately.}$$

From this result we obtain an alternative definition,

$$(3) \qquad\qquad \nabla\,p = \lim_{V \to 0} \frac{1}{V} \int_{(S)} \mathbf{n}p\,dS,$$

where V has been written instead of $d\tau$ and the integral is taken over the surface of the volume V. By $V \to 0$ we mean that the longest dimension of

---

* So named by Sir William Rowan Hamilton, the discoverer of quaternions, from a fancied resemblance to a harp.

the volume V tends to zero.   For example the volume V can always be enclosed in some sphere of diameter l and we then let $l \to 0$.

On this basis we can found a more general definition, namely, that if X is any quantity, scalar or vector,

$$(4) \qquad \nabla X = \lim_{V \to 0} \frac{1}{V} \int_{(S)} \mathbf{n} X \, dS.$$

To show that this is consistent with (1), let l, m, n be the direction cosines of $\mathbf{n}$ so that $\mathbf{n} = \mathbf{i}l + \mathbf{j}m + \mathbf{k}n$.   Then (4) becomes

$$\nabla X = \lim_{V \to 0} \frac{1}{V} \int_{(S)} (\mathbf{i}lX + \mathbf{j}mX + \mathbf{k}nX) \, dS.$$

Suppose the volume V to be split up into slender cylinders whose generators are parallel to the x-axis, and consider the term $\int l\mathbf{i}X \, dS$ for one cylinder of length δx and cross-section ω.   Then for this cylinder $V = \omega \, \delta x$, and $l \, dS = \pm \omega$ on the ends, so that

$$\lim_{V \to 0} \frac{1}{V} \int l\mathbf{i} \times dS = \lim \frac{1}{\omega \, \delta x} \omega \left[\mathbf{i} \left(X + \delta X \right) - \mathbf{i} X \right] = \mathbf{i} \frac{\partial X}{\partial x}.$$

Treating the terms $\mathbf{j}mX$, $\mathbf{k}nX$ similarly, we get

$$(5) \qquad \nabla X = \left(\mathbf{i} \frac{\partial}{\partial x} + \mathbf{j} \frac{\partial}{\partial y} + \mathbf{k} \frac{\partial}{\partial z} \right) X.$$

As particular applications of (5) we can form several combinations.   If ϕ is the velocity potential, we have

$$(6) \qquad \mathbf{q} = - \nabla \phi = - \operatorname{grad} \phi = - \mathbf{i} \frac{\partial \phi}{\partial x} - \mathbf{j} \frac{\partial \phi}{\partial y} - \mathbf{k} \frac{\partial \phi}{\partial z}.$$

Again, for the velocity $\mathbf{q} = \mathbf{i}u + \mathbf{j}v + \mathbf{k}w$, we get

$$(7) \qquad \nabla \mathbf{q} = \operatorname{div} \mathbf{q} = \frac{\partial u}{\partial x} + \frac{\partial v}{\partial y} + \frac{\partial w}{\partial z}.$$

If, further, the motion is irrotational so that $\mathbf{q}$ is given by (6), we have

$$(8) \qquad \nabla \mathbf{q} = - \frac{\partial^2 \phi}{\partial x^2} - \frac{\partial^2 \phi}{\partial y^2} - \frac{\partial^2 \phi}{\partial z^2} = - \nabla^2 \phi,$$

where

$$\nabla^2 = \frac{\partial^2}{\partial x^2} + \frac{\partial^2}{\partial y^2} + \frac{\partial^2}{\partial z^2}$$

is Laplace's operator.

Again, we have the vorticity vector

$$(9) \qquad \boldsymbol{\zeta} = \nabla_{\wedge} \mathbf{q} = \operatorname{curl} \mathbf{q} = \mathbf{i} \left(\frac{\partial w}{\partial y} - \frac{\partial v}{\partial z} \right) + \mathbf{j} \left(\frac{\partial u}{\partial z} - \frac{\partial w}{\partial x} \right)$$
$$+ \mathbf{k} \left(\frac{\partial v}{\partial x} - \frac{\partial u}{\partial y} \right).$$

Combining (9) with (7), where $\boldsymbol{\zeta}$ is to be written for $\mathbf{q}$, we get (9·3)

(10) $\nabla \boldsymbol{\zeta} = \operatorname{div} \boldsymbol{\zeta} = \dfrac{\partial}{\partial x}\left(\dfrac{\partial w}{\partial y} - \dfrac{\partial v}{\partial z}\right) + \dfrac{\partial}{\partial y}\left(\dfrac{\partial u}{\partial z} - \dfrac{\partial w}{\partial x}\right) + \dfrac{\partial}{\partial z}\left(\dfrac{\partial v}{\partial x} - \dfrac{\partial u}{\partial y}\right) = 0,$

since $\partial^2 w/\partial x\,\partial y = \partial^2 w/\partial y\,\partial x$, etc.

If $d\mathbf{r} = \mathbf{i}\,dx + \mathbf{j}\,dy + \mathbf{k}\,dz$ we have the scalar product

(11) $\qquad\qquad (d\mathbf{r}\,\nabla) = dx\dfrac{\partial}{\partial x} + dy\dfrac{\partial}{\partial y} + dz\dfrac{\partial}{\partial z} = d.$

Applied to the pressure p and the velocity $\mathbf{q}$ this gives

(12) $\qquad dp = (d\mathbf{r}\,\nabla)p = dx\dfrac{\partial p}{\partial x} + dy\dfrac{\partial p}{\partial y} + dz\dfrac{\partial p}{\partial z} = d\mathbf{r}\,(\nabla\,p).$

(13) $\qquad \mathbf{i}\,d\mathbf{q} = \mathbf{i}\,(d\mathbf{r}\,\nabla)\mathbf{q} = dx\dfrac{\partial u}{\partial x} + dy\dfrac{\partial u}{\partial y} + dz\dfrac{\partial u}{\partial z}.$

If both sides of (11) are divided by dt we get

(14) $\qquad\qquad (\mathbf{q}\,\nabla) = u\dfrac{\partial}{\partial x} + v\dfrac{\partial}{\partial y} + w\dfrac{\partial}{\partial z},$

a scalar operator which gives the rate of change following the fluid in steady motion, see 21·31 (1).

21·31. The acceleration of an air particle.

Consider the particle whose position vector is $\mathbf{r}$ at time t. Then its velocity depends on both $\mathbf{r}$ and $\mathbf{t}$ so that we have $\mathbf{q} = f(\mathbf{r}, t)$. At time $t + \delta t$ the particle will have moved to the position $\mathbf{r} + \delta\mathbf{r}$ and its velocity will have become $\mathbf{q} + \delta\mathbf{q}$. Therefore

$$\begin{aligned}
\delta\mathbf{q} &= f(\mathbf{r} + \delta\mathbf{r}, t + \delta t) - f(\mathbf{r}, t)\\
&= f(\mathbf{r} + \delta\mathbf{r}, t + \delta t) - f(\mathbf{r}, t + \delta t) + f(\mathbf{r}, t + \delta t) - f(\mathbf{r}, t)\\
&= (\delta\mathbf{r}\,\nabla)f(\mathbf{r}, t + \delta t) + \delta t\,\dfrac{\partial f(\mathbf{r}, t)}{\partial t}.
\end{aligned}$$

Dividing by δt and observing that $\mathbf{q} = d\mathbf{r}/dt$, by proceeding to the limit, we find the acceleration

(1) $\qquad\qquad \dfrac{d\mathbf{q}}{dt} = \dfrac{\partial\mathbf{q}}{\partial t} + (\mathbf{q}\,\nabla)\mathbf{q}.$

Using 21·3 (9), (10) we find after some reduction, which is left to the reader, that

(2) $\qquad\qquad \dfrac{d\mathbf{q}}{dt} = \dfrac{\partial\mathbf{q}}{\partial t} + \tfrac{1}{2}\nabla\,q^2 - \mathbf{q}_{\wedge}\boldsymbol{\zeta}.$

21·4. Gauss's theorem.

Let the closed surface S enclose the volume V and let X be a scalar or vector function of position. Then if $d\tau$ is an element of the volume V, and if dS is an element of the surface S,

$$\int_{(V)} (\nabla X)\,d\tau = -\int_{(S)} \mathbf{n}X\,dS,$$

where **n** is the unit normal vector to dS drawn into the *interior* of the volume enclosed by S. This is Gauss's theorem.

Proof. By drawing three systems of parallel planes the volume V will be divided into elements of volume. If $\delta\tau$ is such an element we shall have, from 21·3 (4),

$$(\nabla X)\delta\tau = -\int_{(\delta\tau)} \mathbf{n}X\, dS,$$

approximately, the integral being taken over the surface of the element $\delta\tau$. The negative sign arises from the fact that in 21·3 the unit normal was drawn *outwards*. Here it is drawn *inwards*.

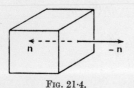

FIG. 21·4.

By summation for all such elements of volume we get

$$\int_{(V)} (\nabla X)\, d\tau = \lim_{\delta\tau\to 0} \Sigma\, (\nabla X)\,\delta\tau = -\Sigma \int_{(\delta\tau)} \mathbf{n}X\, dS.$$

Now at a point on the common boundary of two abutting elements the inward normals to each element are of opposite sign. Thus the surface integrals over boundaries which are shared by two elements of volume cancel, and we are left with the surface integral over S.   Q.E.D.

By giving various values to X we obtain diverse particular forms of Gauss's theorem. Thus the values p, **q**, **ζ**, $_\wedge$**q** give

(1)
$$\int_{(V)} \nabla p\, d\tau = -\int_{(S)} \mathbf{n}p\, dS.$$

(2)
$$\int_{(V)} \nabla \mathbf{q}\, d\tau = -\int_{(S)} \mathbf{n}\mathbf{q}\, dS.$$

(3)
$$\int_{(V)} \nabla \boldsymbol{\zeta}\, d\tau = -\int_{(S)} \mathbf{n}\boldsymbol{\zeta}\, dS.$$

(4)
$$\int_{(V)} \boldsymbol{\zeta}\, d\tau = -\int_{(S)} \mathbf{n}_\wedge\mathbf{q}\, dS.$$

In view of 1·41, (1) states that the resultant thrust due to pressure on the air particles in the volume V is equal to the resultant thrust on the boundary.

As to (3) we have shown, 21·3 (10) that $\nabla \boldsymbol{\zeta} = 0$, so that (3) states that the surface integral of the normal component of the vorticity taken over a closed surface is zero, see 9·31.

Gauss's theorem permits of a particularly striking statement in terms of the following notation. The change of position vector of a point P is denoted by $d\mathbf{P}$, ∇ is denoted by $\partial/\partial\mathbf{P}$ and for **n** dS we use the inwardly directed vector area $d\mathbf{S}$. Then the theorem may be written

$$\int_{(V)} \frac{\partial}{\partial\mathbf{P}} X\, dV = -\int_{(S)} d\mathbf{S}X,$$

where dV is the element of volume previously called $d\tau$.

21·5. The equation of continuity. If we consider a particle of air of infinitesimal volume $d\tau$ and density ρ at time t the mass of this particle does not change as it moves about, and therefore

$$(1) \qquad \frac{d}{dt}(\rho \, d\tau) = 0.$$

This is one form of the *equation of continuity* or *conservation of mass*. If the volume expands, ρ decreases, and vice versa, in such a way that (1) is always satisfied.

Another point of view is the following. Consider a *fixed* closed surface S imagined drawn in the fluid (air). If **n** is the unit inward normal to the element dS the rate at which mass flows into the surface through the boundary is

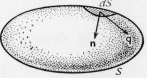

Fig. 21·5.

$$(2) \qquad \int \rho \mathbf{q n} \, dS.$$

The mass of the air within the volume V enclosed by S is

$$(3) \qquad \int_{(V)} \rho \, d\tau.$$

Assuming that no air is created or annihilated within S (no sources or sinks), the mass can only increase by flow through the boundary. Equating (2) to the time rate of increase of (3) we get

$$\frac{\partial}{\partial t} \int_{(V)} \rho \, d\tau = \int_{(S)} \rho \mathbf{q n} \, dS = - \int_{(V)} \nabla(\rho \mathbf{q}) d\tau$$

by Gauss's theorem. Thus

$$\int \left(\frac{\partial \rho}{\partial t} + \nabla(\rho \mathbf{q}) \right) d\tau = 0.$$

Since the surface S can be replaced by any arbitrary closed surface drawn within it, we must have, at every point,

$$(4) \qquad \frac{\partial \rho}{\partial t} + \nabla(\rho \mathbf{q}) = 0,$$

which is another form of the equation of continuity. If the motion is steady, $\partial \rho / \partial t = 0$ and therefore

$$(5) \qquad \nabla(\rho \mathbf{q}) = 0.$$

If the air is in addition regarded as incompressible, we have

$$(6) \qquad \nabla \mathbf{q} = 0,$$

and if the motion is irrotational so that $\mathbf{q} = -\nabla\phi$, we get Laplace's equation (3·311)

$$\nabla^2\phi = \frac{\partial^2\phi}{\partial x^2} + \frac{\partial^2\phi}{\partial y^2} + \frac{\partial^2\phi}{\partial z^2} = 0.$$

Observe that (5) can be written in the form

$$\rho\,\nabla\,\mathbf{q} + (\mathbf{q}\,\nabla)\rho = 0.$$

Using 21·3 (14), in the case of irrotational motion this becomes

$$(7) \qquad -\rho\,\nabla^2\phi + u\frac{\partial\rho}{\partial x} + v\frac{\partial\rho}{\partial y} + w\frac{\partial\rho}{\partial z} = 0.$$

21·6. The equation of motion. Consider the air which at time t occupies the region interior to a *fixed* closed surface S (fig. 21·5) which lies entirely in the fluid (air). By Newton's second law of motion the total force acting on this mass of air is equal to the rate of change of linear momentum. The force is due to

(i) the normal pressure thrusts on the boundary,

(ii) the external force (such as gravity), say $\mathbf{F}$ per unit *mass*. Thus

$$\int_{(S)} p\mathbf{n}\,dS + \int_{(V)} \mathbf{F}\rho\,d\tau = \frac{d}{dt}\int_{(V)} \mathbf{q}\rho\,d\tau.$$

Now,
$$\frac{d}{dt}\int_{(V)} \mathbf{q}\rho\,d\tau = \int_{(V)} \rho\frac{d\mathbf{q}}{dt}\,d\tau + \int \mathbf{q}\frac{d}{dt}(\rho\,d\tau),$$

and the last integral vanishes by 21·5 (1). Also, by Gauss's theorem,

$$\int_{(S)} p\mathbf{n}\,dS = -\int_{(V)} \nabla p\,d\tau.$$

Therefore
$$\int_{(V)} \left(\mathbf{F}\rho - \nabla p - \rho\frac{d\mathbf{q}}{dt}\right) d\tau = 0.$$

Since the volume of integration is entirely arbitrary, the integrand must vanish at every point, and therefore

$$\frac{d\mathbf{q}}{dt} = \mathbf{F} - \frac{1}{\rho}\nabla p,$$

the equation deduced by another method in 9·2.

21·7. Stokes's theorem. Let C be a given closed curve and S a surface * which has C for boundary (like a bowl and its rim). Let $\mathbf{n}$ be the unit vector normal to the element of area dS of S drawn in that sense which is related to the senses of circulation round C and dS by the right-handed screw rule. Then if X is any vector or scalar function of position

$$\int_{(S)} (\mathbf{n}_\wedge\nabla)\,X\,dS = \int_{(C)} d\mathbf{r}X.$$

---

* Such a surface can be conveniently described as a *diaphragm* closing C.

Proof. If by drawing lines across S as in fig. 3·2 (d) we divide S into a network of fine meshes, it is clear that the right-hand integral is equal to the sum of the similar integrals round the individual meshes, for every line which is common to two meshes is described twice in opposite senses, so that its contributions cancel out and we are left with the integral round C.

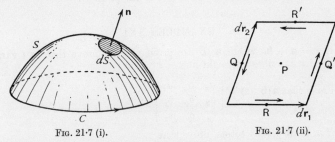

FIG. 21·7 (i).   FIG. 21·7 (ii).

It will therefore be sufficient to prove the theorem for a single mesh. Without loss of generality this single mesh may be taken in the form of a parallelogram whose sides are the infinitesimal vectors $d\mathbf{r}_1$, $d\mathbf{r}_2$. Denote the value of X at the point P by X_P. Then integrating round the mesh, cf. 3·2,

$$\int d\mathbf{r}\, X = d\mathbf{r}_1(X_R - X_{R'}) + d\mathbf{r}_2(X_{Q'} - X_Q),$$

where Q, Q', R, R' are the mid-points of the sides.

If P is the centre of the parallelogram, 21·3 (11) shows that

$$X_{R'} = X_P + \tfrac{1}{2}(d\mathbf{r}_2\,\nabla)X_P, \quad X_R = X_P - \tfrac{1}{2}(d\mathbf{r}_2\,\nabla)X_P,$$

with similar results for X_Q and $X_{Q'}$. Thus round the mesh

$$\int d\mathbf{r}\, X = [-\,d\mathbf{r}_1(d\mathbf{r}_2\,\nabla) + d\mathbf{r}_2(d\mathbf{r}_1\,\nabla)]X_P = [(d\mathbf{r}_{1\wedge}d\mathbf{r}_2)_\wedge\nabla]X_P,$$

by the triple vector product. But from 21·13, $d\mathbf{r}_{1\wedge}d\mathbf{r}_2 = \mathbf{n}dS_P$, for it is the directed area of the parallelogram dS_P. Thus

$$\int d\mathbf{r}\, X = (\mathbf{n}_\wedge\nabla)X\, dS$$

for a single mesh. By summation for all the meshes the theorem follows.

Q.E.D.

As in the case of Gauss's theorem several special forms of this theorem arise by attributing particular forms to X. The form in which we are interested arises when $X = \mathbf{q}$, so that

$$\text{circ } C = \int_{(C)} \mathbf{q}\, d\mathbf{r} = \int_{(S)} (\mathbf{n}_\wedge\nabla)\mathbf{q}\, dS = \int_{(S)} \mathbf{n}(\nabla_\wedge\mathbf{q})dS = \int_{(S)} \mathbf{n}\zeta\, dS,$$

which states that the circulation in any circuit is equal to the integral of the normal component of the vorticity over any diaphragm which closes the circuit.

With the notation given at the end of 21·4 we can write the general form of Stokes's theorem in the following way :

$$\int_{(S)} \left(d\mathbf{S} \wedge \frac{\partial}{\partial \mathbf{P}} \right) X = \int_{(C)} d\mathbf{P}X,$$

where $d\mathbf{P}$ is the change of position vector of the point P.

EXAMPLES XXI

1. If $\mathbf{r} = \mathbf{a} + \mathbf{b}t$, where $\mathbf{a}$, $\mathbf{b}$ are given vectors, prove that as t varies the extremity of $\mathbf{r}$ describes a straight line.

2. Prove that $\mathbf{a}(\mathbf{b} \wedge \mathbf{c}) = \begin{vmatrix} a_x & a_y & a_z \\ b_x & b_y & b_z \\ c_x & c_y & c_z \end{vmatrix}$.

3. If $\mathbf{a}$, $\mathbf{n}$ are given vectors, prove that
$$(\mathbf{r} - \mathbf{a})\mathbf{n} = 0$$
is the equation of a plane.

4. Prove that the triple scalar product $(\mathbf{a} \wedge \mathbf{b})\mathbf{c}$ can be regarded as measuring the volume of a certain parallelepiped.

5. Prove the *cyclic rule* for the triple scalar product, that the product is unaltered if the cyclic order of the vectors is preserved, e.g.
$$(\mathbf{a} \wedge \mathbf{b})\mathbf{c} = \mathbf{a}(\mathbf{b} \wedge \mathbf{c}) = (\mathbf{c} \wedge \mathbf{b})\mathbf{a} = \mathbf{c}(\mathbf{b} \wedge \mathbf{a}).$$

6. Prove that (i) $(\mathbf{a} \wedge \mathbf{b})(\mathbf{c} \wedge \mathbf{d}) = (\mathbf{a}\mathbf{c})(\mathbf{b}\mathbf{d}) - (\mathbf{a}\mathbf{d})(\mathbf{b}\mathbf{c})$
(ii) $\mathbf{a}[\mathbf{b}(\mathbf{c} \wedge \mathbf{d})] - \mathbf{b}[\mathbf{a}(\mathbf{c} \wedge \mathbf{d})] + \mathbf{c}[\mathbf{d}(\mathbf{a} \wedge \mathbf{b})] - \mathbf{d}[\mathbf{c}(\mathbf{a} \wedge \mathbf{b})] = 0.$

7. Prove the *centric rule* for the triple vector product, that the sign changes only with change of the centre vector, e.g.
$$\mathbf{a} \wedge (\mathbf{b} \wedge \mathbf{c}) = (\mathbf{c} \wedge \mathbf{b}) \wedge \mathbf{a} = -\mathbf{a} \wedge (\mathbf{c} \wedge \mathbf{b}).$$

8. Prove that grad $\phi = \mathbf{n} \, \partial\phi/\partial n$, where the differentiation is along the normal $\mathbf{n}$ to the surface $\phi = $ constant.

9. Prove that   (i) $\nabla(\rho\mathbf{q}) = \rho \nabla \mathbf{q} + \mathbf{q}(\nabla \rho)$,
(ii) $\nabla \wedge (\rho\mathbf{q}) = \rho\boldsymbol{\zeta} - \mathbf{q} \wedge \nabla \rho.$

10. Prove that   (i) $\nabla(\boldsymbol{\zeta} \wedge \mathbf{q}) = \mathbf{q}(\nabla \wedge \boldsymbol{\zeta}) - \zeta^2$,
(ii) $(\mathbf{q} \nabla)\mathbf{q} = -\mathbf{q} \wedge \boldsymbol{\zeta} + \frac{1}{2} \nabla \, q^2.$

11. Prove curl (grad ϕ) = 0, and that
$$\nabla^2\mathbf{q} = \nabla(\nabla \mathbf{q}) - \nabla \wedge (\nabla \wedge \mathbf{q}).$$

12. Prove that the velocity components
$$u = \frac{-2xyz}{(x^2 + y^2)^2}, \quad v = \frac{(x^2 - y^2)z}{(x^2 + y^2)^2}, \quad w = \frac{y}{x^2 + y^2}$$
satisfy the equation of continuity for incompressible air, and examine whether the motion is irrotational.

13. Prove that $\nabla(\rho\mathbf{q}) = \rho \nabla \mathbf{q} + \mathbf{q}(\nabla \rho)$, and hence write out the equation of continuity, in cartesian coordinates, for steady motion.

14. Prove that $\dfrac{d\rho}{dt} = \dfrac{\partial\rho}{\partial t} + \mathbf{q}(\nabla\rho)$, and that the equation of continuity may be written $\dfrac{d\rho}{dt} + \rho\,\nabla\,\mathbf{q} = 0$.

15. Prove that a vortex ring always consists of the same fluid particles.

16. In irrotational motion of incompressible air, prove that

$$\frac{p}{\rho} + \tfrac{1}{2}q^2 + \Omega - \frac{\partial\phi}{\partial t}$$

has the same value everywhere at a given instant.

17. Prove that the rate of change of momentum of the air which passes through a closed surface S is

$$\frac{\partial}{\partial t}\int_{(V)}\rho\mathbf{q}\,d\tau + \int_{(S)}\rho\mathbf{q}(\mathbf{nq})\,dS,$$

where $\mathbf{n}$ is the outward unit normal.

18. Prove that $\quad \rho\,\dfrac{d}{dt}\big((\tfrac{1}{2}q^2 + \Omega\big) = -\,\mathbf{q}\,\nabla\,p$, if Ω is independent of t.

19. Obtain the equation of motion in the form

$$\frac{\partial\mathbf{q}}{\partial t} - \mathbf{q}_\wedge\zeta = -\frac{1}{\rho}\,\nabla\,p - \nabla(\Omega + \tfrac{1}{2}q^2).$$

20. For incompressible air in steady motion, prove that

$$\mathbf{q}_\wedge\zeta = \nabla\left(\frac{p}{\rho} + \tfrac{1}{2}q^2 + \Omega\right)$$

and deduce Bernoulli's theorem.

If the motion is irrotational, prove that $\dfrac{p}{\rho} + \tfrac{1}{2}q^2 + \Omega = C$, where C has the same value everywhere.

21. Prove that for incompressible air

$$\frac{d\zeta}{dt} = (\zeta\,\nabla)\,\mathbf{q}.$$

22. Show that $\quad\displaystyle\int_{(V)}\zeta\,d\tau = -\int_{(S)}\mathbf{n}_\wedge\mathbf{q}\,dS.$

23. Obtain the following forms of Gauss's theorem

$$\int_{(V)}\nabla^2\mathbf{q}\,d\tau = -\int_{(S)}(\mathbf{n}\,\nabla)\,\mathbf{q}\,dS,$$

$$\int_{(V)}\nabla^2\phi\,d\tau = -\int_{(S)}(\mathbf{n}\,\nabla)\phi\,dS = -\int\frac{\partial\phi}{\partial n}\,dS.$$

24. Prove Green's theorem that

$$\int_{(V)}(\nabla\,\phi\,\nabla\,\psi)\,d\tau = -\int_{(V)}\phi\,\nabla^2\psi\,d\tau - \int_{(S)}\phi\,\frac{\partial\psi}{\partial n}\,dS$$

$$= -\int_{(V)}\psi\,\nabla^2\phi\,d\tau - \int_{(S)}\psi\frac{\partial\phi}{\partial n}\,dS,$$

and deduce that

$$\int_{(V)}\nabla\,\phi\,.\,\nabla\,\phi\,d\tau = -\int_{(V)}\phi\,\nabla^2\phi\,d\tau - \int\phi\frac{\partial\phi}{\partial n}\,dS.$$

25. Obtain the following particular cases of Stokes's theorem

$$\int_{(S)} (\mathbf{n} \wedge \nabla) \phi \, dS = \int_{(C)} \phi \, d\mathbf{r},$$

$$\int_{(S)} (\mathbf{n} \wedge \nabla) \wedge \mathbf{q} \, dS = \int_{(C)} d\mathbf{r} \wedge \mathbf{q}.$$

Express the second in a form in which the triple vector product does not appear.

26. If **u** is a unit vector, prove geometrically that

$$\mathbf{u}(\mathbf{b} + \mathbf{c}) = \mathbf{u}\mathbf{b} + \mathbf{u}\mathbf{c},$$

and deduce the distributive law for scalar products.

27. By considering the vector product of an arbitrary unit vector **x** and the vector

$$\mathbf{v} = \mathbf{a} \wedge (\mathbf{b} + \mathbf{c}) - \mathbf{a} \wedge \mathbf{b} - \mathbf{a} \wedge \mathbf{c},$$

prove, using the distributive law for scalar products and the theorem of 19·2, that **v** = 0, thereby proving the distributive law for vector products.

INDEX

The references are to pages

A CATALOGUE OF
SELECTED DOVER BOOKS
IN ALL FIELDS OF INTEREST

A CATALOGUE OF SELECTED DOVER
BOOKS IN ALL FIELDS OF INTEREST

RACKHAM'S COLOR ILLUSTRATIONS FOR WAGNER'S RING. Rackham's finest mature work—all 64 full-color watercolors in a faithful and lush interpretation of the *Ring*. Full-sized plates on coated stock of the paintings used by opera companies for authentic staging of Wagner. Captions aid in following complete Ring cycle. Introduction. 64 illustrations plus vignettes. 72pp. 8⅝ x 11¼. 23779-6 Pa. $6.00

CONTEMPORARY POLISH POSTERS IN FULL COLOR, edited by Joseph Czestochowski. 46 full-color examples of brilliant school of Polish graphic design, selected from world's first museum (near Warsaw) dedicated to poster art. Posters on circuses, films, plays, concerts all show cosmopolitan influences, free imagination. Introduction. 48pp. 9⅜ x 12¼. 23780-X Pa. $6.00

GRAPHIC WORKS OF EDVARD MUNCH, Edvard Munch. 90 haunting, evocative prints by first major Expressionist artist and one of the greatest graphic artists of his time: *The Scream, Anxiety, Death Chamber, The Kiss, Madonna*, etc. Introduction by Alfred Werner. 90pp. 9 x 12. 23765-6 Pa. $5.00

THE GOLDEN AGE OF THE POSTER, Hayward and Blanche Cirker. 70 extraordinary posters in full colors, from Maitres de l'Affiche, Mucha, Lautrec, Bradley, Cheret, Beardsley, many others. Total of 78pp. 9⅜ x 12¼. 22753-7 Pa. $5.95

THE NOTEBOOKS OF LEONARDO DA VINCI, edited by J. P. Richter. Extracts from manuscripts reveal great genius; on painting, sculpture, anatomy, sciences, geography, etc. Both Italian and English. 186 ms. pages reproduced, plus 500 additional drawings, including studies for *Last Supper*, Sforza monument, etc. 860pp. 7⅞ x 10¾. (Available in U.S. only) 22572-0, 22573-9 Pa., Two-vol. set $15.90

THE CODEX NUTTALL, as first edited by Zelia Nuttall. Only inexpensive edition, in full color, of a pre-Columbian Mexican (Mixtec) book. 88 color plates show kings, gods, heroes, temples, sacrifices. New explanatory, historical introduction by Arthur G. Miller. 96pp. 11⅜ x 8½. (Available in U.S. only) 23168-2 Pa. $7.50

UNE SEMAINE DE BONTÉ, A SURREALISTIC NOVEL IN COLLAGE, Max Ernst. Masterpiece created out of 19th-century periodical illustrations, explores worlds of terror and surprise. Some consider this Ernst's greatest work. 208pp. 8⅛ x 11. 23252-2 Pa. $5.00

DRAWINGS OF WILLIAM BLAKE, William Blake. 92 plates from Book of Job, *Divine Comedy, Paradise Lost,* visionary heads, mythological figures, Laocoon, etc. Selection, introduction, commentary by Sir Geoffrey Keynes. 178pp. 8⅛ x 11.   22303-5 Pa. $4.00

ENGRAVINGS OF HOGARTH, William Hogarth. 101 of Hogarth's greatest works: *Rake's Progress, Harlot's Progress, Illustrations for Hudibras, Before and After, Beer Street and Gin Lane,* many more. Full commentary. 256pp. 11 x 13¾.   22479-1 Pa. $7.95

DAUMIER: 120 GREAT LITHOGRAPHS, Honore Daumier. Wide-ranging collection of lithographs by the greatest caricaturist of the 19th century. Concentrates on eternally popular series on lawyers, on married life, on liberated women, etc. Selection, introduction, and notes on plates by Charles F. Ramus. Total of 158pp. 9⅜ x 12¼.   23512-2 Pa. $5.50

DRAWINGS OF MUCHA, Alphonse Maria Mucha. Work reveals draftsman of highest caliber: studies for famous posters and paintings, renderings for book illustrations and ads, etc. 70 works, 9 in color; including 6 items not drawings. Introduction. List of illustrations. 72pp. 9⅜ x 12¼. (Available in U.S. only)   23672-2 Pa. $4.00

GIOVANNI BATTISTA PIRANESI: DRAWINGS IN THE PIERPONT MORGAN LIBRARY, Giovanni Battista Piranesi. For first time ever all of Morgan Library's collection, world's largest. 167 illustrations of rare Piranesi drawings—archeological, architectural, decorative and visionary. Essay, detailed list of drawings, chronology, captions. Edited by Felice Stampfle. 144pp. 9⅜ x 12¼.   23714-1 Pa. $7.50

NEW YORK ETCHINGS (1905-1949), John Sloan. All of important American artist's N.Y. life etchings. 67 works include some of his best art; also lively historical record—Greenwich Village, tenement scenes. Edited by Sloan's widow. Introduction and captions. 79pp. 8⅜ x 11¼.
                    23651-X Pa. $4.00

CHINESE PAINTING AND CALLIGRAPHY: A PICTORIAL SURVEY, Wan-go Weng. 69 fine examples from John M. Crawford's matchless private collection: landscapes, birds, flowers, human figures, etc., plus calligraphy. Every basic form included: hanging scrolls, handscrolls, album leaves, fans, etc. 109 illustrations. Introduction. Captions. 192pp. 8⅞ x 11¾.
                    23707-9 Pa. $7.95

DRAWINGS OF REMBRANDT, edited by Seymour Slive. Updated Lippmann, Hofstede de Groot edition, with definitive scholarly apparatus. All portraits, biblical sketches, landscapes, nudes, Oriental figures, classical studies, together with selection of work by followers. 550 illustrations. Total of 630pp. 9⅛ x 12¼.   21485-0, 21486-9 Pa., Two-vol. set $14.00

THE DISASTERS OF WAR, Francisco Goya. 83 etchings record horrors of Napoleonic wars in Spain and war in general. Reprint of 1st edition, plus 3 additional plates. Introduction by Philip Hofer. 97pp. 9⅜ x 8¼.
                    21872-4 Pa. $3.75

THE EARLY WORK OF AUBREY BEARDSLEY, Aubrey Beardsley. 157 plates, 2 in color: *Manon Lescaut, Madame Bovary, Morte Darthur, Salome,* other. Introduction by H. Marillier. 182pp. 8⅛ x 11. 21816-3 Pa. $4.50

THE LATER WORK OF AUBREY BEARDSLEY, Aubrey Beardsley. Exotic masterpieces of full maturity: *Venus and Tannhauser, Lysistrata, Rape of the Lock, Volpone,* Savoy material, etc. 174 plates, 2 in color. 186pp. 8⅛ x 11. 21817-1 Pa. $4.50

THOMAS NAST'S CHRISTMAS DRAWINGS, Thomas Nast. Almost all Christmas drawings by creator of image of Santa Claus as we know it, and one of America's foremost illustrators and political cartoonists. 66 illustrations. 3 illustrations in color on covers. 96pp. 8⅜ x 11¼. 23660-9 Pa. $3.50

THE DORÉ ILLUSTRATIONS FOR DANTE'S DIVINE COMEDY, Gustave Doré. All 135 plates from Inferno, Purgatory, Paradise; fantastic tortures, infernal landscapes, celestial wonders. Each plate with appropriate (translated) verses. 141pp. 9 x 12. 23231-X Pa. $4.50

DORÉ'S ILLUSTRATIONS FOR RABELAIS, Gustave Doré. 252 striking illustrations of *Gargantua and Pantagruel* books by foremost 19th-century illustrator. Including 60 plates, 192 delightful smaller illustrations. 153pp. 9 x 12. 23656-0 Pa. $5.00

LONDON: A PILGRIMAGE, Gustave Doré, Blanchard Jerrold. Squalor, riches, misery, beauty of mid-Victorian metropolis; 55 wonderful plates, 125 other illustrations, full social, cultural text by Jerrold. 191pp. of text. 9⅜ x 12¼. 22306-X Pa. $6.00

THE RIME OF THE ANCIENT MARINER, Gustave Doré, S. T. Coleridge. Dore's finest work, 34 plates capture moods, subtleties of poem. Full text. Introduction by Millicent Rose. 77pp. 9¼ x 12. 22305-1 Pa. $3.00

THE DORE BIBLE ILLUSTRATIONS, Gustave Doré. All wonderful, detailed plates: Adam and Eve, Flood, Babylon, Life of Jesus, etc. Brief King James text with each plate. Introduction by Millicent Rose. 241 plates. 241pp. 9 x 12. 23004-X Pa. $5.00

THE COMPLETE ENGRAVINGS, ETCHINGS AND DRYPOINTS OF ALBRECHT DURER. "Knight, Death and Devil"; "Melencolia," and more—all Dürer's known works in all three media, including 6 works formerly attributed to him. 120 plates. 235pp. 8⅜ x 11¼. 22851-7 Pa. $6.50

MAXIMILIAN'S TRIUMPHAL ARCH, Albrecht Dürer and others. Incredible monument of woodcut art: 8 foot high elaborate arch—heraldic figures, humans, battle scenes, fantastic elements—that you can assemble yourself. Printed on one side, layout for assembly. 143pp. 11 x 16. 21451-6 Pa. $5.00

THE COMPLETE WOODCUTS OF ALBRECHT DURER, edited by Dr. W. Kurth. 346 in all: "Old Testament," "St. Jerome," "Passion," "Life of Virgin," Apocalypse," many others. Introduction by Campbell Dodgson. 285pp. 8½ x 12¼. 21097-9 Pa. $6.95

DRAWINGS OF ALBRECHT DURER, edited by Heinrich Wolfflin. 81 plates show development from youth to full style. Many favorites; many new. Introduction by Alfred Werner. 96pp. 8⅛ x 11. 22352-3 Pa. $4.00

THE HUMAN FIGURE, Albrecht Dürer. Experiments in various techniques—stereometric, progressive proportional, and others. Also life studies that rank among finest ever done. Complete reprinting of *Dresden Sketchbook*. 170 plates. 355pp. 8⅜ x 11¼. 21042-1 Pa. $6.95

OF THE JUST SHAPING OF LETTERS, Albrecht Dürer. Renaissance artist explains design of Roman majuscules by geometry, also Gothic lower and capitals. Grolier Club edition. 43pp. 7⅞ x 10¾ 21306-4 Pa. $2.50

TEN BOOKS ON ARCHITECTURE, Vitruvius. The most important book ever written on architecture. Early Roman aesthetics, technology, classical orders, site selection, all other aspects. Stands behind everything since. Morgan translation. 331pp. 5⅜ x 8½. 20645-9 Pa. $3.75

THE FOUR BOOKS OF ARCHITECTURE, Andrea Palladio. 16th-century classic responsible for Palladian movement and style. Covers classical architectural remains, Renaissance revivals, classical orders, etc. 1738 Ware English edition. Introduction by A. Placzek. 216 plates. 110pp. of text. 9½ x 12¾. 21308-0 Pa. $7.50

HORIZONS, Norman Bel Geddes. Great industrialist stage designer, "father of streamlining," on application of aesthetics to transportation, amusement, architecture, etc. 1932 prophetic account; function, theory, specific projects. 222 illustrations. 312pp. 7⅞ x 10¾. 23514-9 Pa. $6.95

FRANK LLOYD WRIGHT'S FALLINGWATER, Donald Hoffmann. Full, illustrated story of conception and building of Wright's masterwork at Bear Run, Pa. 100 photographs of site, construction, and details of completed structure. 112pp. 9¼ x 10. 23671-4 Pa. $5.00

THE ELEMENTS OF DRAWING, John Ruskin. Timeless classic by great Viltorian; starts with basic ideas, works through more difficult. Many practical exercises. 48 illustrations. Introduction by Lawrence Campbell. 228pp. 5⅜ x 8½. 22730-8 Pa. $2.75

GIST OF ART, John Sloan. Greatest modern American teacher, Art Students League, offers innumerable hints, instructions, guided comments to help you in painting. Not a formal course. 46 illustrations. Introduction by Helen Sloan. 200pp. 5⅜ x 8½. 23435-5 Pa. $3.50

THE ANATOMY OF THE HORSE, George Stubbs. Often considered the great masterpiece of animal anatomy. Full reproduction of 1766 edition, plus prospectus; original text and modernized text. 36 plates. Introduction by Eleanor Garvey. 121pp. 11 x 14¾. 23402-9 Pa. $6.00

BRIDGMAN'S LIFE DRAWING, George B. Bridgman. More than 500 illustrative drawings and text teach you to abstract the body into its major masses, use light and shade, proportion; as well as specific areas of anatomy, of which Bridgman is master. 192pp. 6½ x 9¼. (Available in U.S. only) 22710-3 Pa. $2.50

ART NOUVEAU DESIGNS IN COLOR, Alphonse Mucha, Maurice Verneuil, Georges Auriol. Full-color reproduction of *Combinaisons ornementales* (c. 1900) by Art Nouveau masters. Floral, animal, geometric, interlacings, swashes—borders, frames, spots—all incredibly beautiful. 60 plates, hundreds of designs. 9⅜ x 8-1/16. 22885-1 Pa. $4.00

FULL-COLOR FLORAL DESIGNS IN THE ART NOUVEAU STYLE, E. A. Seguy. 166 motifs, on 40 plates, from *Les fleurs et leurs applications decoratives* (1902): borders, circular designs, repeats, allovers, "spots." All in authentic Art Nouveau colors. 48pp. 9⅜ x 12¼. 23439-8 Pa. $5.00

A DIDEROT PICTORIAL ENCYCLOPEDIA OF TRADES AND INDUSTRY, edited by Charles C. Gillispie. 485 most interesting plates from the great French Encyclopedia of the 18th century show hundreds of working figures, artifacts, process, land and cityscapes; glassmaking, papermaking, metal extraction, construction, weaving, making furniture, clothing, wigs, dozens of other activities. Plates fully explained. 920pp. 9 x 12. 22284-5, 22285-3 Clothbd., Two-vol. set $40.00

HANDBOOK OF EARLY ADVERTISING ART, Clarence P. Hornung. Largest collection of copyright-free early and antique advertising art ever compiled. Over 6,000 illustrations, from Franklin's time to the 1890's for special effects, novelty. Valuable source, almost inexhaustible.
Pictorial Volume. Agriculture, the zodiac, animals, autos, birds, Christmas, fire engines, flowers, trees, musical instruments, ships, games and sports, much more. Arranged by subject matter and use. 237 plates. 288pp. 9 x 12. 20122-8 Clothbd. $13.50

Typographical Volume. Roman and Gothic faces ranging from 10 point to 300 point, "Barnum," German and Old English faces, script, logotypes, scrolls and flourishes, 1115 ornamental initials, 67 complete alphabets, more. 310 plates. 320pp. 9 x 12. 20123-6 Clothbd. $13.50

CALLIGRAPHY (CALLIGRAPHIA LATINA), J. G. Schwandner. High point of 18th-century ornamental calligraphy. Very ornate initials, scrolls, borders, cherubs, birds, lettered examples. 172pp. 9 x 13. 20475-8 Pa. $6.00

ART FORMS IN NATURE, Ernst Haeckel. Multitude of strangely beautiful natural forms: Radiolaria, Foraminifera, jellyfishes, fungi, turtles, bats, etc. All 100 plates of the 19th-century evolutionist's *Kunstformen der Natur* (1904). 100pp. 9⅜ x 12¼.   22987-4 Pa. $4.50

CHILDREN: A PICTORIAL ARCHIVE FROM NINETEENTH-CENTURY SOURCES, edited by Carol Belanger Grafton. 242 rare, copyright-free wood engravings for artists and designers. Widest such selection available. All illustrations in line. 119pp. 8⅜ x 11¼.
23694-3 Pa. $3.50

WOMEN: A PICTORIAL ARCHIVE FROM NINETEENTH-CENTURY SOURCES, edited by Jim Harter. 391 copyright-free wood engravings for artists and designers selected from rare periodicals. Most extensive such collection available. All illustrations in line. 128pp. 9 x 12.
23703-6 Pa. $4.00

ARABIC ART IN COLOR, Prisse d'Avennes. From the greatest ornamentalists of all time—50 plates in color, rarely seen outside the Near East, rich in suggestion and stimulus. Includes 4 plates on covers. 46pp. 9⅜ x 12¼.   23658-7 Pa. $6.00

AUTHENTIC ALGERIAN CARPET DESIGNS AND MOTIFS, edited by June Beveridge. Algerian carpets are world famous. Dozens of geometrical motifs are charted on grids, color-coded, for weavers, needleworkers, craftsmen, designers. 53 illustrations plus 4 in color. 48pp. 8¼ x 11. (Available in U.S. only)   23650-1 Pa. $1.75

DICTIONARY OF AMERICAN PORTRAITS, edited by Hayward and Blanche Cirker. 4000 important Americans, earliest times to 1905, mostly in clear line. Politicians, writers, soldiers, scientists, inventors, industrialists, Indians, Blacks, women, outlaws, etc. Identificatory information. 756pp. 9¼ x 12¾.   21823-6 Clothbd. $40.00

HOW THE OTHER HALF LIVES, Jacob A. Riis. Journalistic record of filth, degradation, upward drive in New York immigrant slums, shops, around 1900. New edition includes 100 original Riis photos, monuments of early photography. 233pp. 10 x 7⅞.   22012-5 Pa. $6.00

NEW YORK IN THE THIRTIES, Berenice Abbott. Noted photographer's fascinating study of city shows new buildings that have become famous and old sights that have disappeared forever. Insightful commentary. 97 photographs. 97pp. 11⅜ x 10.   22967-X Pa. $4.50

MEN AT WORK, Lewis W. Hine. Famous photographic studies of construction workers, railroad men, factory workers and coal miners. New supplement of 18 photos on Empire State building construction. New introduction by Jonathan L. Doherty. Total of 69 photos. 63pp. 8 x 10¾.
23475-4 Pa. $3.00

THE DEPRESSION YEARS AS PHOTOGRAPHED BY ARTHUR ROTH-STEIN, Arthur Rothstein. First collection devoted entirely to the work of outstanding 1930s photographer: famous dust storm photo, ragged children, unemployed, etc. 120 photographs. Captions. 119pp. 9¼ x 10¾.
23590-4 Pa. $5.00

CAMERA WORK: A PICTORIAL GUIDE, Alfred Stieglitz. All 559 illustrations and plates from the most important periodical in the history of art photography, Camera Work (1903-17). Presented four to a page, reduced in size but still clear, in strict chronological order, with complete captions. Three indexes. Glossary. Bibliography. 176pp. 8⅜ x 11¼.
23591-2 Pa. $6.95

ALVIN LANGDON COBURN, PHOTOGRAPHER, Alvin L. Coburn. Revealing autobiography by one of greatest photographers of 20th century gives insider's version of Photo-Secession, plus comments on his own work. 77 photographs by Coburn. Edited by Helmut and Alison Gernsheim. 160pp. 8⅛ x 11.
23685-4 Pa. $6.00

NEW YORK IN THE FORTIES, Andreas Feininger. 162 brilliant photographs by the well-known photographer, formerly with Life magazine, show commuters, shoppers, Times Square at night, Harlem nightclub, Lower East Side, etc. Introduction and full captions by John von Hartz. 181pp. 9¼ x 10¾.
23585-8 Pa. $6.00

GREAT NEWS PHOTOS AND THE STORIES BEHIND THEM, John Faber. Dramatic volume of 140 great news photos, 1855 through 1976, and revealing stories behind them, with both historical and technical information. Hindenburg disaster, shooting of Oswald, nomination of Jimmy Carter, etc. 160pp. 8¼ x 11.
23667-6 Pa. $5.00

THE ART OF THE CINEMATOGRAPHER, Leonard Maltin. Survey of American cinematography history and anecdotal interviews with 5 masters—Arthur Miller, Hal Mohr, Hal Rosson, Lucien Ballard, and Conrad Hall. Very large selection of behind-the-scenes production photos. 105 photographs. Filmographies. Index. Originally Behind the Camera. 144pp. 8¼ x 11.
23686-2 Pa. $5.00

DESIGNS FOR THE THREE-CORNERED HAT (LE TRICORNE), Pablo Picasso. 32 fabulously rare drawings—including 31 color illustrations of costumes and accessories—for 1919 production of famous ballet. Edited by Parmenia Migel, who has written new introduction. 48pp. 9⅜ x 12¼. (Available in U.S. only)
23709-5 Pa. $5.00

NOTES OF A FILM DIRECTOR, Sergei Eisenstein. Greatest Russian filmmaker explains montage, making of Alexander Nevsky, aesthetics; comments on self, associates, great rivals (Chaplin), similar material. 78 illustrations. 240pp. 5⅜ x 8½.
22392-2 Pa. $4.50

HOLLYWOOD GLAMOUR PORTRAITS, edited by John Kobal. 145 photos capture the stars from 1926-49, the high point in portrait photography. Gable, Harlow, Bogart, Bacall, Hedy Lamarr, Marlene Dietrich, Robert Montgomery, Marlon Brando, Veronica Lake; 94 stars in all. Full background on photographers, technical aspects, much more. Total of 160pp. 8⅜ x 11¼. 23352-9 Pa. $5.00

THE NEW YORK STAGE: FAMOUS PRODUCTIONS IN PHOTO-GRAPHS, edited by Stanley Appelbaum. 148 photographs from Museum of City of New York show 142 plays, 1883-1939. *Peter Pan, The Front Page, Dead End, Our Town,* O'Neill, hundreds of actors and actresses, etc. Full indexes. 154pp. 9½ x 10. 23241-7 Pa. $4.50

MASTERS OF THE DRAMA, John Gassner. Most comprehensive history of the drama, every tradition from Greeks to modern Europe and America, including Orient. Covers 800 dramatists, 2000 plays; biography, plot summaries, criticism, theatre history, etc. 77 illustrations. 890pp. 5⅜ x 8½. 20100-7 Clothbd. $10.00

THE GREAT OPERA STARS IN HISTORIC PHOTOGRAPHS, edited by James Camner. 343 portraits from the 1850s to the 1940s: Tamburini, Mario, Caliapin, Jeritza, Melchior, Melba, Patti, Pinza, Schipa, Caruso, Farrar, Steber, Gobbi, and many more—270 performers in all. Index. 199pp. 8⅜ x 11¼. 23575-0 Pa. $6.50

J. S. BACH, Albert Schweitzer. Great full-length study of Bach, life, background to music, music, by foremost modern scholar. Ernest Newman translation. 650 musical examples. Total of 928pp. 5⅜ x 8½. (Available in U.S. only) 21631-4, 21632-2 Pa., Two-vol. set $9.00

COMPLETE PIANO SONATAS, Ludwig van Beethoven. All sonatas in the fine Schenker edition, with fingering, analytical material. One of best modern editions. Total of 615pp. 9 x 12. (Available in U.S. only) 23134-8, 23135-6 Pa., Two-vol. set $13.00

KEYBOARD MUSIC, J. S. Bach. Bach-Gesellschaft edition. For harpsichord, piano, other keyboard instruments. English Suites, French Suites, Six Partitas, Goldberg Variations, Two-Part Inventions, Three-Part Sinfonias. 312pp. 8⅛ x 11. (Available in U.S. only) 22360-4 Pa. $5.50

FOUR SYMPHONIES IN FULL SCORE, Franz Schubert. Schubert's four most popular symphonies: No. 4 in C Minor ("Tragic"); No. 5 in B-flat Major; No. 8 in B Minor ("Unfinished"); No. 9 in C Major ("Great"). Breitkopf & Hartel edition. Study score. 261pp. 9⅜ x 12¼. 23681-1 Pa. $6.50

THE AUTHENTIC GILBERT & SULLIVAN SONGBOOK, W. S. Gilbert, A. S. Sullivan. Largest selection available; 92 songs, uncut, original keys, in piano rendering approved by Sullivan. Favorites and lesser-known fine numbers. Edited with plot synopses by James Spero. 3 illustrations. 399pp. 9 x 12. 23482-7 Pa. $7.95

PRINCIPLES OF ORCHESTRATION, Nikolay Rimsky-Korsakov. Great classical orchestrator provides fundamentals of tonal resonance, progression of parts, voice and orchestra, tutti effects, much else in major document. 330pp. of musical excerpts. 489pp. 6½ x 9¼. 21266-1 Pa. $6.00

TRISTAN UND ISOLDE, Richard Wagner. Full orchestral score with complete instrumentation. Do not confuse with piano reduction. Commentary by Felix Mottl, great Wagnerian conductor and scholar. Study score. 655pp. 8⅛ x 11. 22915-7 Pa. $12.50

REQUIEM IN FULL SCORE, Giuseppe Verdi. Immensely popular with choral groups and music lovers. Republication of edition published by C. F. Peters, Leipzig, n. d. German frontmaker in English translation. Glossary. Text in Latin. Study score. 204pp. 9⅜ x 12¼.

23682-X Pa. $6.00

COMPLETE CHAMBER MUSIC FOR STRINGS, Felix Mendelssohn. All of Mendelssohn's chamber music: Octet, 2 Quintets, 6 Quartets, and Four Pieces for String Quartet. (Nothing with piano is included). Complete works edition (1874-7). Study score. 283 pp. 9⅜ x 12¼.

23679-X Pa. $6.95

POPULAR SONGS OF NINETEENTH-CENTURY AMERICA, edited by Richard Jackson. 64 most important songs: "Old Oaken Bucket," "Arkansas Traveler," "Yellow Rose of Texas," etc. Authentic original sheet music, full introduction and commentaries. 290pp. 9 x 12. 23270-0 Pa. $6.00

COLLECTED PIANO WORKS, Scott Joplin. Edited by Vera Brodsky Lawrence. Practically all of Joplin's piano works—rags, two-steps, marches, waltzes, etc., 51 works in all. Extensive introduction by Rudi Blesh. Total of 345pp. 9 x 12. 23106-2 Pa. $13.50

BASIC PRINCIPLES OF CLASSICAL BALLET, Agrippina Vaganova. Great Russian theoretician, teacher explains methods for teaching classical ballet; incorporates best from French, Italian, Russian schools. 118 illustrations. 175pp. 5⅜ x 8½. 22036-2 Pa. $2.00

CHINESE CHARACTERS, L. Wieger. Rich analysis of 2300 characters according to traditional systems into primitives. Historical-semantic analysis to phonetics (Classical Mandarin) and radicals. 820pp. 6⅛ x 9¼.

21321-8 Pa. $8.95

EGYPTIAN LANGUAGE: EASY LESSONS IN EGYPTIAN HIERO-GLYPHICS, E. A. Wallis Budge. Foremost Egyptologist offers Egyptian grammar, explanation of hieroglyphics, many reading texts, dictionary of symbols. 246pp. 5 x 7½. (Available in U.S. only)

21394-3 Clothbd. $7.50

AN ETYMOLOGICAL DICTIONARY OF MODERN ENGLISH, Ernest Weekley. Richest, fullest work, by foremost British lexicographer. Detailed word histories. Inexhaustible. Do not confuse this with Concise Etymological Dictionary, which is abridged. Total of 856pp. 6½ x 9¼.

21873-2, 21874-0 Pa., Two-vol. set $10.00

A MAYA GRAMMAR, Alfred M. Tozzer. Practical, useful English-language grammar by the Harvard anthropologist who was one of the three greatest American scholars in the area of Maya culture. Phonetics, grammatical processes, syntax, more. 301pp. 5⅜ x 8½.   23465-7 Pa. $4.00

THE JOURNAL OF HENRY D. THOREAU, edited by Bradford Torrey, F. H. Allen. Complete reprinting of 14 volumes, 1837-61, over two million words; the sourcebooks for *Walden*, etc. Definitive. All original sketches, plus 75 photographs. Introduction by Walter Harding. Total of 1804pp. 8½ x 12¼.   20312-3, 20313-1 Clothbd., Two-vol. set $50.00

CLASSIC GHOST STORIES, Charles Dickens and others. 18 wonderful stories you've wanted to reread: "The Monkey's Paw," "The House and the Brain," "The Upper Berth," "The Signalman," "Dracula's Guest," "The Tapestried Chamber," etc. Dickens, Scott, Mary Shelley, Stoker, etc. 330pp. 5⅜ x 8½.   20735-8 Pa. $3.50

SEVEN SCIENCE FICTION NOVELS, H. G. Wells. Full novels. *First Men in the Moon, Island of Dr. Moreau, War of the Worlds, Food of the Gods, Invisible Man, Time Machine, In the Days of the Comet.* A basic science-fiction library. 1015pp. 5⅜ x 8½. (Available in U.S. only)
20264-X Clothbd. $8.95

ARMADALE, Wilkie Collins. Third great mystery novel by the author of *The Woman in White* and *The Moonstone.* Ingeniously plotted narrative shows an exceptional command of character, incident and mood. Original magazine version with 40 illustrations. 597pp. 5⅜ x 8½.
23429-0 Pa. $5.00

MASTERS OF MYSTERY, H. Douglas Thomson. The first book in English (1931) devoted to history and aesthetics of detective story. Poe, Doyle, LeFanu, Dickens, many others, up to 1930. New introduction and notes by E. F. Bleiler. 288pp. 5⅜ x 8½. (Available in U.S. only)
23606-4 Pa. $4.00

FLATLAND, E. A. Abbott. Science-fiction classic explores life of 2-D being in 3-D world. Read also as introduction to thought about hyperspace. Introduction by Banesh Hoffmann. 16 illustrations. 103pp. 5⅜ x 8½.
20001-9 Pa. $1.50

THREE SUPERNATURAL NOVELS OF THE VICTORIAN PERIOD, edited, with an introduction, by E. F. Bleiler. Reprinted complete and unabridged, three great classics of the supernatural: *The Haunted Hotel* by Wilkie Collins, *The Haunted House at Latchford* by Mrs. J. H. Riddell, and *The Lost Stradivarius* by J. Meade Falkner. 325pp. 5⅜ x 8½.
22571-2 Pa. $4.00

AYESHA: THE RETURN OF "SHE," H. Rider Haggard. Virtuoso sequel featuring the great mythic creation, Ayesha, in an adventure that is fully as good as the first book, *She.* Original magazine version, with 47 original illustrations by Maurice Greiffenhagen. 189pp. 6½ x 9¼.
23649-8 Pa. $3.00

UNCLE SILAS, J. Sheridan LeFanu. Victorian Gothic mystery novel, considered by many best of period, even better than Collins or Dickens. Wonderful psychological terror. Introduction by Frederick Shroyer. 436pp. 5⅜ x 8½. 21715-9 Pa. $4.00

JURGEN, James Branch Cabell. The great erotic fantasy of the 1920's that delighted thousands, shocked thousands more. Full final text, Lane edition with 13 plates by Frank Pape. 346pp. 5⅜ x 8½.
23507-6 Pa. $4.00

THE CLAVERINGS, Anthony Trollope. Major novel, chronicling aspects of British Victorian society, personalities. Reprint of Cornhill serialization, 16 plates by M. Edwards; first reprint of full text. Introduction by Norman Donaldson. 412pp. 5⅜ x 8½. 23464-9 Pa. $5.00

KEPT IN THE DARK, Anthony Trollope. Unusual short novel about Victorian morality and abnormal psychology by the great English author. Probably the first American publication. Frontispiece by Sir John Millais. 92pp. 6½ x 9¼. 23609-9 Pa. $2.50

RALPH THE HEIR, Anthony Trollope. Forgotten tale of illegitimacy, inheritance. Master novel of Trollope's later years. Victorian country estates, clubs, Parliament, fox hunting, world of fully realized characters. Reprint of 1871 edition. 12 illustrations by F. A. Faser. 434pp. of text. 5⅜ x 8½. 23642-0 Pa. $4.50

YEKL and THE IMPORTED BRIDEGROOM AND OTHER STORIES OF THE NEW YORK GHETTO, Abraham Cahan. Film *Hester Street* based on *Yekl* (1896). Novel, other stories among first about Jewish immigrants of N.Y.'s East Side. Highly praised by W. D. Howells—Cahan "a new star of realism." New introduction by Bernard G. Richards. 240pp. 5⅜ x 8½. 22427-9 Pa. $3.50

THE HIGH PLACE, James Branch Cabell. Great fantasy writer's enchanting comedy of disenchantment set in 18th-century France. Considered by some critics to be even better than his famous *Jurgen*. 10 illustrations and numerous vignettes by noted fantasy artist Frank C. Pape. 320pp. 5⅜ x 8½. 23670-6 Pa. $4.00

ALICE'S ADVENTURES UNDER GROUND, Lewis Carroll. Facsimile of ms. Carroll gave Alice Liddell in 1864. Different in many ways from final Alice. Handlettered, illustrated by Carroll. Introduction by Martin Gardner. 128pp. 5⅜ x 8½. 21482-6 Pa. $2.00

FAVORITE ANDREW LANG FAIRY TALE BOOKS IN MANY COLORS, Andrew Lang. The four Lang favorites in a boxed set—the complete *Red, Green, Yellow* and *Blue* Fairy Books. 164 stories; 439 illustrations by Lancelot Speed, Henry Ford and G. P. Jacob Hood. Total of about 1500pp. 5⅜ x 8½. 23407-X Boxed set, Pa. $14.00

HOUSEHOLD STORIES BY THE BROTHERS GRIMM. All the great Grimm stories: "Rumpelstiltskin," "Snow White," "Hansel and Gretel," etc., with 114 illustrations by Walter Crane. 269pp. 5⅜ x 8½.
21080-4 Pa. $3.00

SLEEPING BEAUTY, illustrated by Arthur Rackham. Perhaps the fullest, most delightful version ever, told by C. S. Evans. Rackham's best work. 49 illustrations. 110pp. 7⅞ x 10¾.
22756-1 Pa. $2.00

AMERICAN FAIRY TALES, L. Frank Baum. Young cowboy lassoes Father Time; dummy in Mr. Floman's department store window comes to life; and 10 other fairy tales. 41 illustrations by N. P. Hall, Harry Kennedy, Ike Morgan, and Ralph Gardner. 209pp. 5⅜ x 8½.
23643-9 Pa. $3.00

THE WONDERFUL WIZARD OF OZ, L. Frank Baum. Facsimile in full color of America's finest children's classic. Introduction by Martin Gardner. 143 illustrations by W. W. Denslow. 267pp. 5⅜ x 8½.
20691-2 Pa. $3.50

THE TALE OF PETER RABBIT, Beatrix Potter. The inimitable Peter's terrifying adventure in Mr. McGregor's garden, with all 27 wonderful, full-color Potter illustrations. 55pp. 4¼ x 5½. (Available in U.S. only)
22827-4 Pa. $1.10

THE STORY OF KING ARTHUR AND HIS KNIGHTS, Howard Pyle. Finest children's version of life of King Arthur. 48 illustrations by Pyle. 131pp. 6⅛ x 9¼.
21445-1 Pa. $4.00

CARUSO'S CARICATURES, Enrico Caruso. Great tenor's remarkable caricatures of self, fellow musicians, composers, others. Toscanini, Puccini, Farrar, etc. Impish, cutting, insightful. 473 illustrations. Preface by M. Sisca. 217pp. 8⅜ x 11¼.
23528-9 Pa. $6.00

PERSONAL NARRATIVE OF A PILGRIMAGE TO ALMADINAH AND MECCAH, Richard Burton. Great travel classic by remarkably colorful personality. Burton, disguised as a Moroccan, visited sacred shrines of Islam, narrowly escaping death. Wonderful observations of Islamic life, customs, personalities. 47 illustrations. Total of 959pp. 5⅜ x 8½.
21217-3, 21218-1 Pa., Two-vol. set $10.00

INCIDENTS OF TRAVEL IN YUCATAN, John L. Stephens. Classic (1843) exploration of jungles of Yucatan, looking for evidences of Maya civilization. Travel adventures, Mexican and Indian culture, etc. Total of 669pp. 5⅜ x 8½.
20926-1, 20927-X Pa., Two-vol. set $6.50

AMERICAN LITERARY AUTOGRAPHS FROM WASHINGTON IRVING TO HENRY JAMES, Herbert Cahoon, et al. Letters, poems, manuscripts of Hawthorne, Thoreau, Twain, Alcott, Whitman, 67 other prominent American authors. Reproductions, full transcripts and commentary. Plus checklist of all American Literary Autographs in The Pierpont Morgan Library. Printed on exceptionally high-quality paper. 136 illustrations. 212pp. 9⅛ x 12¼.
23548-3 Pa. $7.95

AN AUTOBIOGRAPHY, Margaret Sanger. Exciting personal account of hard-fought battle for woman's right to birth control, against prejudice, church, law. Foremost feminist document. 504pp. 5⅜ x 8½.
20470-7 Pa. $5.50

MY BONDAGE AND MY FREEDOM, Frederick Douglass. Born as a slave, Douglass became outspoken force in antislavery movement. The best of Douglass's autobiographies. Graphic description of slave life. Introduction by P. Foner. 464pp. 5⅜ x 8½.
22457-0 Pa. $5.00

LIVING MY LIFE, Emma Goldman. Candid, no holds barred account by foremost American anarchist: her own life, anarchist movement, famous contemporaries, ideas and their impact. Struggles and confrontations in America, plus deportation to U.S.S.R. Shocking inside account of persecution of anarchists under Lenin. 13 plates. Total of 944pp. 5⅜ x 8½.
22543-7, 22544-5 Pa., Two-vol. set $9.00

LETTERS AND NOTES ON THE MANNERS, CUSTOMS AND CONDITIONS OF THE NORTH AMERICAN INDIANS, George Catlin. Classic account of life among Plains Indians: ceremonies, hunt, warfare, etc. Dover edition reproduces for first time all original paintings. 312 plates. 572pp. of text. 6⅛ x 9¼.
22118-0, 22119-9 Pa.. Two-vol. set $10.00

THE MAYA AND THEIR NEIGHBORS, edited by Clarence L. Hay, others. Synoptic view of Maya civilization in broadest sense, together with Northern, Southern neighbors. Integrates much background, valuable detail not elsewhere. Prepared by greatest scholars: Kroeber, Morley, Thompson, Spinden, Vaillant, many others. Sometimes called Tozzer Memorial Volume. 60 illustrations, linguistic map. 634pp. 5⅜ x 8½.
23510-6 Pa. $7.50

HANDBOOK OF THE INDIANS OF CALIFORNIA, A. L. Kroeber. Foremost American anthropologist offers complete ethnographic study of each group. Monumental classic. 459 illustrations, maps. 995pp. 5⅜ x 8½.
23368-5 Pa. $10.00

SHAKTI AND SHAKTA, Arthur Avalon. First book to give clear, cohesive analysis of Shakta doctrine, Shakta ritual and Kundalini Shakti (yoga). Important work by one of world's foremost students of Shaktic and Tantric thought. 732pp. 5⅜ x 8½. (Available in U.S. only)
23645-5 Pa. $7.95

AN INTRODUCTION TO THE STUDY OF THE MAYA HIEROGLYPHS, Syvanus Griswold Morley. Classic study by one of the truly great figures in hieroglyph research. Still the best introduction for the student for reading Maya hieroglyphs. New introduction by J. Eric S. Thompson. 117 illustrations. 284pp. 5⅜ x 8½.
23108-9 Pa. $4.00

A STUDY OF MAYA ART, Herbert J. Spinden. Landmark classic interprets Maya symbolism, estimates styles, covers ceramics, architecture, murals, stone carvings as artforms. Still a basic book in area. New introduction by J. Eric Thompson. Over 750 illustrations. 341pp. 8⅜ x 11¼.
21235-1 Pa. $6.95

GEOMETRY, RELATIVITY AND THE FOURTH DIMENSION, Rudolf Rucker. Exposition of fourth dimension, means of visualization, concepts of relativity as Flatland characters continue adventures. Popular, easily followed yet accurate, profound. 141 illustrations. 133pp. 5⅜ x 8½.
23400-2 Pa. $2.75

THE ORIGIN OF LIFE, A. I. Oparin. Modern classic in biochemistry, the first rigorous examination of possible evolution of life from nitrocarbon compounds. Non-technical, easily followed. Total of 295pp. 5⅜ x 8½.
60213-3 Pa. $4.00

THE CURVES OF LIFE, Theodore A. Cook. Examination of shells, leaves, horns, human body, art, etc., in *"the* classic reference on how the golden ratio applies to spirals and helices in nature . . . . "—Martin Gardner. 426 illustrations. Total of 512pp. 5⅜ x 8½.
23701-X Pa. $5.95

PLANETS, STARS AND GALAXIES, A. E. Fanning. Comprehensive introductory survey: the sun, solar system, stars, galaxies, universe, cosmology; quasars, radio stars, etc. 24pp. of photographs. 189pp. 5⅜ x 8½. (Available in U.S. only)
21680-2 Pa. $3.00

THE THIRTEEN BOOKS OF EUCLID'S ELEMENTS, translated with introduction and commentary by Sir Thomas L. Heath. Definitive edition. Textual and linguistic notes, mathematical analysis, 2500 years of critical commentary. Do not confuse with abridged school editions. Total of 1414pp. 5⅜ x 8½.
60088-2, 60089-0, 60090-4 Pa., Three-vol. set $18.00

DIALOGUES CONCERNING TWO NEW SCIENCES, Galileo Galilei. Encompassing 30 years of experiment and thought, these dialogues deal with geometric demonstrations of fracture of solid bodies, cohesion, leverage, speed of light and sound, pendulums, falling bodies, accelerated motion, etc. 300pp. 5⅜ x 8½.
60099-8 Pa. $4.00